Lecture Notes in Control and Information Sciences 349

Editors: M. Thoma, M. Morari

Eric Rogers, Krzysztof Galkowski,
David H. Owens

Control Systems Theory and Applications for Linear Repetitive Processes

Springer

Authors

Eric Rogers
School of Electronics
and Computer Science
University of Southampton
Southampton SO17 1BJ
United Kingdom
E-mail: etar@ecs.soton.ac.uk

David H. Owens
Department of Automatic Control and
Systems Engineering
University of Sheffield
Mappin Street
S1 3JD Sheffield
United Kingdom

Krzysztof Galkowski
Institute of Control
and Computation Engineering
The University of Zielona Gora
Podgrna Str. 50
65-246 Zielona Gora
Poland

Library of Congress Control Number: 2006935983

ISSN print edition: 0170-8643
ISSN electronic edition: 1610-7411
ISBN-10 3-540-42663-9 Springer Berlin Heidelberg New York
ISBN-13 978-3-540-42663-9 Springer Berlin Heidelberg New York

Springer is a part of Springer Science+Business Media
springer.com

Typesetting: by the authors and techbooks using a Springer LaTeX macro package

Cover design: *design & production* GmbH, Heidelberg

Printed on acid-free paper SPIN: 10853285 89/techbooks 5 4 3 2 1 0

Preface

Repetitive processes, also termed multipass processes in the early literature, are characterized by a series of sweeps, termed passes, through a set of dynamics where the duration, or length, of each pass is finite. On each pass an output, or pass profile, is produced which acts as a forcing function on, and hence contributes to, the dynamics of the next pass profile. This so-called unit memory property is a special case of the more general situation where it is the previous M passes which contribute to the dynamics of the current one. The positive integer M is termed the memory length and such processes are simply termed non-unit memory.

The concept of a repetitive process was first introduced in the early 1970's as a result of work in The University of Sheffield, UK on the modelling and control of long-wall coal cutting and metal rolling operations. In these applications, productive work is undertaken by a series of passes through a set of dynamics defined over a finite duration, or pass length, which is the first distinguishing feature of a repetitive process. As the process evolves from given initial conditions, an output sequence of pass profiles is produced and it was observed that this could include a first pass profile which had acceptable dynamics along the pass but subsequent passes contained oscillations which grew, or increased in amplitude, severely from pass-to-pass.

Further investigation in the long-wall coal cutting case established that the deterioration in performance after the first pass was due to the effects of the previous pass profile on the production of the current one. In this application, the machine which undertakes the cutting operation rests on the previous pass profile during the production of the current one and its weight alone clearly means that it will most certainly influence the next pass profile, i.e. the output dynamics on any pass acts as a forcing function on, and hence contributes to, the dynamics of the next pass. This is the second distinguishing feature of a repetitive process, i.e. it is possible to generate oscillations which increase in amplitude from pass-to-pass. Such behavior is clearly not acceptable and requires appropriate control action.

Recognizing the unique control problem here, the first approach to the design of control laws was to write down a simplified mathematical model and then make use of standard, termed 1D here, control action. The essence of such an approach is to use a single variable to convert the mathematical model of the process under consideration into that for an equivalent infinite length single pass process in which the relationships between variables are expressed only in terms of the so-called total distance traversed. This then led to the design and evaluation of control schemes for this and other examples such as metal rolling.

The analysis employed in this early work was somewhat application oriented and it was necessary to impose the assumptions that (i) the pass length is 'long' (but finite) and hence the effects of the initial conditions at the start of each pass can be ignored, and (ii) the effects of the previous pass dynamics can be represented by a long delay term. Intuitively, however, the resetting of the initial conditions before the start of each new pass could act as a form of stabilizing action and hence prevent the growth of disturbances. This and the need for a generally applicable control theory led to the development of an alternative approach to stability analysis which does not require the above assumptions and, in particular, takes full account of the interaction between successive pass profiles over the finite pass length. This stability theory is based on an abstract model of the process dynamics in a Banach space setting which includes a wide range of examples with linear dynamics (and a constant pass length) as special cases.

In this abstract model, the critical contribution from the previous pass dynamics to those of the current one is expressed in terms of a bounded linear operator mapping a Banach space into itself and the stability theory is expressed in terms of spectral and induced norm properties of this operator. Hence, unlike the initial approach, this setting provides a rigorous general purpose basis for the control related analysis of linear constant pass length repetitive processes. This is all the more important with the later emergence of other applications and, in particular, those termed algorithmic where adopting a repetitive process setting for analysis either has clear advantages over alternatives or indeed provides the only viable approach.

The stability theory based on the abstract model setting shows that this property for these processes is much more involved than first envisaged. In particular, it shows that the structure of the initial conditions at the start of each new pass is critical to the dynamics which evolve (both along the pass and pass-to-pass) and, critically, they cannot be neglected. Hence, at best, the original approach to the analysis and control of these processes can only be correct under very special circumstances. Moreover, two distinct stability properties can be defined and physically justified, where the essential difference between them is a direct result of the finite pass length.

Given the unique control problem, so-called asymptotic stability demands that the sequence of pass profiles converge to a steady or so-called limit profile which, in turn, is equivalent to demanding bounded-input bounded-output stability (defined in terms of the norm on the underlying function space) over the finite pass length. This, however, does not guarantee that the resulting limit profile has acceptable along the pass dynamics. For example, certain practically relevant sub-classes produce a limit profile which is described by a 1D (differential or discrete) unstable linear state-space model but over a finite duration such a model is guaranteed to produce a bounded response and hence satisfy the definition of asymptotic stability (for repetitive processes).

Stability along the pass removes this difficulty by demanding the bounded-input bounded-output property uniformly, i.e. independent of the pass length. Moreover, asymptotic stability is a necessary condition for stability along the pass and for certain sub-classes of major interest in terms of applications the resulting conditions can be tested by direct application of 1D linear systems tests. Missing, however, is the ability to use these tests, e.g. in the frequency domain, as a basis for control law design.

If the dynamics along the pass are described by a (matrix) discrete linear state equation, it can be shown that stability along the pass (for one particular case of pass initial conditions) of the resulting so-called discrete linear repetitive process is equivalent to bounded-input bounded-output stability of 2D discrete linear systems described by well known and extensively studied state-space models. This, in turn, suggests that the theory for these 2D systems should be directly applicable to discrete linear repetitive processes. Note, however, that this equivalence is only present in the case of the simplest possible boundary conditions and there is no corresponding result for linear repetitive processes whose along the pass dynamics are described by a (matrix) linear differential equation.

By the mid 1990's, the stability theory and associated tests for differential and discrete linear repetitive processes was well developed but there was much yet to be done before their full power in terms of applications could be exploited to the maximum extent. This prompted an expanded research effort into areas of systems theory such as controllability, observability, robust stability, optimal control, and the structure and design of control laws (or controllers) with and without uncertainty in the process description. This monograph gives the results of this work and also its application to, in the main, iterative learning control which is one of the major algorithmic applications for repetitive process systems theory.

The following chapter provides the essential background in terms of examples, their modelling as special cases of the abstract model, the links with certain classes of 2D discrete linear systems and delay differential systems, the development of a 1D equivalent model for the dynamics of discrete linear repetitive processes, and a 2D transfer-function matrix description of the dynamics of differential and discrete processes. The two currently known algorithmic applications for repetitive processes are also introduced by showing how their dynamics fit naturally into the repetitive process setting. This is followed by a chapter giving the abstract model based stability theory and its application in terms of computable tests and (in some relevant cases) the extraction of information concerning expected performance in the presence of stability.

Chapters 4 and 5 give further development of the existing stability theory and tests in two basic directions for the sub-classes of discrete and differential linear repetitive processes which have (currently) the most relevance in terms of applications. This leads to new interpretations of stability in the form of so-called 1D and 2D Lyapunov equations which provide computable information concerning expected performance and also, via Linear Matrix Inequalities (LMIs) and Lyapunov functions, algorithms for control law design to ensure stability and performance. Chapter 6 deals with the case when there is uncertainty in the defining state-space model.

The remaining chapters focus on systems theoretic properties and control law (or controller) design. In Chap. 7, controllability and observability for both differential and discrete linear repetitive processes is treated. As in the theory of 2D/nD discrete linear systems, the situation here is more complex than for 1D linear systems and it is also important to note that some of the properties defined for discrete processes have no 2D linear systems counterparts. In the differential case, the analysis is much less well developed and requires further work to be undertaken.

In Chap. 8, a substantial body of results on control law (or controller) design are developed and illustrative examples given. A major part of these relate to the development of design algorithms which can be computed using LMIs and cover the cases when stability and stability plus performance respectively are required. These control laws are, in general, activated by a combination of current and previous pass information. Moreover, they have a well grounded physical basis, a feature which is not always present in 2D/nD systems. The performance objectives considered include that of forcing the process under control action to be stable along the pass with a resulting limit profile which has acceptable properties as a 1D linear system, which again has a well grounded physical basis.

Linear quadratic optimal control is an obvious approach to the control of the processes considered here, where a cost function can be formed by taking the usual quadratic cost along each pass and then summing over the passes (either for the finite number of passes to be completed or else to infinity). Here it is shown (by a straightforward extension of familiar 1D theory) that such a cost function can be minimized by a state control law which cannot be implemented because it is not causal. It is, however, subsequently shown that a causal solution to this problem does exist but further work is required on the computational aspects.

Chapter 9 deals with control law (or controller) design for robustness and performance. These are the first ever results in this key area and build on those in the previous chapter in terms of the structure of the laws used and LMI based computations. The uncertainty structures considered are expressed in terms of perturbations to the defining process state-space model. This is followed by an H_∞ based design and, in the final section, H_2 and mixed H_2/H_∞ approaches.

Iterative learning control (ILC) is a major application area for repetitive process theory and this is the subject of Chap. 10. The results given range from those previously known, which highlight a major performance trade-off inherent in ILC, through to the very latest analysis supported by experimental results from application to a conveyor system and a gantry robot. Finally, Chap. 11 summarizes the current state of the art and discusses areas for possible future research, where this latter aspect includes both further development of the results reported in this monograph and also extensions to the structure of the models currently considered to capture essential dynamics not included in any of those studied to-date.

Acknowledgements

The research reported in this monograph has benefited greatly from collaborations with a number of PhD students and input from many friends and collaborators worldwide and it would simply be impossible to ensure that we included them all by name here. We do, however, acknowledge John Edwards whose work on the original industrial applications in the 1970's founded this research area and we wish him a long and happy retirement. The PhD research of Artur and Jarek Gramacki, Wojciech Paszke and Bartek Sulikowski (Zielona Gora) contributes significantly to the results of several chapters and the latter two also gave much of their time in constructing the figures and examples etc. Also Lukasz Hladowski (Zielona Gora) has assisted us greatly with, in particular, the referencing and construction of the index. The experimental results in Chap. 10 are based on the PhD research of Tarek Al-Towaim and James Ratcliffe at the University of Southampton under a research programme on ILC theory and experimental verification conducted jointly by the Universities of Southampton and Sheffield. This programme is directed by the first and third authors here together with Paul Lewin (Southampton) and Jari Hatonen (Sheffield). Notker Amann was part of the team (with the first and third authors) that introduced norm optimal control to the community. Financial support from EPSRC, EU, and the Ministry of Scientific Research and Information Technologies of Poland and the Universities of Southampton, Zielona Gora, Sheffield and Wuppertal (where the second author was a Gerhard Mercator Guest Professor in the academic year 2004/05) is gratefully acknowledged. The proof reading skills of Jeffrey Wood have been of enormous assistance in the presentation of this monograph and any errors which remain are the sole responsibility of the authors. Finally, we must thank our families for their support during the period when this monograph was written.

Southampton
Zielona Gora
Sheffield
July 2006

Eric Rogers
Krzysztof Galkowski
David H. Owens

Contents

1 Examples and Representations

Summary. This chapter first introduces the unique features and control problems for repetitive processes by reference to two physical examples – long-wall coal cutting and metal rolling. Two so-called algorithmic examples are considered next, i.e. problem areas where adopting a repetitive process approach to modelling and analysis has clear advantages over alternatives. All these examples are shown to be special cases of the general abstract model in a Banach space setting on which the stability theory for linear repetitive processes is based. Finally, the links at the modelling/structural level with well known 2D discrete and standard, termed 1D in this setting, linear systems are detailed.

1.1 Examples and Control Problems

1.1.1 Long-wall Coal Cutting

The unique characteristic of a repetitive process can be illustrated by considering machining operations where the material or workpiece involved is processed by a sequence of passes of the processing tool. Assuming the pass length $\alpha < \infty$ to be constant, the output vector, or pass profile, $y_k(t)$, $0 \leq t \leq \alpha$ (where t denotes the independent spatial or temporal variable) generated on pass k acts as a forcing function on, and hence contributes to, the dynamics of the next pass profile $y_{k+1}(t)$, $0 \leq t \leq \alpha$, $k \geq 0$.

These processes have their origins in the mining and metal rolling industries where the first to be identified was long-wall coal cutting , which was the most satisfactory, and commonly used, method of mining coal in Great Britain. Even though coal mining in Great Britain is now a much reduced industry in comparison to former times, this example can still be used to illustrate the 'basic mechanics' of a repetitive process and the essential unique control problem. This is treated next, starting with a brief description of the long-wall coal cutting process. We use the notation of the original treatment of this example in [50, 51].

Figures 1.1 and 1.2 illustrate the basic operation of the long-wall coal cutting process in which the coal cutting machine is hauled along the entire length of the coal face riding on the semi-flexible structure of the armored face conveyor, denoted A.F.C., which transports away the coal cut by the rotating drum. In the simplest mode of operation, these machines only cut

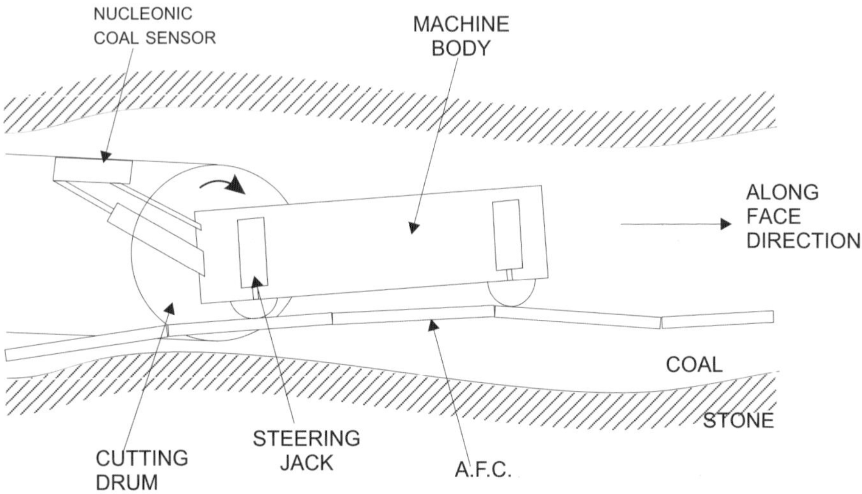

Fig. 1.1. Side elevation of coal cutting machine

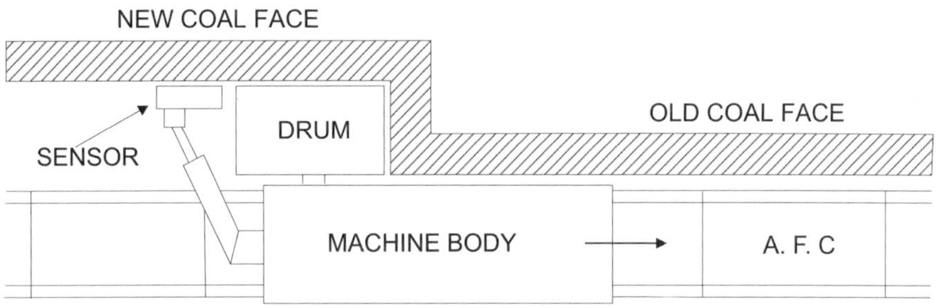

Fig. 1.2. Plan view of coal cutting machine

in one direction, left to right in Figs. 1.1 and 1.2, and they are hauled back in reverse at high speed for the start of each new pass of the coal face.

Between passes, the conveyor is 'snaked' forward using hydraulic rams, as illustrated in Fig. 1.3, so that the machine now rests on the newly cut floor, i.e. the pass profile produced during the previous pass. During the cutting operation, the machine's drum may be raised or lowered with respect to the A.F.C. by using hydraulically operated jacks (illustrated schematically in Fig. 1.1) to tilt the machine body about a datum line on the drum (also termed the face) side. The objective of this operation is the vertical steering of the entire long-wall installation (machine, conveyor and roof support units) to maintain it within the undulating confines of the coal seam (or layer). A nucleonic coal sensor, situated some distance behind the cutting drum, provides the primary control signal by measuring either the floor or ceiling thickness of coal left by the machine (penetration of the stone/coal interface is to be avoided on both economic and safety grounds).

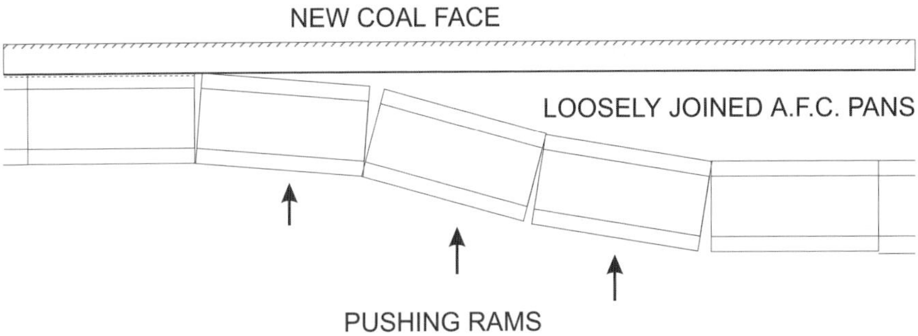

Fig. 1.3. Snaking of conveyor during pushover stage

In order to obtain a simplified mathematical model of this process, consider the idealized side elevation and plan shown in Figs. 1.4 and 1.5 respectively. Here the constants F, R and W represent the feet spacing, drum offset, and width of the machine (and drum) respectively, the variable $J_{k+1}(t)$ represents the controlled drum deflection, $Y_{k+1}(t)$, $e_{k+1}(t)$ denote the coal floor thickness and the height of the A.F.C. above a fixed datum plane respectively, X is the transport delay, or lag, by which the coal floor sensor lags behind the cutting drum, $Z_{k+1}(t)$ denotes the height of the stone/coal interface above the same fixed datum plane as the A.F.C., and $\beta_{k+1}(t)$ denotes the longitudinal tilt of the machine. (The skids labelled A,B,C and D respectively in these last two figures represent the mountings used to fix the machine body to the conveyor and are not relevant to the analysis here.) Suppose also that all angular deflections are small. Then elementary geometrical considerations immediately yield the following description of the coal cutting process dynamics over $0 \le t \le \alpha$, (where α denotes the finite and assumed constant pass length)

$$\begin{aligned} Y_{k+1}(t) + Z_{k+1}(t) &= e_{k+1}(t+R) + W\gamma_{k+1}(t+R) \\ &+ R\beta_{k+1}(t) + J_{k+1}(t) \end{aligned} \tag{1.1}$$

where γ denotes the transverse tilt of the machine.

The transverse and longitudinal tilts of the machine are also those of the supporting conveyor structure and are given by

$$\gamma_{k+1}(t) = \frac{(e_{k+1}(t) - e_k(t))}{W} \tag{1.2}$$

and

$$\beta_{k+1}(t) = \frac{(e_{k+1}(t) - e_{k+1}(t+F))}{F} \tag{1.3}$$

respectively. Suppose also that the A.F.C. moulds itself exactly onto the cut floor on which it rests – the so-called 'rubber conveyor' assumption. Then

$$e_{k+1}(t) = k_2(Y_k(t) + Z_k(t)) \tag{1.4}$$

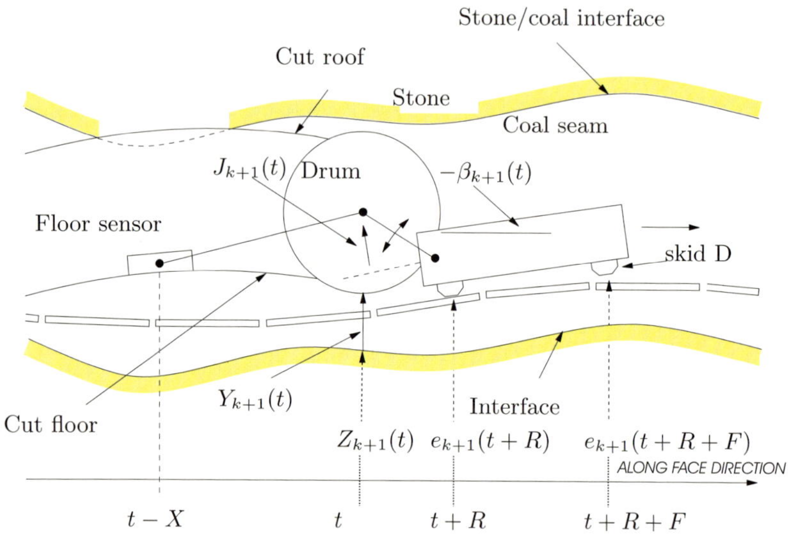

Fig. 1.4. Side elevation with variables labelled

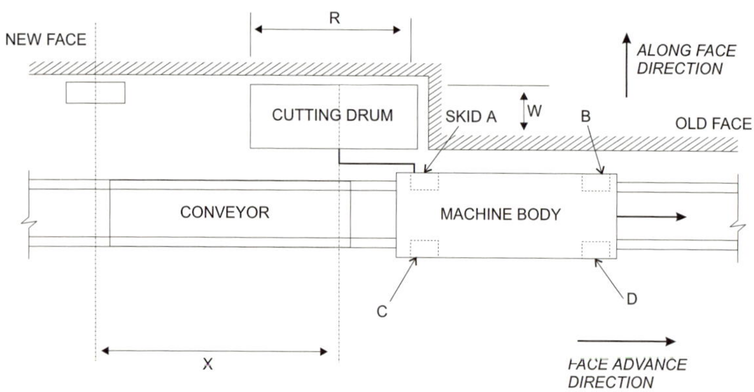

Fig. 1.5. Plan view with variables labelled

where k_2 is a positive real constant. This completes the description of the open-loop system in this case.

One approach to controlling this system is to manipulate the variable $J_{k+1}(t)$ from a delayed measurement of the coal floor thickness $Y_{k+1}(t-X)$. More commonly, however, the roof coal thickness was used since it can be related to $Y_{k+1}(t-X)$ on the assumption that the seam thickness is constant. Suppose also that the sensor and actuator dynamics can be neglected (to a first approximation) and a so-called fixed drum shearer is used, i.e. $R = 0$. Then a possible control law in this case takes the form

$$J_{k+1}(t) = k_1(R_{k+1}(t) - Y_{k+1}(t-X)) - W\gamma_{k+1}(t),\ 0 \le t \le \alpha \tag{1.5}$$

where k_1 is a positive real constant and $R_{k+1}(t)$ is a new external reference vector taken to represent the desired coal thickness on pass $k+1$.

Suppose now, for simplicity, that the variable $Z_{k+1}(t)$ is set equal to zero. Then combining (1.1)–(1.5) yields the following description of the controlled process dynamics over $0 \leq t \leq \alpha$, $k \geq 0$,

$$Y_{k+1}(t) = -k_1 Y_{k+1}(t - X) + k_2 Y_k(t) + k_1 R_{k+1}(t), \; X > 0 \tag{1.6}$$

with assumed pass initial conditions

$$Y_{k+1}(t) = 0, \; -X \leq t \leq 0, \; k \geq 0 \tag{1.7}$$

Figure 1.6 shows the response of this controlled process in the case when $k_1 = 0.8$, $k_2 = 1$, $X = 1.25$, $\alpha = 10$ to a downward step change in $R_{k+1}(t)$ applied at $t = 0$ on each pass, i.e. $R_{k+1}(t) = -1$, $0 \leq t \leq 10$, $k \geq 0$. Note here that the oscillations grow, or increase in amplitude, severely from pass-to-pass (i.e. in the k direction). Consequently the deterioration in performance after the first pass must be due to the effects of the cut floor profile on the previous pass. In other words, the output dynamics on any pass acts (by the basic system geometry) as a forcing function (or disturbance) on, and hence contributes to, the dynamics of the next pass, i.e. the shape of the floor profile produced on the next pass of the cutting machine along the coal face. This interaction between successive pass profile dynamics is a unique characteristic of all repetitive processes and in cases such as that of Fig. 1.6 appropriate control action is clearly required.

If the example under consideration is single-input single-output (SISO) and the dynamics are assumed to be linear, an obvious intuitive approach to

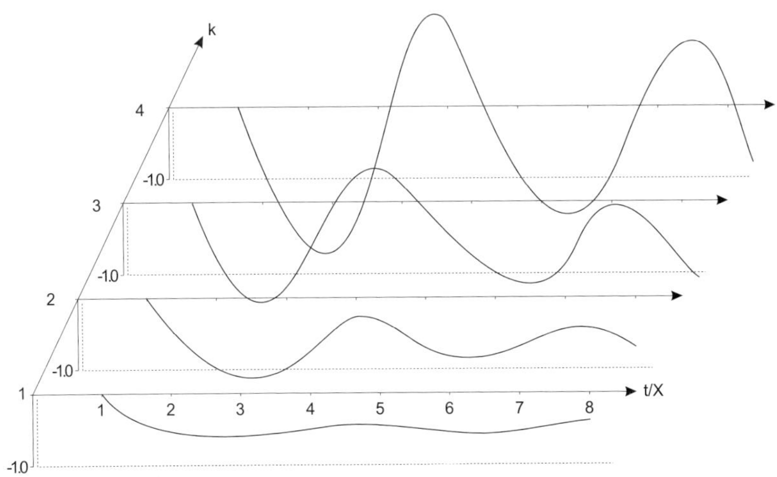

Fig. 1.6. Closed-loop system negative unit step response

stability analysis and control law (or controller) design is to make use of well known classical tools such as the inverse Nyquist criterion. The essence of such an approach is to use the single variable $V = k\alpha + t$ to convert the particular example under consideration into an equivalent infinite length single pass process in which the relationships between variables are expressed only in terms of V, termed the total distance traversed. In particular, a variable, say, $Y_{k+1}(t)$, $k \geq 0$, is identified as a function $Y(V)$ of V defined for $0 \leq V < \infty$.

Applying this approach to (1.6)–(1.7) yields

$$Y(V) = -k_1 Y(V - X) + k_2 Y(V - \alpha) + k_1 R(V) \tag{1.8}$$

and this repetitive process is said to be stable if, and only if, the system of (1.8) is stable in the 1D linear systems sense. Accepting this premise, the original repetitive process dynamics are now amenable to analysis by any of the well known classical (in the 1D sense) techniques. Hence, for example, taking the Laplace transform with respect to V and applying the inverse Nyquist criterion leads to the result that the system is stable as a 1D linear system if, and only if,

$$k_1 < 1 - k_2$$

The above analysis can, at best, only provide initial guidelines for systems analysis and control since it completely ignores the considerable distortion caused to the previous pass profile by the weight of the machine (up to 5 tonnes) as it proceeds along the current pass of the coal seam. This problem is a common feature of a number of known examples of repetitive processes in that dynamic interaction, termed inter-pass smoothing, between passes causes considerable distortion of the previous pass profile with, due to their underlying structure, (potentially) very serious consequences for the future evolution of the process dynamics. Hence if a physically realistic analysis of such examples is to be undertaken then a mathematical means of including this inter-pass smoothing in the process model is required, and it is not at all apparent how this would be achieved in the case when subsequent analysis is to be based on the total distance traversed concept. (See also Sect. 1.2 where it is shown that models of inter-pass smoothing can naturally be included in the abstract model which is the basis for the rigorous stability theory.)

In order to apply classical (frequency domain or otherwise) 1D linear systems theory and control law design techniques to linear repetitive processes it is necessary to make the following assumptions.

- The pass length α is 'long' (but finite) and hence the effects of the initial conditions at the start of each pass can be ignored.
- The effects of the previous pass dynamics can be represented by a long delay term (e.g. $k_2 Y(V - \alpha)$ in (1.8)).

Intuitively, however, the resetting action of the initial conditions on each pass could act as a form of stabilizing action and hence prevent the growth of disturbances (in the case of the long-wall coal cutter these would include, for example, undulations in the floor profile cut during the previous pass). In particular, it is easily shown, using a discretized version of (1.1)–(1.6) with $Y_{k+1}(t)$, $-X \leq t \leq 0$, appropriately chosen, that the initial conditions on each pass have a crucial effect on the performance of the (simplified) long-wall coal cutting dynamics. This, in turn, strongly suggests that for processes with a lag (X) on the current trial, analysis based on the concept of the total distance traversed is valid only in the range

$$k\alpha + X \ll V \ll (k+1)\alpha, \; k \geq 0$$

and only in the following range for processes with no lag in the current pass dynamics

$$k\alpha \ll V \ll (k+1)\alpha, \; k \geq 0$$

Note also that no attempt has been made to use this approach as a basis on which to formulate a general control policy (or strategy) for linear repetitive processes. Instead, attention has been restricted to the problems occurring in a few well known industrial examples.

In summary, therefore, the classically based approach to stability analysis and control law design for linear repetitive processes, as outlined above for the long-wall coal cutting example, is critically limited by the following major factors.

- It completely ignores the effects of the initial conditions on each pass – using the abstract model-based stability theory, see Chap. 2 here, it can be conclusively shown that the structure of these conditions have a critical effect on process stability.
- It does not form the basis for the development of a rigorous generally applicable stability analysis with onward development into the specification and design of control laws.

A major theme of this monograph is that the abstract model-based theory removes these two critical limitations. Prior to introducing this model, however, it is instructive to consider another physical example of a repetitive process in the form of the metal rolling process described next.

1.1.2 Metal Rolling as a Repetitive Process

Metal rolling – see, for example, [52, 61] – is an extremely common industrial process where, in essence, a deformation of a workpiece takes place between two rolls as illustrated in Fig. 1.7.

The first task here is to develop a (simplified but practically feasible) model relating the metal thickness (or gauge) on the current and previous passes through the rolls. These are denoted here by $y_k(t)$ and $y_{k-1}(t)$ respectively and, with reference to Fig. 1.8, the other process variables and physical constants are defined as follows (where we follow the notation in [52, 61])

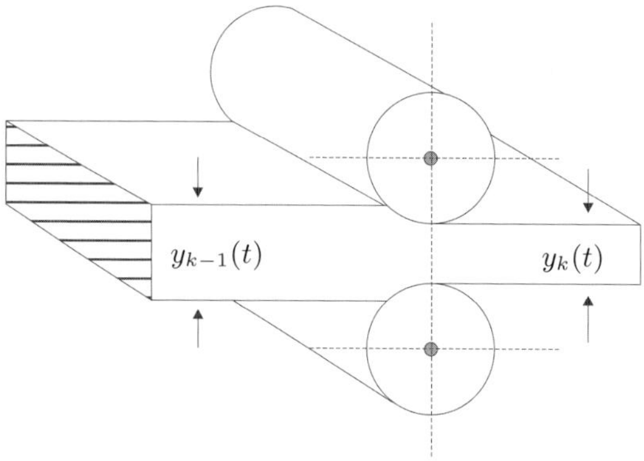

Fig. 1.7. Metal rolling schematic

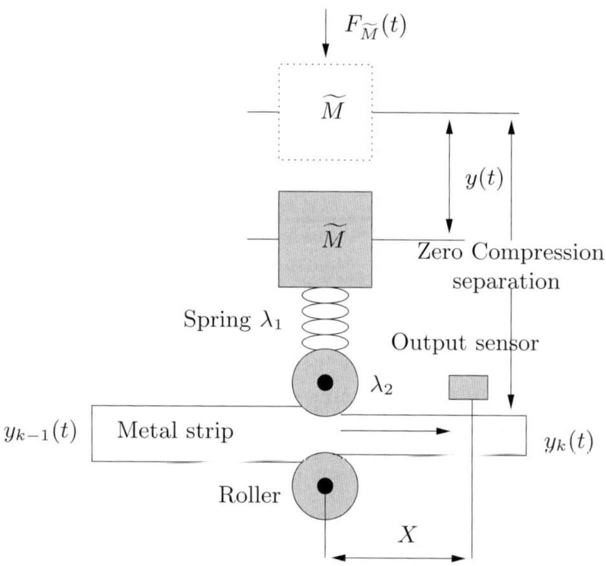

Fig. 1.8. Metal rolling variables

$F_{\widetilde{M}}(t)$ is the force developed by the motor;
$F_s(t)$ is the force developed by the spring;
$\widetilde{M}$ is the lumped mass of the roll-gap adjusting mechanism;
λ_1 is the stiffness of the adjustment mechanism spring;
λ_2 is the hardness of the metal strip;
$\lambda := \frac{\lambda_1 \lambda_2}{\lambda_1 + \lambda_2}$ is the composite stiffness of the metal strip and the roll mechanism;
$y(t)$ is an intermediate variable useful in subsequent analysis.

The force developed by the motor is

$$F_{\widetilde{M}}(t) = F_s(t) + \widetilde{M}\frac{d^2y(t)}{dt^2}$$

and the force developed by the spring is given by

$$F_s(t) = \lambda_1[y(t) + y_k(t)] \tag{1.9}$$

This last force is also applied to the metal strip by the rolls and hence

$$F_s(t) = \lambda_2[y_{k-1}(t) - y_k(t)]$$

The gap-setting motor is conventionally controlled by local feedback of $y(t)$. If proportional plus derivative (PD) action is used to damp the local feedback loop, then

$$F_{\widetilde{M}}(t) = f_a\,[y_d(t) - y(t)] - f_b\frac{dy(t)}{dt}$$

where f_a and f_b are the proportional and derivative gains of the local loop PD controller and $y_d(t)$ denotes the desired value of the motor deflection from the un-stressed position. Also from (1.9)

$$y(t) = \frac{F_s(t)}{\lambda_1} - y_k(t)$$

and, by substituting for $F_s(t)$ in this last equation,

$$y(t) = \frac{\lambda_2}{\lambda_1}y_{k-1}(t) - \frac{\lambda_1 + \lambda_2}{\lambda_1}y_k(t)$$

By obvious substitutions we now have

$$\begin{aligned}\frac{d^2y_k(t)}{dt^2} + 2\zeta\omega_n\frac{dy_k(t)}{dt} + \omega_n^2 y_k(t) \quad &= \quad -\frac{f_a\lambda}{\widetilde{M}\lambda_2}y_d(t) + \frac{\lambda^2}{\widetilde{M}\lambda_2}y_{k-1}(t)\\ &\quad +\frac{\lambda}{\lambda_1}y_k(t)\end{aligned}$$

where $\omega_n^2 := \frac{(f_a+\lambda)}{\widetilde{M}}$ and $2\zeta\omega_n := \frac{f_b}{\widetilde{M}}$. These last two quantities are the undamped natural frequency and damping ratio of the local servo respectively. We will return to this model later in this chapter to illustrate a so-called discrete unit memory linear repetitive process (see Example 1.2.10).

In operation, the work strip can be passed back and forth through a reversing stand, which requires extra power. Hence it is assumed here that the strip is passed repeatedly through a non-reversing single stand, where the roll-gap is reduced for each pass (a process often termed ‘clogging’). This process is, however, ‘slow’ and has a variable pass delay since the stock is usually passed over the top of the rolls. Also, when modelling such behavior links are established between repetitive processes and delay differential systems as shown next.

The thickness of the incoming strip can be related to the actual roll-gap thickness by the following equation

$$y_{k-1}(t) = y_k(t - h_1)$$

where h_1 denotes the pass delay which can be related to the length of the metal strip, which varies from pass-to-pass. This is the so-called inter-pass interaction equation for this process, i.e. it describes the (idealized) dynamics which occur between successive passes.

A commonly used method for controlling the gauge thickness is by proportional feedback control action of the form

$$y_d(t) = -f_c[y_r(t) - y_k(t - h_2)] \tag{1.10}$$

where f_c is the loop gain, $y_r(t)$ is the adjustable reference setting for the desired loop thickness, and h_2 denotes the output sensor measurement delay. This delay is given by $h_2(t) := \frac{X}{v(t)}$, where here X is the distance between the roll-gap and the output sensor and $v(t)$ is the velocity of the metal strip which may also vary from pass-to-pass.

The controlled (or closed-loop) system is this case is modelled by the following forced delay differential equation (obtained after routine manipulations which are omitted here)

$$\frac{d^2 y_k(t)}{dt^2} + f(t) = \frac{c_1 f_a f_c}{\widetilde{M}} y_r(t)$$

where

$$\begin{aligned} f(t) \; := \; & 2\zeta_c \omega_{n_c} \frac{dy_k(t)}{dt} + \omega_{n_c}^2 y_k(t) - c_3 \frac{d^2 y_k(t - h_1)}{dt^2} \\ & -2\zeta_c \omega_{n_c} c_3 \frac{dy_k(t - h_1)}{dt} \\ & - \left(c_3 \omega_{n_c}^2 + \frac{c_2}{\widetilde{M}} \right) y_k(t - h_1) + \frac{c_1 f_a f_c}{\widetilde{M}} y_k(t - h_2) \end{aligned}$$

and

$$\begin{aligned} c_1 &:= \frac{\lambda}{\lambda_2}, \quad c_2 := \lambda c_1, \quad c_3 := \frac{\lambda}{\lambda_1} \\ \omega_{n_c}^2 &:= \frac{f_a + \lambda}{\widetilde{M}}, \quad 2\zeta_c \omega_{n_c} := \frac{f_b}{\widetilde{M}} \end{aligned} \tag{1.11}$$

We will return to this last model in Chap. 7 in connection with the links between repetitive processes and certain classes of delay differential systems.

1.2 A General Abstract Representation

In this section we introduce the abstract model on which the stability theory for linear constant pass length repetitive processes is based. This model was first proposed in [121] with further development in [52] and [149], and its key features in terms of modelling the underlying dynamics are as follows.

- Explicit retention of the effects of the initial conditions on each pass.
- Inclusion of a wide range of linear constant pass length processes as special cases.

Clearly any general model of repetitive processes must, as an essential requirement, include all their unique features. Considering first the most general case, i.e. nonlinear dynamics and a variable pass length, these can be summarized as follows.

- A number of passes, indexed by $k \geq 0$, through a set of dynamics.
- Each pass is characterized by a pass length α_k, and a pass profile $y_k(t)$ defined on $0 \leq t \leq \alpha_k$, where $y_k(t)$ can be a vector or scalar quantity.
- An initial pass profile $y_0(t)$ defined on $0 \leq t \leq \alpha_0$, where α_0 is the initial pass length. The function $y_0(t)$ together with the initial conditions on each pass form the initial, or boundary, conditions for the process.
- Each pass will be subject to its own disturbances and control inputs.
- The process is unit memory, i.e. the dynamics on pass $k+1$ (explicitly) depend only on the independent inputs to that pass and the pass profile on the previous pass k.

Figure 1.9 which illustrates some of these essential features.

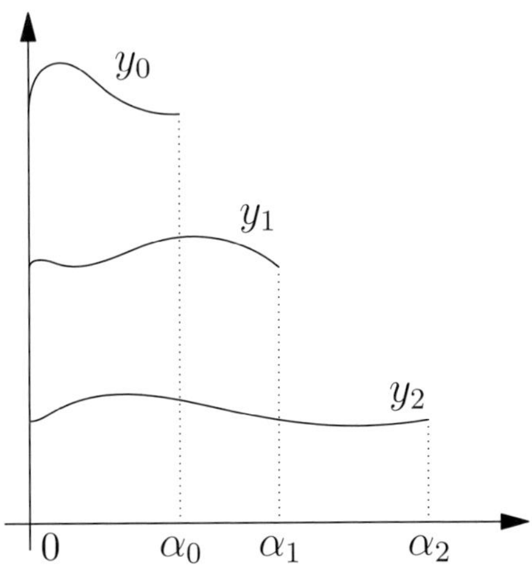

Fig. 1.9. Graphical representation of a sequence of pass profiles

Suppose now that y_k is regarded as a point in a suitably chosen function space. In particular, suppose that $y_k \in E_{\alpha_k}$, $k \geq 0$, where E_{α_k} denotes an appropriately chosen Banach space. Then a general abstract model for repetitive processes can be formulated as a recursion relation of the form

$$y_{k+1} = f_{k+1}(y_k), \ k \geq 0 \tag{1.12}$$

(where f_{k+1} is an abstract mapping of E_{α_k} into $E_{\alpha_{k+1}}$) together with a rule for updating the pass length α_k of the form

$$\alpha_{k+1} = g_{k+1}(\alpha_k, y_{k+1}, y_k), \ k \geq 0 \tag{1.13}$$

Repetitive processes also exist where the current pass profile is a function of the independent inputs to that pass and a finite number $M > 1$ of previous pass profiles. An example is so-called bench mining systems and the integer M is termed the memory length. These processes are designated as 'non-unit memory of length M' or, more simply, 'non-unit memory', and are easily accommodated within the general structure of (1.12)–(1.13). Formally, all that is required is to replace these equations by

$$\begin{aligned} y_{k+1} &= \tilde{f}_{k+1}(y_k, y_{k-1}, \cdots, y_{k+1-M}), \ k \geq 0 \\ \alpha_{k+1} &= \tilde{g}_{k+1}(\alpha_k, \alpha_{k-1}, \cdots, \alpha_{k+1-M}, y_{k+1}, \cdots, y_{k+1-M}), \ k \geq 0 \end{aligned}$$

In actual fact, this last formulation can be avoided by regarding the ordered set $(y_k, y_{k-1}, \cdots, y_{k+1-M})$ as the 'pass profile' in the product space $E_{\alpha_k} \times E_{\alpha_{k-1}} \times \cdots \times E_{\alpha_{k+1-M}}$, i.e.

$$(y_k, y_{k-1}, \cdots, y_{k+1-M}) \in E_{\alpha_k} \times E_{\alpha_{k-1}} \times \cdots \times E_{\alpha_{k+1-M}}$$

Then the two expressions above which define the non-unit memory model become

$$(y_{k+1}, y_k, \cdots, y_{k+2-M}) = (\tilde{f}_{k+1}(y_k, \cdots, y_{k+1-M}), y_k, \cdots, y_{k+2-M})$$

and

$$\alpha_{k+1} = \tilde{g}_{k+1}(\alpha_k, \cdots, \alpha_{k+1-M}, y_{k+1}, \cdots, y_{k+1-M})$$

respectively which have an identical structure to (1.12) and (1.13). Now, however, M points $y_0, y_{-1}, \cdots, y_{1-M}$ are required to define the initial profile.

Any analysis of the abstract model defined above would clearly be a formidable task (with little real applications-oriented progress likely to result). Hence attention has been exclusively focused on processes with a constant pass length (which is 'not unreasonable' to a first approximation in a significant majority of cases encountered), i.e., $\alpha_k = \alpha$, $k \geq 0$.

In the case of processes with linear dynamics, the following definition characterizes a so-called unit memory linear repetitive process in a Banach space setting and forms the basis for onward developments and, in particular, the stability theory.

Definition 1.2.1. *A linear repetitive process of constant pass length $\alpha > 0$ consists of a Banach space E_α, a linear subspace W_α of E_α, and a bounded linear operator L_α mapping E_α into itself (also written $L_\alpha \in B(E_\alpha, E_\alpha)$). The system dynamics are described by linear recursion relations of the form*

$$y_{k+1} = L_\alpha y_k + b_{k+1}, \; k \geq 0 \tag{1.14}$$

where $y_k \in E_\alpha$ is the pass profile on pass k and $b_{k+1} \in W_\alpha$. Here the term $L_\alpha y_k$ represents the contribution from pass k to pass $k+1$ and b_{k+1} represents initial conditions, disturbances and control input effects.

Throughout this monograph we will denote the abstract model of this last definition by S.

In the non-unit memory case, let $L_\alpha^{(j)} \in B(E_\alpha, E_\alpha)$, $1 \leq j \leq M$. Then the abstract representation of a non-unit memory linear repetitive process of memory length M has dynamics described by

$$y_{k+1} = L_\alpha^{(1)} y_k + \cdots + L_\alpha^{(M)} y_{k+1-M} + b_{k+1}, \; k \geq 0$$

where $y_k \in E_\alpha$, $k \geq 1 - M$, $b_{k+1} \in W_\alpha \subset E_\alpha$. Note that this last equation reduces to (1.14) (with $L_\alpha \equiv L_\alpha^{(1)}$) in the case when $M = 1$. Also it can be regarded as a unit memory linear repetitive process S in the product space $E_\alpha^M := E_\alpha \times E_\alpha \times \cdots \times E_\alpha$ (M times) by writing it in the 'companion form' (where I denotes the identity operator on E_α)

$$\begin{bmatrix} y_{k+2-M} \\ \vdots \\ \vdots \\ \vdots \\ y_{k+1} \end{bmatrix} = \begin{bmatrix} 0 & I & 0 & \cdots & 0 \\ 0 & 0 & I & \cdots & 0 \\ 0 & 0 & 0 & \cdots & 0 \\ \vdots & \vdots & \vdots & \ddots & I \\ L_\alpha^{(M)} & L_\alpha^{(M-1)} & L_\alpha^{(M-2)} & \cdots & L_\alpha^{(1)} \end{bmatrix} \begin{bmatrix} y_{k+1-M} \\ \vdots \\ \vdots \\ \vdots \\ y_k \end{bmatrix} + \begin{bmatrix} 0 \\ 0 \\ \vdots \\ 0 \\ b_{k+1} \end{bmatrix}, \; k \geq 0$$

and using the notation

$$L_\alpha := \begin{bmatrix} 0 & I & 0 & \cdots & 0 \\ 0 & 0 & I & \cdots & 0 \\ 0 & 0 & 0 & \cdots & 0 \\ \vdots & \vdots & \vdots & \ddots & I \\ L_\alpha^{(M)} & L_\alpha^{(M-1)} & L_\alpha^{(M-2)} & \cdots & L_\alpha^{(1)} \end{bmatrix} \tag{1.15}$$

Hence results derived for the unit memory case can (in principle at least) be immediately applied to the non-unit memory generalization.

To illustrate the generality of the abstract representation S, the following examples are now considered. Except where stated otherwise, these first arose in (one or more off) [52, 121, 149].

Example 1.2.1. **A delay-difference system.**

The scalar equation over $0 \leq t \leq \alpha,\ k \geq 0$,

$$y_{k+1}(t) = -g_1 y_{k+1}(t - X) + g_2 y_k(t) + g_3 r_{k+1}(t) \tag{1.16}$$

where g_1, g_2 and g_3 are real constants, and $r_{k+1}(t)$ is a new external reference vector taken to represent desired response on pass $k+1$, has already been shown in Sect. 1.1 to arise in the modelling of physical repetitive processes such as long-wall coal cutting. Suppose that the pass initial conditions are of the form

$$y_{k+1}(t) = 0,\ -X \leq t \leq 0,\ k \geq 0$$

Suppose also that $E_\alpha = W_\alpha$ is taken to be the vector space of continuous functions y on $[0, \alpha]$ satisfying the initial condition $y(0) = 0$ and with norm

$$||y|| := \max_{0 \leq t \leq \alpha} |y(t)|$$

Then this example is a special case of S, where the operator L_α is defined by expressing $y_1 = L_\alpha y_0$ in the form

$$\begin{aligned} y_1(t) &= -g_1 y_1(t - X) + g_2 y_0(t),\ 0 \leq t \leq \alpha \\ y_1(t) &= 0,\ -X \leq t \leq 0 \end{aligned}$$

Example 1.2.2. **A differential non-unit memory linear linear repetitive process.**

The state-space model here has the following form over $0 \leq t \leq \alpha,\ k \geq 0$,

$$\begin{aligned} \dot{x}_{k+1}(t) &= Ax_{k+1}(t) + Bu_{k+1}(t) + \sum_{j=1}^{M} B_{j-1} y_{k+1-j}(t) \\ y_{k+1}(t) &= Cx_{k+1}(t) + Du_{k+1}(t) + \sum_{j=1}^{M} D_{j-1} y_{k+1-j}(t) \end{aligned} \tag{1.17}$$

where on pass k, $x_k(t)$ is the $n \times 1$ state vector, $y_k(t)$ is the $m \times 1$ pass profile vector, and $u_k(t)$ is the $r \times 1$ vector of control inputs.

To complete the process description, it is necessary to specify the boundary conditions. The simplest possible form for these is

$$\begin{aligned} x_{k+1}(0) &= d_{k+1},\ k \geq 0 \\ y_{1-j}(t) &= \hat{y}_{1-j}(t),\ 0 \leq t \leq \alpha,\ 1 \leq j \leq M \end{aligned} \tag{1.18}$$

where d_{k+1} is an $n \times 1$ vector with known constant entries and the entries in the $m \times 1$ vectors $\hat{y}_{1-j}(t)$ are known functions of t.

In this case, we choose $E_\alpha = L_2^m[0, \alpha] \cap L_\infty[0, \alpha]$. Then (1.17) and (1.18) define a special case of S under any valid norm $||\cdot||$ with $L_\alpha^{(j)}$, $1 \le j \le M$, defined by the relation

$$(L_\alpha^{(j)} y)(t) := C \int_0^t e^{A(t-\tau)} B_{j-1} y(\tau)\, d\tau + D_{j-1} y(t),\ 0 \le t \le \alpha \tag{1.19}$$

and b_{k+1} over $0 \le t \le \alpha$, $k \ge 0$, by

$$b_{k+1} := Ce^{At} d_{k+1} + C \int_0^t e^{A(t-\tau)} B u_{k+1}(\tau)\, d\tau + D u_{k+1}(t) \tag{1.20}$$

Finally, if the d_{k+1} lie in a subspace W of $\mathbb{R}^n$, $W_\alpha \subset E_\alpha$ can be obtained by evaluating (1.20) for all d_{k+1} and all admissible u_{k+1}.

Example 1.2.3. **A differential unit memory linear repetitive process.** Set $M = 1$ in Example 1.2.2.

The next example is the basis for the application of repetitive process theory to linear iterative learning control analysis and design [126].

Example 1.2.4. **ILC as a differential linear repetitive process.**

Iterative learning control (ILC) is a technique for controlling systems operating in a repetitive (or pass-to-pass) mode with the requirement that a reference trajectory $r(t)$ defined over a finite interval $0 \le t \le T$ is followed to a high precision. Examples of such systems include robotic manipulators that are required to repeat a given task to high precision, chemical batch processes or, more generally, the class of tracking systems. Motivated by human learning, the basic idea of ILC is to use information from previous executions of the task in order to improve performance from pass-to-pass in the sense that the tracking error is sequentially reduced. The objective of ILC schemes is to use their repetitive process structure (i.e. information propagation from pass-to-pass and along a pass) to progressively improve the accuracy with which the core operation under consideration is performed, by updating the control input progressively from pass-to-pass. In common with the ILC literature we will use the word trial instead of pass when considering ILC in this monograph.

Since the original work in the mid 1980's, the general area of ILC has been the subject of intense research effort both in terms of the underlying theory and 'real world' experiments (ILC designs are now routinely supported by the results of experimental verification – see Chap. 9). One of the major streams of research in this general area is based on the fact that the stability theory for linear repetitive processes is directly applicable to a major class of algorithms. Next we introduce this class and its representation as a linear repetitive process.

Commonly used ILC algorithms construct the input to the plant or process from the input used on the last trial plus an additive increment which is typically a function of the past values of the measured output error, i.e. the difference between the achieved output on the current trial and the desired plant output. Suppose that $u_k(t)$ denotes the input to the plant on trial k which is of duration T, i.e. $0 \leq t \leq T$. Suppose also that $e_k(t) = r(t) - y_k(t)$ denotes the current trial error. Then the objective of constructing a sequence of input functions such that the performance achieved is gradually improving with each successive trial can be refined to a convergence condition on the input and error, i.e.

$$\lim_{k \to \infty} ||e_k|| = 0, \ \lim_{k \to \infty} ||u_k - u_\infty|| = 0$$

where $||\cdot||$ is a signal norm in a suitably chosen function space (e.g. $L_2^m[0, \alpha]$) with a norm-based topology and u_∞ is termed the learned control.

This definition of convergent learning is, in effect, a stability problem on a two-dimensional (2D)-product space. As such, it places the analysis of ILC schemes firmly outside standard (or 1D) control theory – although (see Chap. 9) it is still has a significant role to play in certain cases of practical interest. Instead, ILC must be seen in the context of fixed-point problems or, more precisely, repetitive processes. Next we present one particular class of ILC schemes considered in this work (others will be given in Chap. 9) and write the resulting closed-loop (or controlled) system as a differential non-unit memory linear repetitive process.

The plant to be controlled is assumed to be of the following form where $T < \infty$

$$\begin{aligned} \dot{x}_k(t) &= Ax_k(t) + Bu_k(t), \ 0 \leq t \leq T \\ y_k(t) &= Cx_k(t) \end{aligned}$$

Here on trial k, $x_k(t)$ is the $n \times 1$ state vector, $y_k(t)$ is the $m \times 1$ output vector, and $u_k(t)$ is the $r \times 1$ vector of control inputs. Also, without loss of generality, we assume that $x_k(0) = 0, \ k \geq 1$. A member of the class of ILC schemes considered here has the following form which is, in effect, a combination of previous input vectors, the current trial error, and the errors on a finite number of previous trials

$$u_{k+1}(t) = \sum_{j=1}^{M} \alpha_j u_{k+1-j}(t) + \sum_{j=1}^{M} (K_j e_{k+1-j})(t) + (K_0 e_{k+1})(t)$$

In addition to the 'memory' M, the design parameters in this control law are the static scalars $\alpha_j, \ 1 \leq j \leq M$, the linear convolution operator K_0 which describes the current trial error contribution, and the linear convolution operator $K_j, \ 1 \leq j \leq M$, which describes the contribution from the error on trial $k + 1 - j$.

The error dynamics on trial $k+1$ in this case can be written in convolution form as

$$e_{k+1}(t) = r(t) - (Gu_{k+1})(t),\ 0 \le t \le T$$

where

$$(Gu)(t) := C \int_0^t e^{A(t-\tau)} B\, u(\tau)\, d\tau$$

Using this description, it is easily shown that the closed-loop (controlled) error dynamics on trial $k+1$ can be written over $0 \le t \le T$ as

$$\begin{aligned} e_{k+1}(t) &= (I + GK_0)^{-1} \left\{ \sum_{j=1}^{M} ((\alpha_j I - GK_j) e_{k+1-j})(t) \right. \\ &\quad \left. + \left(1 - \sum_{j=1}^{M} \alpha_j\right) r(t) \right\} \end{aligned}$$

(where here $(\cdot)^{-1}$) denotes the inverse convolution operation) or, equivalently, as a differential non-unit memory linear repetitive process which can be written as a special case of the abstract model S in the form

$$\hat{e}_{k+1} = L_\alpha \hat{e}_k + \hat{b},\ k \ge 0$$

where

$$\hat{e}_k(t) := \left[\ e^T_{k+1-M}(t) \quad \cdots \quad \cdots \quad e^T_k(t)\ \right]^T$$

is the so-called error super-vector, and

$$L_\alpha := \begin{bmatrix} 0 & I & 0 & \dots & 0 \\ 0 & 0 & I & \dots & 0 \\ 0 & 0 & 0 & \dots & 0 \\ \vdots & \vdots & \vdots & \ddots & I \\ E_0E_M & E_0E_{M-1} & E_0E_{M-2} & \cdots & E_0E_1 \end{bmatrix}$$

with

$$\begin{aligned} (E_0 y)(t) &:= ((I + GK_0)^{-1} y)(t) \\ (E_j y)(t) &:= ((\alpha_j I - GK_j) y)(t),\ 1 \le j \le M \end{aligned}$$

and

$$\hat{b} := \left[\ 0 \quad 0 \quad \cdots \quad (1 - \sum_{j=1}^{M} \alpha_j) r^T(t)\ \right]^T$$

Finally, suppose that $\hat{e}_k \in E_\alpha$ – a suitably chosen Banach space – and $\hat{b} \in W_\alpha$ – a linear subspace of E_α – and hence the process dynamics can treated as a special case of the abstract model S.

Example 1.2.5. **A differential unit memory linear repetitive process with dynamic boundary conditions.**

The boundary conditions of (1.18) in Example 1.2.2 are the simplest possible and cases exist (see the next example) where they are simply not strong enough to adequately model the underlying process dynamics (even for initial simulation and/or control analysis). Instead, it is necessary to consider a state initial vector sequence which is an explicit function of (in the unit memory case for simplicity) the previous pass profile. One possible form is

$$x_{k+1}(0) = d_{k+1} + \sum_{j=1}^{N} J_j y_k(t_j) \tag{1.21}$$

where d_{k+1} is as in Example 1.2.2, $0 \leq t_1 < t_2 < \cdots < t_N \leq \alpha$, are N sample points along the previous pass, and J_j, $1 \leq j \leq N$, is an $n \times m$ matrix with constant entries.

This process is a special case of S in $E_\alpha = L_2^m[0,\alpha] \cap L_\infty[0,\alpha]$ over $0 \leq t \leq \alpha$ with

$$(L_\alpha y)(t) := Ce^{At}\hat{y} + C\int_0^t e^{A(t-\tau)}B_0 y(\tau)\, d\tau + D_0 y(t)$$

where

$$\hat{y} := \sum_{j=1}^{N} J_j y(t_j)$$

and, over $0 \leq t \leq \alpha$, $k \geq 0$,

$$b_{k+1} := Ce^{At}d_{k+1} + C\int_0^t e^{A(t-\tau)}Bu_{k+1}(\tau)\, d\tau + Du_{k+1}(t)$$

Example 1.2.6. **A class of delay-differential systems modelled as differential unit memory linear repetitive processes.**

A class of delay-differential systems in $\mathbb{R}^n$ can be modelled by the state-space equations

$$\begin{aligned} \dot{x}(t) &= Ax(t) + B_0 x(t-\alpha) + Bu(t),\ t \geq 0 \\ x(t-\alpha) &= x_0(t),\ 0 \leq t \leq \alpha \end{aligned} \tag{1.22}$$

If the delay α is interpreted as a pass length then it is obvious that these systems have certain structural similarities to differential unit memory linear repetitive processes. In particular, introduce the following change of variables over $0 \leq t \leq \alpha$, $k \geq 0$,

$$\begin{aligned} u_{k+1}(t) &:= u(k\alpha + t) \\ x_k(t) &:= x((k-1)\alpha + t) \end{aligned}$$

and define the pass profile as $y_k = x_k,\ k \geq 0$. Then the defining equation (1.22) can be written as a differential unit memory linear repetitive process with state initial vector defined by $x_{k+1}(0) := x_k(\alpha),\ k \geq 0$, i.e. a special case of (1.21) of Example 1.2.5.

The next example, and also Example 1.2.12, arises from the work of Roberts see, for example [139, 140, 141, 142, 143, 144].

Example 1.2.7. **Iterative solution algorithms for nonlinear dynamic optimal control problems modelled as a differential unit memory linear repetitive process.**

The solution of nonlinear optimal control problems often have to be obtained in an iterative manner due to the existence of mixed boundary conditions. This is achieved by specifying an algorithm which updates a trial solution from iteration-to-iteration (or from pass-to-pass). Such an algorithm has a natural 2D systems/repetitive process structure where one direction of information propagation is the time horizon of the dynamic system under investigation and the second is the evolution of the trials (or passes).

A wide range of scenarios could be considered, but here we restrict attention to the case where model-reality differences occur in the dynamic model and both the performance index and the terminal constraints. In particular, we consider so-called dynamic integrated system optimization and parameter estimation, denoted DISOPE, which is a technique for solving optimal control problems where there are differences in structure and parameter values between reality and the model employed in the computations. A representative problem which can be solved by this route in the continuous-time domain is the so-called real optimal control problem, denoted ROP, and defined as follows, where here we (mainly) follow the notation of [140] and hence the superscript $*$ is only used to distinguish variables.

$$\min_{u(t)} J := \min_{u(t)} \left\{ \Phi^*(x(\alpha)) + \int_0^\alpha L^*(x(t), u(t))\, dt \right\} \tag{1.23}$$

such that

$$\dot{x}(t) = f^*(x(t), u(t)),\ x(0) = x_o \tag{1.24}$$

and

$$\Psi^*(x(\alpha)) = 0 \tag{1.25}$$

In this problem defined over the fixed time horizon $t \in [0, \alpha]$, $x(t)$ is the $n \times 1$ state vector, $u(t)$ is the $r \times 1$ control input vector, $\Phi^* : \mathbb{R}^n \to \mathbb{R}$ is the real terminal measure, $L^* : \mathbb{R}^n \times \mathbb{R}^r \to \mathbb{R}$ is the real performance measure function, $f^* : \mathbb{R}^n \times \mathbb{R}^r \to \mathbb{R}^n$ represents the real system state equations, and $\Psi^* : \mathbb{R}^n \to \mathbb{R}^q$ is the real terminal constraint vector.

A so-called dynamic integrated system optimization and parametrization (DISOPE) algorithm obtains the solution of this problem by iterating on the following modified form of a linear quadratic model based optimal control problem (MMOP)

$$\begin{aligned}\hat{u}_k(t), \hat{x}_k(t) &= \arg \min_{u(t),x(t)} \left\{ \frac{1}{2}(x^T(\alpha)Sx(\alpha) + \Gamma_1^T x(\alpha)) \right. \\ &+ \int_0^\alpha \left\{ (\frac{1}{2}(x^T(t)Qx(t) + u^T(t)Ru(t)) \right. \\ &- \lambda_k^T(t)u(t) - \beta_k^T(t)x(t) \\ &+ \left. \left. \frac{1}{2}r_1||u(t) - u_k(t)||^2 + \frac{1}{2}r_2||x(t) - x_k(t)||^2 \right\} dt \right\}\end{aligned} \tag{1.26}$$

subject to

$$\begin{aligned}\dot{x}(t) = Ax(t) + Bu(t) + \gamma(t),\ x(0) = x_o \\ Vx(\alpha) + b + \hat{\gamma} = 0\end{aligned} \tag{1.27}$$

(here $||\cdot||$ denotes the Euclidean norm). Application of well known optimality conditions in this case now yields estimates of the optimal control, state and costate vectors on pass k (iteration (i) in the original work) as

$$\begin{aligned}\hat{u}_k(t) &= \overline{R}^{-1}(-B^T\hat{p}_k(t) + \lambda_k(t) + r_1u_k(t)) \\ \dot{\hat{x}}_k(t) &= A\hat{x}_k(t) + B\hat{u}_k(t) + \gamma_k(t) \\ \dot{\hat{p}}_k(t) &= -\overline{Q}\hat{x}_k(t) - A^T\hat{p}_k(t) + \beta_k(t) + r_2x_k(t)\end{aligned} \tag{1.28}$$

where

$$\begin{aligned}\overline{R} &:= R + r_1I_r \\ \overline{Q} &:= Q + r_2I_n\end{aligned} \tag{1.29}$$

and $p(t)$ is the $n \times 1$ costate vector.

The corresponding mixed boundary conditions are

$$\begin{aligned}\hat{x}_k(0) &= x_o \\ \hat{p}_k(\alpha) &= S\hat{x}_k(\alpha) + \Gamma_1 + [V + \Gamma_2]^T v \\ V\hat{x}_k(\alpha) + b + \hat{\gamma} &= 0\end{aligned} \tag{1.30}$$

In (1.26)–(1.30), Q is an $n \times n$ positive semi-definite matrix (denoted by $\geq$), R is an $r \times r$ positive definite matrix (denoted $>$), V is a $q \times n$ matrix, $b \in \mathbb{R}^q$, v is a Lagrange multiplier, $(\hat{\cdot})_k$ denotes the current solution, and $(\cdot)_k$ the variable value on iteration k. The model parameters $\gamma(t) \in \mathbb{R}^n$ and $\hat{\gamma} \in \mathbb{R}^q$, together with the modifiers $\lambda(t) \in \mathbb{R}^r$ and $\beta(t) \in \mathbb{R}^n$, $\Gamma_1 \in \mathbb{R}^n$, and the $q \times n$ matrix Γ_2 are updated between iterations as follows

$$
\begin{aligned}
\gamma_k(t) &= f^*(x_k(t), u_k(t)) - Ax_k(t) - Bu_k(t) \\
\hat{\gamma}_k &= \Psi^*(x_k(\alpha)) - Vx_k(\alpha) - b \\
\lambda_k(t) &= -\left[\frac{\partial f^*(\cdot)}{\partial u} - B\right]^T p_k(t) - [\nabla_u L^*(\cdot) - Ru_k(t)] \\
\beta_k(t) &= -\left[\frac{\partial f^*(\cdot)}{\partial x} - A\right]^T p_k(t) - [\nabla_x L^*(\cdot) - Qx_k(t)] \\
\Gamma_1 &= \nabla_x \Phi^*(x_k(\alpha)) - Sx_k(\alpha) \\
\Gamma_2 &= \frac{\partial \Psi^*(x_k(\alpha))}{\partial x} - V
\end{aligned} \tag{1.31}
$$

In addition to using the scalars r_1 and r_2, convergence and stability in this algorithm can be regulated useing the following relaxation scheme

$$
\begin{aligned}
u_{k+1}(t) &= u_k(t) + k_u(\hat{u}_k(t) - u_k(t)) \\
x_{k+1}(t) &= x_k(t) + k_x(\hat{x}_k(t) - x_k(t)) \\
p_{k+1}(t) &= p_k(t) + k_p(\hat{p}_k(t) - p_k(t))
\end{aligned}
$$

where k_u, k_x and k_p are scalar gain parameters. Also local convergence and stability properties of the approach can be obtained by considering the special case when the real optimal control problem is also linear with a quadratic performance index and a linear terminal constraint. In the ROP problem defined by (1.23)–(1.25) let

$$
\begin{aligned}
\Phi^*(x(\alpha)) &= \frac{1}{2} x^T(\alpha) S^* x(\alpha) \\
L^*(x(t), u(t)) &= \frac{1}{2}(x^T(t) Q^* x(t) + u^T(t) R^* u(t)) \\
f^*(x(t), u(t)) &= A^* x(t) + B^* u(t) \\
\Psi^*(x(\alpha)) &= V^* x(\alpha) + b^*
\end{aligned}
$$

Then (1.31) becomes

$$
\begin{aligned}
\gamma_k(t) &= (A^* - A)x_k(t) + (B^* - B)u_k(t) \\
\hat{\gamma}_k &= (V^* - V)x_k(\alpha) + b^* - b \\
\lambda_k(t) &= -[B^* - B]^T p_k(t) - (R^* - R)u_k(t) \\
\beta_k(t) &= -[A^* - A]^T p_k(t) - (Q^* - Q)x_k(t) \\
\Gamma_1 &= (S^* - S)x_k(\alpha) \\
\Gamma_2 &= V^* - V
\end{aligned}
$$

Suitable manipulations to eliminate $\gamma_k(t)$, $\lambda_k(t)$, and $\beta_k(t)$ now yields the differential unit memory linear repetitive process state-space model

$$
\begin{aligned}
\dot{\hat{X}}_{k+1}(t) &= \hat{A}\hat{X}_{k+1}(t) + \hat{B}_0 \hat{Y}_k(t) \\
\hat{Y}_{k+1}(t) &= \hat{C}\hat{X}_{k+1}(t) + \hat{D}_0 \hat{Y}_k(t)
\end{aligned} \tag{1.32}
$$

where

$$\hat{X}_{k+1}(t) := \begin{bmatrix} \hat{x}_k(t) \\ \hat{p}_k(t) \end{bmatrix}, \quad \hat{Y}_k(t) := \begin{bmatrix} u_k(t) \\ x_k(t) \\ p_k(t) \end{bmatrix}$$

and

$$\begin{aligned} \hat{A} &:= \begin{bmatrix} A & -B\overline{R}^{-1}B^T \\ -\overline{Q} & -A^T \end{bmatrix}, \quad \hat{C} := \begin{bmatrix} 0 & -k_u\overline{R}^{-1}B^T \\ k_x I_n & 0 \\ 0 & k_p I_n \end{bmatrix} \\ \hat{B}_0 &:= \begin{bmatrix} (B^* - B\overline{R}^{-1}R^*) & (A^* - A) & B\overline{R}^{-1}(B - B^*)^T \\ 0 & (\overline{Q} - Q^*) & (A - A^*)^T \end{bmatrix} \\ \hat{D}_0 &:= \begin{bmatrix} (I_r - k_u\overline{R}^{-1}R^*) & 0 & k_u\overline{R}^{-1}(B - B^*)^T \\ 0 & (1 - k_u)I_n & 0 \\ 0 & 0 & (1 - k_p)I_n \end{bmatrix} \end{aligned}$$

Eliminating Γ_1, Γ_2, and $\hat{\gamma}$ shows that the solution of (1.32) is subject to the mixed boundary conditions

$$\begin{aligned} \hat{x}_k(0) &= x_o \\ \hat{p}_k(\alpha) &= S\hat{x}_k(\alpha) + (S^* - S)x_k(\alpha) + (V^*)^T v \end{aligned}$$

and

$$V\hat{x}_k(\alpha) + (V^* - V)x_k(\alpha) + b^* = 0$$

Also routine algebraic manipulations now give the following expression for the initial costate vector

$$\begin{aligned} \hat{p}(0) &= -\tilde{\psi}_{22}(\alpha,0)^{-1}\left\{\overline{\psi}_{21}(\alpha,0)\hat{x}_k(0) + \int_0^\alpha \overline{\psi}_2(\alpha,\tau)B_0\hat{Y}_k(\tau)\,d\tau \right. \\ &+ \left. E_0\hat{Y}_k(\alpha) + b_0\right\} \end{aligned}$$

where

$$\begin{aligned} \overline{\psi}_{21}(\alpha,0) &= \tilde{\psi}_{21}(\alpha,0) + (V^*)^T\left[V\psi_{12}(\alpha,0)\tilde{\psi}_{22}^{-1}(\alpha,0)(V^*)^T\right]^{-1} \\ &\times\ V\overline{\psi}_{11}(\alpha,0) \\ \overline{\psi}_2(\alpha,\tau) &= \tilde{\psi}_2(\alpha,\tau) + (V^*)^T\left[V\psi_{12}(\alpha,0)\tilde{\psi}_{22}^{-1}(\alpha,0)(V^*)^T\right]^{-1} \\ &\quad \times V\overline{\psi}_1(\alpha,\tau) \end{aligned}$$

$$\begin{aligned} E_0 &= \begin{bmatrix} 0 & \mathcal{H} & 0 \end{bmatrix} \\ \mathcal{H} &= -(S^* - S) + (V^*)^T[V\psi_{12}(\alpha,0)\tilde{\psi}_{22}^{-1}(\alpha,0)(V^*)^T]^{-1}\overline{V} \\ b_0 &= (V^*)^T\left[V\psi_{12}(\alpha,0)\tilde{\psi}_{22}(\alpha,0)^{-1}(V^*)^T\right]^{-1} b^* \end{aligned}$$

with

$$\begin{aligned}
\overline{\psi}_{11}(\alpha,0) &= \psi_{11}(\alpha,0) - \psi_{12}(\alpha,0)\tilde{\psi}_{22}^{-1}(\alpha,0)\tilde{\psi}_{21}(\alpha,0) \\
\overline{\psi}_{1}(\alpha,\tau) &= \psi_{1}(\alpha,\tau) - \psi_{12}(\alpha,0)\tilde{\psi}_{22}^{-1}(\alpha,0)\tilde{\psi}_{2}(\alpha,\tau) \\
\overline{V} &= V^* - V + V\psi_{12}(\alpha,0)\tilde{\psi}_{22}^{-1}(\alpha,0)(S^* - S)
\end{aligned}$$

and

$$\begin{aligned}
\tilde{\psi}_{22}(\alpha,0) &= \psi_{22}(\alpha,0) - S\psi_{12}(\alpha,0) \\
\tilde{\psi}_{21}(\alpha,0) &= \psi_{21}(\alpha,0) - S\psi_{11}(\alpha,0) \\
\tilde{\psi}_{2}(\alpha,\tau) &= \psi_{2}(\alpha,\tau) - S\psi_{1}(\alpha,\tau)
\end{aligned}$$

where

$$e^{\hat{A}(\alpha-\tau)} = \begin{bmatrix} \psi_{11}(\alpha,\tau) & \psi_{12}(\alpha,\tau) \\ \psi_{21}(\alpha,\tau) & \psi_{22}(\alpha,\tau) \end{bmatrix} =: \begin{bmatrix} \psi_{1}(\alpha,\tau) \\ \psi_{2}(\alpha,\tau) \end{bmatrix}$$

The combined state initial vector can now be written as

$$\begin{aligned}
\hat{X}_{k+1}(0) &:= \begin{bmatrix} \hat{x}_k(0) \\ \hat{p}_k(0) \end{bmatrix} = F_0\hat{x}_k(0) + \tilde{E}_0\hat{Y}_k(\alpha) \\
&+ \int_0^\alpha W(\alpha,\tau)\hat{Y}_k(\tau)\, d\tau + \tilde{b}_0
\end{aligned}$$

where

$$\begin{aligned}
F_0 &:= \begin{bmatrix} I_n \\ -\tilde{\psi}_{22}^{-1}(\alpha,0)\overline{\psi}_{21}(\alpha,0) \end{bmatrix} \\
\tilde{E}_0 &:= \begin{bmatrix} 0 \\ -\tilde{\psi}_{22}^{-1}(\alpha,0)E_0 \end{bmatrix} \\
W(\alpha,\tau) &:= \begin{bmatrix} 0 \\ -\tilde{\psi}_{22}^{-1}(\alpha,0)\overline{\psi}_{2}(\alpha,\tau)B_0 \end{bmatrix}
\end{aligned}$$

and

$$\tilde{b}_0 := \begin{bmatrix} 0 \\ -\tilde{\psi}_{22}^{-1}(\alpha,0)b_0 \end{bmatrix}$$

Example 1.2.8. **A differential unit memory linear repetitive process with inter-pass smoothing effects.**

One possible method [149] of modelling the effects of inter-pass smoothing on the dynamics of a process is to assume that the pass profile at any point t on pass $k+1$ is a function of the state and input vectors at this point and the complete pass profile on pass k. One candidate representation in the unit memory case (with $D = 0,\ D_0 = 0$ for simplicity) is

$$\begin{aligned}
\dot{x}_{k+1}(t) &= Ax_{k+1}(t) + Bu_{k+1}(t) + B_0\int_0^\alpha K(t,\tau)\, y_k(\tau)\, d\tau \\
y_{k+1}(t) &= Cx_{k+1}(t),\ 0 \le t \le \alpha,\ k \ge 0
\end{aligned}$$

with (for simplicity) $x_{k+1}(0) = d_{k+1}$, $k \geq 0$. In this representation, the inter-pass interaction term $\int_0^\alpha K(t,\tau)y_k(\tau)\,d\tau$ represents a 'smoothing out' of the previous pass profile in a manner governed by the properties of the kernel $K(t,\tau)$. Note also that the particular choice of $K(t,\tau) = \delta(t-\tau)I_m$, where $\delta(\cdot)$ denotes the Dirac delta function, reduces this model to that of Example 1.2.3 (with $D = 0$, $D_0 = 0$).

It is easily verified that the construction of Example 1.2.3 is also valid for this model, i.e. it is a special case of S over $0 \leq t \leq \alpha$ with

$$(L_\alpha y)(t) := C\int_0^t e^{A(t-\tau)}B_0\int_0^\alpha K(\tau,t')y(t')dt'\,d\tau$$

and, over $0 \leq t \leq \alpha$, $k \geq 0$,

$$b_{k+1} := Ce^{At}d_{k+1} + C\int_0^t e^{A(t-t')}Bu_{k+1}(t')dt'$$

Example 1.2.9. **A discrete non-unit memory linear repetitive process.** This is the natural discrete analogue of the differential non-unit memory linear repetitive process state-space model introduced in Example 1.2.2 and has the form

$$\begin{aligned} x_{k+1}(p+1) &= Ax_{k+1}(p) + Bu_{k+1}(p) + \sum_{j=1}^{M} B_{j-1}y_{k+1-j}(p) \\ y_{k+1}(p) &= Cx_{k+1}(p) + Du_{k+1}(p) + \sum_{j=1}^{M} D_{j-1}y_{k+1-j}(p) \end{aligned} \quad (1.33)$$

where on pass k, $x_k(p)$ is the $n \times 1$ state vector, $y_k(p)$ is the $m \times 1$ pass profile vector, and $u_k(p)$ is the $r \times 1$ vector of control inputs. The simplest possible set of boundary conditions for this process are given by

$$\begin{aligned} x_{k+1}(0) &= d_{k+1},\ k \geq 0 \\ y_{1-j}(p) &= \hat{y}_{1-j}(p),\ 1 \leq j \leq M,\ 0 \leq p \leq \alpha, \end{aligned} \quad (1.34)$$

where d_{k+1} is an $n \times 1$ vector with known constant entries, and the entries in the $m \times 1$ vectors $\hat{y}_{1-j}(p)$, $1 \leq j \leq M$, are known functions of p over $0 \leq p \leq \alpha$.

In this case, set $E_\alpha = \ell_2^m[0,\alpha]$ – the space of all real $m \times 1$ vectors of length α (corresponding to $p = 1, 2, \cdots, \alpha$). Then it follows immediately that this model is a special case of S over $1 \leq p \leq \alpha$ with

$$(L_\alpha^{(j)}y)(p) := \sum_{h=0}^{p-1} CA^{p-1-h}B_0y(h) + D_{j-1}y(p)$$

and over $1 \leq p \leq \alpha$, $k \geq 0$,

$$b_{k+1} := CA^p d_{k+1} + \sum_{h=0}^{p-1} CA^{p-1-h} Bu_{k+1}(h) + Du_{k+1}(p)$$

Example 1.2.10. **A discrete unit memory linear repetitive process.** Set $M = 1$ in Example 1.2.9.

One physical example here is in the analysis of the metal rolling process of the previous section where, as in [178], we wish to study the effects of choosing $F_{\widetilde{M}}$ as the system input. Then the governing differential equation is

$$\ddot{y}_k(t) + \frac{\lambda}{\widetilde{M}} y_k(t) = \frac{\lambda}{\lambda_1} \ddot{y}_{k-1}(t) + \frac{\lambda}{\widetilde{M}} y_{k-1}(t) - \frac{\lambda}{\widetilde{M}\lambda_1} F_{\widetilde{M}}(t)$$

Suppose now that T denotes the sampling period used. Then on replacing differentiation by backward difference we obtain the difference equation approximation

$$\begin{aligned} y_{k+1}(t) &= a_1 y_{k+1}(t-T) + a_2 y_{k+1}(t-2T) + a_3 y_k(t) \\ &+ a_4 y_k(t-T) + a_5 y_k(t-2T) + bF_{\widetilde{M}} \end{aligned}$$

where

$$\begin{aligned} a_1 &:= \frac{2\widetilde{M}}{\lambda T^2 + \widetilde{M}},\ a_2 := -\frac{\widetilde{M}}{\lambda T^2 + \widetilde{M}},\ a_3 := \frac{\lambda}{\lambda T^2 + \widetilde{M}} \left(T^2 + \frac{\widetilde{M}}{\lambda_1} \right) \\ a_4 &:= -\frac{2\lambda\widetilde{M}}{\lambda_1(\lambda T^2 + \widetilde{M})},\ a_5 := \frac{\lambda\widetilde{M}}{\lambda_1(\lambda T^2 + \widetilde{M})},\ b := -\frac{\lambda T^2}{\lambda_2(\lambda T^2 + \widetilde{M})} \end{aligned}$$

Now set $t = pT$ and write $y_k(p) = y_k(pT)$. Then it follows immediately that the dynamics here can be written as a discrete unit memory linear repetitive process of the form (1.33)–(1.34) with $M = 1$ where

$$\begin{aligned} x_k(p) &:= \begin{bmatrix} y_k(p-1) & y_k(p-2) & y_{k-1}(p-1) & y_{k-1}(p-2) \end{bmatrix}^T \\ u_k(p) &:= \widetilde{F}(p) \end{aligned}$$

and

$$A := \begin{bmatrix} a_1 & a_2 & a_4 & a_5 \\ 1 & 0 & 0 & 0 \\ 0 & 0 & 0 & 0 \\ 0 & 0 & 1 & 0 \end{bmatrix},\ B := \begin{bmatrix} b \\ 0 \\ 0 \\ 0 \end{bmatrix},\ B_0 := \begin{bmatrix} a_3 \\ 0 \\ 0 \\ 0 \end{bmatrix}$$

$$C := \begin{bmatrix} a_1 & a_2 & a_4 & a_5 \end{bmatrix},\ D := b,\ D_0 := a_3$$

Example 1.2.11. **ILC as a discrete linear repetitive process**

This is simply the discrete analogue of Example 1.2.4 and hence the details are omitted here.

Example 1.2.12. **Iterative solution algorithms for discrete nonlinear optimal control problems as discrete linear repetitive processes.**

This is the equivalent discrete time problem to that of Example 1.2.7. Hence the details are omitted here and a detailed treatment can be found in [141].

Example 1.2.13. **A discrete unit memory linear repetitive process with dynamic boundary conditions.**

As in Example 1.2.10 with

$$x_{k+1}(0) = d_{k+1} + \sum_{j=1}^{N} J_j y_k(p_j)$$

where $0 \leq p_1 < p_2 < \cdots < p_N \leq \alpha$ are N sample points along the previous pass and J_j, $1 \leq j \leq N$, is an $n \times m$ matrix with constant entries. It is a simple exercise to show that this model is also a special case of S in $E_\alpha = \ell_2^m[0, \alpha]$ and hence the details are omitted. One case of particular interest is (the most general) case when

$$x_{k+1}(0) = d_{k+1} + \sum_{j=1}^{\alpha-1} J_j y_k(j) \tag{1.35}$$

1.3 2D and 1D Discrete Linear Systems Equivalent Models

Discrete linear repetitive processes share certain structural similarities with well known and extensively studied 2D quarter plane causal discrete linear systems. Also there are obvious structural links between this sub-class of linear repetitive processes and 1D discrete linear time invariant systems. In this section we investigate these links by addressing the following question where, for ease of presentation, attention is restricted to the unit memory case:

Do there exist 2D Roesser/Fornasini-Marchesini and/or 1D discrete linear time invariant systems equivalent state-space model descriptions of the dynamics of discrete linear repetitive processes?

The modelling, or representation, of 2D discrete linear systems is somewhat more involved than in the 1D case. At a basic level, the types of representations possible can be classified according to whether or not

(i) an input/output structure; and
(ii) latent (auxiliary) variables

are included in addition to system variables.

As in the 1D case, state-space models are a very important class of internal representations and several versions exist. The most commonly used, and extensively studied, are those due to Roesser [145] and Fornasini-Marchesini [62] respectively. The Roesser state-space model (omitting the output equation which has no role in this work) has the structure

$$\begin{aligned} x_h(i+1,j) &= A_1x_h(i,j) + A_2x_v(i,j) + B_1u(i,j) \\ x_v(i,j+1) &= A_3x_h(i,j) + A_4x_v(i,j) + B_2u(i,j) \end{aligned} \tag{1.36}$$

In this model i and j are the positive integer valued horizontal and vertical coefficients, x_h is the $n \times 1$ vector of horizontally transmitted information, x_v is the $m \times 1$ vector of vertically transmitted information, and u is the $r \times 1$ vector of control inputs.

In the Fornasini-Marchesini model, the state vector is not split into horizontal and vertical components. Again, the output equation is not required in this work and, with $z(i,j)$ denoting the appropriately dimensioned state vector, the basic model of this type has the structure

$$\begin{aligned} z(i+1,j+1) &= A_5z(i+1,j) + A_6z(i,j+1) + A_7z(i,j) \\ &+ B_3u(i+1,j) + B_4u(i,j+1) \end{aligned} \tag{1.37}$$

where, as in (1.36), u is the (appropriately dimensioned) control input vector.

These two model structures are not completely independent and in the majority of cases every Fornasini-Marchesini model can be transformed into a Roesser model and vice-versa. This particular aspect of 2D systems theory has been the subject of much research effort but for the purposes of the work reported here, the key point to note is that, on deleting the control input terms for simplicity, the Roesser model can be written in terms of $z(i,j) := \left[x_h^T(i,j), y_v^T(i,j)\right]^T$ as

$$z(i+1,j+1) = \begin{bmatrix} A_1 & A_2 \\ 0 & 0 \end{bmatrix} z(i,j+1) + \begin{bmatrix} 0 & 0 \\ A_3 & A_4 \end{bmatrix} z(i+1,j)$$

which is a special case of (1.37) with no input terms and no term in $z(i,j)$.

An essential difference between discrete linear repetitive processes of the form defined by Example 1.2.10 and 2D discrete linear systems described by the Roesser (and hence Fornasini-Marchesini) state-space model is that in the former case information propagation in one of the two separate directions, along the pass, only occurs over the fixed finite duration of the pass length. Hence, despite initial structural similarities (essentially replace A, B_0, C, D_0 with A_1, A_2, A_3, A_4 respectively and treat the state vector and pass profile as horizontally and vertically transmitted information respectively), (1.36) is not an equivalent 2D linear systems Roesser state-space model description of the dynamics of discrete linear repetitive processes. The existence of such an 'equivalence' has been extensively investigated in the past and has only led to a result which shows that stability tests can be interchanged between these two areas (see [24] and Chap. 2) instead of a basis on which to also address other key systems theoretic questions for discrete linear repetitive processes, such as what exactly does controllability of these processes mean? (In answering this general question (see Chap. 6) other differences with 2D discrete linear systems will emerge and likewise when we come to control law design (see Chaps. 7 and 8).)

To develop a 'fully equivalent' Roesser/Fornasini-Marchesini state-space model description of the dynamics of discrete linear repetitive processes [68], it is necessary to use the so-called augmented state vector defined as

$$\mathcal{X}(k,p) := \left[\begin{array}{cc} x_k^T(p) & y_k^T(p) \end{array}\right]^T$$

and set $u(k,p) = u_k(p)$. In this case the dynamics of a discrete unit memory linear repetitive process of the form defined by Example 1.2.10 can be written as

$$E\mathcal{X}(k+1,p+1) = A_8\mathcal{X}(k+1,p) + A_9\mathcal{X}(k,p) + B_5u(k+1,p) \tag{1.38}$$

where

$$\begin{aligned} E &:= \begin{bmatrix} I_n & 0 \\ 0 & 0 \end{bmatrix}, \; A_8 := \begin{bmatrix} A & 0 \\ C & -I_m \end{bmatrix} \\ A_9 &:= \begin{bmatrix} 0 & B_0 \\ 0 & D_0 \end{bmatrix}, \; B_5 := \begin{bmatrix} B \\ D \end{bmatrix} \end{aligned} \tag{1.39}$$

This is a singular version of the Fornasini-Marchesini state-space model of (1.37) with $A_6 = 0$, $B_4 = 0$.

A considerable volume of literature now exists on the analysis of 2D discrete linear systems described by singular versions of the Roesser and Fornasini-Marchesini state-space models. In Chap. 6 of this monograph we will use elements of this theory (together with the results which follow next) to develop a transition matrix (or fundamental matrix sequence) for the dynamics of discrete unit memory linear repetitive processes written in the form of (1.38)–(1.39). This, in turn, leads to a general response formula for these processes and the definition and characterization of local reachability and controllability properties. In what follows next, we investigate what other types of 2D models can be obtained for these processes by the application of suitable transformations.

In 2D linear systems theory it is known that state variable transformations can be applied to both regular (also known as standard or non-singular in the literature) or singular Fornasini-Marchesini state-space models to yield equivalent models with the Roesser structure. To detail this further, consider, for example, the case of (1.37) when $A_7 = 0$, i.e. the so-called first-order form. Then in this case the Roesser model results when we enforce the partition defined by

$$
\begin{aligned}
A_5 &= \begin{bmatrix} * & * \\ 0 & 0 \end{bmatrix}, \; A_6 = \begin{bmatrix} 0 & 0 \\ * & * \end{bmatrix} \\
B_3 &= \begin{bmatrix} * \\ 0 \end{bmatrix}, \; B_4 = \begin{bmatrix} 0 \\ * \end{bmatrix}
\end{aligned}
$$

where here $*$ denotes a possibly non-zero entry.

In the case of discrete unit memory linear repetitive processes of the form defined by Example 1.2.10, the resulting Fornasini-Marchesini state-space model matrices of (1.39) do not satisfy the above conditions and hence this method cannot be applied. Instead, introduce

$$
\begin{aligned}
\eta(k,p) &:= E\mathcal{X}(k,p+1) - A_8\mathcal{X}(k,p) \\
\gamma(k,p) &:= \eta(k,p) - B_5 u(k,p)
\end{aligned} \tag{1.40}
$$

where the role of the second equation here is to avoid the undesirable feature of a shift in the input vector. Substitution of (1.40) into (1.38) immediately yields the following singular Roesser state-space model interpretation of the dynamics of discrete unit memory linear repetitive processes

$$
\begin{aligned}
\begin{bmatrix} I_{n+m} & 0 \\ 0 & E \end{bmatrix} \begin{bmatrix} \gamma(k+1,p) \\ \mathcal{X}(k,p+1) \end{bmatrix} &= \begin{bmatrix} 0 & A_9 \\ I_{n+m} & A_8 \end{bmatrix} \begin{bmatrix} \gamma(k,p) \\ \mathcal{X}(k,p) \end{bmatrix} \\
&+ \begin{bmatrix} 0 \\ B_5 \end{bmatrix} u(k,p)
\end{aligned} \tag{1.41}
$$

In this model, both the state $\{x\}$ and the pass profile $\{y\}$, which form the augmented state vector $\mathcal{X}$, constitute the information transmitted in one direction. (Compare this with the earlier claim that the pass profile $\{y\}$ alone constitutes information propagated in one of the two separate directions (from pass-to-pass)). Note also that the state dimension has been doubled with respect to that of (1.38).

A key property of the state-space model of the dynamics of discrete unit memory linear repetitive processes of the form defined by Example 1.2.10 is that it is regular. This, in turn, suggests that a regular Roesser state-space model interpretation of their dynamics can also be developed.

Introduce the following transformations into the state-space model of Example 1.2.10

$$
\begin{aligned}
\eta(k,p) &:= x(k,p+1) - Ax(k,p) - Bu(k,p) \\
\mu(k,p) &:= y(k,p) - Cx(k,p) - D_0 u(k,p)
\end{aligned}
$$

where, to conform with the 2D systems literature, $x(k,p) := x_k(p)$, $y(k,p) := y_k(p)$, $u(k,p) := u_k(p)$. Then the result can be written as

$$\begin{bmatrix} x(k,p+1) \\ \mu(k+1,p) \\ \eta(k+1,p) \end{bmatrix} = \begin{bmatrix} A & 0 & I_n \\ D_0C & D_0 & 0 \\ B_0C & B_0 & 0 \end{bmatrix} \begin{bmatrix} x(k,p) \\ \mu(k,p) \\ \eta(k,p) \end{bmatrix} + \begin{bmatrix} B \\ D_0D \\ B_0D \end{bmatrix} u(k,p) \tag{1.42}$$

i.e., a regular Roesser model. Note also that the state dimension here is $2n+m$ as opposed to $2(n+m)$ for the singular Roesser model of (1.41).

Consider now the case when D_0 is nonsingular and define the so-called restricted state vector for a discrete unit memory linear repetitive process of the form defined by Example 1.2.10 as

$$\hat{\mathcal{X}}(k,p) := \begin{bmatrix} x^T(k,p) & \mu^T(k,p) \end{bmatrix}^T$$

Hence in this case

$$\eta(k,p) = B_0 D_0^{-1} \mu(k,p)$$

and the following restricted 2D state-space model of the Roesser type is obtained for the dynamics of processes described by Example 1.2.10

$$\begin{bmatrix} x(k,p+1) \\ \mu(k+1,p) \end{bmatrix} = \begin{bmatrix} A & B_0D_0^{-1} \\ D_0C & D_0 \end{bmatrix} \begin{bmatrix} x(k,p) \\ \mu(k,p) \end{bmatrix} + \begin{bmatrix} B \\ D_0D \end{bmatrix} u(k,p)$$

In the 2D linear systems case (and also nD where $n \geq 3$, there has been work reported over the years on the construction of 1D linear systems state-space equivalent models for the underlying dynamics, resulting in, for example, the so-called 'wave advance' models. One of the reasons advocated for using these models in analysis is that the local state formulation of the Roesser/Fornasini-Marchesini state-space models is not a very convenient framework in which to address problems of controllability and observability. Here we briefly review the wave advance model developed in [136] (see also [12] for the special case of 2D systems) for nD systems (using mainly the notation of this first reference) and then proceed to develop an equivalent 1D linear systems state-space model description for the dynamics of discrete linear repetitive processes with static and dynamic boundary conditions. These models will be extensively used in the analysis of Chaps. 6 and 7.

Following [12], denote the independent variable set by $\mu = \{(i,j) : i,j = 0,1,\cdots,q\}$ and consider the bivariate difference equation

$$x(i+1,j+1) = \mathcal{J}_1 x(i,j+1) + \mathcal{J}_2 x(i+1,j) + \mathcal{J}_3 u(i,j+1) + \mathcal{J}_4 u(i+1,j) \tag{1.43}$$

with x, u vector-valued. Then, for example, given knowledge of $x(i,j)$ on the set $\{(0,4),(1,3),(2,2),(3,1),(4,0)\}$, (1.43) computes $x(i,j)$ on the set $(1,4),(2,3),(3,2),(4,1)\}$, i.e. in this sense information advances in a 'wavefront' across x and hence the designation 'wave advance'.

Suppose now that ψ is a column vector whose entries are computed as

$$\begin{aligned}
\psi(0) &= x(0,0) \\
\psi(1) &= \begin{bmatrix} x^T(1,0) & x^T(0,1) \end{bmatrix}^T \\
&\vdots \\
\psi(\tilde{q}) &= \begin{bmatrix} x^T(\tilde{q},0) & x^T(\tilde{q}-1,1) & \cdots & x^T(0,\tilde{q}) \end{bmatrix}^T
\end{aligned}$$

(where $\tilde{q} \le q$) and for convenience we number the entries in ψ from 0. Consider now the jth component of $\psi(\tilde{q}+1)$, i.e.

$$\psi_j(\tilde{q}+1) = x(\tilde{q}+1-j,j),\ j = 0,1,\cdots,\tilde{q}$$

Then from (1.43) we have that

$$\begin{aligned}
\psi_j(\tilde{q}+1) &= \mathcal{J}_1 x(\tilde{q}-j,j) + \mathcal{J}_2 x(\tilde{q}+1-j,j-1) + \mathcal{J}_3 u(\tilde{q}-j,j) \\
&+ \mathcal{J}_4 u(\tilde{q}+1-j,j-1)
\end{aligned}$$

and hence

$$\psi(\tilde{q}+1) = \mathcal{J}_1\psi_j(\tilde{q}) + \mathcal{J}_2\psi_{j-1}(\tilde{q}) + \mathcal{J}_3 v_j(\tilde{q}) + \mathcal{J}_4 v_{j-1}(\tilde{q})$$

where

$$v(\tilde{q}) := \begin{bmatrix} u^T(\tilde{q},0) & u^T(\tilde{q}-1,1) & \cdots & u^T(0,\tilde{q}) \end{bmatrix}^T$$

Now define the following $(\tilde{q}+2) \times (\tilde{q}+1)$ block matrices

$$A(\tilde{q}) := \begin{bmatrix} \mathcal{J}_1 & 0 & 0 & \cdots & 0 \\ \mathcal{J}_2 & \mathcal{J}_1 & 0 & \cdots & 0 \\ 0 & \mathcal{J}_2 & \mathcal{J}_1 & \cdots & 0 \\ 0 & 0 & \mathcal{J}_2 & \ddots & 0 \\ \vdots & \vdots & \vdots & \ddots & \mathcal{J}_1 \\ 0 & 0 & 0 & \cdots & \mathcal{J}_2 \end{bmatrix},\quad B(\tilde{q}) = \begin{bmatrix} \mathcal{J}_3 & 0 & 0 & \cdots & 0 \\ \mathcal{J}_4 & \mathcal{J}_3 & 0 & \cdots & 0 \\ 0 & \mathcal{J}_4 & \mathcal{J}_3 & \cdots & 0 \\ 0 & 0 & \mathcal{J}_4 & \ddots & 0 \\ \vdots & \vdots & \vdots & \ddots & \mathcal{J}_3 \\ 0 & 0 & 0 & \cdots & \mathcal{J}_4 \end{bmatrix}$$

Then we can write (1.43) in the form

$$\psi(\tilde{q}+1) = A(\tilde{q})\psi(\tilde{q}) + B(\tilde{q})v(\tilde{q}),\ \tilde{q} = 0,1,\cdots \tag{1.44}$$

Using this last equation, we have the conclusion that the stationary 2D linear systems bivariate difference equation (1.43) (which is structurally similar to the Fornasini-Marchesini state-space model) can be replaced by a dynamically equivalent 1D time-varying linear systems state-space model. In particular, the matrices and vectors in (1.44) expand as $\tilde{q}$ increases. This fact alone has greatly reduced the value of the 1D equivalent model in 2D linear systems theory. As shown below, however, we can construct an equivalent 1D state-space model for discrete unit memory linear repetitive processes with either static or dynamic boundary conditions which is 'time-invariant', i.e. the matrices and vectors involved are of constant dimensions and have constant entries.

Consider first a process of the form defined by Example 1.2.10. Then the equivalent 1D model can be obtained (see, for example, [67]) by a number of routes. Of these, the simplest is to employ a simple 'forward transformation' of the pass profile vector followed by a change of variable in the pass number. In particular, introduce

$$v_k(p) := y_{k-1}(p),\ 0 \le p \le \alpha - 1,\ k \ge 0$$

and $l := k + 1$. Also define the so-called global pass profile, state, and input super-vectors respectively as

$$Y(l) := \begin{bmatrix} v_l(0) \\ v_l(1) \\ \vdots \\ v_l(\alpha-1) \end{bmatrix},\ X(l) := \begin{bmatrix} x_l(1) \\ x_l(2) \\ \vdots \\ x_l(\alpha) \end{bmatrix},\ U(l) := \begin{bmatrix} u_l(0) \\ u_l(2) \\ \vdots \\ u_l(\alpha-1) \end{bmatrix}$$

Given this setting, it follows immediately that the dynamics of discrete unit memory linear repetitive processes of the form defined by Example 1.2.10 are equivalently described by the 1D discrete linear systems state-space model given next where $l \ge 1$

$$\begin{aligned} Y(l+1) &= \mathfrak{B}_1 Y(l) + \mathfrak{B}_2 U(l) + \mathfrak{B}_3 x_l(0) \\ X(l) &= \mathfrak{B}_4 Y(l) + \mathfrak{B}_5 U(l) + \mathfrak{B}_6 x_l(0) \end{aligned} \tag{1.45}$$

with associated boundary conditions

$$\begin{aligned} x_l(0) &= d_l,\ l \ge 1 \\ v_1(p) &= y_0(p),\ 0 \le p \le \alpha - 1 \end{aligned} \tag{1.46}$$

The matrices $\mathfrak{B}_1$, $\mathfrak{B}_2$, $\mathfrak{B}_3$, $\mathfrak{B}_4$, $\mathfrak{B}_5$ and $\mathfrak{B}_6$ are of dimensions $m\alpha \times m\alpha$, $m\alpha \times r\alpha$, $m\alpha \times n$, $n\alpha \times m\alpha$, $n\alpha \times r\alpha$, and $n\alpha \times n$ respectively and are defined as follows

$$\mathfrak{B}_1 := \begin{bmatrix} D_0 & 0 & 0 & \dots & 0 \\ CB_0 & D_0 & 0 & \dots & 0 \\ CAB_0 & CB_0 & D_0 & \dots & 0 \\ \vdots & \vdots & \vdots & \ddots & \vdots \\ CA^{\alpha-2}B_0 & CA^{\alpha-3}B_0 & CA^{\alpha-4}B_0 & \dots & D_0 \end{bmatrix}$$

$$\mathfrak{B}_2 := \begin{bmatrix} D & 0 & 0 & \dots & 0 \\ CB & D & 0 & \dots & 0 \\ CAB & CB & D & \dots & 0 \\ \vdots & \vdots & \vdots & \ddots & \vdots \\ CA^{\alpha-2}B & CA^{\alpha-3}B & CA^{\alpha-4}B & \dots & D \end{bmatrix}, \; \mathfrak{B}_3 := \begin{bmatrix} C \\ CA \\ CA^2 \\ \vdots \\ CA^{\alpha-1} \end{bmatrix}$$

$$\mathfrak{B}_4 := \begin{bmatrix} B_0 & 0 & \dots & 0 \\ AB_0 & B_0 & \dots & 0 \\ \vdots & \vdots & \ddots & \vdots \\ A^{\alpha-1}B_0 & A^{\alpha-2}B_0 & \dots & B_0 \end{bmatrix}$$

$$\mathfrak{B}_5 := \begin{bmatrix} B & 0 & \dots & 0 \\ AB & B & \dots & 0 \\ \vdots & \vdots & \ddots & \vdots \\ A^{\alpha-1}B & A^{\alpha-2}B & \dots & B \end{bmatrix}, \; \mathfrak{B}_6 := \begin{bmatrix} A \\ A^2 \\ \vdots \\ A^{\alpha} \end{bmatrix}$$

To develop the 1D equivalent model in the case of dynamic boundary conditions , we (see also [72]) consider the most general version of the state initial vector sequence of Example 1.2.13 which has the form (1.35) or, in terms of l,

$$x_l(0) = d_l + \sum_{j=0}^{\alpha-1} J_j v_l(j), \; l \geq 1$$

In this case, with the super-vectors defined as before, the 1D equivalent model takes the form

$$\begin{aligned} Y(l+1) =& \mathfrak{B}_7 Y(l) + \mathfrak{B}_2 U(l) + \mathfrak{B}_3 x_l(0) \\ X(l) =& \mathfrak{B}_8 Y(l) + \mathfrak{B}_5 U(l) + \mathfrak{B}_6 x_l(0) \end{aligned} \tag{1.47}$$

where

$$\begin{aligned} \mathfrak{B}_7 &:= \mathfrak{B}_1 + \mathfrak{B}_3 J \\ \mathfrak{B}_8 &:= \mathfrak{B}_4 + \mathfrak{B}_6 J \end{aligned} \tag{1.48}$$

and

$$J := \begin{bmatrix} J_0 & J_1 & \cdots & J_{\alpha-1} \end{bmatrix} \tag{1.49}$$

In Chap. 6 we will use this 1D model to characterize so-called pass profile controllability in terms of rank tests on matrices with constant entries. Note also that this 1D equivalent model extends in a natural manner (see [72]) to the non-unit memory processes of Example 1.2.9.

1.4 2D Transfer-Function and Related Representations

In 1D linear systems theory, the transfer-function (or transfer-function matrix) and similar representations, such as matrix fraction and matrix factorization descriptions, play a central role. For 2D/nD linear systems/repetitive processes, it also is possible to define such descriptions for differential and discrete linear repetitive processes and (potentially) use them to solve control related problems. Here we develop such representations, which are necessarily defined as a function of two complex variables, starting with the 2D transfer-function matrix representations. The results of this section are, in the main, from [146] (differential) and [147] (discrete) respectively.

Consider first a discrete unit memory linear repetitive process of the form defined by Example 1.2.10 with zero boundary conditions. Also since the state on pass 0 plays no role, it is convenient to re-label the state vector as $x_{k+1}(p) \mapsto x_k(p)$ (keeping of course the same interpretation). Now let z_1 denote the shift operator along the pass applied e.g. to $x_k(p)$ as follows

$$x_k(p) := z_1 x_k(p+1)$$

and z_2 the pass-to-pass shift operator applied e.g. to $y_k(p)$ as follows

$$y_k(p) := z_2 y_{k+1}(p)$$

Then the process dynamics can be written in the form

$$Y(z_1, z_2) = G(z_1, z_2) U(z_1, z_2)$$

where $Y(z_1, z_2)$ and $U(z_1, z_2)$ denote the result of applying these shift operators to the pass profile and input vectors respectively and $G(z_1, z_2)$ is the 2D transfer-function matrix given by

$$G(z_1, z_2) := \begin{bmatrix} 0 & I_m \end{bmatrix} \begin{bmatrix} I_n - z_1 A & -z_1 B_0 \\ -z_2 C & I_m - z_2 D_0 \end{bmatrix}^{-1} \begin{bmatrix} B \\ D \end{bmatrix} \tag{1.50}$$

or, after some routine manipulations,

$$G(z_1, z_2) = \left(I_m - G_1(z_1^{-1}) z_2\right)^{-1} G_0(z_1^{-1})$$

where

$$\begin{aligned} G_0(z_1^{-1}) &:= C(z_1^{-1} I_n - A)^{-1} B + D \\ G_1(z_1^{-1}) &:= C(z_1^{-1} I_n - A)^{-1} B_0 + D_0 \end{aligned}$$

The 2D transfer-function matrix for a discrete non-unit memory linear repetitive process of the form defined by Example 1.2.9 with zero boundary conditions can be derived by the same route and hence the details are omitted here. One form of the final result is

$$G(z_1, z_2) := \left(I_m - \sum_{j=1}^{M} G_j(z_1^{-1}) z_2^j\right)^{-1} G_0(z_1^{-1})$$

where $G_0(z_1^{-1})$ is given above and

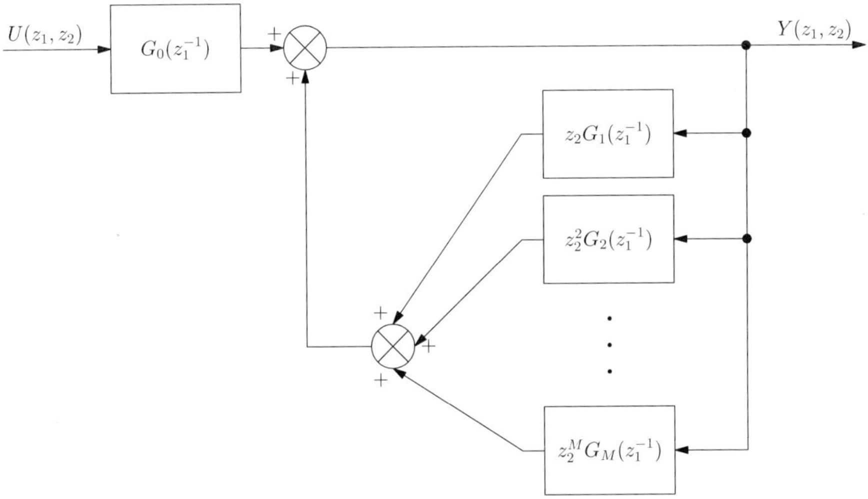

Fig. 1.10. Block diagram of 2D transfer-function matrix description – discrete case

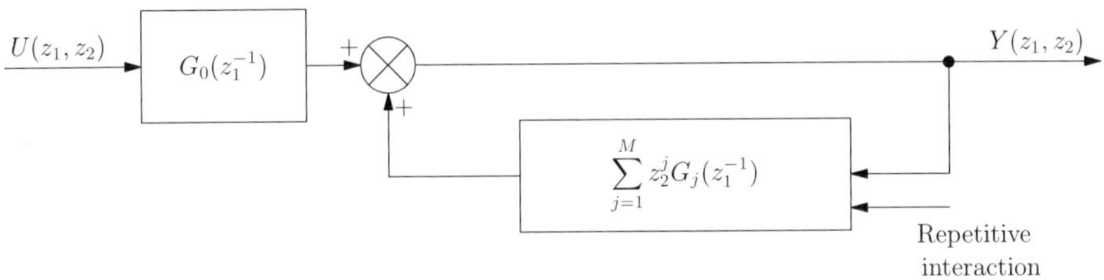

Fig. 1.11. Equivalent representation of the previous block diagram

$$G_j(z_1^{-1}) := C(z_1^{-1}I_n - A)^{-1}B_{j-1} + D_{j-1},\ 1 \le j \le M \tag{1.51}$$

This leads to the block diagram interpretation of Fig. 1.10 or 1.11. Either of these diagrams illustrate that the process dynamics in this case can be interpreted as a dynamic pre-compensator followed by a positive feedback loop with unity gain in the forward path and dynamic elements in the feedback loop. These feedback elements are the repetitive interaction. Note also that, in effect, the transfer-function matrix $G_0(z_1^{-1})$ describes the contribution of the current pass input acting alone to the current pass profile (i.e. pass $k+1$) and the transfer-function matrix $G_j(z_1^{-1})$, $1 \le j \le M$, the contribution of pass profile $k+1-j$ acting alone. This latter fact will be of particular interest in terms of the 1D Lyapunov equation based analysis of Chap. 3 and (in the unit memory case for illustration) we can describe the contribution from the previous pass dynamics using

$$y_{k+1}(z_1^{-1}) = G_1(z_1^{-1})y_k(z_1^{-1}),\ k \ge 0$$

From the structure of $G(z_1, z_2)$ and, in particular, (1.50) a natural definition of a characteristic polynomial for the unit memory case is

$$\mathcal{C}(z_1, z_2) := \det\left(\begin{bmatrix} I_n - z_1 A & -z_1 B_0 \\ -z_2 C & I_m - z_2 D_0 \end{bmatrix}\right) \tag{1.52}$$

Note here that if the previous pass terms are deleted from the defining state-space model then this characteristic polynomial reduces to that for 1D discrete linear systems. Also it will be shown in the next chapter that $\mathcal{C}(z_1, z_2)$ provides a complete characterization of stability (so-called stability along the pass). This in turn leads to a so-called 2D Lyapunov equation interpretation of stability (Chap. 3) and then (Chaps. 7 and 8) to extremely powerful Linear Matrix Inequality (LMI) based algorithms for the design of control laws (or controllers) for stability and performance.

In the case of differential linear repetitive processes of the form defined by Example 1.2.3 with zero boundary conditions we employ the Laplace transform s in the along the pass direction and the shift operator z_2 in the pass-to-pass direction, defined as per the discrete case above, and then the process dynamics can be written in the form

$$Y(s, z_2) = G(s, z_2)U(s, z_2)$$

and the 2D transfer-function matrix $G(s, z_2)$ is given by

$$G(s, z_2) := \begin{bmatrix} 0 & I_m \end{bmatrix} \begin{bmatrix} sI_n - A & -B_0 \\ -z_2 C & I_m - z_2 D_0 \end{bmatrix}^{-1} \begin{bmatrix} B \\ D \end{bmatrix}$$

or, after some routine manipulations,

$$G(s, z_2) = (I_m - G_1(s) z_2)^{-1} G_0(s) \tag{1.53}$$

where

$$\begin{aligned} G_0(s) &:= C(sI_n - A)^{-1}B + D \\ G_1(s) &:= C(sI_n - A)^{-1}B_0 + D_0 \end{aligned}$$

The 2D transfer-function matrix for a differential non-unit memory linear repetitive process of the form defined by Example 1.2.2 with zero boundary conditions can be derived by the same route and hence the details are omitted here. One form of the final result is

$$G(s, z_2) = \left(I_m - \sum_{j=1}^{M} G_j(s) z_2^j\right)^{-1} G_0(s)$$

where $G_0(s)$ is given above and

$$G_j(s) := C(sI_n - A)^{-1}B_{j-1} + D_{j-1},\ 1 \leq j \leq M \tag{1.54}$$

This 2D transfer-function matrix has an identical block diagram representation to that of Fig. 1.10 or 1.11 for its discrete counterpart.

The characteristic polynomial for the unit memory case is

$$\mathcal{C}(s, z_2) := \det\left(\begin{bmatrix} sI_n - A & -B_0 \\ -z_2 C & I_m - z_2 D_0 \end{bmatrix}\right) \tag{1.55}$$

This characteristic polynomial will pay an identical role to that above for the discrete processes in the analysis of subsequent chapters.

Remark 1.4.1. The finite pass length does not cause a problem in the application of the z_1 and s transforms for the discrete and differential cases respectively, provided the variables considered are suitably extended from $[0, \alpha]$ to $[0, \infty]$ and here we have assumed that this has been done. (This fact will also be required in various other places in this monograph.)

It is also possible to develop a Rosenbrock system matrix (see, for example [95, 159] in the 1D case) representation of the dynamics of differential and discrete linear repetitive processes of the forms defined by Examples 1.2.2 and 1.2.9 respectively, where here we limit attention to the corresponding unit memory versions (all other cases follow as natural generalizations). This immediately raises the question; what is meant by coprimeness in the context of linear repetitive processes/nD linear systems? The answer, as summarized next, is (as expected) somewhat more complex than in the 1D case and requires the following background results (given here for the general nD case). Except where indicated, the analysis which follows can be found in [89, 90] or the relevant references cited in these works.

To proceed, we will require some notation not previously introduced in this chapter. In particular, we let $H(\underline{z})$ denote a $q_1 \times q_2$ n-dimensional (nD) polynomial matrix with entries in $\underline{z} \in F[\underline{z}]$, where $\underline{z} := (z_1, \cdots, z_n)$ is the set of indeterminates and F is the coefficient ring ($\mathbb{R}$ or $\mathbb{C}$ here).

A square polynomial matrix $U(\underline{z})$ with elements in $F[\underline{z}]$ is said to be unimodular over the ring $F[\underline{z}]$ if the elements of the inverse $U^{-1}(\underline{z})$ are polynomials over $F[\underline{z}]$. Equivalently, a unimodular matrix in $F[\underline{z}]$ is one whose determinant is a unit of the coefficient ring F, i.e. an element of F with multiplicative inverse also in F.

Elementary matrices are a sub-class of unimodular matrices over a ring and play a crucial role in 1D linear systems theory, where all unimodular matrices can be formed as a product of elementary matrices. For general rings, $F[z_1, z_2, \cdots, z_n]$, $n \geq 2$, not all unimodular matrices can be formed as a product of elementary matrices. (Examples to illustrate this fact are available.)

This fact arises since there is no division algorithm in $F[z_1, z_2, \cdots, z_n]$. The existence of a division algorithm for Euclidian polynomial rings forms the basis for the algorithmic derivation of many canonical forms and solution techniques at the heart of 1D linear systems theory, e.g. the Smith form and the solution of 1D polynomial equations. In the case of $n \geq 2$ progress is possible by noting that any polynomial ring can be regarded as a subring of a larger ring with a division algorithm. The exact mechanism is to favor one of the indeterminates and consider elements of the ring to be polynomial in this indeterminate with coefficients rational in the others. If, for example, z_n is the favored indeterminate, the resulting ring is denoted by $F(z_1, z_2, \cdots, z_{n-1})[z_n]$.

A set of polynomials, say $a_i(z_1)$, $1 \leq i \leq h$, in one indeterminate are said to be factor coprime provided there is no value $\tilde{z}_1 \in \mathbb{C}$ such that they are all identically zero. If such a value exists, then $z_1 - \tilde{z}_1$ is a factor of all the polynomials in the set. In the case of $n \geq 2$, this no longer holds (examples to illustrate this are very easy to generate) and hence it is necessary to distinguish between zero coprimeness and factor coprimeness. The following fundamental result, termed Hilbert's Nullstellensatz , is of critical importance in this general area.

Lemma 1.4.1. *Let f be a polynomial in $F[z_1, z_2, \cdots, z_n]$, which vanishes at all zeros common to the polynomials $f_1, f_2, \cdots, f_c$. Then*

$$f^b = a_1 f_1 + a_2 f_2 + \cdots + a_c f_c$$

is valid for some integer b and $a_1, a_2, \cdots, a_c \in F[z_1, z_2, \cdots, z_n]$ (and conversely).

In 1D linear systems theory, a polynomial matrix with rank degeneracies can be viewed as the product of two polynomial matrices, one with full rank and the other containing the rank degeneracies. The first matrix here is termed relatively prime and the second a greatest common divisor. Consider now a $q_1 \times q_2$ nD polynomial matrix $A(\underline{z})$ with $q_1 \leq q_2$. Then there are three different notions of primeness for this matrix, termed factor, minor and zero respectively and are defined as follows.

Definition 1.4.1. *Let $A(\underline{z})$ be a $q_1 \times q_2$ nD polynomial matrix with $q_1 \leq q_2$. Then*

(i) $A(\underline{z})$ is factor (left) coprime if all factorizations have the form

$$A(\underline{z}) = Q(\underline{z})\hat{A}(\underline{z})$$

where $\hat{A}(\underline{z})$ is a polynomial matrix of dimension $q_1 \times q_2$ and the $q_1 \times q_1$ polynomial matrix $Q(\underline{z})$ is unimodular.

(ii) $A(\underline{z})$ is minor (left) coprime if all $q_1 \times q_2$ minors of this matrix form a factor coprime set of polynomials, i.e. they have no common factor other than a unit of F.

(iii) $A(\underline{z})$ is zero (left) coprime if $\exists$ no n-tuple $\underline{z} \in \mathbb{C}^n$ which is a zero of all $q_1 \times q_1$ minors of $A(\underline{z})$.

Note 1.4.1. A matrix $A(\underline{z})$ with $q_1 \geq q_2$ is defined to be factor/minor/zero (right) coprime if $A^T(\underline{z})$ is factor/minor/zero (left) coprime respectively.

The following result [179] is known as coprimeness equivalence.

Theorem 1.4.1. *For $n = 1$ all three definitions of coprimeness are the same, for $n = 2$ minor and factor coprimeness are equivalent and zero coprimeness is a different notion, and for $n \geq 3$ none of the definitions of coprimeness are equivalent. Also, for any n*

$$\text{zero coprime} \Rightarrow \text{minor coprime} \Rightarrow \text{factor coprime}$$

We also need primeness for two matrices as per the following definition in [179].

Definition 1.4.2. *Let $A(\underline{z})$ and $B(\underline{z})$ be two nD polynomial matrices of dimensions $q_1 \times q_2$ and $q_1 \times q_3$ respectively with $q_2 + q_3 \geq q_1 \geq 1$, and let*

$$C(\underline{z}) = [A(\underline{z})\ B(\underline{z})]$$

Then these matrices are said to be

(i) zero (left) coprime if there exists no n-tuple $\underline{z} \in \mathbb{C}^n$ which is a zero of all the $q_1 \times q_1$ minors of $C(\underline{z})$.
(ii) minor (left) coprime if these minors are relatively prime.
(iii) factor (left) coprime if in any polynomial decomposition

$$C(\underline{z}) = C_1(\underline{z})C_2(\underline{z})$$

where if $C_1(\underline{z})$ is square then this matrix must also be elementary.

Note 1.4.2. Zero right coprimeness etc is defined in an analogous manner (by applying the matrix transpose) and hence the details are omitted here.

Two of these three definitions of coprimeness can be characterized by necessary and sufficient conditions. These Bezout identities for zero and minor coprimeness are stated next (for the left case, with the right case following by applying the matrix transpose) and are fundamental to some of the results developed in Chap. 6.

Theorem 1.4.2. *Two nD polynomial matrices $A(\underline{z})$ and $B(\underline{z})$ of dimensions $q_1 \times q_2$ and $q_1 \times q_3$ respectively with $q_2 + q_3 \geq q_1 \geq 1$ are*

(i) zero (left) coprime if, and only if, $\exists$ two polynomial matrices $X(\underline{z})$ and $Y(\underline{z})$ of dimensions $q_2 \times q_1$ and $q_3 \times q_1$ respectively, such that

$$A(\underline{z})X(\underline{z}) + B(\underline{z})Y(\underline{z}) = I_{q_1}$$

(ii) minor (left) coprime if, and only if, $\exists$ $q_2 \times q_1$ polynomial matrices $X_1(\underline{z}), \cdots, X_n(\underline{z})$ and $q_3 \times q_1$ polynomial matrices $Y_1(\underline{z}), \cdots, Y_n(\underline{z})$ such that

$$\begin{aligned} A(\underline{z})X_1(\underline{z}) + B(\underline{z})Y_1(\underline{z}) &= \psi_1(\underline{z}_1^c)I_{q_1} \\ A(\underline{z})X_2(\underline{z}) + B(\underline{z})Y_2(\underline{z}) &= \psi_2(\underline{z}_2^c)I_{q_1} \\ &\vdots \\ A(\underline{z})X_n(\underline{z}) + B(\underline{z})Y_n(\underline{z}) &= \psi_n(\underline{z}_n^c)I_{q_1} \end{aligned}$$

where $\psi_i(\underline{z}_i^c)$ is a polynomial in the other $n-1$ indeterminates.

This result alone means that generalizing, for example, 1D Rosenbrock system matrix theory to nD linear systems is a non-trivial task. In the special case of differential and discrete linear repetitive processes of the form defined by Examples 1.2.2 and 1.2.9 respectively, substantial progress is possible – see Chap. 6. The Rosenbrock system matrix for these processes is now introduced, starting with the differential unit memory case (Example 1.2.3).

Write

$$\begin{aligned} G_0(s) &= D^{-1}(s)N(s) \in \mathbb{R}^{m\times r}(s) \\ G_1(s) &= \overline{N}(s)\overline{D}^{-1}(s) \in \mathbb{R}^{m\times m}(s) \end{aligned} \tag{1.56}$$

where $N(s), D(s)$ are (zero) left coprime, and $\overline{N}(s), \overline{D}(s)$ are (zero) right coprime. In this case, the coprimeness type is termed zero because all the matrices involved are polynomial in one indeterminate and hence all three definitions of 2D coprimeness are equivalent. Hence the 2D transfer-function matrix in this case can be written as

$$\begin{aligned} G(s,z_2) &= \overline{D}(s)\left[\overline{D}(s) - z_2\overline{N}(s)\right]^{-1} D^{-1}(s)N(s) \\ &= \overline{D}(s)\left[D(s)(\overline{D}(s) - z_2\overline{N}(s))\right]^{-1} N(s) \end{aligned} \tag{1.57}$$

and a Rosenbrock system matrix over $\mathbb{R}[s,z_2]$ giving rise to $G(s,z_2)$ is the following $(m+m)\times(m+r)$ polynomial matrix

$$P(s,z_2) := \begin{bmatrix} D(s)\left[\overline{D}(s) - z_2\overline{N}(s)\right] & N(s) \\ -\overline{D}(s) & 0 \end{bmatrix} \tag{1.58}$$

Finally, the extension to the other cases is straightforward and hence the details are omitted here.

2 Stability – Theory, Tests and Performance Bounds

Summary. This chapter presents the stability theory for the Banach space based abstract model of Definition 1.2.1, together with the results of applying it to the various forms of differential and discrete linear repetitive processes introduced in the previous chapter. Also the question of testing the resulting stability conditions is addressed by a number of routes. These lead in most cases to computationally feasible tests and, in one case of applications interest, computable bounds on expected performance.

2.1 Asymptotic Stability

As illustrated by Fig. 1.6 for the long-wall coal cutting process model, the unique undesirable feature/control problem for repetitive processes is the possible presence in the output sequence of pass profiles $\{y_k\}_{k\geq 1}$ of oscillations that increase in amplitude from pass-to-pass (i.e. in the k direction in the notation for variables used here). Hence the natural intuitive definition of asymptotic stability is to demand that, given any initial profile y_0 and any disturbance sequence $\{b_k\}_{k\geq 1}$ which 'settles down' to a steady disturbance b_∞ as $k \to \infty$, the sequence $\{y_k\}_{k\geq 1}$ generated by the abstract model S of Definition 1.2.1 'settles down' to a steady, or so-called limit, profile as $k \to \infty$. This idea is illustrated in Fig. 2.1. (Except where other references are given, the material in this section can be found in [52, 121], or [149].)

The major drawback of such an approach is that it does not explicitly include the intuitive idea that asymptotic stability should be retained if the model is perturbed slightly due to modelling errors or simulation approximations. Consequently the following definition of asymptotic stability is preferred since it ensures that the 'set of stable systems' is open (in a well defined sense) in the class of all linear repetitive processes.

Definition 2.1.1. *A linear repetitive process S of constant finite pass length $\alpha > 0$ is said to be asymptotically stable if $\exists$ a real scalar $\delta > 0$ such that, given any initial pass profile y_0 and any strongly convergent disturbance sequence $\{b_k\}_{k\geq 1} \subset W_\alpha$, the sequence $\{y_k\}_{k\geq 1}$ generated by the perturbed process*

$$y_{k+1} = (L_\alpha + \gamma)y_k + b_{k+1}, \ k \geq 0 \tag{2.1}$$

converges strongly to a limit profile $y_\infty \in E_\alpha$ whenever $||\gamma|| \leq \delta$.

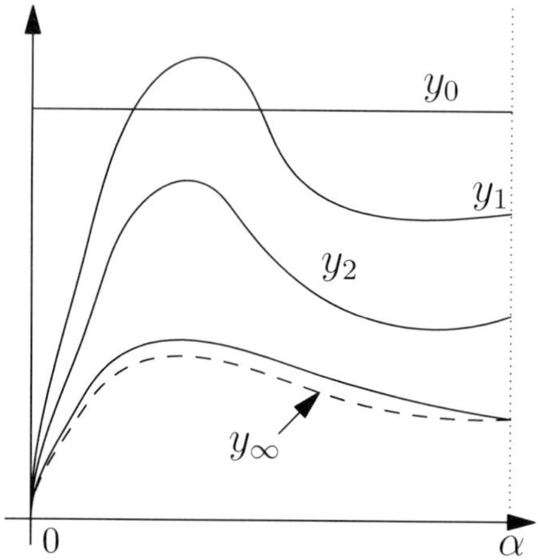

Fig. 2.1. Schematic illustration of asymptotic stability

Note: y_∞ does, of course, depend on γ, y_0, and $\{b_k\}_{k\geq 1}$.

To justify the use of the term 'asymptotic stability' in the above definition, consider the case when $b_{k+1} = 0,\ k \geq 0$, which is strongly convergent to zero. Hence, from the definition, asymptotic stability requires that the solution sequence $\{y_k\}_{k\geq 1}$ generated by (2.1) converges strongly to a limit profile y_∞ for each y_0 and model perturbation γ. Then the solution of (2.1) can be written as

$$y_k = (L_\alpha + \gamma)^k y_0,\ k \geq 0 \tag{2.2}$$

and consider the case when $\gamma = \frac{L_\alpha \delta}{||L_\alpha||}$ to yield $||\gamma|| = \delta$. Hence it follows immediately that the sequence $(L_\alpha + \gamma)^k y_0,\ k \geq 0$, is strongly convergent and hence bounded for each $y_0 \in E_\alpha$.

Application of the well known Banach-Steinhaus theorem now states that $\exists$ a real number $M_\alpha > 0$ such that

$$||(L_\alpha + \gamma)^k|| \leq M_\alpha,\ k \geq 0$$

(where (as before) $||\cdot||$ is also used to denote the induced norm) or, equivalently,

$$\left(1 + \frac{\delta}{||L_\alpha||}\right)^k ||L_\alpha^k|| \leq M_\alpha,\ k \geq 0 \tag{2.3}$$

Consider now the case of $\gamma = 0$ (the 'real' (or unperturbed) system) and introduce

$$\lambda_\alpha := \left(1 + \frac{\delta}{||L_\alpha||}\right)^{-1} < 1 \tag{2.4}$$

Then (2.3) takes the form

$$||L_\alpha^k|| \leq M_\alpha \lambda_\alpha^k,\ k \geq 0 \tag{2.5}$$

and using (2.2) it follows that

$$||y_k|| = ||L_\alpha^k y_0|| \leq ||L_\alpha^k||\ ||y_0|| \leq M_\alpha \lambda_\alpha^k ||y_0||$$

Hence, in the absence of disturbances, the output sequence $\{y_k\}_{k\geq 1}$ converges strongly to zero for all initial profiles. Physically, this requires that the effects of the initial profile are rapidly attenuated after a 'large' number of passes at a geometric rate.

The characterization of asymptotic stability of S is in terms of the spectral radius of the bounded linear operator L_α whose definition is as follows.

Definition 2.1.2. *A complex number λ is said not to be a spectral value of L_α if, and only if, the bounded linear operator $\lambda I - L_\alpha$, where I is the identity operator in E_α, has range dense in E_α and a bounded inverse $(\lambda I - L_\alpha)^{-1}$. The set, $\sigma(L_\alpha)$, of all spectral values of L_α is termed the spectrum of L_α and its spectral radius is defined to be the finite positive number*

$$r(L_\alpha) := \sup_{\lambda \in \sigma(L_\alpha)} |\lambda|$$

or, equivalently,

$$r(L_\alpha) = \lim_{k\to\infty} ||L_\alpha^k||^{\frac{1}{k}} \tag{2.6}$$

The proof of Theorem 2.1.1 below for asymptotic stability of the general case of S requires the following results stated here without proof since these follow immediately on using standard results and concepts from functional analysis. (For background on the required functional analysis see, for example, [43] or [170] and for their first use in repetitive process analysis [121].)

Lemma 2.1.1. *Consider the linear operator $L_\alpha \in B(E_\alpha, E_\alpha)$. Then we have the following results.*

(a) Suppose that $||L_\alpha|| < 1$. Then the operator $I - L_\alpha$ has a bounded inverse which can be expressed in terms of the absolutely convergent power series

$$(I - L_\alpha)^{-1} = \sum_{j=1}^{\infty} L_\alpha^{j-1}$$

and

$$||(I - L_\alpha)^{-1}|| \leq \frac{1}{1 - ||L_\alpha||}$$

(b) If $\lambda < r(L_\alpha) < 1$, then

$$\beta(\lambda) := \sup_{|z|\geq\lambda} ||(zI - L_\alpha)^{-1}|| < \infty$$

(c) If $r(L_\alpha) < 1$ then $\exists$ real scalars $\delta > 0$ and $\lambda \in (r(L_\alpha), 1)$ such that $r(L_\alpha + \gamma) < \lambda$ whenever $||\gamma|| \leq \delta$.

(d) With the notation of (c) above, suppose that $r(L_\alpha) < 1$ and $||\gamma|| \leq \delta$. Then $\exists$ a real number $M_\alpha(\gamma) > 0$ such that

$$||(L_\alpha + \gamma)^k|| \leq M_\alpha(\gamma)\lambda^k, \; k \geq 0$$

(e) Again with the above notation, suppose that $r(L_\alpha) < 1$ and $||\gamma|| \leq \delta$. Suppose also that the sequence $\{b_k\}_{k\geq 1}$ converges strongly to $b_\infty \in E_\alpha$. Then

$$\lim_{k\to\infty} \sum_{j=1}^{k} (L_\alpha + \gamma)^{j-1}(b_{k+1-j} - b_\infty) = 0$$

(the limit being interpreted in the sense of the norm).

(f) Suppose that $r(L_\alpha) < 1$ and $||\gamma|| \leq \delta$. Suppose also that the sequence $\{b_k\}_{k\geq 1}$ converges strongly to $b_\infty \in E_\alpha$. Then the power series

$$H := \sum_{j=1}^{k} (L_\alpha + \gamma)^{j-1} b_\infty$$

is absolutely convergent as $k \to \infty$.

Theorem 2.1.1. *The linear repetitive process S of constant pass length $\alpha > 0$ is asymptotically stable if, and only if,*

$$r(L_\alpha) < 1$$

Note 2.1.1. For ease of presentation, we will drop the expression of constant pass length $\alpha > 0$ from the statements of definitions, theorems and lemmas etc from this point onwards.

Proof. Suppose that S is asymptotically stable. Then use of (2.4) and (2.5) in (2.6) yields

$$r(L_\alpha) = \lim_{k\to\infty} ||L_\alpha^k||^{\frac{1}{k}} \leq \lambda_\alpha \lim_{k\to\infty} M_\alpha^{\frac{1}{k}} = \lambda_\alpha < 1 \tag{2.7}$$

Conversely, suppose that $r(L_\alpha) < 1$ and write the solution of (2.1) in the form

$$\begin{aligned} y_k &= (L_\alpha + \gamma)^k y_0 + \sum_{j=1}^{k} (L_\alpha + \gamma)^{j-1} b_{k+1-j} \\ &= (L_\alpha + \gamma)^k y_0 + \sum_{j=1}^{k} (L_\alpha + \gamma)^{j-1}(b_{k+1-j} - b_\infty) + \sum_{j=1}^{k} (L_\alpha + \gamma)^{j-1} b_\infty \end{aligned}$$

Then, since $||(L_\alpha + \gamma)^k y_0|| \leq M_\alpha(\gamma)\lambda^k ||y_0||$ by (d) of Lemma 2.1.1, it is clear that $(L_\alpha + \gamma)^k y_0 \to \infty$ as $k \to \infty$. Also it is easily verified using (e) and (f) of Lemma 2.1.1 that the sequence $\{y_k\}_{k\geq 1}$ converges strongly to $y_\infty \in E_\alpha$ where

$$y_\infty := \lim_{k\to\infty} y_k = \sum_{j=1}^{\infty}(L_\alpha + \gamma)^{j-1} b_\infty \tag{2.8}$$

and the proof is complete. □

This last result provides a necessary and sufficient condition for asymptotic stability which, as shown later in this section, can be applied to many of the special cases given in Chap. 1 to produce computationally feasible stability tests. It does not, however, provide any directly useful information as to the expected response of an asymptotically stable example to a given input sequence. In particular, it does not provide such information on what happens to the sequence of pass profiles as $k \to \infty$. The limit profile defined and characterized next removes this deficiency.

Definition 2.1.3. *Suppose that the linear repetitive process S is asymptotically stable and let $\{b_k\}_{k\geq 1}$ be a disturbance sequence that converges strongly to a disturbance b_∞. Then the strong limit*

$$y_\infty := \lim_{k\to\infty} y_k$$

is termed the limit profile corresponding to $\{b_k\}_{k\geq 1}$.

Theorem 2.1.2. *Suppose that the linear repetitive process S is asymptotically stable and let $\{b_k\}_{k\geq 1}$ be a disturbance sequence that converges strongly to a disturbance b_∞. Then the limit profile corresponding to this disturbance sequence is the unique solution of the linear equation*

$$y_\infty = L_\alpha y_\infty + b_\infty \tag{2.9}$$

Proof. Using (2.8) with $\gamma = 0$ yields a power series representation of y_∞ which satisfies (2.9). The uniqueness of y_∞ follows immediately on writing (2.9) in the form $(I - L_\alpha)y_\infty = b_\infty$ and noting, by asymptotic stability, that $r(L_\alpha) < 1$ and hence $(I - L_\alpha)$ has a bounded inverse in E_α. Equivalently, (2.9) has a unique solution which can be written in the form $y_\infty = (I - L_\alpha)^{-1} b_\infty$. □

Corollary 2.1.1. *The limit profile y_∞ is independent of the initial pass profile y_0 and the direction of approach to b_∞.*

Note 2.1.2. Formally, (2.9) can be obtained from (2.1) with $\gamma = 0$ by replacing all variables by their strong limits.

To be of use in a particular application, the abstract results of Theorems 2.1.1 and 2.1.2 must be converted (if possible) into a suitable computational form. No general rules exist for this task, other than the obvious necessity to compute the spectral values of L_α and hence their moduli. Note also that severe difficulties could arise if the space E_α and the operator L_α have a complex structure. As shown below, however, this task is possible for (most of) the processes considered in this monograph where the approach to the spectral calculations used here is to consider the equation

$$(zI - L_\alpha)y = \eta \tag{2.10}$$

and construct necessary and sufficient conditions on the complex scalar z to ensure that (i) a solution exists $\forall\, \eta \in E_\alpha$, and (ii) that this solution is bounded in the sense that $||y|| \leq K_0||\eta||$ for some real scalar $K_0 > 0$ and $\forall\, \eta \in E_\alpha$. This yields the following results.

Theorem 2.1.3. *The linear repetitive process S generated by a model of the form defined by Example 1.2.1 is asymptotically stable if, and only if,*

$$|g_2| < 1$$

Proof. In this particular case, (2.10) can be written in the form

$$zy(t) - w(t) = \eta(t) \tag{2.11}$$

where $w(t)$ is defined by

$$\begin{aligned} w(t) &= -g_1 w(t-X) + g_2 y(t),\ 0 \leq t \leq \alpha \\ w(t) &= 0,\ -X \leq t \leq 0 \end{aligned} \tag{2.12}$$

Also by assumption $X > 0$ and hence it is possible to choose an integer $h \geq 1$ such that $(h-1)X < \alpha \leq hX$ and, after some routine manipulation, (2.11)–(2.12) can be written in matrix form as

$$z\begin{bmatrix} y(t) \\ y(t+X) \\ \vdots \\ y(t+(h-1)X) \end{bmatrix} - \begin{bmatrix} w(t) \\ w(t+X) \\ \vdots \\ w(t+(h-1)X) \end{bmatrix} = \begin{bmatrix} \eta(t) \\ \eta(t+X) \\ \vdots \\ \eta(t+(h-1)X) \end{bmatrix}$$

for $0 \leq t \leq X$. Eliminating the w variables now gives

$$(zI_h - K)\begin{bmatrix} y(t) \\ y(t+X) \\ \vdots \\ y(t+(h-1)X) \end{bmatrix} = \begin{bmatrix} \eta(t) \\ \eta(t+X) \\ \vdots \\ \eta(t+(h-1)X) \end{bmatrix},\ 0 \leq t \leq X \tag{2.13}$$

where the $h \times h$ matrix K is lower triangular with elements defined as follows

$$\begin{aligned} K_{i,i} &= g_2,\ 1 \leq i \leq h \\ K_{i+1,j+1} &= K_{i,j},\ 1 \leq i,j \leq h-1,\ j < i \end{aligned}$$

Choosing $z \neq g_2$, it is clear that (2.13) has a solution at each point $t \in [0, X]$. Moreover, it is easily shown that $y(0) = 0$ and that $y(t)$ is continuous on $0 \leq t \leq \alpha$. Also, consider, without loss of generality, the norm $||x|| := \max_{1 \leq j \leq n} |x_j|$ in $\mathbb{R}^h$ and apply this to (2.13) to yield (here again $|| \cdot ||$ is also used to denote the induced norm)

$$||y|| \leq ||(zI_h - K)^{-1}|| \; ||\eta||$$

Hence the only candidate for a spectral value of L_α is $z = g_2$. In this case, $(zI_h - K)$ is singular and it is a simple exercise to construct an η such that (2.13) has no solution. This, in turn, means that $\sigma(L_\alpha) = \{g_2\}$, $r(L_\alpha) = |g_2|$ and the proof is complete. □

Suppose now that the condition in Theorem 2.1.3 holds and a strongly convergent sequence $\{r_k\}_{k \geq 1}$ is applied (with limit $r_\infty(t)$). Then a representation of the corresponding limit profile can be obtained by (see Theorem 2.1.2) replacing all variables in the defining equation (1.16) by their corresponding strong limits. This yields

$$y_\infty(t) = -g_1 y_\infty(t - X) + g_2 y_\infty(t) + g_3 r_\infty(t), \; 0 \leq t \leq \alpha$$

or, after routine manipulation,

$$y_\infty(t) = -\frac{g_1}{1 - g_2} y_\infty(t - X) + \frac{g_3}{1 - g_2} r_\infty(t), \; 0 \leq t \leq \alpha$$

Theorem 2.1.4. *The linear repetitive process S generated by a differential non-unit memory linear repetitive process of the form defined by Example 1.2.2 is asymptotically stable if, and only if,*

$$r(\tilde{D}) < 1 \tag{2.14}$$

where

$$\tilde{D} := \begin{bmatrix} 0 & I & 0 & \dots & 0 \\ 0 & 0 & I & \dots & 0 \\ 0 & 0 & 0 & \dots & 0 \\ \vdots & \vdots & \vdots & \ddots & I \\ D_{M-1} & D_{M-2} & D_{M-3} & \cdots & D_0 \end{bmatrix} \tag{2.15}$$

Proof. The abstract equation

$$(zI - L_\alpha) \begin{bmatrix} y_1 \\ y_2 \\ y_3 \\ \vdots \\ y_M \end{bmatrix} = \begin{bmatrix} \eta_1 \\ \eta_2 \\ \eta_3 \\ \vdots \\ \eta_M \end{bmatrix} \in E_\alpha^M$$

in this particular case, where L_α is defined by (1.15) and (1.19), can be written out as

$$\begin{aligned} zy_j(t) &= y_{j+1}(t) + \eta_j(t),\ 1 \le j \le M-1 \\ \dot{x}(t) &= Ax(t) + \sum_{j=1}^{M} B_{j-1}\, y_{M+1-j}(t),\ x(0) = 0 \\ zy_M(t) &= Cx(t) + \sum_{j=1}^{M} D_{j-1}\, y_{M+1-j}(t) + \eta_M(t),\ 0 \le t \le \alpha \end{aligned} \tag{2.16}$$

Using induction, it is easily seen from the first equation in (2.16) that

$$y_j(t) = z^{j-M} y_M(t) + \sum_{h=j}^{M-1} z^{j-h-1}\eta_h(t),\ 1 \le j \le M-1 \tag{2.17}$$

and using this last expression to eliminate $y_1, y_2, \cdots, y_{M-1}$ from the differential equation in (2.16) yields, over $0 \le t \le \alpha$,

$$\begin{aligned} \dot{x}(t) &= Ax(t) + \sum_{j=1}^{M} B_{j-1} z^{1-j} y_M(t) + \sum_{j=2}^{M} \sum_{i=M+1-j}^{M-1} B_{j-1}\, z^{M-j-i}\eta_i(t) \\ zy_M(t) &= Cx(t) + \sum_{j=1}^{M} D_{j-1} z^{1-j} y_M(t) \\ &+ \sum_{j=2}^{M} \sum_{i=M+1-j}^{M-1} D_{j-1} z^{M-j-i}\eta_i(t) + \eta_M(t),\ x(0) = 0 \end{aligned} \tag{2.18}$$

Now introduce the polynomial matrix

$$P(z) := zI_m - D_0 - z^{-1}D_1 - \ldots - z^{1-M}D_{M-1} \tag{2.19}$$

and suppose that $P(z)$ is nonsingular. Then (2.18) takes the form

$$\begin{aligned} \dot{x}(t) &= \left(A + \sum_{j=1}^{M} B_{j-1} z^{1-j} P^{-1}(z) C\right) x(t) \\ &+ \left(\sum_{j=1}^{M} B_{j-1} z^{1-j}\right) P^{-1}(z) \left(\eta_M(t) \right. \\ &+ \left. \sum_{i=1}^{M-1} \sum_{j=M+1-i}^{M} D_{j-1} z^{M-i-j} \eta_i(t) \right) \\ &+ \sum_{i=1}^{M-1} \sum_{j=M+1-i}^{M} B_{j-1} z^{M-i-j} \eta_i(t) \end{aligned}$$

$$y_M(t) = P^{-1}(z)\left(Cx(t) + \eta_M(t) + \sum_{i=1}^{M-1}\sum_{j=M+1-i}^{M} D_{j-1}z^{M-i-j}\eta_i(t)\right)$$

and it is clear in this case that $zI - L_\alpha$ has a bounded inverse, and hence the spectrum of L_α is a subset of the roots of $\det P(z) = 0$. Also set $t = 0$ in (2.18) and consider the case when $\eta_j(t) = 0,\ 1 \le j \le M-1$. Then

$$P(z)y_M(0) = \eta_M(0)$$

which has no solution if $\eta_M(t)$ is not in the range of $P(z)$, i.e. $zI - L_\alpha$ cannot be surjective if $\det P(z) = 0$.

At this stage, we have shown that $\sigma(L_\alpha)$ is this case is equal to the set of solutions of $\det P(z) = 0$. Hence $r(L_\alpha) < 1$ if, and only if, (2.14) holds. □

Corollary 2.1.2. *The linear repetitive process S generated by a differential unit memory linear repetitive process of the form defined by Example 1.2.3 is asymptotically stable if, and only if,* $r(D_0) < 1$.

Note 2.1.3. It is (at first) somewhat surprising to find that asymptotic stability of differential linear repetitive processes is essentially independent of the system matrices and, in particular, independent of the eigenvalues of the matrix A. This is an immediate consequence of the fact that the pass length α is finite, and will change drastically when the case of $\alpha \to \infty$ is considered later in this chapter.

Suppose now that the condition in Theorem 2.1.4 holds and a strongly convergent control input sequence $\{u_k\}_{k\ge1}$ is applied. Then a representation of the corresponding limit profile $y_\infty(t)$ can, as for the case of Theorem 2.1.3, be obtained by replacing all variables in the defining state-space model by their strong limits. The result over $0 \le t \le \alpha$ is

$$\begin{aligned}\dot{x}_\infty(t) &= Ax_\infty(t) + Bu_\infty(t) + \hat{B}y_\infty(t)\\ y_\infty(t) &= Cx_\infty(t) + Du_\infty(t) + \hat{D}y_\infty(t)\\ x_\infty(0) &= d_\infty\end{aligned} \tag{2.20}$$

where

$$\hat{B} := \sum_{j=1}^{M} B_{j-1}, \quad \hat{D} := \sum_{j=1}^{M} D_{j-1} \tag{2.21}$$

and d_∞ is the strong limit of the sequence $\{d_k\}_{k\ge1}$. Alternatively, since asymptotic stability ensures that $\det(I_m - \hat{D}) \ne 0$,

$$\begin{aligned}\dot{x}_\infty(t) &= (A + \hat{B}(I_m - \hat{D})^{-1}C)x_\infty(t) + (B + \hat{B}(I_m - \hat{D})^{-1}D)u_\infty(t)\\ y_\infty(t) &= (I_m - \hat{D})^{-1}Cx_\infty(t) + (I_m - \hat{D})^{-1}Du_\infty(t)\end{aligned} \tag{2.22}$$

Setting $M = 1$ in (2.20)–(2.22) gives the corresponding results for the unit memory case.

Consider now asymptotic stability and the corresponding limit profiles for the discrete processes of Examples 1.2.9 and 1.2.10 respectively. Then the following results are stated without proof since these follow from identical arguments to those used in establishing their differential counterparts.

Theorem 2.1.5. *The linear repetitive process S generated by a discrete non-unit memory linear repetitive process of the form defined by Example 1.2.9 is asymptotically stable if, and only if, $r(\tilde{D}) < 1$, where the matrix $\tilde{D}$ is defined by (2.15).*

Corollary 2.1.3. *The linear repetitive process S generated by a discrete unit memory linear repetitive process of the form defined by Example 1.2.10 is asymptotically stable if, and only if, $r(D_0) < 1$.*

If the condition in Theorem 2.1.5 holds, the corresponding limit profile state-space model in response to the strongly convergent input sequence $\{u_k\}_{k\geq 1}$ is given by

$$\begin{aligned} x_\infty(p+1) &= (A+\hat{B}(I_m-\hat{D})^{-1}C)x_\infty(p)+(B+\hat{B}(I_m-\hat{D})^{-1}D)u_\infty(p) \\ y_\infty(p) &= (I_m-\hat{D})^{-1}Cx_\infty(p)+(I_m-\hat{D})^{-1}Du_\infty(p) \\ x_\infty(0) &= d_\infty \end{aligned}$$

where $\hat{B}$ and $\hat{D}$ are again defined by (2.21) and (as per the differential case) d_∞ is the strong limit of the sequence $\{d_k\}_{k\geq 1}$. Setting $M=1$ in this last equation gives the corresponding unit memory result.

The limit profiles for both differential and discrete non-unit memory linear repetitive processes are described by 1D differential and discrete linear time invariant state-space models respectively. Hence if the process under consideration is asymptotically stable then its repetitive dynamics can, after a 'sufficiently large' number of passes have elapsed, be described by a 1D linear systems state-space model. The implications of this fact in terms of the control of these processes will be considered in Chap. 7. In Chap. 9, the application of these results to the ILC problem of Example 1.2.4 will be detailed.

Now we consider asymptotic stability of differential and discrete linear repetitive processes subject to dynamic boundary conditions. Consider first the differential case of Example 1.2.5. Then we have the following result with $D_0 = 0$ for simplicity.

Theorem 2.1.6. *[125] Suppose that the pair $\{A, B_0\}$ is controllable and $D_0 = 0$ for simplicity of presentation. Then the linear repetitive process S generated by a differential unit memory linear repetitive process of the form defined by Example 1.2.5 is asymptotically stable if, and only if,*

$$r(L_\alpha) = \max\{0, \sup\{|z| : z \neq 0 \;\&\; \det(zI_n - M(z)) = 0\}\} < 1$$

where

$$M(z) := \sum_{j=1}^{N} J_j C e^{\hat{A}(z)t_j}$$

and

$$\hat{A}(z) := A + z^{-1}B_0C,\ z \neq 0$$

Proof. To evaluate $r(L_\alpha)$ we examine the problem of solving (2.10), i.e. $(zI - L_\alpha)y = \eta$, for $y \in E_\alpha$ when $\eta \in E_\alpha$. In particular, we construct conditions on z such that the map $\eta \mapsto y$ is defined and bounded in E_α. If $f = L_\alpha y$ then $zy - f = \eta \Rightarrow y = z^{-1}(f + \eta)$ and the relationship between f and y is described by

$$\begin{aligned} \dot{x}(t) &= Ax(t) + B_0 y(t) \\ f(t) &= Cx(t),\ x(0) = \sum_{j=1}^{N} J_j y(t_j) \end{aligned}$$

Using $y = z^{-1}(f + \eta)$ we now have that

$$x(t) = e^{\hat{A}(z)t}\left(\sum_{j=1}^{N} z^{-1}J_j\left[Cx(t_j) + \eta(t_j)\right]\right) + \int_0^t e^{\hat{A}(z)(t-\tau)}z^{-1}B_0\eta(\tau)\,d\tau \tag{2.23}$$

The existence of a solution y to $(zI - L_\alpha)y = \eta$ is equivalent to the consistency of (2.23) for $x(t)$ at $t_1, t_2, \cdots, t_N$. (Note: Matching $y(t)$ at $t_1, t_2, \cdots, t_N$ ensures that $zy - f = \eta$ between these points.) Also a solution exists if, and only if,

$$\begin{aligned} zx(t_h) &- \sum_{j=1}^{N} e^{\hat{A}(z)t_h} J_j C x(t_j) = \sum_{j=1}^{N} e^{\hat{A}(z)t_h} J_j \eta(t_j) \\ &+ \int_0^{t_h} e^{\hat{A}(z)(t_h-\tau)} B_0 \eta(\tau)\,d\tau \end{aligned} \tag{2.24}$$

Now let $g_h(z)$, $1 \le h \le N$, denote the right-hand side of this last equation and define

$$g(z) := \left[\ g_1^T(z)\ \ \cdots\ \ g_N^T(z)\ \right]^T$$

Also let $\hat{M}(z)$ denote the $nN \times nN$ matrix with $n \times n$ block entries $\hat{M}_{hj}(z)$ where

$$\hat{M}_{hj}(z) := e^{\hat{A}(z)t_h} J_j C$$

Hence (2.24) can be written as

$$\left[zI_{nM} - \hat{M}(z)\right] X_s = g(z) \tag{2.25}$$

where $X_s := \left[\begin{array}{ccc} x^T(t_1) & \dots & x^T(t_N) \end{array}\right]^T$.

It is now routine to show that a sufficient condition for the existence of a unique solution X_s of (2.25) is that $\det(zI_{nN} - \hat{M}(z)) \neq 0$. In this situation, $z \notin \sigma(L_\alpha)$. Also it is easy to show that this non-singularity condition is necessary by using the controllability assumption to construct an f such that there is no solution to $\left[zI_{nN} - \hat{M}(z)\right] X_s = g(z)$, and hence no solution to $(zI - L_\alpha)y = f$ for that choice of f.

Finally, write

$$\hat{M}(z) = \left[\begin{array}{c} e^{\hat{A}(z)t_1} \\ e^{\hat{A}(z)t_2} \\ \vdots \\ e^{\hat{A}(z)t_N} \end{array}\right] \left[\begin{array}{cccc} J_1C & J_2C & \cdots & J_NC \end{array}\right]$$

and note that $\det(zI_{nN} - \hat{M}(z)) = z^{n(N-1)} \det(zI_n - M(z))$. The result is now proved by noting that

$$\sigma(L_\alpha) = \{z : z \neq 0,\ \det(zI_n - M(z)) = 0\} \cup \{0\}$$

□

Further simplification (reduction in dimension) is possible in some special case, e.g. the following.

Corollary 2.1.4. *Consider the linear repetitive process S generated by a differential unit memory linear repetitive process of the form defined by Example 1.2.5 (with $D_0 = 0$ for simplicity of presentation) in the special case when $J_j = \widehat{J}T_j$, $1 \leq j \leq N$, where the $n \times m$ matrix $\widehat{J}$ and the $m \times m$ matrices T_j, $1 \leq j \leq N$, have constant entries. Then S in this case is asymptotically stable if, and only if, all solutions of*

$$\det\left(zI_m - \sum_{j=1}^{N} T_j C e^{\hat{A}(z)t_j} \widehat{J}\right) = 0$$

have modulus strictly less than unity.

In general, Theorem 2.1.6 shows that the property of asymptotic stability for differential linear repetitive processes is critically dependent on the structure of $x_{k+1}(0)$, $k \geq 0$. Suppose also that this sequence is incorrectly modelled as in the process of Example 1.2.3 instead of as a special case of Example 1.2.5. Then the process under consideration could well be interpreted as asymptotically stable when in actual fact it is asymptotically unstable.

The limit profile in the case when the condition of Theorem 2.1.6 holds is given by

$$\begin{aligned}\dot{x}_\infty(t) &= (A + B_0 C)x_\infty(t) + Bu_\infty(t)\\ y_\infty(t) &= Cx_\infty(t)\\ x_\infty(0) &= (I_n - M(1))^{-1}d_\infty\end{aligned}$$

where the matrix inverse involved exists by asymptotic stability.

Consider now the delay-differential case of Example 1.2.6. Then in this case the condition for asymptotic stability follows as a special case of the last result, i.e. when $N = 1$ and $t_1 = \alpha$. In which case we have asymptotic stability if, and only if, all roots of

$$\det\left(zI_n - e^{(A+z^{-1}B_0)\alpha}\right) = 0 \tag{2.26}$$

have modulus strictly less than unity. Also set $z = e^{s\alpha}$. Then (2.26) is equivalent to the requirement that

$$\det\left(sI_n - A - B_0 e^{-s\alpha}\right) \neq 0,\ Re(s) \geq 0 \tag{2.27}$$

i.e. the repetitive process concept of asymptotic stability coincides with the normal condition for stability of this class of delay differential systems.

In the case of the optimal control application of Example 1.2.7, we have the following result (where, without loss of generality, we consider the case when $x_o = 0$, and also $b = 0$ and $b^* = 0$).

Theorem 2.1.7. *[140] The linear repetitive process S generated by Example 1.2.7 is asymptotically stable if, and only if, all solutions of*

$$\det\left(I_{2n} - \tilde{E}_0 F(\lambda)e^{(\hat{A}+\hat{B}_0F(\lambda))\alpha} - \int_0^\alpha W(\alpha,\tau)F(\lambda)e^{(\hat{A}+\hat{B}_0F(\lambda))\tau}\,d\tau\right) = 0 \tag{2.28}$$

where

$$F(\lambda) := (\lambda I_{2n+r} - \hat{D}_0)^{-1}\hat{C},$$

satisfy $|\lambda| < 1$.

Proof. Following the proof (and notation) of Theorem 1 in [140] we can write

$$\begin{aligned}\dot{x}(t) &= \hat{A}x(t) + \hat{B}_0 y(t)\\ \lambda y(t) &= \hat{C}x(t) + \hat{D}_0 y(t)\\ x(0) &= \tilde{E}_0 y(\alpha) + \int_0^\alpha W(\alpha,\tau)y(\tau)\,d\tau\end{aligned}$$

Hence

$$\begin{aligned}\dot{x}(t) &= (\hat{A} + \hat{B}_0 F(\lambda))x(t)\\ y(t) &= (\lambda I_{2n+r} - \hat{D}_0)^{-1}\hat{C}x(t)\end{aligned}$$

and therefore

$$\begin{aligned} x(t) &= e^{(\hat{A}+\hat{B}_0 F(\lambda))t} x(0) \\ y(t) &= F(\lambda) e^{(\hat{A}+\hat{B}_0 F(\lambda))t} x(0) \end{aligned}$$

It now follows that

$$\begin{aligned} x(0) &= \tilde{E}_0 F(\lambda) \left\{ e^{(\hat{A}+\hat{B}_0 F(\lambda))\alpha} x(0) \right. \\ &+ \left. \int_0^\alpha W(\alpha, \tau) F(\lambda) e^{(\hat{A}+\hat{B}_0 F(\lambda))\tau} \, d\tau \right\} x(0) \end{aligned}$$

and hence the spectral values of L_α in this case are given by (2.28). The result follows immediately. □

Consider now the case of a discrete unit memory linear repetitive process with dynamic boundary conditions. Then we have the following result.

Theorem 2.1.8. *The linear repetitive process S generated by a discrete unit memory linear repetitive process of the form defined by Example 1.2.13 with state initial vector sequence (1.35) is asymptotically stable if, and only if, all roots of*

$$\det\left(zI_n - \sum_{j=0}^{\alpha-1} J_j z (zI_m - D_0)^{-1} C (A + B_0 (zI_m - D_0)^{-1} C)^j\right) = 0 \quad (2.29)$$

have modulus strictly less than unity.

Proof. Omitted since it follows identical steps to that of Theorem 2.1.6 (See also Sect. III of [156].) □

The limit profile resulting from this last result is (with $D = 0$ for simplicity) described over $0 \leq p \leq \alpha$ by

$$\begin{aligned} x_\infty(p+1) &= (A + B_0 (I_m - D_0)^{-1} C) x_\infty(p) + B u_\infty(p) \\ y_\infty(p) &= (I_m - D_0)^{-1} C x_\infty(p) \\ x_\infty(0) &= \left(I_n - \sum_{j=1}^{\alpha-1} J_j C (A + B_0 C)^j\right)^{-1} d_\infty \end{aligned}$$

where d_∞ is the strong limit of $\{d_k\}_{k\geq 1}$.

It is also possible to interpret asymptotic stability of S in bounded-input bounded-output (BIBO) terms. This is detailed next and is a key step in developing the stronger concept of stability along the pass in the next section. The formal definition is as follows which demands that the BIBO property is also retained in the presence of small modelling errors.

Definition 2.1.4. *The linear repetitive process S is said to be BIBO stable over the finite pass length α if $\exists$ a real scalar $\delta > 0$ such that, given any initial pass profile y_0 and any disturbance sequence $\{b_k\}_{k\geq 1}$ bounded in norm (i.e. $||b_k|| \leq c_1$ for some $c_1 \geq 0$ and $\forall\, k \geq 1$), the output sequence generated by the perturbed process (2.1) is bounded in norm whenever $||\gamma|| \leq \delta$.*

The following result establishes the equivalence of the apparently different stability concepts defined by Definitions 2.1.1 and 2.1.4 respectively.

Theorem 2.1.9. *The linear repetitive process S is BIBO stable if, and only if, it is asymptotically stable.*

Proof. Suppose that S is BIBO stable. Then consideration of the bounded sequence $b_k = 0$, $k \geq 1$, and $\gamma = \frac{\delta\, L_\alpha}{||L_\alpha||}$ leads to the boundedness of (2.2) (the dynamics of S) and hence to (2.5) where λ_α is defined by (2.4). It now follows from (2.7) and Theorem 2.1.1 that asymptotic stability holds.

Conversely, suppose that S is asymptotically stable (and hence $r(L_\alpha) < 1$) and that the disturbance sequence is bounded, i.e. $||b_k|| \leq c_1$, $k \geq 1$. Also write the solution of (2.1) in the form

$$y_k = (L_\alpha + \gamma)^k y_0 + \sum_{j=1}^{k} (L_\alpha + \gamma)^{j-1} b_{k+1-j}$$

and hence (using Lemma 2.1.1 (d))

$$\begin{aligned} ||y_k|| &\leq ||(L_\alpha + \gamma)^k||\; ||y_0|| + \sum_{j=1}^{k} ||(L_\alpha + \gamma)^{j-1}||\; ||b_{k+1-j}|| \\ &\leq M_\alpha(\gamma) \left(\lambda^k ||y_0|| + \frac{c_1}{1-\lambda} \right) \\ &\leq M_\alpha(\gamma) \left(||y_0|| + \frac{c_1}{1-\lambda} \right) \end{aligned}$$

whenever $||\gamma|| \leq \delta$. It now follows immediately that S is BIBO stable. □

2.2 Stability Along the Pass

Under well defined conditions, asymptotic stability of S, i.e. BIBO stability over the finite pass length α, guarantees the existence of a limit profile. The fact that this property is not uniform with respect to α means that the resulting limit profile could produce unacceptable along the pass dynamics. A simple illustration of this fact is the following SISO differential unit memory linear repetitive process over $0 \leq t \leq \alpha$, $k \geq 0$, where β is a real scalar

$$\begin{aligned}
\dot{x}_{k+1}(t) &= -x_{k+1}(t) + u_{k+1}(t) + (1+\beta)y_k(t) \\
y_{k+1}(t) &= x_{k+1}(t) \\
x_{k+1}(0) &= 0
\end{aligned} \tag{2.30}$$

This process is asymptotically stable ($D_0 = 0$) and the resulting limit profile ($M = 1$ in (2.22) and evaluate in terms of (2.30)) is described over $0 \leq t \leq \alpha$ by

$$\dot{y}_\infty(t) = \beta y_\infty(t) + u_\infty(t),\ y_\infty(0) = 0$$

Suppose now that the above process is subject to the control input sequence $\{u_k(t)\}_{k\geq 1} = 1$ with $y_0(t) = 0,\ 0 \leq t \leq \alpha$. Then over $0 \leq t \leq \alpha$,

$$\begin{aligned}
y_1(t) &= 1 - e^{-t} \\
y_\infty(t) &= \frac{1}{\beta}(e^{\beta t} - 1)
\end{aligned}$$

Hence, despite the fact the first pass profile $y_1(t)$ is a quite acceptable 'classical' response to the unit step input $u_1(t) = 1$, the resulting limit profile can have totally unacceptable along the pass dynamic characteristics. For example, if $\beta > 0$, then the limit profile grows exponentially and can be said to be 'unstable along the pass' in an obvious intuitive sense.

The natural definition of stability along the pass for cases such as that illustrated by (2.30), is to demand that the limit profile is stable in the 1D sense as $\alpha \to \infty$ (i.e. $\beta < 0$ in (2.30)). Unfortunately, however, this intuitively appealing idea does not apply in a simple manner to processes such as those described by Example 1.2.8 (inter-pass smoothing). Consequently the definition of stability along the pass of the abstract process S is expressed in terms of the rate of approach to the limit profile as $\alpha \to \infty$. This requires the following preliminary result where ,unless other references are given, the material in this section can again be found in [52, 121], or [149].

Lemma 2.2.1. *Suppose that the linear repetitive process S is asymptotically stable and the constant disturbance sequence $b_{k+1} = b_\infty$, $k \geq 1$, generates the limit profile y_∞. Then $\exists$ real scalars $M_\alpha > 0$ and $\lambda_\alpha \in (0,1)$ such that*

$$||y_k - y_\infty|| \leq M_\alpha \lambda_\alpha^k \left\{ ||y_0|| + \frac{||b_\infty||}{1-\lambda_\alpha} \right\},\ k \geq 0$$

Proof. Since $b_{k+1} = b_\infty$, $k \geq 0$, the solution of (2.1) with $\gamma = 0$ can be written as

$$y_k = L_\alpha^k y_0 + \sum_{j=1}^{k} L_\alpha^{j-1} b_\infty$$

and, using (2.8) with $\gamma = 0$, the limit profile can be expressed as

$$y_\infty = \sum_{j=1}^{\infty} L_\alpha^{j-1} b_\infty$$

Hence the 'error' $y_k - y_\infty$ can be written in the form

$$y_k - y_\infty = L_\alpha^k y_0 - \sum_{j=k+1}^{\infty} L_\alpha^{j-1} b_\infty,\ k \geq 0$$

or, taking the norm to obtain a (numerical) estimate of convergence,

$$||y_k - y_\infty|| \leq ||L_\alpha^k||\ ||y_0|| + \sum_{j=k+1}^{\infty} ||L_\alpha^{j-1}||\ ||b_\infty||$$

The result now follows on applying (2.5) to this last equation. □

Remark 2.2.1. In effect, this last result states that under asymptotic stability the output sequence $\{y_k\}_{k\geq 1}$ generated by S approaches the resulting limit profile y_∞ at a geometric rate governed by the scalar λ_α. This result, together with Definition 2.2.1 below, plays a significant part later in this chapter when the subject is the development of computable performance bounds.

Given this last result, the following definition of stability along the pass is expressed in terms of finite bounds on the scalars M_α and λ_α as $\alpha \to \infty$. Its effective action is to demand that the rate of approach of the output sequence to the limit profile has a guaranteed geometric upper bound independent of the pass length.

Definition 2.2.1. *The linear repetitive process S is said to be stable along the pass if $\exists$ finite real scalars $M_\infty > 0$ and $\lambda_\infty \in (0,1)$, independent of α, which, for each constant disturbance sequence $b_{k+1} = b_\infty$, $k \geq 0$, ensure that the output sequence satisfies*

$$||y_k - y_\infty|| \leq M_\infty \lambda_\infty^k \left\{ ||y_0|| + \frac{||b_\infty||}{1-\lambda_\infty} \right\},\ k \geq 0 \tag{2.31}$$

Despite its well defined physical meaning, the above definition of stability along the pass is not in an appropriate form for the development of conditions under which this property holds. A more useful definition in this respect is implied by the next result which then leads to necessary and sufficient conditions for this property.

Lemma 2.2.2. *The linear repetitive process S is stable along the pass if, and only if, $\exists$ finite real scalars $M_\infty > 0$ and $\lambda_\infty \in (0,1)$, independent of α, such that the following inequality holds $\forall\, \alpha > 0$*

$$||L_\alpha^k|| \leq M_\infty \lambda_\infty^k,\ k \geq 0 \tag{2.32}$$

Proof. Suppose that S is stable along the pass and set $b_\infty = 0$ in (2.31) where y_∞ is the corresponding limit profile. Then this inequality reduces to $||y_k|| \leq M_\infty \lambda_\infty^k ||y_0||, \forall \alpha > 0$, and, since y_0 is arbitrary and $y_k = L_\alpha^k\, y_0$, (2.32) follows trivially. □

Theorem 2.2.1. *The linear repetitive process S is stable along the pass if, and only if,*

(a)

$$r_\infty := \sup_{\alpha>0} r(L_\alpha) < 1 \tag{2.33}$$

and

(b)

$$M_0 := \sup_{\alpha>0} \sup_{|z|\geq\lambda} ||(zI - L_\alpha)^{-1}|| < \infty \tag{2.34}$$

for some real number $\lambda \in (r_\infty, 1)$.

Proof. To establish necessity, note from Lemma 2.2.2 that $\forall\, \alpha > 0$,

$$r(L_\alpha) = \lim_{k\to\infty} ||L_\alpha^k||^{\frac{1}{k}} \leq M_\infty^{\frac{1}{k}} \lambda_\infty = \lambda_\infty < 1$$

and hence $r_\infty \leq \lambda_\infty < 1$. Suppose also that λ is any number in the range $\lambda_\infty < \lambda < 1$. Then it is clear that $(zI - L_\alpha)^{-1}$ can be expressed as an absolutely convergent power series in z^{-1} for $|z| \geq \lambda$ and hence $\forall\, \alpha > 0$,

$$\begin{aligned} ||(zI - L_\alpha)^{-1}|| &= |z^{-1}| \left\| \sum_{j=0}^{\infty} \frac{1}{z^j} L_\alpha^j \right\| \\ &\leq \frac{1}{\lambda} \sum_{j=0}^{\infty} \frac{1}{\lambda^j} ||L_\alpha^j|| \\ &\leq \frac{M_\infty}{\lambda} \sum_{j=0}^{\infty} \left(\frac{\lambda_\infty}{\lambda}\right)^j \\ &= \frac{M_\infty}{\lambda\,(1 - \frac{\lambda_\infty}{\lambda})} \end{aligned}$$

To prove sufficiency, suppose that $r_\infty < 1$ and consider the contour $\widehat{M}$ in the complex plane defined by the relation $\widehat{M} := \{z : |z| = \lambda\}$. Also exploit standard results from operator calculus in Banach spaces to write L_α^k as the contour integral

$$L_\alpha^k = \frac{1}{2\pi\imath} \int_{\widehat{M}} z^k (zI - L_\alpha)^{-1}\, dz$$

where $z = \lambda e^{\imath\theta}$ and $\imath$ denotes the 'square root of minus one'. Then taking norms and using (2.34) yields

$$\begin{aligned} ||L_\alpha^k|| &\leq \frac{1}{2\pi} \left\| \int_0^{2\pi} \lambda^k e^{\imath k\theta} (\lambda e^{\imath\theta} - L_\alpha)^{-1} \lambda \imath e^{\imath\theta}\, d\theta \right\| \\ &\leq \frac{1}{2\pi} \int_0^{2\pi} \lambda^{k+1} ||(\lambda e^{\imath\theta} - L_\alpha)^{-1}||\, d\theta \\ &\leq (M_0\lambda)\lambda^k \end{aligned}$$

This verifies (2.34) with $M_\infty = M_0\lambda$ and $\lambda = \lambda_\infty$ and the proof is complete. □

Note 2.2.1. It is possible to show [121] that (2.34) can be relaxed slightly to

$$M_0 := \sup_{\alpha>0} \sup_{|z|=\lambda} ||(zI - L_\alpha)^{-1}|| < \infty$$

Also (2.33) is equivalent to asymptotic stability $\forall\, \alpha > 0$. This is one reason for retaining the separate identities of (a) and (b) of Theorem 2.2.1, despite the fact that the latter does imply the former. Another reason is that in a number of cases considered to date (including those considered below) (a) has proved easier (in relative terms) to interpret than (b).

The 'boundedness' condition (b) of Theorem 2.2.1 is equivalent to the requirement that $\exists\, \lambda \in (r_\infty, 1)$ such that (2.10), i.e.

$$(zI - L_\alpha)y = \eta \tag{2.35}$$

has a uniformly bounded, with respect to α, solution y for all choices of $\eta \in E_\alpha$ satisfying $\sup_{\alpha>0} ||\eta|| < \infty$ and $\forall\, |z| \geq \lambda$. Also it is clear that, in general, this condition could prove very difficult to interpret. However, for the following cases from Sect. 1.2 the results given next are obtained.

Theorem 2.2.2. *The linear repetitive process S generated by the model of Example 1.2.1 is stable along the pass if, and only if,*

(a)

$$|g_2| < 1 \tag{2.36}$$

and

(b)

$$|g_1| < 1 - |g_2| \tag{2.37}$$

Proof. In the case of (a), it follows immediately from the proof of Theorem 2.1.3 that the spectrum of L_α in this case is independent of α and hence $r(L_\alpha) = |g_2|$, $\forall \alpha > 0$. To prove (b), first note that (2.35) in this case can be written as

$$y(t) = -\frac{zg_1}{z - g_2} y(t - X) + \frac{(g_1\eta(t - X) + \eta(t))}{z - g_2}$$

Hence it follows immediately that (2.34) holds if, and only if,

$$\sup_{|z|\geq\lambda} \left| \frac{-zg_1}{z - g_2} \right| < 1 \tag{2.38}$$

for some $\lambda \in (|g_2|, 1)$. Elementary graphical considerations now reduce (2.38) to (2.37) and the proof is complete. □

Note 2.2.2. In the above theorem (2.36) is a consequence of (2.37) and hence testing for stability along the pass reduces to the simple task of checking this latter condition.

In this particular case, some indication of the equivalence of the concepts of stability along the pass and the classical analysis of Sect. 1.1 can be obtained by considering the 'characteristic equation'

$$\rho_c(s) := 1 + g_1 e^{-sX} - g_2 e^{-s\alpha} = 0 \tag{2.39}$$

as $\alpha \to \infty$. In particular, we will show below that (2.37) is necessary and sufficient for all solutions of (2.39) to have strictly negative real parts for all choices of $\alpha > 0$.

First note that if (2.37) (and hence (2.36)) holds then

$$|1 + g_1 e^{-sX}| \geq 1 - |g_1| > |g_2| \geq |g_2 e^{-s\alpha}|$$

for all $Re(s) \geq 0$ and for all α, i.e. $\rho_c(s) \neq 0$ for these values of s and α.

Conversely, if $\rho_c(s) \neq 0$, $\mathrm{Re}(s) \geq 0$, and $\alpha > 0$, let q be a real positive parameter and set

$$s = \begin{cases} \frac{(q+\imath\pi)}{X}, & g_1 > 0 \\ \frac{(q+2\imath\pi)}{X}, & g_1 < 0 \end{cases}$$

Also let α be an even positive integer multiple of X if g_2 is positive, an odd integer multiple of X if $g_1 > 0$ and $g_2 < 0$, and an odd integer multiple of $\frac{X}{2}$ if both $g_1 < 0$ and $g_2 < 0$. It then follows that

$$\rho_c(s) = 1 - |g_1| e^{-q} - |g_2| e^{-Re(\alpha s)} \neq 0, \; q \geq 0$$

from which (2.37) follows trivially.

In the case of differential non-unit memory linear repetitive processes of the form defined by Example 1.2.2 we have the following result.

Theorem 2.2.3. *The linear repetitive process S generated by a differential non-unit linear repetitive process of the form defined by Example 1.2.2 is stable along the pass if, and only if,*

(a)

$$r_\infty = \sup \{|z| : \det P(z) = 0\} < 1$$

and

(b) $\exists$ real numbers $\epsilon > 0$ and $\lambda \in (r_\infty, 1)$ such that

$$\det \left(sI_n - A - \sum_{j=1}^{M} B_{j-1} z^{1-j} P^{-1}(z) C \right) \neq 0 \tag{2.40}$$

for all complex numbers z, s satisfying $|z| \geq \lambda$ and $Re(s) > -\epsilon$, where $P(z)$ is defined by (2.19).

Proof. To prove (a), note from Theorem 2.1.4 that $r(L_\alpha)$ is independent of α and hence the result. In the case of (b), first note again that (b) of the general result Theorem 2.2.1 is equivalent to the existence of the uniform bound defined in terms of (2.35) which, in this particular case, is equivalent to the existence of such a bound on the solutions of (2.16) as $\alpha \to \infty$. Given (2.17), this is equivalent to the existence of such a bound on the solutions of (2.18) as $\alpha \to \infty$. Also this last condition is equivalent to (b) above since the fact that all the coefficients of $\eta_1, \eta_2, \cdots, \eta_M$ in (2.18) are bounded in any region $|z| \geq \lambda$ with $\lambda > r_\infty$ reduces the boundedness condition to a stability condition on the matrix

$$\Psi(z) := A + \sum_{j=1}^{M} B_{j-1} z^{1-j} P^{-1}(z) C$$

for $|z| \geq \lambda$. The fact that this must be a strong stability condition follows from a routine argument (based some controllability and observability assumptions) which is omitted here since it is not central to the subsequent use of this result (for full details, see the proof of Theorem 3.3.5 in [149]). $\square$

Setting $M = 1$ in this last result gives the following corollary for differential unit memory linear repetitive processes.

Corollary 2.2.1. *Suppose that the pair $\{A, B_0\}$ is controllable and the pair $\{C, A\}$ is observable. Then the linear repetitive process S generated by a differential unit memory linear repetitive process of the form defined by Example 1.2.3 is stable along the pass if, and only if,*

(a) all eigenvalues of the matrix D_0 have modulus strictly less than unity, and
(b) $\exists$ real numbers $\epsilon > 0$ and $\lambda \in (r(D_0), 1)$ such that all eigenvalues of the matrix $A + B_0(zI_m - D_0)^{-1}C$ lie to the left of the line $Re(s) = -\epsilon$, for all $|z| \geq \lambda$.

In the case of the discrete processes, the following results are obtained (in an analogous manner to their differential counterparts).

Theorem 2.2.4. *The linear repetitive process S generated by a discrete non-unit memory linear repetitive process of the form defined by Example 1.2.9 is stable along the pass if, and only if,*

(a)

$$r_\infty = \sup \{|z| : \det P(z) = 0\} < 1$$

and

(b) $\exists$ real numbers $\epsilon > 0$ and $\lambda \in (r_\infty, 1)$ such that

$$\det \left(z_1 I_n - A - \sum_{j=1}^{M} B_{j-1} z^{1-j} P^{-1}(z) C \right) \neq 0$$

for all complex numbers z_1, z satisfying $|z_1| \geq 1 - \epsilon$ and $|z| \geq \lambda$, where $P(z)$ is again defined by (2.19).

Corollary 2.2.2. *Suppose that the pair $\{A, B_0\}$ is controllable and the pair $\{C, A\}$ is observable. Then the linear repetitive process S generated by a discrete unit memory linear repetitive process of the form defined by Example 1.2.10 is stable along the pass if, and only if,*

(a) all eigenvalues of the matrix D_0 have modulus strictly less than unity, and
(b) $\exists$ real numbers $\epsilon > 0$ and $\lambda \in (r(D_0), 1)$ such that all eigenvalues of the matrix $A + B_0(zI_m - D_0)^{-1}C$ have modulus less than or equal to $1 - \epsilon$ for all choices of $|z| \geq \lambda$.

A major difficulty with these conditions is that they are not computationally feasible, i.e. they do not lead to implementable stability tests. This key area is the subject of the next section for which the following equivalent sets of necessary and sufficient conditions for stability along the pass (Theorem 3.3.7 in [149]) are a starting point.

Theorem 2.2.5. *The conditions of Theorem 2.2.3 are equivalent to*

(a) all eigenvalues of the block companion matrix $\tilde{D}$ of (2.15) have modulus strictly less than unity, i.e.

$$r(\tilde{D}) < 1$$

(b)

$$\det(sI_n - A) \neq 0, \ Re(s) \geq 0$$

and

(c) all eigenvalues of the block companion transfer-function matrix

$$G(s) := \begin{bmatrix} 0 & I_m & 0 & \cdots & 0 \\ 0 & 0 & I_m & \cdots & 0 \\ \vdots & \vdots & \vdots & \ddots & \vdots \\ 0 & 0 & 0 & \cdots & I_m \\ G_M(s) & G_{M-1}(s) & G_{M-2}(s) & \cdots & G_1(s) \end{bmatrix}$$

have modulus strictly less than unity for $s = \imath\omega, \forall\omega \geq 0$, where (see (1.54))

$$G_j(s) = C(sI_n - A)^{-1}B_0 + D_{j-1}, \ 1 \leq j \leq M \tag{2.41}$$

Setting $M = 1$ in this last theorem generates the unit memory case.

Corollary 2.2.3. *The conditions of Corollary 2.2.1 are equivalent to*

(a)

$$r(D_0) < 1 \text{ and } \det(sI_n - A) \neq 0, \ Re(s) \geq 0$$

and

(b) all eigenvalues of the transfer-function matrix $G_1(s)$ defined by setting $M = 1$ in (2.41) have modulus strictly less than unity for $s = \imath\omega$, $\forall\, \omega \geq 0$.

The corresponding results for the discrete cases are as follows.

Theorem 2.2.6. *The conditions of Theorem 2.2.4 are equivalent to*

(a)

$$r(\tilde{D}) < 1$$

where $\tilde{D}$ is again defined by (2.15),

(b)

$$r(A) < 1$$

and

(c) all eigenvalues of the block-companion transfer-function matrix

$$G(z_1^{-1}) := \begin{bmatrix} 0 & I_m & 0 & \cdots & 0 \\ 0 & 0 & I_m & \cdots & 0 \\ \vdots & \vdots & \vdots & \ddots & \vdots \\ 0 & 0 & 0 & \cdots & I_m \\ G_M(z_1^{-1}) & G_{M-1}(z_1^{-1}) & G_{M-2}(z_1^{-1}) & \cdots & G_1(z_1^{-1}) \end{bmatrix}$$

where (see (1.51)),

$$G_j(z_1^{-1}) = C(z_1^{-1} I_n - A)^{-1} B_{j-1} + D_j,\ 1 \leq j \leq M \qquad (2.42)$$

have modulus strictly less than unity $\forall\, |z_1^{-1}| = 1$.

Corollary 2.2.4. *The conditions of Corollary 2.2.2 are equivalent to*

(a)

$$r(D_0) < 1,\ r(A) < 1$$

and,

(b) all eigenvalues of the transfer-function matrix $G_1(z_1^{-1})$ defined by setting $M = 1$ in (2.42) have modulus strictly less than unity $\forall\, |z_1^{-1}| = 1$.

It is also possible to express stability along the pass of processes described by Examples 1.2.3 and 1.2.10 in terms of their characteristic polynomials defined by (1.55) and (1.52) respectively. These results are stated next (consult the cited reference for the proof in each case).

Theorem 2.2.7. *[153] Suppose that the pair $\{A, B_0\}$ is controllable and the pair $\{C, A\}$ is observable. Then the linear repetitive process S generated by a differential unit memory linear repetitive process of the form defined by Example 1.2.3 is stable along the pass if, and only if, its characteristic polynomial $\mathcal{C}(s, z_2)$ defined by (1.55) satisfies*

$$\mathcal{C}(s, z_2) \neq 0 \text{ in } \mathbb{I} \times \mathbb{U} \qquad (2.43)$$

where $\mathbb{I} := \{s : Re(s) \geq 0\}$ and $\mathbb{U} := \{z_2 : |z_2| \leq 1\}$.

Theorem 2.2.8. *[152] Suppose that the pair $\{A, B_0\}$ is controllable and the pair $\{C, A\}$ is observable. Then the linear repetitive process S generated by a discrete unit memory linear repetitive process of the form defined by Example 1.2.10 is stable along the pass if, and only if, its characteristic polynomial $\mathcal{C}(z_1, z_2)$ defined by (1.52) satisfies*

$$\mathcal{C}(z_1, z_2) \neq 0 \text{ in } \mathbb{U}_c^2 \tag{2.44}$$

where $\mathbb{U}_c^2 := \{(z_1, z_2) : |z_1| \leq 1,\ |z_2| \leq 1\}$ denotes the closed unit bidisc.

As noted in Chap. 1, these last two results will lead to a so-called 2D Lyapunov equation interpretation of stability along the pass (in the next two chapters) and then (Chaps. 7 and 8) to extremely powerful LMI based algorithms for the design of control laws for stability and performance. (See also [24] where the links between stability along the pass of discrete (unit memory) linear repetitive processes and BIBO stability of 2D linear systems described by the Roesser model were first established.)

In the case of a differential unit memory process of the form defined by Example 1.2.5, there are two possible cases which need to be considered. The first of these is that as $\alpha \to \infty$ we keep N and t_j fixed and the second is that as $\alpha \to \infty$, we allow $N \to \infty$ and $t_j \to \infty$. Of these cases, the first is more practically relevant and the task now is to examine the equation $(zI - L_\alpha)y = \eta$ in E_α where $\alpha \geq t_N$ and $\eta \in E_\alpha$. We require that:

(i) $\forall\, \alpha \geq t_N$, $r(L_\alpha) < 1$, which, due to the assumptions on N and t_j holds if, and only if, Theorem 2.1.6 holds; and
(ii) $\exists\, \lambda \in (r_\infty, 1)$ such that the map $f \mapsto y$ is defined and uniformly bounded with respect to $\alpha \geq t_N$ and $|z| \geq \lambda$.

We now have the following result.

Theorem 2.2.9. *[125] Suppose that the pair $\{A, B_0\}$ is controllable and the pair $\{C, A\}$ is observable. Then the linear repetitive process S generated by a differential unit memory linear repetitive process of the form defined by Example 1.2.5 is stable along the pass if, and only if,*

(a) the condition of Theorem 2.1.6 holds,
(b) all eigenvalues of the matrix A have strictly negative real parts, and
(c)

$$\sup_{\omega \geq 0} r(G_{dc}(\imath\omega)) < 1 \tag{2.45}$$

where

$$G_{dc}(s) := C(sI_n - A)^{-1}B_0$$

Proof. The necessity of (a) follows as $r(L_\alpha) \leq r_\infty < 1$ requires asymptotic stability on finite intervals. To prove the necessity of (b) it is necessary to consider unbounded intervals. Write $[0, \infty) = [0, \alpha_0] \cup [\alpha_0, \infty)$ with $\alpha_0 > t_N$ and let $f \in E_\infty$. Then (b) of Theorem 2.2.1 shows that $y \in E_\infty$. Also by the controllability assumption we can choose η such that $x(\alpha_0)$ in (2.23) for this case is arbitrary and the resultant response (i.e. $y(t)$ generated by the solution of (2.23)) on $[\alpha_0, \infty)$ is then given by $y(t) = Ce^{\hat{A}(z)(t-\alpha_0)}x(\alpha_0))$. The uniform boundedness of the response, and the observability assumption, now establish that $\hat{A}(z)$ is Hurwitz (i.e. all eigenvalues have strictly negative real parts) for all choices of complex z in the range $\{z : |z| \geq \lambda\}$. The necessity of (b) now follows from this fact, the multivariable Nyquist criterion and the matrix identity

$$\det(sI_n - \hat{A}(z)) = \frac{\det(sI_n - A)\det(zI_m - G_{dc}(s))}{z^m}$$

To prove sufficiency, first note that (a) and the fact that the spectrum of L_α is independent of α ensures that $r_\infty < 1$ for any $\alpha > t_N$. Now consider the equation $(zI - L_\alpha)y = \eta$. Condition (a) then ensures that the map $(zI - L_\alpha)^{-1} : \eta \mapsto y$ in $L_2^m[0, \alpha_0] \cap L_\infty[0, \alpha_0]$ is uniformly bounded on any interval $[0, \alpha_0]$ with $\alpha_0 > t_N$. Condition (b) shows that the corresponding matrix $\hat{A}(z)$ is bounded and stable for all complex z in the set $\{z : |z| \geq \lambda\}$ where λ is any point in the non-empty set $((\sup_{\omega \geq 0} r(G_{dc}(\imath\omega)), 1)$. Hence for these values of z $\exists \tilde{M} > 0$ and $\epsilon > 0$ such that $||e^{\hat{A}(z)t}|| \leq \tilde{M}e^{-\epsilon t}$. The map $\eta \mapsto y$ in $L_2^m[\alpha_0, \alpha) \cap L_\infty[\alpha_0, \alpha)$ is hence also uniformly bounded over the infinite range $\alpha \geq \alpha_0$ and $\{z : |z| \geq \lambda\}$. Condition (b) of Theorem 2.2.1 follows by combining these results. □

In the case of the delay-differential processes of Example 1.2.6, the following result is obtained as a special case of Theorem 2.2.9.

Theorem 2.2.10. *The linear repetitive process S generated by a differential unit memory linear repetitive process of the form defined by Example 1.2.6 is stable along the pass if, and only if, (2.27) holds, all eigenvalues of the matrix A have strictly negative real parts, and (2.45) holds.*

In the case of discrete unit memory processes of the form of defined by Example 1.2.13 we have the following result whose proof is omitted here since it follows identical steps (with routine changes where appropriate) to that of Theorem 2.2.9.

Theorem 2.2.11. *The linear repetitive process S generated by a discrete unit memory linear repetitive process of the form defined by Example 1.2.13 (with state initial vector sequence defined by (1.35)) is stable along the pass if, and only if, the condition of Theorem 2.1.8 holds, $r(A) < 1$, and all eigenvalues of the transfer-function matrix $G_1(z_1^{-1})$ obtained by setting $M = 1$ in (2.42) have modulus strictly less than unity $\forall\, |z_1^{-1}| = 1$.*

2.3 Stability Tests and Performance Bounds

Consider first the problem of determining the stability along the pass properties of a differential non-unit memory linear repetitive process of the form defined by Example 1.2.2. Then it is clear that the three conditions of Theorem 2.2.5 can be tested by direct application of appropriate 1D linear systems stability tests. In particular, (a) and (b) of this result are straightforward and, with the emphasis on computer aided analysis, (c) can be tested by plotting the continuous curves, or characteristic loci, generated by the eigenvalues $g_i(s)$, $1 \leq i \leq mM$, of $G(s)$, $s = \imath\omega$, $\forall\, \omega \geq 0$, in the complex plane and checking whether or not they lie completely inside the unit circle.

This last test is particularly simple in the SISO unit memory case when (c) reduces to

$$|G_1(\imath\omega)| < 1, \;\forall\, \omega \geq 0 \tag{2.46}$$

or, equivalently, the frequency response (i.e. Nyquist) plot of $G_1(s)$ lies completely inside the unit circle in the complex plane. Also a physical interpretation of these conditions is possible where (a) and (b) are to expected intuitively. In particular, suppose that $x_{k+1}(0) = 0$, $k \geq 0$, and $u_{k+1}(t) = 0$, $0 \leq t \leq \alpha$, $k \geq 0$, then $y_k(0) = D_0^k y_0(0)$. Hence asymptotic stability means that the sequence of initial pass profile responses $\{y_k(0)\}_{k\geq 1}$ does not become unbounded (in a well defined sense) as $k \to \infty$.

In the case of (b) in the result under consideration, it is easily seen that this ensures that the contribution to the pass profile from the current pass input is uniformly bounded with respect to the pass length. As the example of (2.30) demonstrates, however, this is only a necessary condition for stability along the pass. In the case of (c), a little manipulation in the case of a zero initial state vector and control input sequences, enables the process to be described by

$$y_{k+1}(s) = G_1(s)y_k(s), \; k \geq 0$$

Setting $s = \imath\omega$ gives $y_k(\imath\omega) = G_1^k(\imath\omega)y_0(\imath\omega)$, $k \geq 0$. Hence (noting (2.46)) this condition is equivalent to the requirement that each frequency component of the initial profile is attenuated from pass-to-pass.

For discrete linear repetitive processes of the form defined by Examples 1.2.9 and 1.2.10 respectively, it is clear that the above discussion transfers directly with, in effect, the imaginary axis in the complex plane replaced by the boundary of the unit circle in the same plane. Hence the details are omitted here. For the discrete processes of Example 1.2.13, the following result shows that stability along the pass can also be determined by direct application of 1D linear systems tests. Its effective action is to replace the condition of Theorem 2.1.8 by, in effect, the requirement that the state matrix $\mathfrak{B}_7$ of the 1D equivalent model of (1.47)–(1.49) is stable in the 1D discrete sense.

Theorem 2.3.1. *[156] The linear repetitive process S generated a discrete unit memory linear repetitive process of the form defined by Example 1.2.13* *with state initial vector sequence (1.35) is stable along the pass if, and only if,*

(a) $r(\mathfrak{B}_7) < 1$ *and* $r(A) < 1$; *and*
(b) all eigenvalues of the transfer-function matrix $G_1(z_1^{-1})$ *obtained by setting* $M = 1$ *in (2.42) have modulus strictly less than unity* $\forall\,|z_1^{-1}| = 1$.

Proof. It is clearly enough to prove that $r(\mathfrak{B}_7) < 1$ is equivalent to the condition of Theorem 2.1.8. First consider (2.29) and note that, by application of suitable block elementary operations which leave the determinant invariant, that this condition is equivalent to

$$\det \begin{pmatrix} I_n & -W & \cdots & 0 & 0 \\ 0 & I_n & \cdots & 0 & 0 \\ \vdots & \vdots & \ddots & \ddots & \vdots \\ 0 & 0 & \cdots & I_n & -W \\ -W_{\alpha-1} & -W_{\alpha-2} & \cdots & -W_1 & zI_n - W_0 \end{pmatrix} \neq 0, \forall\,|z| \geq 1 \quad (2.47)$$

where

$$\begin{aligned} W &:= A + B_0(zI_m - D_0)^{-1}C \\ W_j &:= J_j z(zI_m - D_0)^{-1}C,\ 0 \leq j \leq \alpha - 1 \end{aligned}$$

Now multiply the first $\alpha - 1$ block rows by the scalar z (which does not add any non-zero roots to the resulting polynomial) and then perform the following block elementary operations.

(i) Multiply the last block row from the left by the matrix A and add the result to the preceding block row.
(ii) Multiply the penultimate row of the matrix resulting from (i) by A and add it to the block row above.
(iii) Continue in this manner until (ii) is applied to the first two block rows.

These last operations show that (2.47) is equivalent to

$$\det \begin{pmatrix} zI_n - X_{1,1} & -X_{1,2} & \cdots & -X_{1,\alpha} \\ \vdots & \vdots & \ddots & \vdots \\ -X_{\alpha-1,1} & -X_{\alpha-1,2} & \cdots & -X_{\alpha-1,\alpha} \\ -X_{\alpha,1} & -X_{\alpha,2} & \cdots & zI_n - X_{\alpha,\alpha} \end{pmatrix} \neq 0,\ \forall\,|z| \geq 1 \quad (2.48)$$

where for $i, j = 1, 2, \cdots, \alpha$,

$$X_{i,j} := \begin{cases} A^{\alpha-i}J_{\alpha-j}\psi(z)C, & i \geq j \\ A^{\alpha-i}J_{\alpha-j}\psi(z)C - A^{j-i-1}B_0\psi(z)C, & i < j \end{cases}$$

and $\psi(z) := z(zI_m - D_0)^{-1}$. Now we can write (2.48) in the form

$$\det(zI_{\alpha m} - PR) \neq 0,\ \forall\,|z| \geq 1$$

where P is of dimension $\alpha n \times \alpha m$ and R is an $\alpha m \times \alpha m$ diagonal matrix with each diagonal entry equal to C. Also $\det(zI - PR) = \det(zI - RP)$, and hence (2.48) is equivalent to

$$\det \begin{pmatrix} zI_m - Z_{1,1} & -Z_{1,2} & \cdots & -Z_{1,\alpha} \\ \vdots & \vdots & \ddots & \vdots \\ -Z_{\alpha-1,1} & -Z_{\alpha-1,2} & \cdots & -Z_{\alpha-1,\alpha} \\ -Z_{\alpha,1} & -Z_{\alpha,2} & \cdots & zI_m - Z_{\alpha,\alpha} \end{pmatrix} \neq 0, \ \forall\, |z| \geq 1$$

where for $i, j = 1, 2, \cdots, \alpha$,

$$Z_{i,j} := \begin{cases} CA^{\alpha-i} J_{\alpha-j} \psi(z), & i \geq j \\ CA^{\alpha-i} J_{\alpha-j} \psi(z) - CA^{j-i-1} B_0 \psi(z), & i < j \end{cases}$$

Now, extract the matrix $\psi(z)$ from each block row on the right (and note that $r(D_0) < 1$ by assumption), and the proof is completed by applying appropriate block row and column permutations. □

One possible difficulty with this last result is numerical problems arising from the (potentially large) dimensions of the matrix $\mathfrak{B}_7$. The following result, whose proof is trivial and hence omitted here, gives an alternative set of conditions which could be more appropriate (e.g. in computation terms) for a given case.

Theorem 2.3.2. *The linear repetitive process S generated by a discrete unit memory linear repetitive process of the form defined by Example 1.2.13 with pass state initial vector sequence (1.35) is stable along the pass if, and only if,*

(a) all eigenvalues of the transfer-function matrix

$$\tilde{G}(z) := J(zI_{\alpha m} - \mathfrak{B}_7)^{-1} \mathfrak{B}_3$$

where $\mathfrak{B}_7$, $\mathfrak{B}_3$, and J are as in (1.47)–(1.49), have modulus strictly less than unity $\forall\, |z| = 1$.

(b) $r(A) < 1$; and

(c) (b) of Theorem 2.3.1 holds.

In the differential case of Example 1.2.5, no real progress is possible and this condition will have to analyzed numerically. Example 1.2.6 is considered again in Chap. 7. In the case of the optimal control application of Example 1.2.7, the structure of the matrix $\hat{A}$ means that it can never have all its eigenvalues in the open-left half of the complex plane. Hence this is a case where asymptotic stability is all that can be achieved. Finally, in the case of Example 1.2.8 (inter-pass smoothing), it is not possible to explicitly characterize the spectral values in manner which leads to stability tests – a problem which is returned to later in this section.

Consider again the results of Theorems 2.2.5 and 2.2.6. Then a major deficiency, relative to the 1D case, is that the 'Nyquist-like' tests which can be used to examine the stability along the pass conditions do not also provide any 'meaningful indicators' as to expected performance achievable for a stable along the pass example. In particular, it is clear that computable information concerning the following aspects of the performance of a stable example should be extremely useful for, say, control law design studies.

(a) The rate of approach of the output sequence of pass profiles $\{y_k\}_{k\geq 1}$ to the resulting limit profile y_∞.
(b) The error $y_k - y_\infty$ on any pass $k \geq 1$.

In the remainder of this section we develop an approach to stability testing which yields such information (at the expense of using sufficient but not necessary conditions for stability along the pass) at, in effect, no extra computational cost.

The analysis below requires some preliminary results which are given next. Also no loss of generality arises in restricting attention to the unit memory case. Finally, we show that the resulting tests, unlike their counterparts above, can also be extended in a straightforward manner to processes of the form defined by Example 1.2.8. The results in this section for the differential cases are from [148, 150] for processes of the form defined by Example 1.2.3 and [155] for those defined by Example 1.2.8. The discrete process results are from [151].

The following result is central to the stability analysis which follows. Its proof is well known in the literature and is hence omitted here.

Lemma 2.3.1. *Suppose that f is a scalar continuous function defined on $[0, \infty]$ and of bounded variation on any finite interval $[0, T]$. Then the total variation of f on $[0, T]$ can be written as*

$$N_T(f) := |f(0+)| + \sum_{j=1}^{h} |f(t_j) - f(t_{j-1})| + |f(T) - f(t_h)|$$

where $0 = t_0 < t_1 < t_2 \cdots$ are the local maxima and minima of f on the extended half line and h is defined by the relation $t_h \leq T < t_{h+1}$. In the case of $T = \infty$

$$N_\infty(f) := \sup_{T\geq 0} N_T(f) = \lim_{T\to\infty} N_T(f)$$

Figure 2.2 illustrates the computation of $N_T(f)$.

Some extra notation is required before stating the next result, starting with $[\cdot]_p$ which has a dual role here depending on whether it applies to a matrix or a matrix of operators. In the former case, $[F]_p$ denotes the matrix where each element is formed by taking the absolute value of the corresponding element in F; in the latter case , $[F]_p$ denotes the matrix where

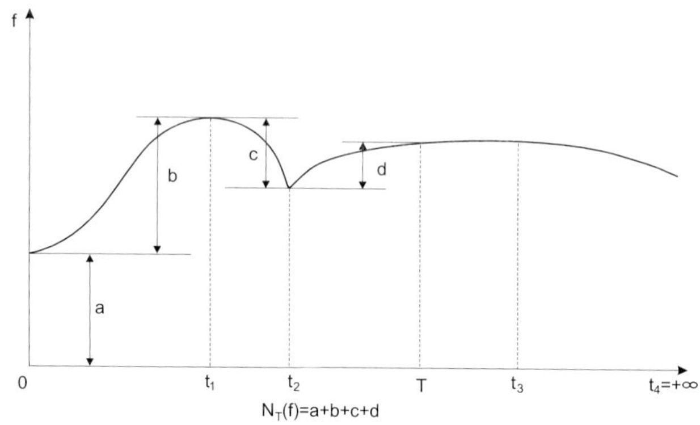

Fig. 2.2. Illustrating the computation of $N_T(f)$

each element is formed by taking the induced norm of the corresponding element in F. Also, if y is the $m \times 1$ vector, then $[y]_p$ is the column vector with each element equal to the L_∞ norm of the corresponding entry in y. If the elements in y are denoted by y_i, $1 \leq i \leq m$, then the norm of $[y]_p$ is given by $||[y]_p|| := \max_i \; |y_i|$. The same symbol is also used (as previously in this work) to denote the matrix norm induced by this vector norm, i.e. $||[F]_p|| = \max_i \sum_j f_{ij}$, where f_{ij} denotes the (i,j) element in $[F]_p$. Also we need the element-wise partial ordering on (compatible) matrices defined by $A_{ij} \leq B_{ij}$ if, and only if, the elements of these matrices, denoted by the subscript i, j, satisfy $A_{ij} \leq B_{ij}\ \forall\, i$ and j.

Lemma 2.3.2. *Given $\alpha > 0$, suppose that the operator*

$$(Ky)(t) := \int_0^\alpha \mathbb{K}(t,\tau)y(\tau)\, d\tau$$

maps $L_\infty^m(0,\alpha)$ into itself. Then

$$[K]_p \leq \sup_{t \geq 0} \int_0^t [\mathbb{K}(t,\tau)]_p\, d\tau \tag{2.49}$$

where the supremum is taken with respect to the element-wise partial ordering $\leq$ defined above. Also equality holds in (2.49) if the kernel $\mathbb{K}(t,\tau)$ is positive (pointwise) at $\alpha = \infty$.

Proof. This follows from straightforward arguments based on the definition of each element of $[\cdot]_p$. Hence the details are omitted. □

Note 2.3.1. The supremum in (2.49) is assumed to be finite.

Return now to Example 1.2.3 and suppose that all eigenvalues of the matrix A has strictly negative real parts – a necessary condition for stability along the pass and, for simplicity, $x_{k+1}(0) = 0,\ k \geq 0$. Consider also the 1D differential linear system in $L_\infty(0,\infty)$ defined in terms of the state-space quadruple $\{A, B_0, C, D_0\}$ of such a repetitive process by the convolution operator L_0,

$$w(t) := (L_0 y)(t) = C\int_0^t e^{A(t-\tau)} B_0 y(\tau)\, d\tau + D_0 y(t)$$

Suppose also that $W(t)$ denotes the step-response matrix of this system, i.e.

$$W(t) := C\int_0^t e^{A\tau} B_0\, d\tau + D_0,\ t \geq 0$$

Then it follows immediately that

$$[w]_p \leq [L_0]_p [y]_p$$

Also, each element in the matrix $[L_0]_p$ is the total variation of the corresponding entry in $W(t)$ on $[0,\infty]$ and (as discussed below) is easily computed or estimated from experimental data.

It now follows immediately that in the case of a differential unit memory linear repetitive process of the form defined by Example 1.2.3 the bounded linear operator L_α of S satisfies

$$[L_\alpha]_p \leq [L_0]_p,\ \forall\, \alpha > 0 \tag{2.50}$$

The following result now expresses stability along the pass here in terms of $[L_0]_p$.

Theorem 2.3.3. *The linear repetitive process S generated by a differential unit memory linear repetitive process of the form defined by Example 1.2.3 is stable along the pass if*

$$r([L_0]_p) < 1 \tag{2.51}$$

Proof. It is required to show that (2.51) is a sufficient condition for (a) and (b) of Theorem 2.2.1 to hold for this case, where the proof for (a) follows immediately from application of the standard fact that for square matrices P and Q of the same dimensions

$$0 \leq [P]_p \leq Q \Rightarrow r(P) \leq r([P]_p) \leq r(Q) \tag{2.52}$$

To prove (b), first note that if $r(L_\alpha) < |z|$ then $z^{-1}(I - \frac{L_\alpha}{z})^{-1}$ can be represented by the absolutely convergent power series

$$\frac{1}{z}\left(I - \frac{L_\alpha}{z}\right)^{-1} = \frac{1}{z}\left(I + \frac{L_\alpha}{z} + \left(\frac{L_\alpha}{z}\right)^2 + \cdots\right)$$

Use of some obvious norm-like properties of $[\cdot]_p$ applied to matrices, and (2.50), yields

$$[(zI - L_\alpha)^{-1}]_p \leq \frac{1}{|z|}\sum_{j=0}^{\infty}\frac{[L_\alpha^j]_p}{|z|^j} \leq \frac{1}{|z|}\sum_{j=0}^{\infty}\frac{[L_\alpha]_p^j}{|z|^j} \leq \frac{1}{|z|}\sum_{j=0}^{\infty}\frac{[L_0]_p^j}{|z|^j} \tag{2.53}$$

Now consider the last term in (2.53) and note that $r([L_0]_p) < 1 \Rightarrow \exists$ a real number $\lambda \in (r([L_0]_p), 1)$ and hence

$$\begin{aligned} \frac{1}{|z|}\sum_{j=0}^{\infty}\frac{[L_0]_p^j}{|z|^j} &\leq \frac{1}{\lambda}\sum_{j=0}^{\infty}\frac{[L_0]_p^j}{\lambda^j} \\ &= (\lambda I_m - [L_0]_p)^{-1} =: \tilde{M} < \infty \end{aligned} \tag{2.54}$$

Combining (2.53) and (2.54) yields

$$[(zI - L_\alpha)^{-1}]_p \leq \tilde{M},\ \forall\, \alpha > 0,\ \forall\, |z| \geq \lambda \tag{2.55}$$

Finally, taking the norm of (2.55) now leads immediately to the conclusion that

$$\sup_{\alpha>0}\ \sup_{\substack{|z|\geq\lambda \\ z\in\mathbb{C}}} ||([(zI - L_\alpha)^{-1}]_p)|| = \sup_{\alpha>0}\ \sup_{\substack{|z|\geq\lambda \\ z\in\mathbb{C}}} ||(zI - L_\alpha)^{-1}|| \leq ||\tilde{M}||$$

and the proof is complete. □

As an alternative to the test of Theorem 2.3.3, compute $||([L_0]_p)||$. Then using the spectral radius inequality

$$r([L_0]_p) \leq ||([L_0]_p)||$$

and (2.52) yields

$$r(L_\alpha) \leq r([L_\alpha]_p) \leq r([L_0]_p) \leq ||([L_0]_p)||,\ \forall\, \alpha > 0$$

Hence $||([L_0]_p)|| < 1$ is a sufficient condition for stability along the pass of processes described by Example 1.2.3. This result is stated formally as follows.

Theorem 2.3.4. *The linear repetitive process S generated by a differential unit memory linear repetitive process of the form defined by Example 1.2.3 is stable along the pass if*

$$||([L_0]_p)|| < 1 \tag{2.56}$$

Note 2.3.2. Unlike the test of Theorem 2.3.3, this last test always produces an inconclusive result in the non-unit memory case. This is due to the structure of L_α in this case which means that (2.56) never holds.

Suppose now that (2.51) holds and the control input sequence applied $\{u_k\}_{k\geq 1}$ is constant in the pass-to-pass (i.e. k) direction (e.g. a unit step input applied at $t = 0$ on each pass), and hence $b_k = b_\infty$, $k \geq 1$, in the abstract model S. Then the following result is the basis of the computable performance information available.

Theorem 2.3.5. *Suppose that the linear repetitive process S generated by a differential unit memory linear repetitive process of the form defined by Example 1.2.3 is stable along the pass and the condition of Theorem 2.3.3 holds. Suppose also that the control sequence applied is constant in the pass-to-pass direction and hence $b_k = b_\infty$, $k \geq 1$, in the abstract model S. Then, $\forall\, \alpha > 0$, $\exists$ an $m \times m$ matrix $\tilde{Q} \geq 0$ and a real scalar $\gamma \in (r([L_0]_p), 1)$ such that the error $y_k - y_\infty$ satisfies*

$$[y_k - y_\infty]_p \leq \tilde{Q}\gamma^k([y_0]_p + (I_m - [L_0]_p)^{-1}[b_\infty]_p),\ k \geq 0 \tag{2.57}$$

Proof. In the case when $b_k = b_\infty$, $k \geq 1$, it follows immediately that the error $y_k - y_\infty$ is given by

$$y_k - y_\infty = L_\alpha^k y_0 - \sum_{j=k+1}^{\infty} L_\alpha^{j-1} b_\infty$$

and hence

$$[y_k - y_\infty]_p \leq [L_0^k]_p \left([y_0]_p + \sum_{j=k+1}^{\infty} [L_0]_p^{j-1-k} [b_\infty]_p \right)$$

(where again use has been made of (2.52) and some obvious 'norm-like' properties of $[\cdot]_p$ applied to matrices). Given $r([L_0]_p) < 1$, $(I_m - [L_0]_p)^{-1}$ exists and can be written in the form

$$(I_m - [L_0]_p)^{-1} = \sum_{j=k+1}^{\infty} [L_0]_p^{j-1-k} \tag{2.58}$$

It now remains to prove that $\exists$ an $m \times m$ matrix $\tilde{Q} \geq 0$ and a scalar $\gamma \in (r([L_0]_p), 1)$ such that

$$[L_0^k]_p \leq \tilde{Q}\gamma^k,\ k \geq 0$$

This follows immediately since $r([L_0]_p) < 1$ enables us to choose real numbers $\tilde{\eta} > 0$ and $\gamma \in (r([L_0]_p), 1)$ such that

$$||([L_0^k]_p)|| \leq \tilde{\eta}\gamma^k,\ k \geq 0$$

Also $[L_0^k]_p \leq \hat{Q}$ clearly holds, where $\hat{Q}$ is an $m \times m$ matrix with each element equal to $||([L_0^k]_p)||$. The result (2.57) now follows from (2.58) and defining $\tilde{Q}$ as the $m \times m$ matrix with each element equal to $\tilde{\eta}$. □

Return now to the rate of approach of the output sequence $\{y_k\}_{k\geq 1}$ to y_∞ under stability along the pass, and select γ as any number in the known range $(r([L_0]_p), 1)$. Then Theorem 2.3.5 shows that $\{y_k\}_{k\geq 1}$ approaches y_∞ at a geometric rate governed by this computable number. Note also that a number of refinements of this basic result can (in principle) be obtained by use of appropriate filtering operations on, in particular, the elements of $W(t)$. The details are omitted here and can be found in [150].

The following result provides an alternative to Theorem 2.3.5 (in the unit memory case only).

Theorem 2.3.6. *Suppose that the linear repetitive process S generated by a differential unit memory linear repetitive process of the form defined by Example 1.2.3 is stable along the pass and the condition of Theorem 2.3.4 holds. Suppose also that the control input sequence applied is constant in the pass-to-pass direction and hence $b_k = b_\infty$, $k \geq 1$, in S. Then $\forall \alpha > 0$, the error $y_k - y_\infty$ satisfies*

$$||y_k - y_0|| \leq ||([L_0]_p)||^k \left(||y_0|| + \frac{||b_\infty||}{1 - ||([L_0]_p)||} \right), \ k \geq 0$$

Proof. Simply take the norm of (2.57). □

In effect, this last result states that $\{y_k\}_{k\geq 0}$ approaches y_∞ at a geometric rate governed by the computable scalar $||([L_0]_p)|| < 1$. As with Theorem 2.3.5, a number of refinements of this basic result can (in principle) be obtained by use of appropriate filtering operations on, in particular, the elements of $W(t)$. Details of these can again be found in [150].

Consider now the problem of obtaining bounds for the error $y_k - y_\infty$, $k \geq 0$, where, for simplicity, the initial profile y_0 is set to zero. Then Theorem 2.3.5 states that

$$[y_k - y_\infty]_p \leq M_k [b_\infty]_p \tag{2.59}$$

where, for notational convenience, we introduce the matrix

$$M_k := (I_m - [L_0]_p)^{-1} [L_0]_p^k$$

Suppose that $[b_\infty]_p$ is computed as detailed below and let $y_k^i(t)$ and $y_\infty^i(t)$ denote the ith output channel of $y_k(t)$ and $y_\infty(t)$ respectively, $1 \leq i \leq m$. Then it follows immediately from (2.59) that

$$[y_k(t) - y_\infty(t)]_p \leq [y_k - y_\infty]_p \leq M_k [b_\infty]_p, \ t \geq 0$$

or equivalently that $y_k^i(t)$ lies in the 'band' defined by

$$y_\infty^i(t) - m_k^i \leq y_k^i(t) \leq y_\infty^i(t) + m_k^i, \ t \geq 0 \tag{2.60}$$

where m_k^i denotes the ith entry in the $m \times 1$ vector $M_k[b_\infty]_p$. This band has a simple graphical interpretation and is obviously compatible with computer-aided analysis. Also its width from pass-to-pass is, in effect, governed by the relationship

$$M_{k+1} = M_k[L_0]_p$$

With the initial profile y_0 again set equal to zero, and from the definition of $||\cdot||$, it follows immediately from Theorem 2.3.6 that

$$|y_k^i(t) - y_\infty^i(t)| \leq ||y_k - y_\infty|| \leq \frac{||([L_0]_p)||^k \ ||b_\infty||}{1 - ||([L_0]_p)||}, \ t \geq 0$$

Suppose also that $||b_\infty||$ is computed as detailed below and define γ_k as

$$\gamma_k := \frac{||([L_0]_p)||^k \ ||b_\infty||}{1 - ||([L_0]_p)||}$$

Then it follows immediately that $y_k^i(t)$ lies in the band defined by

$$y_\infty^i(t) - \gamma_k \leq y_k^i(t) \leq y_\infty^i(t) + \gamma_k, \quad (t \geq 0)$$

which again has a simple graphical interpretation.

Theorem 2.3.6 yields a single band for all channels. Hence, in comparison to Theorem 2.3.5, it analyzes the 'worst' channel rather than individual channels. It does, however, provide sharper information in the sense that the width of its error band decreases at a geometric rate governed by the computable scalar $||([L_0]_p)||$.

To compute $[b_\infty]_p$, and hence $||b_\infty||$, first note that b_∞ is the response to $u_\infty = u_k$, $\forall\ k \geq 1$, of the stable (recall the assumption on the eigenvalues of the matrix A) 1D differential linear system defined by the state-space quadruple $\{A, B, C, D\}$. Suppose also that $b_\infty(t)$ is available from simulation studies on this system in response to $u_\infty(t)$. Then it follows immediately that

$$[b_\infty]_p = \begin{bmatrix} b_1 & \cdots & b_m \end{bmatrix}^T$$

where

$$b_i := \sup_{t \geq 0} |b_\infty^i(t)|, \ 1 \leq i \leq m$$

Also

$$||b_\infty|| = \max_{1 \leq i \leq m} b_i$$

An alternative to the individual channel bands of Theorem 2.3.5 exists in terms of an upper bound on $[b_\infty]_p$ which, although more conservative in general, could prove computationally more attractive in certain cases of practical interest. To develop this, first suppose the availability of the step-response matrix $W_b(t)$ of $b_\infty := L_a u_\infty$ where L_a is the convolution operator associated with the 1D differential linear system defined by the state-space quadruple $\{A, B, C, D\}$. Then with $[L_a]_p$ computed as before,

$$[b_\infty]_p \leq [L_a]_p [b_\infty]_p$$

where

$$[u_\infty]_p := \begin{bmatrix} u^1 & \cdots & u^r \end{bmatrix}^T$$

with

$$u^i := \sup_{t \geq 0} |u^i_\infty(t)|$$

in which $u^i_\infty(t)$ denotes the ith channel of $u_\infty(t)$. Hence

$$M_k [b_\infty]_p \leq M_k [L_a]_p [b_\infty]_p =: \tilde{m}_k$$

and the following band is generally more conservative than that of (2.60)

$$y^i_\infty(t) - \tilde{m}^i_k \leq y^i_k(t) \leq y^i_\infty(t) + \tilde{m}^i_k \tag{2.61}$$

where $\tilde{m}^i_k$ denotes the ith entry in $\tilde{m}_k$.

In common with the 1D linear systems case, it is to be expected that the dynamic response of a particular example will be judged in terms of appropriate benchmark choices of the control input vector. Consider, therefore, the ith channel of $y_k(t)$. Then one obvious choice is

$$\tilde{u}_\infty(s) = \frac{e_i}{s}$$

where the tilde denotes the Laplace transform and e_i is the $r \times 1$ column vector with all entries equal to zero except for the ith, which is equal to one.

Equivalently, a unit step is applied at $t = 0$ on each pass in this channel with all others identically zero. Also

$$u^j := \begin{cases} 1, & i = j \\ 0, & i \neq j \end{cases}$$

and this is one practically relevant case where the error bands of (2.61) may prove computationally more attractive than those of (2.60).

Return now to the single error band resulting from Theorem 2.3.6, and note that

$$||b_\infty|| \leq ||L_a|| \; ||u_\infty||$$

where

$$||u_\infty|| = \max_{1 \leq i \leq r} u^i$$

Hence the following band is generally more conservative than that of (2.61)

$$y^i_\infty(t) - \hat{\gamma}^k \leq y^i_k(t) \leq y^i_\infty(t) + \hat{\gamma}^k \tag{2.62}$$

where

$$\hat{\gamma}^k := \frac{||([L_0]_p)||^k \, ||L_a|| \, ||u_\infty||}{1 - ||([L_0]_p)||}$$

Finally, if u_∞ is specified as above, then $||u_\infty|| = 1$ and again this is one practically relevant case where (2.62) may prove more computationally attractive than (2.61).

A unique feature of the analysis so far in this section is that it extends in a natural manner to the case when inter-pass smoothing of the form of Example 1.2.8 is present. In particular, write the operator L_α in this case in the form

$$\begin{aligned} y(t) &= (L_\alpha y)(t) \\ &= C \int_0^t e^{A(t-\tau)} B_0 v(\tau) \, d\tau \end{aligned}$$

where

$$v(t) := \int_0^\alpha K(t,\tau) y(\tau) \, d\tau$$

Then it follows immediately that

$$[y]_p \le [L_0]_p \, [v]_p$$

and hence that

$$[L_\alpha]_p \le L_I =: [L_0]_p \, [K]_p, \; \forall \, \alpha > 0$$

where $[K]_p$ is given by Lemma 2.3.2. It is also clear that all of the analysis to-date in this section generalizes in a natural manner and hence the details are omitted.

Consider now the actual application of the results given so far in this section. Then the key task is the application of Lemma 2.3.1. This is also illustrated in Fig. 2.2 and is clearly compatible with computer aided analysis. Note also that

$$\lim_{T \to \infty} |N_\infty(f) - N_T(f)| = 0 \tag{2.63}$$

and consequently $N_\infty(f)$ can be accurately estimated using data on a 'long enough' domain $0 \le t \le T$. On such an interval, the continuity of $N_T(f)$ as a function of the stationary points $t_1, t_2, \cdots$ implies that it is insensitive to errors in their estimation. If $f(t)$ is contaminated by noise, $N_T(f)$ must be evaluated by inspection of $f(t) + n(t)$ where, under the assumption of a 'sufficiently large' signal to noise ratio, the stationary points of $f(t)$ can be estimated to 'reasonable' accuracy by visual smoothing of the recorded response $f(t) + n(t)$. This, together with (2.63), leads to the conclusion that the estimation of $N_\infty(f)$ is a 'robust operation' in many cases of practical interest.

There are a number of special cases where a closed form for $[L_0]_p$ can be obtained. Here we only note the (potentially) most common use of these which arises when all entries of $W(t)$ are monotonic and sign-definite. In this case

$$[L_0]_p = [W(\infty)]_p \tag{2.64}$$

which is particularly easy to compute. Alternatively, if the matrices in the state-space model giving rise to $W(t)$ are assumed known, it follows immediately that

$$[L_0]_p = [-CA^{-1}B_0 + D_0]_p \tag{2.65}$$

The analysis so far in this section generalizes in a natural manner to the relevant discrete cases. Hence the details are omitted here with the note that the following is the equivalent result for the total variation. (Note also that the proof of this result does not have the complications caused by continuity in the differential case.)

Lemma 2.3.3. *Suppose that the sequence defined by $f(0) = d$ and*

$$f(h) := \sum_{j=1}^{h} g(j), \ h \geq 1$$

is bounded with local maxima or minima at sample instants $1 \leq h_1 < h_2 < \cdots$ in the extended positive integers. Then, with $h_0 = 0$,

$$N_h(f) = |d| + \sum_{j=1}^{h} |g(j)|$$

where

$$N_h(f) := |f(0)| + \sum_{j=1}^{\tilde{h}(h)} |f(h_j) - f(h_{j-1})| + |f(h) - f(\tilde{h}(h))|$$

and

$$N_\infty(f) := \sup_{h \geq 0} N_h(f)$$

where $\tilde{h}(h)$ is the largest integer $h_1, h_2, \cdots$ satisfying $h_j < h$.

2.4 Stability of Discrete Processes via 2D Spectral Methods

In this section we use results from [58] for 2D discrete linear systems to obtain stability along the pass conditions for discrete unit memory linear repetitive processes of the form defined by Example 1.2.10 in terms of the properties of matrices with constant entries. The basis of these tests is the fact that stability along the pass can (by the result of Theorem 2.2.8) be characterized in terms of the zero manifolds of the characteristic polynomial $\mathcal{C}(z_1, z_2)$ of (1.52) written in the form

$$\mathcal{C}(z_1, z_2) = \det(I_{n+m} - z_1\widehat{A}_1 - z_2\widehat{A}_2)$$

where

$$\widehat{A}_1 := \begin{bmatrix} A & B_0 \\ 0 & 0 \end{bmatrix}, \quad \widehat{A}_2 := \begin{bmatrix} 0 & 0 \\ C & D_0 \end{bmatrix} \tag{2.66}$$

In particular, stability along the pass can be characterized in terms of the zero manifolds of this polynomial which we denote by the set

$$\prod(\widehat{A}_1, \widehat{A}_2) := \{(z_1, z_2) \in \mathbb{C} \times \mathbb{C} : \det(I_{n+m} - z_1\widehat{A}_1 - z_2\widehat{A}_2) = 0\} \tag{2.67}$$

Now we have the following results (where $\mathbb{U}_c^2$ denotes the closed unit bidisc (as defined in Theorem 2.2.8)).

Theorem 2.4.1. *The linear repetitive process S generated by a discrete unit memory linear repetitive process of the form defined by Example 1.2.10 is stable along the pass if, and only if,*

$$\mathbb{U}_c^2 \cap \prod(\widehat{A}_1, \widehat{A}_2) = \emptyset \text{ (the null set)}$$

Also we have the equivalent characterization.

Theorem 2.4.2. *The linear repetitive process S generated by a discrete unit memory linear repetitive process of the form defined by Example 1.2.10 is stable along the pass if, and only if,*

$$r(\widehat{A}_1 + e^{\imath\omega}\widehat{A}_2) < 1, \ \forall\, \omega \in (0, 2\pi)$$

Proof. Omitted since it follows identical steps to that in [63] for 2D discrete linear systems described by the Fornasini-Marchesini state-space model. □

An obvious necessary condition for stability along the pass resulting from this last result is that $\widehat{A}_1 \pm \widehat{A}_2$ must both be stable in the 1D discrete linear systems sense. Also this last result suggests the following definition.

Definition 2.4.1. *The 2D spectral radius of the matrix pair $(\widehat{A}_1, \widehat{A}_2)$ is defined as*

$$\mu(\widehat{A}_1, \widehat{A}_2) := \sup_{\omega \in (0, 2\pi)} r(\widehat{A}_1 + e^{\imath\omega}\widehat{A}_2)$$

Also it is clear that S is stable along the pass if, and only if, the 2D spectral radius $\mu(\widehat{A}_1, \widehat{A}_2) < 1$.

In order to proceed, recall (see, for example, [83]) that the numerical range (also known as the field values) of a $q \times q$ matrix F is defined as

$$V(F) := \{x^*Fx : x^*x = 1, \; x \in \mathbb{C}^q\}$$

where here $*$ denotes the complex conjugate transpose operation, and its associated numerical radius is given by

$$v(F) := \sup\{|\tilde{q}| : \tilde{q} \in V(F)\}$$

The numerical range of F is closed, bounded and convex. Also it can be shown that the spectrum, $\sigma(F)$, of F satisfies

$$\sigma(F) \subset V(F), \; F \neq 0$$

Hence

$$r(F) \leq v(F)$$

This last fact has led to the suggestion that the numerical radius is a measure of the stability of matrices and, in particular, the following well known result (see [16] in the 2D linear systems case) expresses stability along the pass of discrete unit memory linear repetitive processes of the form defined by Example 1.2.10 solely in terms of matrices with constant entries.

Theorem 2.4.3. *The linear repetitive process S generated by a discrete unit memory linear repetitive process of the form defined by Example 1.2.10 is stable along the pass if*

$$v(\widehat{A}_1) + v(\widehat{A}_2) < 1$$

Note, however, that this result can be very conservative. For example, if

$$\widehat{A}_1 = \begin{bmatrix} \frac{3}{4} & 0 \\ 0 & 0 \end{bmatrix}, \quad \widehat{A}_2 = \begin{bmatrix} 0 & 0 \\ 0 & \frac{3}{4} \end{bmatrix} \tag{2.68}$$

then the condition in Theorem 2.4.3 does not hold since $v(\widehat{A}_1) = v(\widehat{A}_2) = \frac{3}{4}$. This process is, however, easily shown to be stable along the pass.

A possible reason for the conservative nature of Theorem 2.4.3 is that it is based on individual, rather than mutual, properties of $\widehat{A}_1$ and $\widehat{A}_2$. This has led to the concept of the generalized numerical range which is defined as follows.

Definition 2.4.2. *The generalized numerical range $V_g(\widehat{A}_1, \widehat{A}_2)$ is defined as*

$$V_g(\widehat{A}_1, \widehat{A}_2) := \{(x^*\widehat{A}_1x, \; x^*\widehat{A}_2x) : x^*x = 1, \; x \in \mathbb{C}^{n+m}\}$$

The associated generalized numerical radius is of the form

$$v_g(\widehat{A}_1, \widehat{A}_2) := \sup\{|q_1| + |q_2| : (q_1, q_2) \in V_g(\widehat{A}_1, \widehat{A}_2)\}$$

Also if an algorithm is available for computing the numerical range/numerical radius then it can be trivially extended to compute the generalized numerical range/radius.

It is clear that the numerical range of a matrix is closely related to its spectrum. Similarly, the generalized numerical range is connected to the generalized eigenvalue problem $\det(\widehat{A}_1 - \lambda \widehat{A}_2)$, i.e.

$$q_2 \widehat{A}_1 x = q_1 \widehat{A}_2 x,\ q_1, q_2 \in \mathbb{C},\ (q_1, q_2) \neq (0,0),\ x \in \mathbb{C}^{n+m},\ x^* x = 1$$

The generalized eigenvalues are given by

$$\lambda = \frac{q_1}{q_2}$$

but, numerically, it is more convenient to compute the pair (q_1, q_2). Note, however, that the pair (q_1, q_2) is not unique and certain conventions are followed in numerical analysis to make them relatively unique. These involve choosing the pair (q_1, q_2) such that $q_i = x^* \widehat{A}_i x,\ i = 1, 2$.

The generalized spectrum $\sigma_g(\widehat{A}_1, \widehat{A}_2)$ and generalized spectral radius are defined as follows.

Definition 2.4.3. *The generalized spectrum $\sigma_g(\widehat{A}_1, \widehat{A}_2)$ is defined in the case of $q_1, q_2 \in \mathbb{C}$ as*

$$\sigma_g(\widehat{A}_1, \widehat{A}_2) := \left\{ (x^* \widehat{A}_1 x, x^* \widehat{A}_2 x) : q_2 \widehat{A}_1 x = q_1 \widehat{A}_2 x,\ x^* x = 1,\ x \in \mathbb{C}^{n+m} \right\}$$

The generalized spectral radius is defined as

$$r_g(\widehat{A}_1, \widehat{A}_2) := \sup\{|q_1| + |q_2| : (q_1, q_2) \in \sigma_g(\widehat{A}_1, \widehat{A}_2)\}$$

Note also that the generalized numerical range is induced by all $x : x^* x = 1$ but the generalized spectrum is based on a subset of such x. This fact establishes the following result.

Lemma 2.4.1. *The generalized numerical range and the generalized spectrum satisfy*

$$\sigma_g(\widehat{A}_1, \widehat{A}_2) \subset V_g(\widehat{A}_1, \widehat{A}_2),\ \widehat{A}_1 \neq 0,\ \widehat{A}_2 \neq 0$$

The zero manifold $\prod(\widehat{A}_1, \widehat{A}_2)$ is (from (2.67)) defined by

$$\det(I_{n+m} - z_1 \widehat{A}_1 - z_2 \widehat{A}_2) = 0$$

and hence for any $(z_1, z_2) \in \prod(\widehat{A}_1, \widehat{A}_2)\ \exists\, x$ such that

$$(I_{n+m} - z_1 \widehat{A}_1 - z_2 \widehat{A}_2) x = 0,\ x^* x = 1$$

Pre-multiplying this last equation by x^* now gives

$$1 - q_1 z_1 - q_2 z_2 = 0$$

where

$$q_i = x^* \widehat{A}_i x,\ i = 1, 2$$

i.e. the pair $(q_1, q_2) \in V_g(\widehat{A}_1, \widehat{A}_2)$ and we have, in effect, proved the following result.

Theorem 2.4.4. *The zero manifold* $\prod(\widehat{A}_1, \widehat{A}_2)$ *is a subset of the linear manifold defined by*

$$Z(\widehat{A}_1, \widehat{A}_2) := \{(z_1, z_2) : q_1 z_1 + q_2 z_2 = 1,\ (q_1, q_2) \in V_g(\widehat{A}_1, \widehat{A}_2)\}$$

This next theorem establishes the importance of the manifold $Z(\widehat{A}_1, \widehat{A}_2)$ induced by the generalized numerical range in terms of the stability along the pass.

Theorem 2.4.5. *The linear repetitive process S generated by a discrete unit memory linear repetitive process of the form defined by Example 1.2.10 is stable along the pass if*

$$v_g(\widehat{A}_1, \widehat{A}_2) < 1 \tag{2.69}$$

Proof. Suppose that (2.69) holds. Then

$$|q_1| + |q_2| < 1,\ (q_1, q_2) \in V_g(\widehat{A}_1, \widehat{A}_2)$$

and hence (z_1, z_2) does not belong to $\mathbb{U}_c^2$ when $q_1 z_1 + q_2 z_2 = 1$. Hence the manifold $Z(\widehat{A}_1, \widehat{A}_2)$, and therefore the zero manifold $\prod(\widehat{A}_1, \widehat{A}_2)$, does not have zeros in $\mathbb{U}_c^2$. Hence stability along the pass holds. □

Return now to the process generating (2.68). Then in this case it is easy to show that

$$V_g(\widehat{A}_1, \widehat{A}_2) = \left\{\left(\frac{3}{4}h,\ \frac{3}{4}(1-h)\right) : h \in [0, 1]\right\}$$

and hence

$$v_g(\widehat{A}_1, \widehat{A}_2) = \frac{3}{4}$$

Hence, using this route, we conclude that this process is stable along the pass. This demonstrates that Theorem 2.4.5 is less conservative than Theorem 2.4.3.

The generalized spectrum $\sigma_g(\widehat{A}_1, \widehat{A}_2)$ is a subset of the generalized numerical range $V_g(\widehat{A}_1, \widehat{A}_2)$. Hence we have the following result.

Lemma 2.4.2. *The following relationship holds*

$$r_g(\widehat{A}_1, \widehat{A}_2) \le v_g(\widehat{A}_1, \widehat{A}_2)$$

The generalized spectral radius can be computed by solving the generalized eigenvalue problem discussed above. If it is greater than unity, then the generalized numerical radius is greater than unity and hence the stability test arising from Theorem 2.4.5 is inconclusive.

Lemma 2.4.3. *The 2D spectral radius is not greater than the generalized numerical radius, i.e.*

$$\mu(\widehat{A}_1, \widehat{A}_2) \le v_g(\widehat{A}_1, \widehat{A}_2)$$

Proof. Suppose that $\lambda(\imath\omega)$ is the eigenvalue of the matrix $F(\imath\omega) := \widehat{A}_1 + e^{\imath\omega}\widehat{A}_2$ with the largest modulus and let $f(\imath\omega)$ be the corresponding normalized eigenvector for a particular value of ω, i.e.

$$F(\imath\omega)f(\imath\omega) = \lambda(\imath\omega)f(\imath\omega)$$

or, on pre-multiplying this last equation by $f^*(\imath\omega)$,

$$\rho_1(\imath\omega) + e^{\imath\omega}\rho_2(\imath\omega) = \lambda(\imath\omega)$$

where $\rho_i(\imath\omega) := f^*(\imath\omega)\widehat{A}_i f(\imath\omega),\ i = 1, 2$. Hence

$$r(F(\imath\omega)) = |\lambda(\imath\omega)|$$

and therefore

$$\begin{aligned} \mu(\widehat{A}_1, \widehat{A}_2) &= \sup_{\omega\in(0,2\pi)} |\lambda(\imath\omega)| \\ &\leq \sup_{\omega\in(0,2\pi)} |\rho_1(\imath\omega) + e^{\imath\omega}\rho_2(\imath\omega)| \\ &\leq \sup_{\omega\in(0,2\pi)} (|\rho_1(\imath\omega)| + |\rho_2(\imath\omega)|) \end{aligned}$$

Since $(\rho_1(\imath\omega), \rho_2(\imath\omega)) \in V_g(\widehat{A}_1, \widehat{A}_2)$, the result follows. □

One of the potential benefits of using the generalized numerical range is that phase information can also be used in stability testing. Some initial analysis on this topic for 2D discrete linear systems can be found in [58] and its interpretation/extension to discrete linear repetitive processes (of the form defined by Example 1.2.10) is clearly an area for further research.

3 Lyapunov Equations for Discrete Processes

Summary. In this chapter we investigate the role of Lyapunov equations in the stability related analysis of discrete linear repetitive processes. There are two types of Lyapunov equation which, in common with the 2D systems literature, are termed 1D and 2D respectively. The key difference here is that in the latter case the matrices involved have constant entries whereas those in the former have entries which are functions of a complex variable. Moreover, the 1D equation gives necessary and sufficient conditions for stability along the pass whereas the 2D equation gives, in general, sufficient but not necessary conditions. The role of these equations in the extraction of computable measures of the performance of stable processes is also investigated.

3.1 The 1D Lyapunov Equation

Note 3.1.1. From this point onwards, a real symmetric positive definite (respectively negative definite) matrix, say U_1, will be denoted by $U_1 > 0$ (respectively $U_1 < 0$). Also for symmetric matrices of the same dimensions, say U_2 and U_3, we use the notation $U_2 \geq U_3$ (respectively $U_2 \leq U_3$) to denote the case when $U_2 - U_3$ is positive semi-definite (respectively $U_2 - U_3$ is negative semi-definite).

The basic result [154] for discrete linear repetitive processes of the form defined by Example 1.2.10 is as follows. This result makes use of the 2D transfer-function matrix description of the process dynamics and, in particular, $G_1(z_1^{-1})$, obtained from (2.42) by setting $M = 1$, which describes the contribution of the previous pass dynamics to those of the current one.

Theorem 3.1.1. *Consider the linear repetitive process S generated by a discrete unit memory linear repetitive process of the form defined by Example 1.2.10 with the pairs $\{A, B_0\}$ and $\{C, A\}$ controllable and observable respectively. Then S is stable along the pass if, and only if,*

(a)

$$r(D_0) < 1, \quad r(A) < 1 \tag{3.1}$$

and

(b) ∃ a rational polynomial matrix solution $P(z_1^{-1})$ of the Lyapunov equation

$$G_1^T(z_1)P(z_1^{-1})G_1(z_1^{-1}) - P(z_1^{-1}) = -I_m \tag{3.2}$$

bounded in an open neighborhood of $\mathbb{C}_{uc} := \{z_1^{-1} : |z_1^{-1}| = 1\}$ with the properties that

(i) $P(z_1^{-1}) = P^T(z_1)$
(ii) $\beta_2^2 I_m \geq P(z_1^{-1}) = P^T(z_1) \geq \beta_1^2 I_m \;\forall\, z_1^{-1} \in \mathbb{C}_{uc}$ *for some choices of real scalars* $\beta_i \geq 1,\ i = 1, 2$.

Proof. If S is stable along the pass then the conditions of (3.1) are necessary and hence $G_1(z_1^{-1})$ is bounded in an open neighborhood of $\mathbb{C}_{uc}$. Also by (b) of Corollary 2.2.4, it follows that for each $z_1^{-1} \in \mathbb{C}_{uc}$, ∃ a unique positive definite Hermitian (from this point onwards such a matrix is denoted by PDH) solution $P(z_1^{-1})$ of (3.2). A routine continuity and compactness argument now shows the necessity of the condition listed under (ii) above and also extends the solution to an open neighborhood of $\mathbb{C}_{uc}$.

To complete the proof, it remains to show that (a) and (b) here imply that the spectral radius of $G_1(z_1^{-1})$ is strictly less than unity $\forall\, z_1^{-1} \in \mathbb{C}_{uc}$. This is trivially verified pointwise from (3.2). □

In computational (or testing) terms, only $\mathbb{C}_{uc}$ need be considered, the extension off this curve being achieved (if necessary) by analytic continuation means. Note also that the numbers β_2 and β_1 play no role in stability analysis but, together with $P(z_1^{-1})$, they are the key to obtaining bounds on expected system performance as shown next.

Suppose that the particular example under consideration is stable along the pass. Then well known factorization techniques enable $P(z_1^{-1})$ to be written as

$$P(z_1^{-1}) = F^T(z_1)F(z_1^{-1})$$

where $F(z_1^{-1})$ is both stable and minimum phase. Hence it has a stable minimum phase inverse.

Given these facts, return to the process under consideration and consider the case when the current pass input terms are deleted. Then as described in Sect. 1.4, the process dynamics can be written as

$$y_{k+1}(z_1^{-1}) = G_1(z_1^{-1})y_k(z_1^{-1}),\ k \geq 0$$

Also let

$$\hat{y}_k(z_1^{-1}) := F(z_1^{-1})y_k(z_1^{-1}),\ k \geq 0$$

denote 'filtered' (by properties of $F(z_1^{-1})$) outputs. Then the following result gives bounds on expected performance.

Theorem 3.1.2. *Suppose that the linear repetitive process S generated by a discrete unit memory linear repetitive process of the form defined by Example 1.2.10 is stable along the pass. Then $\forall\, k \geq 0$,*

$$||\hat{y}_{k+1}||^2_{\ell_2^m(0,\infty)} = ||\hat{y}_k||^2_{\ell_2^m(0,\infty)} - ||y_k||^2_{\ell_2^m(0,\infty)} \tag{3.3}$$

and hence the filtered output (pass profile) sequence $\{||\hat{y}_k||_{\ell_2^m(0,\infty)}\}_{k\geq 0}$ *is strictly monotonically decreasing to zero and satisfies, for* $k \geq 0$, *the inequality*

$$||\hat{y}_k||_{\ell_2^m(0,\infty)} \leq \lambda^k ||\hat{y}_0||_{\ell_2^m(0,\infty)}$$

where

$$\lambda := (1 - \beta_2^{-2})^{\frac{1}{2}} < 1$$

Also the actual output sequence $\{||y_k||_{\ell_2^m(0,\infty)}\}_{k\geq 0}$ *is bounded by*

$$||y_k||_{\ell_2^m(0,\infty)} \leq M_{1d}\lambda^k ||y_0||_{\ell_2^m(0,\infty)}$$

where

$$M_{1d} := \beta_2\beta_1^{-1} \geq 1$$

Proof. Consider the case of $z_1^{-1} \in \mathbb{C}_{uc}$ in the Lyapunov equation (3.2) and pre- and post-multiply this equation by $y_k^T(z_1)$ and $y_k(z_1^{-1})$ respectively. Then on integrating around the closed curve $\mathbb{C}_{uc}$, Parseval's identity yields (3.3). Using the inequality

$$\beta_1^2 I_m \leq P(z_1^{-1}) \leq \beta_2^2 I_m, \quad |z_1^{-1}| = 1$$

with $\hat{y}_k = F y_k$, yields

$$\beta_1^2 ||y_k||^2_{\ell_2^m(0,\infty)} \leq ||\hat{y}_k||^2_{\ell_2^m(0,\infty)} \leq \beta_2^2 ||y_k||^2_{\ell_2^m(0,\infty)}$$

and then

$$||\hat{y}_{k+1}||^2_{\ell_2^m(0,\infty)} \leq ||\hat{y}_k||^2_{\ell_2^m(0,\infty)} - \beta_2^{-2} ||\hat{y}_k||^2_{\ell_2^m(0,\infty)}$$

The required properties of $\{||\hat{y}_k||_{\ell_2^m(0,\infty)}\}_{k\geq 0}$ and $\{||y_k||_{\ell_2^m(0,\infty)}\}_{k\geq 0}$ now follow as simple consequences of this result. □

This result provides the following computable information (see below) on the convergence to the resulting limit profile of a stable along the pass discrete unit memory linear repetitive process of the form defined by Example 1.2.10.

- The sequence of filtered outputs $\{\hat{y}_k\}_{k\geq 0}$ consists of monotone signals converging to zero at a computable geometric rate in $\ell_2^m(0,\infty)$.
- The actual output sequence $\{y_k\}_{k\geq 0}$ converges to zero at the same geometric rate but this is no longer necessarily monotonic. Further, the deviation from monotonicity is described by the parameter M_{1d}, computed from the solution of the 1D Lyapunov equation.

Next we consider the solution of the 1D Lyapunov equation of Theorem 3.1.1.

As a preliminary step, it is convenient to introduce at this stage the so-called augmented plant matrix for discrete unit memory linear repetitive processes of the form defined by Example 1.2.10 as

$$\Phi := \begin{bmatrix} A & B_0 \\ C & D_0 \end{bmatrix} \tag{3.4}$$

This matrix also plays a central role in the development of a 2D Lyapunov equation with constant coefficients for these processes (see the next section). Its first use in the current context is to provide the following set of easily tested necessary conditions for stability along the pass.

Lemma 3.1.1. *Consider the linear repetitive process S generated by a discrete unit memory linear repetitive process of the form defined by Example 1.2.10. Then the following are necessary conditions for stability along the pass.*

(a) the conditions listed under (a) of Theorem 3.1.1 hold; and
(b) $r(\Phi) < 1$

Proof. Clearly only (b) needs to be established. This follows immediately from the condition of Theorem 2.2.8 on considering the case when $z_1 = z_2 = 1$ in the characteristic polynomial. □

Return now to (3.2) and suppose that the rational polynomial solution matrix $P(z_1^{-1})$, $z_1^{-1} \in \mathbb{C}_{uc}$, has been obtained. Also write z_1^{-1} in polar form as $z_1^{-1} = e^{\imath\theta}$, $\theta \in [0, 2\pi]$, and let $h_j(\theta)$ denote the jth principal minor of the solution matrix. Then the $h_j(\theta)$ are functions of the real single variable θ over the closed interval $[0, 2\pi]$. Consequently the following set of conditions [154] are, in effect, a systematic test procedure for Theorem 3.1.1. (See [106] for the corresponding result in the case of 2D discrete linear systems described by the Roesser model.)

Lemma 3.1.2. *The conditions of Theorem 3.1.1 are equivalent to the following:*

(a) $r(D_0) < 1, \quad r(A) < 1, \quad r(\Phi) < 1;$ *and*
(b) $h_j(\theta) > 0, \quad \forall\, \theta \in [0, 2\pi],\ 1 \le j \le m$

To illustrate how this last result can be used in a 'synthesis type' role, consider the SISO case with $|D_0| < 1$ (asymptotically stable). Suppose also that A has the eigenvalue-eigenvector decomposition

$$\Lambda := T_2^{-1} A T_2 = \text{diag}\{\lambda_i\}_{1 \le i \le n}$$

and let λ_i, $1 \le i \le n$, be real and such that $r(A) < 1$. Also introduce $T = \text{diag}\,\{T_2, 1\}$ and transform the augmented plant matrix Φ to

$$\tilde{\Phi} = T^{-1} \Phi T = \begin{bmatrix} \Lambda & A_2 \\ A_3 & D_0 \end{bmatrix}$$

where

$$\begin{aligned} A_2 &:= T_2^{-1}B_0 = \begin{bmatrix} d_1 & d_2 & \cdots & d_n \end{bmatrix}^T \\ A_3 &:= CT_2 = \begin{bmatrix} f_1 & f_2 & \cdots & f_n \end{bmatrix} \end{aligned}$$

and introduce

$$\delta_i := d_i f_i, \ 1 \leq i \leq n \tag{3.5}$$

Then, as shown next, (b) of Lemma 3.1.2 holds for the case when all the δ_i have the same sign if, and only if,

$$\max\{|G_1(1)|, \ |G_1(-1)|\} < 1 \tag{3.6}$$

To prove (3.6), note that in this case

$$h_1(\theta)|1 - G_1(z_1^{-1})|^2 = 1, \ z_1^{-1} = e^{\imath\theta}$$

and hence $h_1(\theta) > 0$, $\theta \in [0, 2\pi]$, if, and only if, $|G_1(z_1^{-1})| < 1$ and necessity is immediate. For sufficiency, note that

$$|G_1(z_1^{-1})| = \left| D_0 + \sum_{i=1}^{n} \frac{\delta_i}{z_1^{-1} - \lambda_i} \right| = \left| D_0 + \mathrm{sgn}(\delta_1) \sum_{i=1}^{n} g_i(z_1^{-1}) \right|$$

where

$$g_i(z_1^{-1}) := \frac{|\delta_i|}{z_1^{-1} - \lambda_i}, \ 1 \leq i \leq n$$

A simple argument now shows that each of these functions maps the unit circle in the z_1^{-1}-plane onto a circle centered on the real axis. Hence this maximum value can only occur when $z_1^{-1} = 1$ or -1 and (3.6) follows immediately and we have established the following result.

Theorem 3.1.3. *Consider a discrete unit memory linear repetitive process of the form defined by Example 1.2.10 which is SISO and for which A is a diagonalizable matrix with real eigenvalues λ_i, $1 \leq i \leq n$. Suppose also that all δ_i defined by (3.5) have the same sign. Then the linear repetitive process S generated in this case is stable along the pass if, and only if,*

(a) $|D_0| < 1, \ |\lambda_i| < 1, \ 1 \leq i \leq n;$ *and*
(b) $\max\{|D_0 + C(I_n - A)^{-1}B_0|, \ |D_0 - C(I_n - A)^{-1}B_0|\} < 1$

If the δ_i defined by (3.5) have different signs the following corollary gives a sufficient condition for stability along the pass.

Corollary 3.1.1. *Consider a discrete unit memory linear repetitive process of the form defined by Example 1.2.10 which is SISO and for which A is a diagonalizable matrix with real eigenvalues λ_i, $1 \leq i \leq n$. Suppose also that the numbers δ_i, $1 \leq i \leq q$, have the same sign and that so do δ_i, $q+1 \leq i \leq n$. Then the linear repetitive process S generated in this case is stable along the pass if* *(a)* $|D_0| < 1,\ |\lambda_i| < 1,\ 1 \leq i \leq n$; *and*

(b) $\max\{\gamma_1 + \hat{\gamma}_1,\ \gamma_1 + \hat{\gamma}_2,\ \gamma_2 + \hat{\gamma}_1,\ \gamma_2 + \hat{\gamma}_2\} < 1$, *where*

$$\begin{aligned} \gamma_1 &:= \left| D_0 + \sum_{i=1}^{q} \frac{|\delta_i|}{1-\lambda_i} \right|, \quad \gamma_2 := \left| D_0 - \sum_{i=1}^{q} \frac{|\delta_i|}{1+\lambda_i} \right| \\ \hat{\gamma}_1 &:= \left| \sum_{i=q+1}^{n} \frac{|\delta_i|}{1-\lambda_i} \right|, \quad \hat{\gamma}_2 := \left| \sum_{i=q+1}^{n} \frac{|\delta_i|}{1+\lambda_i} \right| \end{aligned}$$

Proof. We have that

$$\begin{aligned} |D_0 + C(z_1^{-1}I_n - A)^{-1}B_0| &= \left| D_0 + \sum_{i=1}^{q} \frac{\delta_i}{z_1^{-1} - \lambda_i} + \sum_{i=q+1}^{n} \frac{\delta_i}{z_1^{-1} - \lambda_i} \right| \\ &\leq \max_{|z_1^{-1}|=1} \left| D_0 + \operatorname{sgn}(\delta_1) \sum_{i=1}^{q} \frac{|\delta_i|}{z_1^{-1} - \lambda_i} \right| \\ &+ \max_{|z_1^{-1}|=1} \left| \sum_{i=q+1}^{n} \frac{|\delta_i|}{z_1^{-1} - \lambda_i} \right| \end{aligned}$$

□

Note: Further refinements of the result given in Corollary 3.1.1 are possible. For example, a more accurate estimate can be obtained by combining any two terms in $\{\delta_i\}_{1 \leq i \leq n}$ having opposite signs.

Return now to the conditions of Lemma 3.1.2 and, in particular, (b) of this result. Then in the general case it is clear that constructing the $h_j(\theta)$ and testing each of them for the required property is only feasible in a restricted number of special cases. Also in the search for alternatives the most appealing, from an applications standpoint, would be expressed in terms of matrices with constant entries. Next we give one such test [154] based on the use of the matrix Kronecker product, which is denoted by $\otimes$. (See [3] for the corresponding analysis in the case of 2D discrete linear systems described by the Roesser model.)

Basically, what is required here is to show that there exists a PDH matrix $P(e^{\imath\theta})$ which solves the following equivalent version of (3.2) for all $\theta \in [0, 2\pi]$.

$$P(e^{\imath\theta}) - G_1^T(e^{-\imath\theta})P(e^{\imath\theta})G_1(e^{\imath\theta}) = I_m$$

or, using $\otimes$,

$$\left(I_{m^2} - G_1^T(e^{-\imath\theta}) \otimes G_1^T(e^{\imath\theta})\right) S_{so}\left[P(e^{\imath\theta})\right] = S_{so}\left[I_{m^2}\right]$$

where $S_{so}[\cdot]$ denotes the stacking operator. The following is now a new set of conditions for stability along the pass of the discrete unit memory linear repetitive processes under consideration here.

Theorem 3.1.4. *Consider the linear repetitive process S generated by a discrete unit memory linear repetitive process of the form defined by Example 1.2.10. Then S is stable along the pass if, and only if,*

(a) $r(D_0) < 1,\ r(A) < 1$
(b) $P = P(e^{\imath\theta_0})$, *the solution of*

$$P - G_1^T(e^{-\imath\theta_0})PG_1(e^{\imath\theta_0}) = I_m$$

is positive definite for some $\theta_0 \in [0, 2\pi]$; *and*
(c)

$$\det\left(I_{m^2} - G_1^T(e^{-\imath\theta}) \otimes G_1^T(e^{\imath\theta})\right) \neq 0,\ \forall\, \theta \in [0, 2\pi] \tag{3.7}$$

Proof. It clearly suffices to show that (b) and (c) here are, together, equivalent to (b) of Theorem 3.1.1 and, in particular, the solution of the 1D Lyapunov equation (3.2) on $\mathbb{C}_{uc}$. First note, therefore, that (3.7) guarantees the existence of a unique solution, $P(e^{\imath\theta})$, of (3.2). Also for this matrix to be PDH it is required that its eigenvalues remain positive for all $\theta \in [0, 2\pi]$. This is equivalent to (b) and (c) here since the fact that eigenvalues of $P(e^{\imath\theta})$ are real continuous functions of θ means that they will always be positive if $P(e^{\imath\theta}) > 0$ for an arbitrary value of θ, say θ_0, and (3.7) holds. □

Consider now the problem of testing Theorem 3.1.4 for a given example. Then, in effect, the remaining difficulty is (c). The following result now expresses (3.2) in terms of a condition defined by matrices with constant entries.

Theorem 3.1.5. *Consider the linear repetitive process S generated by a discrete unit memory linear repetitive process of the form defined by Example 1.2.10. Then S is stable along the pass if, and only if,*

(a) $r(D_0) < 1,\ r(A) < 1$
(b)

$$r(G_1(e^{\imath\theta_0})) < 1$$

for some $\theta_0 \in [0, 2\pi]$; *and*
(c)

$$\det\left(\lambda^2 X_1 + \lambda X_2 + X_3\right) \neq 0,\ \forall\, \lambda \in \mathbb{C}_{uc} \tag{3.8}$$

where

$$X_1 := \begin{bmatrix} 0 & 0 & 0 & 0 \\ 0 & 0 & 0 & 0 \\ 0 & 0 & 0 & 0 \\ 0 & 0 & 0 & -A^T \otimes I_n \end{bmatrix} \tag{3.9}$$

$$X_2 := \begin{bmatrix} 0 & B_0^T \otimes D_0^T & 0 & B_0^T \otimes B_0^T \\ 0 & -A^T \otimes I_m & 0 & 0 \\ 0 & 0 & I_{nm} & 0 \\ 0 & 0 & 0 & I_{n^2} + A^T \otimes A^T \end{bmatrix} \tag{3.10}$$

$$X_3 := \begin{bmatrix} I_{m^2} - D_0^T \otimes D_0^T & 0 & I_m \otimes B_0^T & 0 \\ C^T \otimes I_m & I_{nm} & 0 & 0 \\ D_0^T \otimes C^T & 0 & -I_m \otimes A^T & 0 \\ C^T \otimes C^T & 0 & 0 & -I_n \otimes A^T \end{bmatrix} \tag{3.11}$$

Proof. It is easily seen that (b) here is equivalent to (b) of Theorem 3.1.4 and hence it remains to prove that (c) of each of these two theorems are equivalent. Consider, therefore,

$$\begin{aligned} g(e^{\imath\theta}) &:= \det\left(I_{m^2} - G_1^T(e^{-\imath\theta}) \otimes G_1^T(e^{\imath\theta})\right) \\ &= \det\left(I_{m^2} - D_0^T \otimes D_0^T - e^{\imath\theta}\left(B_0^T \otimes D_0^T\right)\right. \\ &\quad \times \left(\left(I_n - e^{\imath\theta} A^T\right) \otimes I_m\right)^{-1} \left(C^T \otimes I_m\right) \\ &\quad - \left(I_m \otimes B_0^T\right)\left(I_m \otimes \left(I_n e^{\imath\theta} - A^T\right)\right)^{-1}\left(D_0^T \otimes C^T\right) \\ &\quad \left. -e^{\imath\theta}\left(B_0^T \otimes B_0^T\right) h(e^{\imath\theta})^{-1}\left(C^T \otimes C^T\right)\right) \end{aligned}$$

where

$$h(e^{\imath\theta}) := -\left(A^T \otimes I_n\right) e^{2\imath\theta} + \left(I_{n^2} + A^T \otimes A^T\right) e^{\imath\theta} - \left(I_n \otimes A^T\right)$$

Hence

$$g(e^{\imath\theta}) = \det\left(I_{m^2} - D_0^T \otimes D_0^T - UV^{-1}W\right) \tag{3.12}$$

where

$$U := \begin{bmatrix} e^{\imath\theta}\left(B_0^T \otimes D_0^T\right) & \left(I_m \otimes B_0^T\right) & e^{\imath\theta}\left(B_0^T \otimes B_0^T\right) \end{bmatrix}$$

$$V := \begin{bmatrix} I_{nm} - e^{\imath\theta}\left(A^T \otimes I_m\right) & 0 & 0 \\ 0 & e^{\imath\theta} I_{nm} - I_m \otimes A^T & 0 \\ 0 & 0 & h(e^{\imath\theta}) \end{bmatrix}$$

$$W := \begin{bmatrix} C^T \otimes I_m \\ D_0^T \otimes C^T \\ C^T \otimes C^T \end{bmatrix}$$

Also $r(A) < 1$ from (a) of the theorem implies that $\forall\, \theta \in [0, 2\pi]$

$$\det\left(I_{nm} - e^{\imath\theta}\left(A^T \otimes I_m\right)\right) \det\left(e^{\imath\theta} I_{nm} - \left(I_m \otimes A^T\right)\right) \det h(e^{\imath\theta}) \neq 0 \tag{3.13}$$

Now pre-multiplying (3.12) by the left-hand side of (3.13) and using a standard determinant equality we obtain that

$$g(e^{\imath\theta}) \neq 0, \ \forall\, \theta \in [0, 2\pi]$$

is equivalent to

$$\tilde{g}(e^{\imath\theta}) \neq 0, \ \forall\, \theta \in [0, 2\pi] \tag{3.14}$$

where

$$\tilde{g}(e^{\imath\theta}) := \begin{bmatrix} I_{m^2} - D_0^T \otimes D_0^T & U \\ W & V \end{bmatrix}$$

Finally, it is easy to see that (3.14) is identical to (3.8). □

At this stage, the stability testing problem has been reduced to testing (or determining) the eigenvalue locations of three constant matrices and those of a second order matrix polynomial. The difficulty with the latter condition is that the matrix X_1 given in (3.9) is singular and hence the required eigenvalues cannot be obtained directly using existing software packages. In what follows, it will be shown that this last condition can be replaced by an equivalent one which is easily checked. This is a two stage task to deal with the cases when the matrix A is nonsingular and singular respectively.

In the case when $\det A \neq 0$, we have the following result.

Theorem 3.1.6. *Consider the linear repetitive process S generated by a discrete unit memory linear repetitive process of the form defined by Example 1.2.10 with* $\det A \neq 0$. *Then S is stable along the pass if, and only if,*

(a) $r(D_0) < 1, \quad r(A) < 1$
(b) $r(G_1(e^{\imath\theta_0})) < 1$ *for some* $\theta_0 \in [0, 2\pi]$; *and*
(c)

$$\det\left(\lambda Y_1 + Y_0\right) \neq 0, \ \forall\, \lambda \in \mathbb{C}_{uc} \tag{3.15}$$

where

$$Y_0 := \begin{bmatrix} I_{nm} & -\left(C^T \otimes I_m\right) Y_2 & 0 & 0 \\ 0 & Y_3 & 0 & 0 \\ 0 & 0 & 0 & -I_{n^2} \\ 0 & Y_4 & Y_5 & Y_6 \end{bmatrix}$$

$$Y_1 := \begin{bmatrix} Y_7 & 0 & Y_8 & 0 \\ Y_9 & I_{nm} & Y_{10} & 0 \\ 0 & 0 & I_{n^2} & 0 \\ Y_{11} & 0 & Y_{12} & I_{n^2} \end{bmatrix}$$

and

$$\begin{aligned}
Y_2 &:= \left(I_{m^2} - D_0^T \otimes D_0^T\right)^{-1}\left(I_m \otimes B_0^T\right) \\
Y_3 &:= -\left(I_m \otimes A^T\right) - \left(D_0^T \otimes C^T\right) Y_2 \\
Y_4 &:= \left(A^T \otimes I_n\right)^{-1}\left(C^T \otimes C^T\right) Y_2 \\
Y_5 &:= \left(A^T \otimes I_n\right)^{-1}\left(I_n \otimes A^T\right) \\
Y_6 &:= -\left(A^T \otimes I_n\right)^{-1}\left(I_{n^2} + A^T \otimes A^T\right) \\
Y_7 &:= -\left(A^T \otimes I_m\right) - \left(C^T \otimes I_m\right)\left(I_{m^2} - D_0^T \otimes D_0^T\right)^{-1}\left(B_0^T \otimes D_0^T\right) \\
Y_8 &:= -\left(C^T \otimes I_m\right)\left(I_{m^2} - D_0^T \otimes D_0^T\right)^{-1}\left(B_0^T \otimes B_0^T\right) \\
Y_9 &:= -\left(D_0^T \otimes C^T\right)\left(I_{m^2} - D_0^T \otimes D_0^T\right)^{-1}\left(B_0^T \otimes D_0^T\right) \\
Y_{10} &:= -\left(D_0^T \otimes C^T\right)\left(I_{m^2} - D_0^T \otimes D_0^T\right)^{-1}\left(B_0^T \otimes B_0^T\right) \\
Y_{11} &:= \left(A^T \otimes I_n\right)^{-1}\left(C^T \otimes C^T\right)\left(I_{m^2} - D_0^T \otimes D_0^T\right)^{-1}\left(B_0^T \otimes D_0^T\right) \\
Y_{12} &:= \left(A^T \otimes I_n\right)^{-1}\left(C^T \otimes C^T\right)\left(I_{m^2} - D_0^T \otimes D_0^T\right)^{-1}\left(B_0^T \otimes B_0^T\right)
\end{aligned}$$

Proof. We need to show that (c) here is equivalent to (c) of Theorem 3.1.5. To proceed, consider again (3.8) which, since $\det A \neq 0$ by assumption, can be expressed as

$$\begin{aligned}
\det\left(\lambda^2 X_1 + \lambda X_2 + X_3\right) &= \det\left(-A^T \otimes I_n\right) \\
&\times \det\left(\begin{bmatrix} \lambda H_{11} + H_{10} & \lambda H_{21} \\ H_{30} & \lambda^2 I_{n^2} + \lambda H_{41} + H_{40} \end{bmatrix}\right)
\end{aligned}$$

where

$$\begin{aligned}
H_{10} &:= \begin{bmatrix} I_{m^2} - D_0^T \otimes D_0^T & 0 & I_m \otimes B_0^T \\ C^T \otimes I_m & I_{nm} & 0 \\ D_0^T \otimes C^T & 0 & -I_m \otimes A^T \end{bmatrix} \\
H_{11} &:= \begin{bmatrix} 0 & B_0^T \otimes D_0^T & 0 \\ 0 & -A^T \otimes I_m & 0 \\ 0 & 0 & I_{nm} \end{bmatrix}, \quad H_{21} := \begin{bmatrix} B_0^T \otimes B_0^T \\ 0 \\ 0 \end{bmatrix} \\
H_{30} &:= \begin{bmatrix} -\left(A^T \otimes I_n\right)^{-1}\left(C^T \otimes C^T\right) & 0 & 0 \end{bmatrix}
\end{aligned}$$

$$H_{40} := \left(A^T \otimes I_n\right)^{-1}\left(I_n \otimes A^T\right), \quad H_{41} := -\left(A^T \otimes I_n\right)^{-1}\left(I_{n^2} + A^T \otimes A^T\right)$$

Hence we now have that (3.8) in this case is equal to

$$\det\left(\lambda^2 I_{n^2} + \lambda H_{41} + H_{40}\right)\hat{H} \neq 0, \ \forall \in \mathbb{C}_{uc} \tag{3.16}$$

where

$$\hat{H} := \left(\lambda H_{11} + H_{10} - \lambda H_{21}\left(\lambda^2 I_{n^2} + \lambda H_{41} + H_{40}\right)^{-1} H_{30}\right)$$

Applying appropriate algebraic operations now shows that (3.16) is equal to

$$\det\left(\begin{bmatrix} \lambda H_{11}+H_{10} & \lambda H_{21} & 0 \\ 0 & \lambda I_{n^2} & -I_{n^2} \\ H_{30} & H_{40} & \lambda I_{n^2}+H_{41} \end{bmatrix}\right) \neq 0,\ \forall\, \lambda \in \mathbb{C}_{uc} \qquad (3.17)$$

Now rewrite (3.17) in the form

$$\det\left(\begin{bmatrix} \hat{H}_{10} & \lambda \hat{H}_{21}+\hat{H}_{20} \\ \hat{H}_{30} & \lambda \hat{H}_{41}+\hat{H}_{40} \end{bmatrix}\right) \neq 0,\ \forall\, \lambda \in \mathbb{C}_{uc}$$

where

$$\begin{aligned}
\hat{H}_{10} &:= I_{m^2} - D_0^T \otimes D_0^T, \qquad \hat{H}_{20} := \begin{bmatrix} 0 & I_m \otimes B_0^T & 0 & 0 \end{bmatrix} \\
\hat{H}_{21} &:= \begin{bmatrix} B_0^T \otimes D_0^T & 0 & B_0^T \otimes B_0^T & 0 \end{bmatrix} \\
\hat{H}_{30} &:= \begin{bmatrix} C^T \otimes I_m \\ D_0^T \otimes C^T \\ 0 \\ -\left(A^T \otimes I_n\right)^{-1}\left(C^T \otimes C^T\right) \end{bmatrix}, \quad \hat{H}_{41} := \begin{bmatrix} -A^T \otimes I_m & 0 \\ 0 & I_{2n^2+nm} \end{bmatrix} \\
\hat{H}_{40} &:= \begin{bmatrix} I_{nm} & 0 & 0 & 0 \\ 0 & -I_m \otimes A^T & 0 & 0 \\ 0 & 0 & 0 & -I_{n^2} \\ 0 & 0 & \left(A^T \otimes I_n\right)^{-1}\left(I_n \otimes A^T\right) & -\hat{X} \end{bmatrix} \\
\hat{X} &:= -\left(A^T \otimes I_n\right)^{-1}\left(I_{n^2} + A^T \otimes A^T\right)
\end{aligned}$$

This last step shows that (3.17) is equivalent to

$$\det \hat{H}_{10}\ \det\left(\left(\hat{H}_{41} - \hat{H}_{30}\hat{H}_{10}^{-1}\hat{H}_{21}\right)\lambda + \hat{H}_{40} - \hat{H}_{30}\hat{H}_{10}^{-1}\hat{H}_{20}\right) \neq 0$$

$\forall\, \lambda \in \mathbb{C}_{uc}$, where

$$\begin{aligned}
\hat{H}_{41} - \hat{H}_{30}\hat{H}_{10}^{-1}\hat{H}_{21} &= Y_1 \\
\hat{H}_{40} - \hat{H}_{30}\hat{H}_{10}^{-1}\hat{H}_{20} &= Y_0
\end{aligned}$$

and the proof is complete. □

The conditions of this last theorem can be easily tested using existing software. In particular, they require tests on the eigenvalue locations of three matrices with constant entries and the solution of a well known generalized eigenvalue problem.

In the case when $\det A = 0$, we have the following result.

Theorem 3.1.7. *Consider the linear repetitive process S generated by a discrete unit memory linear repetitive process of the form defined by Example 1.2.10 with* $\det A = 0$. *Then S is stable along the pass if, and only if,*

(a) (a) and (b) of Theorem 3.1.6 hold; and
(b)

$$\det\left(\lambda\widetilde{W}_1 + \widetilde{W}_2\right) \neq 0, \ \forall\, \lambda \in \mathbb{C}_{uc} \tag{3.18}$$

where

$$\widetilde{W}_1 := \begin{bmatrix} \hat{W}_{11} & \hat{W}_{21} & 0 \\ 0 & I_{n_\lambda} & 0 \\ \Lambda^{-1}\hat{W}_{31} & 0 & I_{n_\lambda} \end{bmatrix}$$
$$\widetilde{W}_2 := \begin{bmatrix} \hat{W}_{10} & \hat{W}_{20} & 0 \\ 0 & 0 & -I_{n_\lambda} \\ \Lambda^{-1}\hat{W}_{30} & \Lambda^{-1}\hat{W}_{40} & \Lambda^{-1}\hat{W}_{41} \end{bmatrix}$$

and

$$-T^{-1}\left(A^T \otimes I_n\right)T = \begin{bmatrix} 0 & 0 \\ 0 & \Lambda \end{bmatrix} \tag{3.19}$$

where n_λ denotes the rank of A, $\det \Lambda \neq 0$ and $\det T \neq 0$. The matrices $\hat{W}_{10}, \cdots, \hat{W}_{41}$ are obtained by a conformal partition of the left-hand side of the following equation

$$\begin{bmatrix} I & 0 \\ 0 & T^{-1} \end{bmatrix}\left(\lambda^2 X_1 + \lambda X_2 + X_3\right)\begin{bmatrix} I & 0 \\ 0 & T \end{bmatrix}$$
$$= \begin{bmatrix} \lambda\hat{W}_{11} + \hat{W}_{10} & \lambda\hat{W}_{21} + \hat{W}_{20} \\ \lambda\hat{W}_{31} + \hat{W}_{30} & \lambda^2\Lambda + \lambda\hat{W}_{41} + \hat{W}_{40} \end{bmatrix} \tag{3.20}$$

Proof. It is clearly only required to show that (b) here is equivalent to (c) of Theorem 3.1.5. Consider first (3.19) and note that there always exist nonsingular Λ and T such that this equation holds. One possible choice is Λ as a matrix in Jordan form with all non-zero eigenvalues of $-\left(A^T \otimes I_n\right)$ and T the matrix of all eigenvectors. With X_i, $1 \leq i \leq 3$, defined by (3.9)–(3.11) respectively, this implies that the decomposition of (3.20) always exists. Hence (c) of Theorem 3.1.5 is equivalent to the determinant of the right-hand side of (3.20) being non-zero $\forall\, \lambda \in \mathbb{C}_{uc}$. Finally, a similar approach to that used in the second-half of the proof of Theorem 3.1.6 shows that this condition is equivalent to (c) of Theorem 3.1.5. □

The stability along the pass conditions of Theorems 3.1.4–3.1.7 consist of one or more simple 1D stability conditions and a somewhat more involved determinant condition. One means of assessing the advantages arising from using Theorems 3.1.6 and 3.1.7 over Theorem 3.1.5 and, in turn, Theorem 3.1.4, is the maximal order of the determinant which defines the last condition in each of these results.

Suppose, therefore, that Λ is an $n_\Lambda \times n_\Lambda$ matrix and note that in Theorem 3.1.4 and, in particular, (3.7), the determinant of a matrix of size

m^2 with entries of the order of n^2 leads to the maximal determinant order of m^2n^2. For Theorem 3.1.5, (3.8), the determinant of a second order matrix polynomial of size $(n+m)^2$ leads to the maximal order of $2(n+m)^2$. For Theorem 3.1.6, (3.15), the maximal order is $2n(n+m)$ and in Theorem 3.1.7, (3.18) it is $(n+m)^2 + n_{\Lambda}$.

To apply Theorem 3.1.6, it is necessary to invert two matrices of dimensions $m^2 \times m^2$ and $n^2 \times n^2$ respectively but despite this the maximal order of the determinant has decreased and, as illustrated in Example 3.1.1 below, this leads to a less involved test. The decrease in the maximal order of the determinant from Theorem 3.1.5 to Theorem 3.1.7 is less than that from Theorem 3.1.5 to Theorem 3.1.6, but for Theorem 3.1.7 the only matrix inversion required is of a diagonal matrix and it can, unlike Theorem 3.1.6, deal with the case when $\det A = 0$. Finally, as noted above, the condition in Theorem 3.1.4 has the disadvantage that it cannot be tested by existing software due to the fact that the matrix X_1 is singular.

Next we give two examples to highlight the above results.

Example 3.1.1. As a comparative exercise, consider the case when n (current pass state vector dimension) is 6 and m (the pass profile vector dimension) is 2 and also $\det A \neq 0$. Then the order of the determinant in (3.7) of Theorem 3.1.4 is 144, the order of the determinant in (3.8) of Theorem 3.1.5 is 128, and the order of the determinant in (3.15) of Theorem 3.1.6 is 32.

Example 3.1.2. Consider the following special case of the model of Example 1.2.10

$$
\begin{aligned}
x_{k+1}(p+1) &= \begin{bmatrix} -0.4 & 0.02 \\ -0.6 & 0.03 \end{bmatrix} x_{k+1}(p) + Bu_{k+1}(p) \\
&+ \begin{bmatrix} 0.1423 & 0 \\ -0.0342 & 0 \end{bmatrix} y_k(p) \\
y_{k+1}(p) &= \begin{bmatrix} 0.3895 & 0.03895 \\ 0 & 0 \end{bmatrix} x_{k+1}(p) + Du_{k+1}(p) \\
&+ \begin{bmatrix} -0.5 & 0.75 \\ 0 & 0 \end{bmatrix} y_k(p)
\end{aligned}
$$

In this case $\det D_0 = \det A = 0$, and hence Theorem 3.1.7 has to be used. Here

$$
\begin{aligned}
|\lambda_i(D_0)| &= \{\ 0, \quad 0.5\ \} \\
|\lambda_i(A)| &= \{\ 0, \quad 0.37\ \}
\end{aligned}
$$

Also for $\theta_0 = 1$,

$$
|\lambda_i(G_1(e^{\imath\theta_0}))| = \{\ 0, \quad 0.4694\ \}
$$

In the case of the final condition, the fact that rank $A \otimes I_2 =$ dimension of $\Lambda = 2$ means that

$$\begin{gathered}\hat{W}_{11},\ \hat{W}_{10} \text{ are } 14 \times 14 \text{ matrices}\\ \hat{W}_{21},\ \hat{W}_{20} \text{ are } 14 \times 2 \text{ matrices}\\ \hat{W}_{31},\ \hat{W}_{30} \text{ are } 2 \times 14 \text{ matrices}\\ \Lambda,\ \hat{W}_{41} \text{ and } \hat{W}_{40} \text{ are } 2 \times 2 \text{ matrices}\end{gathered}$$

and hence $\widetilde{W}_1$ and $\widetilde{W}_2$ are 18×18 matrices. Also the condition

$$\det\left(\lambda\widetilde{W}_1 + \widetilde{W}_2\right) = 0$$

is satisfied for λ which are either infinity or has an absolute value of less than 10^{-5}. Hence by Theorem 3.1.7 this example is stable along the pass.

3.2 The 2D Lyapunov Equation

3.2.1 Stability and the 2D Lyapunov Equation

Suppose that $r(D_0) < 1$ (and $r(A) < 1$) hold. Then it follows immediately from the result of Theorem 2.2.8 that a discrete unit memory linear repetitive process of the form defined by Example 1.2.10 is stable along the pass if, and only if,

$$\det\left(I_m - z_2 G_1(z_1^{-1})\right) \neq 0,\ \forall\ |z_1^{-1}| = 1 \text{ and } |z_2| \leq 1 \tag{3.21}$$

(where $G_1(z_1^{-1})$ is obtained by setting $M = 1$ in (1.51)). The so-called 2D Lyapunov equation in 2D/nD systems theory was first developed [134] as a condition for the multivariate characteristic polynomial of a 2D/nD differential linear system to be strictly Hurwitz, i.e. no zeros in the region $\text{Re}(s_i) \geq 0,\ i = 1, 2, \cdots, n$. This result was then extended to the discrete case using the 2D/nD double bilinear transformation.

Consider again the augmented plant matrix Φ of (3.4). Then the 2D Lyapunov equation condition for stability along the pass asserts that this property holds if, and only if, $\exists$ matrices $W_1 > 0$, $W_2 > 0$, and $Q > 0$ of dimensions $n \times n$, $m \times m$, and $(n+m) \times (n+m)$ respectively such that

$$\Phi^T W \Phi - W = -Q \tag{3.22}$$

where W is the direct sum, denoted $\oplus$, of W_1 and W_2, i.e.

$$W := \begin{bmatrix} W_1 & 0 \\ 0 & W_2 \end{bmatrix}$$

In the case of 2D discrete linear systems, it known [11] that the 2D Lyapunov equation is, in general, only a sufficient condition for stability. Below we show that the same is true for the 2D Lyapunov equation (3.22). This requires the following definitions and preliminary results [10] for strictly bounded real matrices (in the square case).

Definition 3.2.1. *Let $S(\gamma)$ be a square matrix of real rational functions in the complex variable γ. Then this matrix is said to be strictly bounded real (SBR) provided*

(a) all poles of $S(\gamma)$ lie in $|\gamma| < 1$; and

(b) $I - S^T(e^{-\imath\omega})S(e^{\imath\omega}) > 0$, $\forall\, \omega \in [0, 2\pi]$.

The conditions under (a) and (b) in this last definition can be reduced to conditions on the matrices of a minimal state-space realization of $S(\gamma)$. In particular, suppose that $S(\gamma)$ has a minimal realization given by the quadruple $\{F, G, H, J\}$ such that

$$S(\gamma) = H^T(\gamma I - F)^{-1}G + J \tag{3.23}$$

Then the following is the so–called strictly bounded real lemma.

Lemma 3.2.1. *Suppose that $S(\gamma)$ is a square matrix of real rational functions in γ with minimal state-space realization defined by $\{F, G, H, J\}$ as per (3.23). Then this matrix is SBR if, and only if, $\exists$ a matrix $\widetilde{P} > 0$ such that*

$$Q_1 := \begin{bmatrix} \widetilde{P} - F^T\widetilde{P}F - HH^T & -F^T\widetilde{P}G - HJ \\ -(F^T\widetilde{P}G + HJ)^T & I - J^TJ - G^T\widetilde{P}G \end{bmatrix} > 0 \tag{3.24}$$

In what follows, we develop the necessary and sufficient condition for the existence of matrices $W = W_1 \oplus W_2 > 0$ and $Q > 0$ satisfying the 2D Lyapunov equation (3.22). This result is based on Lemma 3.2.1 and will subsequently related to the stability along the pass condition of (3.21). Also an example is given which shows that this condition is not sufficient for the existence of solution matrices for the associated 2D Lyapunov equation.

Theorem 3.2.1. *Consider a discrete linear repetitive process of the form defined by Example 1.2.10 and suppose that, for some nonsingular $m \times m$ matrix T, $\tilde{G}_1(z_1^{-1})$ defined by*

$$\tilde{G}_1(z_1^{-1}) := TG_1(z_1^{-1})T^{-1} = T\left[C(z_1^{-1}I_n - A)^{-1}B_0 + D_0\right]T^{-1} \tag{3.25}$$

is SBR. Then $\exists$ matrices $W = W_1 \oplus W_2 > 0$ and $Q > 0$ which satisfy the 2D Lyapunov equation (3.22). Conversely, if (3.22) holds for $W = W_1 \oplus W_2 > 0$ and $Q > 0$, $\exists$ a nonsingular matrix T such that $\tilde{G}_1(z_1^{-1})$ is SBR.

Proof. This is well known in the literature for 2D discrete linear systems but it is instructive to also give it here, starting with that for sufficiency.

Since T is nonsingular, the assumptions made here imply that the quadruple $\{A, B_0T^{-1}, TC, TD_0T^{-1}\}$ is a minimal realization of $\tilde{G}_1(z_1^{-1})$. Also by Lemma 3.2.1, it follows that $\exists$ a matrix $\widetilde{P} > 0$ such that

$$Q_1 := \begin{bmatrix} H_1 & H_2 \\ H_2^T & H_3 \end{bmatrix} > 0$$

where

$$\begin{aligned} H_1 &:= \widetilde{P} - A^T \widetilde{P} A - C^T T^T T C \\ H_2 &:= -A^T \widetilde{P} B_0 T^{-1} - C^T T^T T D_0 T^{-1} \\ H_3 &:= I_m - T^{-T} D_0^T T^T T D_0 T^{-1} - T^{-T} B_0^T \widetilde{P} B_0 T^{-1} \end{aligned}$$

Now pre- and post-multiply the expression above for Q_1 by

$$\begin{bmatrix} I_n & 0 \\ 0 & T^T \end{bmatrix} \text{ and } \begin{bmatrix} I_n & 0 \\ 0 & T \end{bmatrix}$$

respectively, to yield

$$\begin{bmatrix} I_n & 0 \\ 0 & T^T \end{bmatrix} Q_1 \begin{bmatrix} I_n & 0 \\ 0 & T \end{bmatrix} = \begin{bmatrix} \widetilde{P} & 0 \\ 0 & T^T T \end{bmatrix} - \Phi^T \begin{bmatrix} \widetilde{P} & 0 \\ 0 & T^T T \end{bmatrix} \Phi$$

i.e. the 2D Lyapunov equation (3.22) with $W := \widetilde{P} \oplus T^T T$ and $Q := [I_n \oplus T^T]^T Q_1 [I_n \oplus T]$.

Now suppose that (3.22) holds and set $T = W_2^{\frac{1}{2}}, F = A, G = B_0 T^{-1}, H^T = TC$ and $J = TD_0T^{-1}$. Then multiply (3.22) from the left and right by

$$\begin{bmatrix} I_n & 0 \\ 0 & T^{-1} \end{bmatrix}$$

to yield (3.24) with $\widetilde{P} = W_1$ and

$$Q_1 = \begin{bmatrix} I_n & 0 \\ 0 & T^{-1} \end{bmatrix} Q \begin{bmatrix} I_n & 0 \\ 0 & T^{-1} \end{bmatrix}$$

and the SBR property of $\tilde{G}_1(z_1^{-1})$ now follows from Lemma 3.2.1. □

Now let $\lambda_i\{G_1(e^{\imath\omega})\}$, $1 \le i \le m$, denote the eigenvalues of $G(z_1^{-1})$ when $z_1^{-1} = e^{\imath\omega}$, $\omega \in [0, 2\pi]$. Then (3.21) for stability along the pass requires that

$$|\lambda_i\{G_1(e^{\imath\omega})\}| < 1, \ 1 \le i \le m, \ \forall\, \omega \in [0, 2\pi]$$

or, equivalently, for any nonsingular $m \times m$ matrix T

$$|\lambda_i\{\tilde{G}_1(e^{\imath\omega})\}| = |\lambda_i\{TG_1(e^{\imath\omega})T^{-1}\}| < 1, \ 1 \le i \le m, \ \forall\, \omega \in [0, 2\pi] \quad (3.26)$$

The following result now shows that the SBR property implies that (3.21) holds but, as illustrated by an example below, the converse is not true.

Lemma 3.2.2. *If $\tilde{G}_1(z_1^{-1})$ defined by (3.25) is SBR then (3.26) holds.*

Proof. If $\tilde{G}_1(e^{\imath\omega})$ is SBR, it follows that

$$I_m - \tilde{G}_1^T(e^{-\imath\omega})\tilde{G}_1(e^{\imath\omega}) > 0, \ \forall\, \omega \in [0, 2\pi]$$

which implies that

$$|\lambda_i\{\tilde{G}_1(e^{\imath\omega})\}| < 1, \ \forall\, \omega \in [0, 2\pi], \ 1 \leq i \leq m$$

□

To demonstrate that the converse of Lemma 3.2.2 is not true, consider the following example [21] (where the current pass input vector terms have been deleted since they do not influence stability along the pass).

Example 3.2.1. Consider the following special case of the model of Example 1.2.10

$$\begin{aligned} x_{k+1}(p+1) &= \begin{bmatrix} -2.81 & 1 & 0 & 0 & 0 & 0 \\ -2.657 & 0 & 1 & 0 & 0 & 0 \\ -0.845 & 0 & 0 & 0 & 0 & 0 \\ 0 & 0 & 0 & 2.81 & 1 & 0 \\ 0 & 0 & 0 & -2.657 & 0 & 1 \\ 0 & 0 & 0 & 0.845 & 0 & 0 \end{bmatrix} x_{k+1}(p) \\ &+ \begin{bmatrix} 0.028 & 0 \\ 0.008 & 0 \\ 0.012 & 0 \\ 0 & 0.028 \\ 0 & -0.008 \\ 0 & 0.012 \end{bmatrix} y_k(p) \\ y_{k+1}(p) &= \begin{bmatrix} 0 & 0 & 0 & 1 & 0 & 0 \\ 1 & 0 & 0 & 0 & 0 & 0 \end{bmatrix} x_{k+1}(p) + \begin{bmatrix} 0.5 & 0.007 \\ -0.007 & 0.5 \end{bmatrix} y_k(p) \end{aligned}$$

First note that the necessary conditions $r(D_0) < 1$, $r(A) < 1$, and $r(\Phi) < 1$, for stability along the pass hold and also $G_1(z_1^{-1})$ can be written in the form

$$G_1(z_1^{-1}) = \begin{bmatrix} 0.5 & G_a(z_1^{-1}) \\ G_b(z_1^{-1}) & 0.5 \end{bmatrix} \tag{3.27}$$

where

$$\begin{aligned} G_a(z_1^{-1}) &:= \frac{0.028(z_1^{-1})^2 - 0.008z_1^{-1} + 0.012}{(z_1^{-1})^3 - 2.81(z_1^{-1})^2 + 2.657z_1^{-1} - 0.845} + 0.007 \\ G_b(z_1^{-1}) &:= \frac{0.028(z_1^{-1})^2 + 0.008z_1^{-1} + 0.012}{(z_1^{-1})^3 + 2.81(z_1^{-1})^2 + 2.657z_1^{-1} + 0.845} - 0.007 \end{aligned} \tag{3.28}$$

Also it is easily verified, by evaluating $|G_a(e^{\imath\omega})G_b(e^{\imath\omega})| \ \forall\, \omega \in [0, 2\pi]$, that

$$|G_a(e^{\imath\omega})G_b(e^{\imath\omega})| < 0.007$$

and hence the eigenvalues of $G_1(e^{\imath\omega})$ are close to 0.5, i.e. this example is stable along the pass.

To show that $G_1(z_1^{-1})$ is not SBR, it is required to show that there is no $m \times m$ nonsingular matrix T such that $\tilde{G}_1(z_1^{-1}) = TG_1(z_1^{-1})T^{-1}$ satisfies

$$I_m - \tilde{G}_1^T(e^{-\imath\omega})\tilde{G}_1(e^{\imath\omega}) > 0, \; \forall\, \omega \in [0, 2\pi] \tag{3.29}$$

This is done by assuming that such a matrix T does exist and then establishing a contradiction.

Suppose, therefore, that $\widetilde{P} := T^T T$ and then (3.29) can be rewritten as

$$\widetilde{P} - G_1^T(e^{-\imath\omega})\widetilde{P}G_1(e^{\imath\omega}) > 0, \; \forall\, \omega \in [0, 2\pi]$$

Next we show that there exists no matrix $\widetilde{P} > 0$ such that the following two conditions hold

$$\widetilde{P} - G_1^T(1)\widetilde{P}G_1(1) > 0 \tag{3.30}$$

and

$$\widetilde{P} - G_1^T(-1)\widetilde{P}G_1(-1) > 0 \tag{3.31}$$

Using (3.27) and (3.28) we have that

$$G_1(1) = \begin{bmatrix} 0.5 & \beta \\ -\zeta & 0.5 \end{bmatrix}, \quad G_1(-1) = \begin{bmatrix} 0.5 & -\beta \\ \zeta & 0.5 \end{bmatrix}$$

where $\beta = 16.007$ and $\zeta = 0.000435$. Since $\widetilde{P}$ is symmetric, denote its elements by $\widetilde{p}_{11}$, $\widetilde{p}_{12} = \widetilde{p}_{21}$, and $\widetilde{p}_{22}$ where no loss of generality arises from setting $\widetilde{p}_{11} = 1$. Hence since $\widetilde{P} > 0$, we have that $|\widetilde{p}_{12}| < \sqrt{\widetilde{p}_{22}}$, and we can proceed to show that for (3.30) to hold we must have that

$$\sqrt{\widetilde{p}_{22}} > \frac{2|\beta|}{3} \approx 10 \tag{3.32}$$

Similarly, we find that for (3.31) to hold we must have that

$$\sqrt{\widetilde{p}_{22}} < \frac{3}{2|\beta|} \approx 0.1$$

This last condition clearly contradicts (3.32) and hence there is no nonsingular matrix T such that (3.29) holds.

The following result can now be stated.

Theorem 3.2.2. *Consider the linear repetitive process S generated by a discrete unit memory linear repetitive process of the form defined by Example 1.2.10. Then S is stable along the pass if $\exists$ matrices $W_1 > 0, W_2 > 0$ and $Q > 0$ such that the 2D Lyapunov equation*

$$\Phi^T W \Phi - W = -Q \tag{3.33}$$

holds with $W = W_1 \oplus W_2$.

There are a number of special cases when the 2D Lyapunov equation condition is both necessary and sufficient for stability along the pass. The most important of these (arguably) is when the process is SISO and we have the following result.

Corollary 3.2.1. *If the process under consideration is SISO then the 2D Lyapunov equation condition of Theorem 3.2.2 is necessary and sufficient for stability along the pass.*

Proof. In the SISO case $\tilde{G}_1(z_1^{-1})$ is a scalar function in z_1^{-1} and consequently the two scalar conditions

$$|\lambda_1\{\tilde{G}_1(e^{\imath\omega})\}| < 1, \ \forall\, \omega \in [0, 2\pi]$$

and

$$1 - \tilde{G}_1(e^{-\imath\omega})\tilde{G}_1(e^{\imath\omega}) > 0, \ \forall\, \omega \in [0, 2\pi]$$

are equivalent. □

A second case when the 2D Lyapunov equation is both necessary and sufficient is when the augmented plant matrix Φ is normal (see [55] for the case of 2D discrete linear systems described by the Roesser model).

Corollary 3.2.2. *Suppose that the process under consideration has an augmented plant matrix which is normal, i.e. $\Phi^T\Phi = \Phi\Phi^T$. Then the 2D Lyapunov equation condition of Theorem 3.2.2 is necessary and sufficient for stability along the pass.*

Proof. Necessity is obvious. To prove sufficiency, first note that $r(\Phi) < 1$ (a necessary condition for stability along the pass) means that Φ satisfies the 1D discrete linear systems Lyapunov equation, i.e.

$$\Phi^T W \Phi - W < 0 \tag{3.34}$$

for some $W > 0$. Also if λ_i, $1 \le i \le n+m$, denote the eigenvalues of Φ then, since Φ is normal,

$$\Phi = R\mathrm{diag}\{\lambda_i\}_{1 \le i \le n+m} R^* = R\Lambda R^* \tag{3.35}$$

where here $*$ denotes the complex conjugate transpose operation, and R is the orthogonal eigenvector matrix of Φ. Substitution of (3.35) and $W = I_{n+m}$ into the left–hand side of (3.34) now yields

$$\Phi^T W \Phi - W = R\left(\mathrm{diag}\{|\lambda_i|^2\}_{1 \le i \le n+m} - I_{n+m}\right) R^* =: R\left(|\Lambda|^2 - I_{n+m}\right) R^*$$

where (since $r(\Phi) < 1$ by assumption) $|\lambda_i|^2 < 1$, $1 \le i \le n+m$. Equivalently the 1D linear systems Lyapunov equation (3.34) holds under the choice of $W = I_{n+m}$, and hence also, since I_{n+m} is block diagonal under any partition, the 2D Lyapunov equation for stability along the pass holds. □

In the case of discrete linear repetitive processes of the form defined by Example 1.2.10, but not for other classes of 2D discrete linear systems, it is possible [74] to use the 2D Lyapunov equation to extract computable information on the rate of convergence to the resulting limit profile of the pass profile sequence of a stable along the pass example. Suppose, therefore, that the 2D Lyapunov equation holds for the example under consideration which is operating under zero control input sequence and introduce

$$\begin{aligned}||x_{k+1}(p+1)||^2_{W_1} &:= x^T_{k+1}(p+1)W_1x_{k+1}(p+1)\\ ||y_{k+1}(p)||^2_{W_2} &:= y^T_{k+1}(p)W_2y_{k+1}(p)\end{aligned}$$

Then, using (3.33),

$$\begin{aligned}||x_{k+1}(p+1)||^2_{W_1} + ||y_{k+1}(p)||^2_{W_2} &- ||x_k(p)||^2_{W_1} - ||y_k(p)||^2_{W_2}\\ &= -||x_k(p)||^2_{I_n} - ||y_k(p)||^2_{I_m} \quad (3.36)\end{aligned}$$

Now, assuming that $x_{k+1}(0) = 0,\ k \geq 0$, introduce for $j \geq 0$,

$$\begin{aligned}||x_{j+1}||^2_{W_1} &:= \sum_{p=0}^{\infty} ||x_{j+1}(p+1)||^2_{W_1}\\ ||y_{j+1}||^2_{W_2} &:= \sum_{p=0}^{\infty} ||y_{j+1}(p)||^2_{W_2}\end{aligned}$$

Then applying these summations to (3.36) gives

$$\begin{aligned}||x_{k+1}||^2_{W_1} + ||y_{k+1}||^2_{W_2} &= ||x_k||^2_{W_1-I_n} + ||y_k||^2_{W_2-I_m}\\ &\leq \lambda(||x_k||^2_{W_1} + ||y_k||^2_{W_2})\end{aligned}$$

where λ is a positive constant which is defined as follows.

Assume that

$$\begin{aligned}W_1 - I_n &\leq \lambda_1 W_1\\ W_2 - I_m &\leq \lambda_2 W_2\end{aligned}$$

where λ_1 and λ_2 are positive real scalars. Then

$$||(x_{k+1}, y_{k+1})||^2 := ||x_{k+1}||^2_{W_1} + ||y_{k+1}||^2_{W_2} \leq \lambda(||(x_k, y_k)||^2) \quad (3.37)$$

where $\lambda = \max(\lambda_1, \lambda_2)$. Moreover, the process is stable along the pass if $\lambda < 1$ which is always guaranteed to be true if $W_1 > I_n$, $W_2 > I_m$. Hence we have geometric convergence to zero in the pass-to-pass direction.

It is also possible to compute the rate λ using the fact that if a square matrix, say $\hat{W}$, satisfies $\hat{W} = \hat{W}^T \geq I$ then $\exists\, \hat{\lambda} \in [0, 1) : \hat{W} - I \leq \hat{\lambda}\hat{W}$. Also, with $\overline{\sigma}(\cdot)$ denoting the maximum singular value of its matrix argument, it is easily shown that

$$\hat{W} - I \leq \hat{W} - \frac{1}{\overline{\sigma}^2(\hat{W})}\hat{W} = \left(1 - \frac{1}{\overline{\sigma}^2(\hat{W})}\right)\hat{W}$$

Hence in the case of (3.37)

$$\lambda = \max\left(1 - \frac{1}{\overline{\sigma}^2(W_i)}\right),\ i = 1, 2$$

3.2.2 An Alternative 2D Lyapunov Equation

It is possible to develop an alternative Lyapunov equation (see [87] for the Fornasini-Marchesini 2D discrete linear systems state-space model case) to that of the previous sub–section which will be of major use in terms of the robustness properties of discrete unit memory linear repetitive processes (see Chap. 5). The starting point is the fact that stability along the pass is equivalent to the requirement that the characteristic polynomial $\mathcal{C}(z_1, z_2)$ (see Theorem 2.2.8) satisfies

$$\mathcal{C}(z_1, z_2) \neq 0 \text{ in } \mathbb{U}_c^2$$

or, equivalently,

$$\det\left(I_{n+m} - z_1\widehat{A}_1 - z_2\widehat{A}_2\right) \neq 0 \text{ in } \mathbb{U}_c^2$$

where $\widehat{A}_1$ and $\widehat{A}_2$ are defined by (2.66) (and $\mathbb{U}_c^2$ denotes the closed unit bidisc).

Now we have the following result.

Theorem 3.2.3. *Consider the linear repetitive process S generated by a discrete unit memory linear repetitive process of the form defined by Example 1.2.10. Then S is stable along the pass if $\exists$ an $(n+m) \times (n+m)$ matrix $\widehat{P} > 0$ such that*

$$Q = \begin{bmatrix} \delta_1\widehat{P} & 0 \\ 0 & \delta_2\widehat{P} \end{bmatrix} - \widehat{A}^T\widehat{P}\widehat{A} > 0 \tag{3.38}$$

for any $\delta_i > 0$, $i = 1, 2$, $\delta_1 + \delta_2 = 1$, and $\widehat{A} := \begin{bmatrix} \widehat{A}_1 & \widehat{A}_2 \end{bmatrix}$.

Proof. Suppose that (3.38) holds and assume that

$$\det\left(I_{n+m} - z_1\widehat{A}_1 - z_2\widehat{A}_2\right) = 0 \tag{3.39}$$

for some $(z_1, z_2) \in \mathbb{C}^2$. Then $\exists$ an $(n+m) \times 1$ vector $x \neq 0$ such that

$$\left(I_{n+m} - z_1\widehat{A}_1 - z_2\widehat{A}_2\right)x = 0$$

or

$$x = \left(z_1 \widehat{A}_1 + z_2 \widehat{A}_2\right) x \tag{3.40}$$

Use of (3.38) and (3.40) now yields

$$\begin{aligned} x^* \widehat{P} x &= x^* \begin{bmatrix} z_1^* I_{n+m} & z_2^* I_{n+m} \end{bmatrix} \widehat{A}^T \widehat{P} \widehat{A} \begin{bmatrix} z_1 I_{n+m} \\ z_2 I_{n+m} \end{bmatrix} x \\ &= \left(\delta_1 |z_1|^2 + \delta_2 |z_2|^2\right) x^* P x \\ &\quad - x^* \begin{bmatrix} z_1^* I_{n+m} & z_2^* I_{n+m} \end{bmatrix} Q \begin{bmatrix} z_1 I_{n+m} \\ z_2 I_{n+m} \end{bmatrix} x \end{aligned} \tag{3.41}$$

Hence

$$\left(\delta_1 |z_1|^2 + \delta_2 |z_2|^2 - 1\right) x^* P x = x^* \begin{bmatrix} z_1^* I_{n+m} & z_2^* I_{n+m} \end{bmatrix} Q \begin{bmatrix} z_1 I_{n+m} \\ z_2 I_{n+m} \end{bmatrix} x$$

The facts that $\widehat{P} > 0$ and $Q > 0$ now give

$$\begin{aligned} x^* \widehat{P} x &> 0 \\ x^* \begin{bmatrix} z_1^* I_{n+m} & z_2^* I_{n+m} \end{bmatrix} Q \begin{bmatrix} z_1 I_{n+m} \\ z_2 I_{n+m} \end{bmatrix} x &> 0 \end{aligned}$$

and using these in (3.41) yields

$$\delta_1 |z_1|^2 + \delta_2 |z_2|^2 - 1 > 0$$

or

$$\frac{|z_1|^2}{\frac{1}{\delta_1}} + \frac{|z_2|^2}{\frac{1}{\delta_2}} > 1$$

Hence the region in the $(|z_1|, |z_2|)$-plane in which (3.39) holds is that outside the ellipse

$$\frac{|z_1|^2}{\frac{1}{\delta_1}} + \frac{|z_2|^2}{\frac{1}{\delta_2}} = 1$$

Suppose now that

$$\det\left(I_{n+m} - z_1 \widehat{A}_1 - z_2 \widehat{A}_2\right) \neq 0 \tag{3.42}$$

for $(z_1, z_2) \in \mathbb{C}^2$ such that

$$\frac{|z_1|^2}{\frac{1}{\delta_1}} + \frac{|z_2|^2}{\frac{1}{\delta_2}} \leq 1$$

Suppose also that (3.42) is valid for $(z_1, z_2) \in \mathbb{U}_0^2$, where

$$\mathbb{U}_0^2 := \{(z_1, z_2) : \delta_1 |z_1|^2 + \delta_2 |z_2|^2 \leq 1\}$$

Then it is clear that $\mathbb{U}_c^2 \subset \mathbb{U}_0^2$ and hence if (3.42) is valid for $(z_1, z_2) \in \mathbb{U}_0^2$ then it is also valid for $(z_1, z_2) \in \mathbb{U}_c^2$, and the proof is complete. □

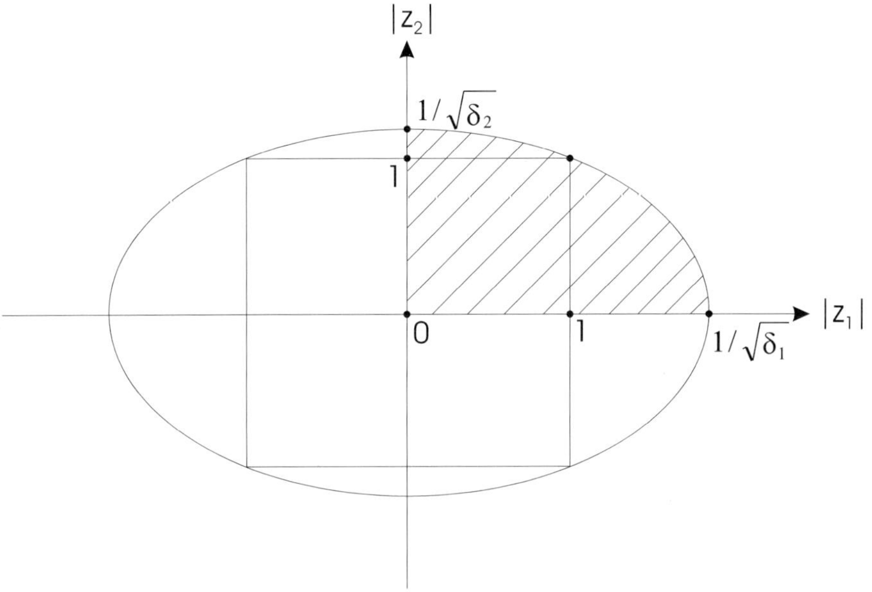

Fig. 3.1. Stability region

Figure 3.1 shows a schematic illustration of the region such that $\mathcal{C}(z_1, z_2) \neq 0$ if Theorem 3.2.3 holds.

It follows immediately from Theorem 3.2.3 that stable along the pass examples with $\det\left(I_{n+m} - z_1\widehat{A}_1 - z_2\widehat{A}_2\right) = 0$ in the shaded regions S_1 and S_2 in Fig. 3.2 do not satisfy the 2D Lyapunov equation defined by this result. In particular, it is not possible to find admissible δ_i, $i = 1, 2$, such that the solutions of $\det\left(I_{n+m} - z_1\widehat{A}_1 - z_2\widehat{A}_2\right) = 0$ in both the areas S_1 and S_2 are outside the corresponding ellipse. This means that there is a (potentially large) number of examples whose stability along the pass properties cannot be determined by application of the 2D Lyapunov equation condition of Theorem 3.2.3. We return to this last problem below but first we give the following result, obtained by imposing the constraint that $\delta_1 = \delta_2 = \frac{1}{2}$ in Theorem 3.2.3.

Theorem 3.2.4. *Consider the linear repetitive process S generated by a discrete unit memory linear repetitive process of the form defined by Example 1.2.10. Then the 2D Lyapunov equation condition of Theorem 3.2.3 holds under* $\delta_1 = \delta_2 = \frac{1}{2}$ *if, and only if, the* $(n+m) \times 2(n+m)$ *matrix* $\widehat{A}$ *can be decomposed as*

$$\widehat{A} = T^{-1}\widetilde{R}\begin{bmatrix} \Sigma & 0 \end{bmatrix}\widetilde{S}\begin{bmatrix} T & 0 \\ 0 & T \end{bmatrix} \tag{3.43}$$

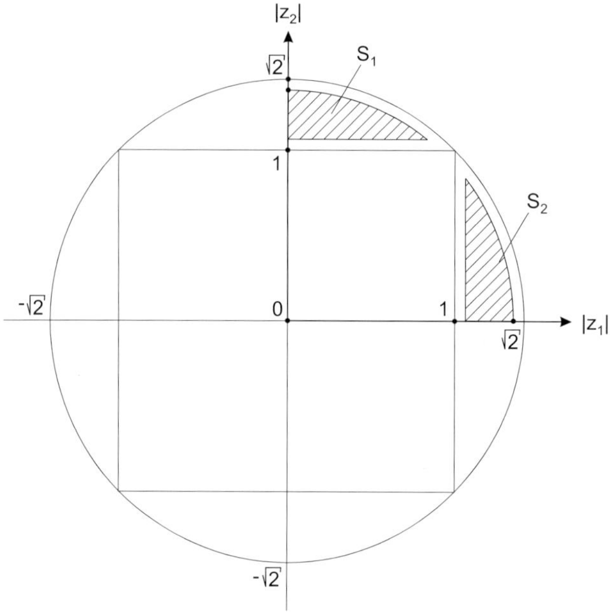

Fig. 3.2. Modified stability region

where $\widetilde{R}$ *and* $\widetilde{S}$ *are orthogonal matrices with dimensions* $(n+m) \times (n+m)$ *and* $2(n+m) \times 2(n+m)$ *respectively, the* $(n+m) \times (n+m)$ *matrix* T *is nonsingular, and*

$$\Sigma := \operatorname{diag}\{r_i\}_{1 \le i \le n+m}, \ |r_i| < \frac{1}{\sqrt{2}}, \ 1 \le i \le n+m$$

Proof. To prove that this condition is sufficient, assume that (3.43) is valid, and choose $\widehat{P} = T^T T > 0$. Then when $\delta_1 = \delta_2 = \frac{1}{2}$,

$$\begin{aligned} Q &= \begin{bmatrix} \delta_1 \widehat{P} & 0 \\ 0 & \delta_2 \widehat{P} \end{bmatrix} - \widehat{A}^T \widehat{P} \widehat{A} \\ &= \begin{bmatrix} T^T & 0 \\ 0 & T^T \end{bmatrix} \widetilde{S}^T \begin{bmatrix} \delta_1 I_{n+m} - \Sigma^2 & 0 \\ 0 & \delta_2 I_{n+m} \end{bmatrix} \widetilde{S} \begin{bmatrix} T & 0 \\ 0 & T \end{bmatrix} \end{aligned}$$

Clearly $Q > 0$ and this implies that the condition stated in Theorem 3.2.3 holds.

To prove necessity, suppose that $\exists$ an $(n+m) \times (n+m)$ matrix $\widehat{P} > 0$ such that (3.38) holds when $\delta_1 = \delta_2 = \frac{1}{2}$. Then since $\widehat{P} > 0$, an $(n+m) \times (n+m)$ matrix T exists such that $\widehat{P} = T^T T$. Hence using (3.38) we have that

$$\begin{bmatrix} \sqrt{\delta_1} T^T & 0 \\ 0 & \sqrt{\delta_2} T^T \end{bmatrix} \begin{bmatrix} \sqrt{\delta_1} T & 0 \\ 0 & \sqrt{\delta_2} T \end{bmatrix} - \widehat{A}^T T^T T \widehat{A} > 0$$

or, equivalently,

$$\begin{bmatrix} \delta_1 I_{n+m} & 0 \\ 0 & \delta_2 I_{n+m} \end{bmatrix} - \widetilde{A}^T \widetilde{A} > 0 \tag{3.44}$$

where

$$\begin{aligned} \widetilde{A} &:= \begin{bmatrix} \widetilde{A}_1 & \widetilde{A}_2 \end{bmatrix} \\ \widetilde{A}_1 &:= T\widehat{A}_1 T^{-1} \\ \widetilde{A}_2 &:= T\widehat{A}_2 T^{-1} \end{aligned}$$

Now write the singular value decomposition of $\widetilde{A}$ as

$$\widetilde{A} = \widetilde{R} \begin{bmatrix} \Sigma & 0 \end{bmatrix} \widetilde{S}$$

where $\widetilde{R}$, Σ, and $\widetilde{S}$ are defined as in (3.43). Then on substituting this last expression into (3.44) we obtain

$$\begin{bmatrix} \delta_1 I_{n+m} & 0 \\ 0 & \delta_2 I_{n+m} \end{bmatrix} - \widetilde{S}^T \begin{bmatrix} \Sigma^2 & 0 \\ 0 & 0 \end{bmatrix} \widetilde{S} > 0$$

or, since $\delta_1 = \delta_2 = \frac{1}{2}$,

$$\begin{bmatrix} \delta_1 I_{n+m} - \Sigma^2 & 0 \\ 0 & \delta_2 I_{n+m} \end{bmatrix} > 0$$

This last result implies that $|r_i| < \frac{1}{\sqrt{2}}$, $1 \le i \le n+m$, and hence $\widehat{A}$ can be decomposed into the form given in (3.43). □

In what follows, we develop a less conservative version (see [110] for the case of 2D discrete linear systems described by the Fornasini-Marchesini model) of the 2D Lyapunov equation sufficient condition for stability along the pass given in Theorem 3.2.3. The starting point is the fact that a symmetric positive definite (or positive semi–definite) matrix, say H, can always be written in the form $H = U^T \Gamma U$, where U is an orthogonal matrix and, in the positive definite case, $\Gamma = \text{diag}\{\sigma_1, \cdots, \sigma_{n+m}\}$ with $\sigma_i > 0$, $1 \le i \le n+m$. Also if we write $\Gamma^{\frac{1}{2}} = \text{diag}\{\sigma_1^{\frac{1}{2}}, \cdots, \sigma_{n+m}^{\frac{1}{2}}\}$, then $H = H^{\frac{T}{2}} H^{\frac{1}{2}}$, where $H^{\frac{1}{2}} = \Gamma^{\frac{1}{2}} U$ and $H^{\frac{T}{2}} = \left(H^{\frac{1}{2}}\right)^T$. Such an $H^{\frac{1}{2}}$ is often referred to as the (nonsymmetric) square root of H.

Using the above fact, the 2D Lyapunov equation of (3.38) can be written as

$$Q = \begin{bmatrix} \widehat{P}^{\frac{T}{2}} \left(\delta_1 I_{n+m}\right) \widehat{P}^{\frac{1}{2}} & 0 \\ 0 & \widehat{P}^{\frac{T}{2}} \left(\delta_2 I_{n+m}\right) \widehat{P}^{\frac{1}{2}} \end{bmatrix} - \widehat{A}^T \widehat{P}^{\frac{T}{2}} I_{n+m} \widehat{P}^{\frac{1}{2}} \widehat{A}$$

Below, we consider the 2D Lyapunov equation

$$Q = \begin{bmatrix} \widehat{P}^{\frac{T}{2}} W_1 \widehat{P}^{\frac{1}{2}} & 0 \\ 0 & \widehat{P}^{\frac{T}{2}} W_2 \widehat{P}^{\frac{1}{2}} \end{bmatrix} - \widehat{A}^T \widehat{P}^{\frac{T}{2}} \widehat{R} \widehat{P}^{\frac{1}{2}} \widehat{A} \tag{3.45}$$

for appropriately dimensioned matrices $\widehat{P} > 0, W_1 > 0, W_2 > 0$, and $\widehat{R} > 0$. Note also that (3.45) reduces to (3.38) when $W_1 = \delta_1 I_{n+m}$, $W_2 = \delta_2 I_{n+m}$ and $\widehat{R} = I_{n+m}$.

Theorem 3.2.5. *Consider the linear repetitive process S generated by a discrete unit memory linear repetitive process of the form defined by Example 1.2.10. Then S is stable along the pass if $\exists$ matrices $\widehat{P} > 0, W_1 > 0, W_2 > 0$, and $\widehat{R} > 0$ such that the matrix Q defined by (3.45) satisfies $Q > 0$ and*

$$\widehat{R} - W_1 - W_2 \geq 0 \tag{3.46}$$

Proof. This is by contradiction. Suppose, therefore, that the conditions of the theorem hold but the process under consideration is unstable along the pass. Then $\exists\, (z_1, z_2) \in \mathbb{U}_c^2$ such that

$$\det\left(I_{n+m} - z_1 \widehat{A}_1 - z_2 \widehat{A}_2\right) = 0 \tag{3.47}$$

Hence $\exists\, v \neq 0$ such that

$$v = \widehat{A} \begin{bmatrix} z_1 I_{n+m} \\ z_2 I_{n+m} \end{bmatrix} v \tag{3.48}$$

Using (3.45) and (3.48) we now have that

$$\begin{aligned} v^* \widehat{P}^{\frac{T}{2}} \widehat{R} \widehat{P}^{\frac{1}{2}} v &= v^* \begin{bmatrix} z_1^* I_{n+m} & z_2^* I_{n+m} \end{bmatrix} \begin{bmatrix} \widehat{P}^{\frac{T}{2}} W_1 \widehat{P}^{\frac{1}{2}} & 0 \\ 0 & \widehat{P}^{\frac{T}{2}} W_2 \widehat{P}^{\frac{1}{2}} \end{bmatrix} \\ &\quad \times \begin{bmatrix} z_1 I_{n+m} \\ z_2 I_{n+m} \end{bmatrix} v - v_c^* Q v_c \end{aligned}$$

and $v_c^* := v^* \begin{bmatrix} z_1^* I_{n+m} & z_2^* I_{n+m} \end{bmatrix}$. Therefore

$$v^* \widehat{P}^{\frac{T}{2}} \left(\widehat{R} - |z_1|^2 W_1 - |z_2|^2 W_2\right) \widehat{P}^{\frac{1}{2}} v = -v_c^* Q v_c \tag{3.49}$$

It now follows from (3.47) that $(z_1, z_2) \neq 0$, and hence $v_c \neq 0$. Hence $Q > 0$ implies that the right–hand side of (3.49) is negative.

Note now that $|z_1| \leq 1$, $|z_2| \leq 1$, and the positive semi–definite condition of (3.46) imply that $\widehat{R} - |z_1|^2 W_1 - |z_2|^2 W_2 \geq 0$. Hence the left–hand side of (3.49) is nonnegative and this leads to a contradiction and the proof is complete. □

The following corollaries now follow from Theorem 3.2.5.

Corollary 3.2.3. *The linear repetitive process S generated by a discrete unit memory linear repetitive process of the form defined by Example 1.2.10 is stable along the pass if ∃ a matrix $\widehat{P} > 0$ such that*

$$Q = \begin{bmatrix} \widehat{P}^{\frac{T}{2}} W_1 \widehat{P}^{\frac{1}{2}} & 0 \\ 0 & \widehat{P}^{\frac{T}{2}} W_2 \widehat{P}^{\frac{1}{2}} \end{bmatrix} - \widehat{A}^T \widehat{P} \widehat{A} > 0$$

where $W_1 = U^T \Sigma_1 U$, $W_2 = U^T \Sigma_2 U$, U is orthogonal, and

$$\begin{aligned} \Sigma_1 &:= \operatorname{diag}\{\sigma_{1,1}, \cdots, \sigma_{1,n+m}\} \\ \Sigma_2 &:= \operatorname{diag}\{\sigma_{2,1}, \cdots, \sigma_{2,n+m}\} \end{aligned}$$

$$\left. \begin{aligned} &\sigma_{1,j} > 0 \\ &\sigma_{2,j} > 0 \\ &\sigma_{1,j} + \sigma_{2,j} = 1, \ \ 1 \le j \le n+m \end{aligned} \right\} \tag{3.50}$$

Proof. Set $\widehat{R} = I_{n+m}$ and then $I_{n+m} - W_1 - W_2 = 0$. The result now follows on using Theorem 3.2.5. □

Corollary 3.2.4. *The linear repetitive process S generated by a discrete unit memory linear repetitive process of the form defined by Example 1.2.10 is stable along the pass if ∃ a matrix $\widehat{P} > 0$ such that*

$$Q = \begin{bmatrix} \widehat{P}^{\frac{T}{2}} W_1 \widehat{P}^{\frac{1}{2}} & 0 \\ 0 & \widehat{P}^{\frac{T}{2}} W_2 \widehat{P}^{\frac{1}{2}} \end{bmatrix} - \widehat{A}^T \widehat{P} \widehat{A} > 0$$

where (the elements in) $W_i := \operatorname{diag}\{\sigma_{i,1}, \cdots, \sigma_{i,n+m}\}$, $i = 1, 2$, satisfy (3.50).

Proof. Choose $U = I_{n+m}$. □

Corollary 3.2.5. *The result of Theorem 3.2.3 is recovered by setting $\widehat{R} = I_{n+m}$, $W_1 = \delta_1 I_{n+m}$, $W_2 = \delta_2 I_{n+m}$.*

The following result, whose proof is omitted here since it is essentially identical to that in the 2D discrete systems literature, shows that $\widehat{R} = I_{n+m}$ in the corollaries above does not destroy the generality of Theorem 3.2.5.

Theorem 3.2.6. *The sufficient conditions of Theorem 3.2.5 are equivalent to the existence of matrices $\widehat{P} > 0$, $W_1 > 0$, and $W_2 > 0$ such that*

$$Q = \begin{bmatrix} \widehat{P}^{\frac{T}{2}} W_1 \widehat{P}^{\frac{1}{2}} & 0 \\ 0 & \widehat{P}^{\frac{T}{2}} W_2 \widehat{P}^{\frac{1}{2}} \end{bmatrix} - \widehat{A}^T \widehat{P} \widehat{A} > 0 \tag{3.51}$$

and

$$I_{n+m} - W_1 - W_2 \ge 0 \tag{3.52}$$

In Chap. 5 it will be shown that the 2D Lyapunov equation is closely related to the definition of a Lyapunov function for processes of the form defined by Example 1.2.10 which is subsequently used in control law design (Chaps. 7 and 8).

3.2.3 Solving the 2D Lyapunov Equation

Consider first the 2D Lyapunov equation of Theorem 3.2.2. Then one approach to obtaining solution algorithms for 2D discrete linear systems described by the Roesser model is given in [2] and here we detail how these case be applied in the discrete linear repetitive process case. To begin, first note that Definition 3.2.1 describes the SBR property in terms of frequency–domain constraints. Lemmas 3.2.3 and 3.2.4 below give two alternative formulations of the so-called DTSBR lemma and it is these which will form the basis of solution algorithms.

Lemma 3.2.3. *Let $S(\gamma)$ be a square matrix of real rational functions in the complex variable γ defined from a minimal state-space realization by (3.23). Then $S(\gamma)$ is SBR if, and only if, $\exists$ real matrices L and $\widetilde{L}$, a matrix $\widetilde{P} > 0$, and a constant β, $|\beta| < 1$, such that*

$$\begin{aligned} F^T\widetilde{P}F - \beta^2\widetilde{P} + HH^T &= -LL^T \\ G^T\widetilde{P}G - \beta^2 I + J^TJ &= -\widetilde{L}^T\widetilde{L} \\ F^T\widetilde{P}G + HJ &= -L\widetilde{L} \end{aligned}$$

Lemma 3.2.4. *Let $S(\gamma)$ be a square matrix of real rational functions in the complex variable γ defined from a minimal state-space realization by (3.23). Then $S(\gamma)$ is SBR if, and only if, $\exists$ real matrices L and $\widetilde{L}$, a matrix $\widetilde{P} > 0$, a full rank matrix N, and a constant β, $|\beta| < 1$, such that*

$$\begin{aligned} F^T\widetilde{P}F - \widetilde{P} + HH^T &= -LL^T - NN^T \\ G^T\widetilde{P}G - \beta^2 I + J^TJ &= -\widetilde{L}^T\widetilde{L} \\ F^T\widetilde{P}G + HJ &= -L\widetilde{L} \end{aligned}$$

Return now to the 2D Lyapunov equation. Then the first solution algorithm given below starts from choosing Q to be of the form

$$Q = \begin{bmatrix} \widetilde{L}^T \\ L \end{bmatrix} \begin{bmatrix} \widetilde{L} & L^T \end{bmatrix} + (1-\beta^2)\begin{bmatrix} W_1 & 0 \\ 0 & W_2 \end{bmatrix}$$

Under this choice, the 2D Lyapunov equation (3.33) can be rewritten in the following form

$$A^TW_1A - \beta^2W_1 + C^TW_2C = -\widetilde{L}^T\widetilde{L} \tag{3.53}$$

$$D_0^TW_2D_0 - \beta^2W_2 + B_0^TW_1B_0 = -LL^T \tag{3.54}$$

$$A^TW_1B_0 + C^TW_2D_0 = -\widetilde{L}^TL^T \tag{3.55}$$

which is equivalent to the conditions of Lemma 3.2.3 under obvious substitutions.

Conversely, suppose that Q is taken to be of the form

$$Q = \begin{bmatrix} \widetilde{L}^T \\ L \end{bmatrix} \begin{bmatrix} \widetilde{L} & L^T \end{bmatrix} + \begin{bmatrix} (1-\beta^2)W_1 & 0 \\ 0 & NN^T \end{bmatrix}$$

Then the 2D Lyapunov equation gives

$$A^T W_1 A - \beta^2 W_1 + C^T W_2 C = -\widetilde{L}^T \widetilde{L} \tag{3.56}$$

$$D_0^T W_2 D_0 - W_2 + B_0^T W_1 B_0 = -LL^T - NN^T \tag{3.57}$$

$$A^T W_1 B_0 + C^T W_2 D_0 = -\widetilde{L}^T L^T \tag{3.58}$$

which is equivalent to the conditions of Lemma 3.2.4 under obvious substitutions.

It now follows immediately that solution matrices $W > 0$ and $Q > 0$ can be obtained by solving either of the following routes

1. Find $\{W_1, W_2, L, \widetilde{L}, \beta\}$ where $W_1 > 0$, $W_2 > 0$, L and $\widetilde{L}$ are real and $|\beta| < 1$ such that (3.53)–(3.55) and are satisfied.
2. Find $\{W_1, W_2, L, \widetilde{L}, N, \beta\}$ where $W_1 > 0$, $W_2 > 0, \widetilde{L}, N$ full rank, L real, $|\beta| < 1$, such that (3.56)–(3.58) are satisfied.

Hence it is clear that obtaining positive definite solutions to the 2D Lyapunov equation is equivalent to solving the DTSBR lemma. This leads to solution algorithms based on 1D spectral factorization and an algebraic Riccati equation respectively which are identical to those for a 2D linear system described by the Roesser model. Hence the details are omitted here and a detailed treatment can be found in [2]. We postpone further discussion of solution algorithms for this 2D Lyapunov equation until we have considered the same problem for the alternative equation also developed in this chapter.

Consider now the alternative version of the 2D Lyapunov equation and, in particular, Theorem 3.2.6. Then in what follows we consider the following problem.

Given $\widehat{A} = \begin{bmatrix} \widehat{A}_1 & \widehat{A}_2 \end{bmatrix}$, find matrices $\widehat{P} > 0$, $W_1 > 0$, and $W_2 > 0$ such that Q defined by (3.51) satisfies $Q > 0$ and (3.52) holds.

Here we follow the analysis in [110] and, as a starting point, we need the following result that relates the existence of such $\widehat{P}$, W_1 and W_2 to a norm minimization problem.

Theorem 3.2.7. *There exist $\widehat{P} > 0$, $W_1 > 0$, and $W_2 > 0$ such that (3.52) holds and Q of (3.51) is positive definite if, and only if,*

$$\min \left\| \tilde{A} \begin{bmatrix} V_1 & 0 \\ 0 & V_2 \end{bmatrix} \right\| < 1 \tag{3.59}$$

subject to the constraints that T, V_1, and V_2 are nonsingular and

$$I - V_1^{-T} V_1^{-1} - V_2^T V_2^{-1} \geq 0$$

where here $||\cdot||$ denotes the induced Euclidean norm of its matrix argument.

Proof. Suppose first that matrices $\widehat{P} > 0$, $W_1 > 0$, and $W_2 > 0$ exist such that Q in (3.51) is positive definite and (3.52) holds. Also write

$$W_1 = V_1^{-T} V_1^{-1}, \; W_2 = V_2^{-T} V_2^{-1}$$

and then on setting $T^{-1} = \widehat{P}^{\frac{1}{2}}$ we can write (3.51) and (3.52) as

$$\begin{bmatrix} V_1^T T^T & 0 \\ 0 & V_2^T T^T \end{bmatrix} Q \begin{bmatrix} TV_1 & 0 \\ 0 & TV_2 \end{bmatrix} = I - \begin{bmatrix} V_1^T & 0 \\ 0 & V_2^T \end{bmatrix} \tilde{A}^T \tilde{A} \begin{bmatrix} V_1 & 0 \\ 0 & V_2 \end{bmatrix} \tag{3.60}$$

and

$$I - V_1^{-T} V_1^{-1} - V_2^{-T} V_2^{-1} \geq 0 \tag{3.61}$$

respectively.

Since $Q > 0$, (3.60) implies that

$$\left\| \tilde{A} \begin{bmatrix} V_1 & 0 \\ 0 & V_2 \end{bmatrix} \right\| < 1 \tag{3.62}$$

and hence (3.59) holds.

Conversely, if (3.59) holds then $\exists$ nonsingular T, V_1, V_2 satisfying (3.62) and this means that

$$I - V_1^{-T} V_1^{-1} - V_2^{-T} V_2^{-1} \geq 0 \tag{3.63}$$

and

$$\hat{Q} := I - \begin{bmatrix} V_1^T & 0 \\ 0 & V_2^T \end{bmatrix} \tilde{A}^T \tilde{A} \begin{bmatrix} V_1 & 0 \\ 0 & V_2 \end{bmatrix} > 0 \tag{3.64}$$

Now set $\widehat{P} = T^{-T} T^{-1}$, $W_1 = V_1^{-T} V_1^{-1}$, $W_2 = V_2^{-T} V_2^{-1}$. Then (3.64) implies that Q of (3.51) satisfies $Q > 0$ where here

$$Q = \begin{bmatrix} T^{-T} V_1^{-T} & 0 \\ 0 & T^{-T} V_2^{-T} \end{bmatrix} \hat{Q} \begin{bmatrix} V_1^{-1} T^{-1} & 0 \\ 0 & V_2^{-1} T^{-1} \end{bmatrix}$$

and (3.63) gives (3.52). □

There are a range of methods available to solve the optimization problem of (3.59), but here we only consider two special cases where V_1 and V_2 assume diagonal forms. This leads to much simplified unconstrained minimization problems.

In the first of these cases, the form of V_1 and V_2 is as follows

$$V_1 = \operatorname{diag}\left\{\frac{1}{\nu_{1i}}\right\}, \quad V_2 = \operatorname{diag}\left\{\frac{1}{\nu_{2i}}\right\}$$

where $\nu_{1i} > 0$, $\nu_{2i} > 0$, $1 \le i \le n+m$ and, to satisfy (3.63),

$$\nu_{1i}^2 + \nu_{2i}^2 = 1, \; 1 \le i \le n+m$$

This last constraint implies that $\nu_{1i} \in (0,1)$ and $\nu_{2i} = (1-\nu_{1i})^{\frac{1}{2}}$. Also the hyperbolic tangent transformation

$$v_{1i} := \frac{1+\tanh \nu_i}{2}, \; 1 \le i \le n+m$$

can be used to satisfy the constraint $\nu_{1i} \in (0,1)$, where $\nu_i \in (-\infty, \infty)$.

The minimization problem to be solved here has, with the above substitutions, been simplified to

$$\text{minimize} \left\| \tilde{A} \begin{bmatrix} V_1 & 0 \\ 0 & V_2 \end{bmatrix} \right\| \tag{3.65}$$

over T nonsingular and $-\infty < \nu_1, \; \ldots, \nu_{n+m} < \infty$, where $1 \le i \le n+m$,

$$\begin{aligned} V_1 &= \operatorname{diag}\left\{\frac{2}{1+\tanh \nu_i}\right\} \\ V_2 &= \operatorname{diag}\left\{\frac{2}{[4-(1+\tanh \nu_i)^2]^{\frac{1}{2}}}\right\} \end{aligned}$$

Note here that although the state transformation matrix T is required to be nonsingular, the problem can still be described as unconstrained since (i) the set of points in the parameter space which results in a singular T is zero (in measure theory terms) and hence the chance of encountering such a point in the parameter space is extremely small, and (ii) during the minimization itself, updating a parameter vector in such a way that it is closer to a singular T is always discouraged as an almost singular T tends to increase the value of the norm in (3.65) due to the presence of T^{-1} in $\tilde{A}$. Hence in this norm minimization problem, the use of a 'good' optimization algorithm would be highly unlikely to lead to a solution which corresponds to an almost singular T.

The second algorithm starts from V_1 and V_2 of the following form

$$V_1 = \frac{1}{\nu_1} I, \quad V_2 = \frac{1}{\nu_2} I \tag{3.66}$$

with $\nu_i > 0$, $i = 1,2$, and, to satisfy (3.61), $\nu_1^2 + \nu_2^2 = 1$. Hence $\nu_1 \in (0,1)$ and $\nu_2 = (1-\nu_1^2)^{\frac{1}{2}}$. In a similar manner to the previous algorithm, the hyperbolic tangent transformations

$$\nu_1 := \frac{1 + \tanh \nu}{2}, \quad \nu_2 := \frac{1}{2} \left[4 - (1 + \tanh \nu)^2\right]^{\frac{1}{2}} \tag{3.67}$$

where $\nu \in (-\infty, \infty)$, can be employed such that the minimization problem

$$\text{minimize} \left\| \tilde{A} \begin{bmatrix} V_1 & 0 \\ 0 & V_2 \end{bmatrix} \right\| \tag{3.68}$$

over T nonsingular and $-\infty < \nu < +\infty$, with V_i, $i = 1, 2$, specified as above, is essentially unconstrained. Also it is easy to see that, since $W_1 = V_1^{-T} V_1^{-1} = \nu_1^2 I_{n+m}$ and $W_2 = V_2^{-T} V_2^{-1} = \nu_2^2 I_{n+m}$, (3.68) with (3.66) and (3.67) provides a feasible approach to solving the 2D Lyapunov equation under consideration.

Overall, the solution algorithms given in this section are somewhat involved and it is not at all clear how they would perform numerically. In Chap. 8 LMI based alternatives which can then be used in control law design (based on the assumption of a perfect process model and under certain well defined uncertainty structures respectively). The major mechanism for this is through appropriately defined Lyapunov functions.

4 Lyapunov Equations for Differential Processes

Summary. In this chapter we investigate the role of Lyapunov equations in the stability related analysis of differential linear repetitive processes. As for the discrete processes of the previous chapter, there are two types of Lyapunov equation, again termed 1D and 2D respectively. Also it is shown that the Lyapunov equation approach can be extended to deal with one case of dynamic boundary conditions.

4.1 The 1D Lyapunov Equation

The basic result [124] for differential unit memory linear repetitive processes of the form defined by Example 1.2.3 is as follows. This result makes use of the 2D transfer-function matrix description of the process dynamics and, in particular, $G_1(s)$ obtained by setting $M = 1$ in (2.41) which describes the contribution of the previous pass dynamics to those of the current one.

Theorem 4.1.1. *Consider the linear repetitive process S generated by a differential unit memory linear repetitive process of the form defined by Example 1.2.3. Suppose also that the pair $\{A, B_0\}$ is controllable and the pair $\{C, A\}$ is observable. Then S is stable along the pass if, and only if,*

(a)

$$r(D_0) < 1, \ \det(sI_n - A) \neq 0, \ \mathrm{Re}(s) \geq 0 \tag{4.1}$$

and

(b) $\exists$ a rational polynomial matrix solution $P(s)$ of the Lyapunov equation

$$G_1^T(-s)P(s)G_1(s) - P(s) = -I_m \tag{4.2}$$

bounded in an open neighborhood of the imaginary axis in the complex plane with the properties that

(i) $P(s) = P^T(-s)$

(ii) $\beta_2^2 I_m \geq P(\imath\omega) = P^T(-\imath\omega) \geq \beta_1^2 I_m$, $\forall\, \omega \geq 0$, for some choice of real scalars $\beta_i \geq 1$, $i = 1, 2$.

Proof. If S is stable along the pass then the conditions of (4.1) are necessary and hence $G_1(s)$ is bounded in an open neighborhood of the imaginary axis. Also by (c) of Corollary 2.2.3 (i.e. all eigenvalues of $G_1(s)$ have modulus strictly less than unity, $s = \imath\,\omega, \forall \omega \geq 0$) $\exists$ a unique positive definite Hermitian (PDH) solution matrix $P(\imath\omega)$ of the equation

$$G_1^T(-\imath\omega)P(\imath\omega)G_1(\imath\omega) - P(\imath\omega) = -I_m \tag{4.3}$$

Consequently a solution of (4.2) exists on the imaginary axis. Also

$$\lim_{|s|\to\infty} G_1(s) = D_0$$

and it follows immediately from uniqueness and continuity of solutions that

$$\lim_{|\omega|\to\infty} P(\imath\omega) = P_\infty \tag{4.4}$$

where $P_\infty = P_\infty^T \geq 0$ is real and solves

$$D_0^T P_\infty D_0 - P_\infty = -I_m \tag{4.5}$$

The fact that $r(D_0) < 1$ now yields $P_\infty > 0$ and hence $P(\imath\omega)$ is bounded $\forall\,\omega \geq 0$. Further, the necessity of property (ii) above follows immediately from a continuity and compactness argument.

It now remains to extend the solution to an open neighborhood of the imaginary axis. This is achieved by regarding the Lyapunov equation as a linear matrix equation in $\mathbb{C}^{n+m}$, with nonsingular (in some neighborhood of the imaginary axis) coefficient matrix whose entries are rational polynomials in the complex variable s. The fact that a solution $P(s)$ exists, is unique and rational, and satisfies (ii) above is then a trivial consequence of standard linear algebraic arguments.

Finally, it remains to show that (a) and (b) here imply that the spectral radius of $G_1(\imath\omega)$ is strictly less than unity $\forall\,\omega$. This is trivially verified pointwise from (4.2). □

As in the discrete case of the previous chapter, the numbers $\beta_i,\ i = 1, 2$ play no role in stability analysis but, together with $P(s)$, they are the key to obtaining bounds on expected performance. This is developed next.

Suppose that the example under consideration is stable along the pass. Then standard factorization techniques enable $P(s)$ to be written as

$$P(s) = F^T(-s)F(s)$$

where, without loss of generality,

$$\lim_{|s|\to\infty} F(s) = P_\infty^{\frac{1}{2}}$$

and the matrix on the right-hand side of this equation is the unique, symmetric and positive definite square root of P_∞ of (4.4) and (4.5).

Given these facts, return to the process under consideration and suppose that the current pass input terms are deleted. Then as described in Sect. 1.4 (for the discrete case), the process dynamics can be written as

$$y_{k+1}(s) = G_1(s)y_k(s), \; k \geq 0$$

Also let

$$\hat{y}_k(s) := F(s)y_k(s), \; k \geq 0$$

denote 'filtered' (by properties of $F(s)$) outputs. Then the following result (again from [124]) gives bounds on expected performance.

Theorem 4.1.2. *Suppose that the linear repetitive process S generated by a differential unit memory linear repetitive process of the form defined by Example 1.2.3 is stable along the pass. Then, $\forall\, k \geq 0$,*

$$||\hat{y}_{k+1}||^2_{L_2^m(0,\infty)} = ||\hat{y}_k||^2_{L_2^m(0,\infty)} - ||y_k||^2_{L_2^m(0,\infty)} \tag{4.6}$$

and hence the filtered output sequence $\{||\hat{y}_k||_{L_2^m(0,\infty)}\}_{k\geq 0}$ is strictly monotonically decreasing to zero and satisfies, for $k \geq 0$, the inequality

$$||\hat{y}_k||_{L_2^m(0,\infty)} \leq \lambda^k ||\hat{y}_0||_{L_2^m(0,\infty)}$$

where

$$\lambda := (1 - \beta_2^{-2})^{\frac{1}{2}} < 1$$

Also the actual output sequence $\{||y_k||_{L_2^m(0,\infty)}\}_{k\geq 0}$ is bounded by

$$||y_k||_{L_2^m(0,\infty)} \leq M_{1d}\lambda^k ||y_0||_{L_2^m(0,\infty)}$$

where

$$M_{1d} := \beta_2 \beta_1^{-1} \geq 1$$

Proof. Set $s = \imath\omega$ in the Lyapunov equation (4.2), and then pre-and post-multiply it by $y_k^T(-\imath\omega)$ and $y_k(\imath\omega)$ respectively. Then on integrating over the frequency interval $[0, \infty]$, Parseval's identity yields (4.6). Using the inequality

$$\beta_1^2 I_m \leq P(\imath\omega) \leq \beta_2^2 I_m, \; \omega \geq 0$$

with $\hat{y}_k = F y_k$, yields

$$\beta_1^2 ||y_k||^2_{L_2^m(0,\infty)} \leq ||\hat{y}_k||^2_{L_2^m(0,\infty)} \leq \beta_2^2 ||y_k||^2_{L_2^m(0,\infty)}$$

and hence

$$||\hat{y}_{k+1}||^2_{L_2^m(0,\infty)} \leq ||\hat{y}_k||^2_{L_2^m(0,\infty)} - \beta_2^{-2} ||\hat{y}_k||^2_{L_2^m(0,\infty)}$$

The required properties of $\{||\hat{y}_k||_{L_2^m(0,\infty)}\}_{k\geq 0}$ and $\{||y_k||_{L_2^m(0,\infty)}\}_{k\geq 0}$ now follow as simple consequences of these results. □

This result provides the following computable information on the convergence to the resulting limit profile of a stable along the pass differential unit memory linear repetitive process of the form defined by Example 1.2.3.

- The sequence of filtered outputs $\{\hat{y}\}_{k\geq 0}$ consists of monotonic signals converging to zero at a computable geometric rate in $L_2^m(0,\infty)$.
- The actual output sequence $\{y\}_{k\geq 0}$ converges to zero at the same geometric rate but is no longer necessarily monotonic. Further the deviation from monotonicity is described by the parameter M_{1d}, computed from the solution of the 1D Lyapunov equation.

As per the discrete case of the previous chapter, we now develop methods for solving the 1D Lyapunov equation. These solution procedures are either explicit, i.e. solve for $P(\imath\omega)$ and then test it for the PDH property, or implicit, i.e. avoid the need to solve for $P(\imath\omega)$.

Consider first the explicit route where in computational, or testing, terms only the imaginary axis of the complex plane needs to be considered in the 1D Lyapunov equation, the extension off this curve being achieved by analytic continuation means (if required). Hence if the conditions listed under (a) of Theorem 4.1.1 hold, the essential task is to solve for $P(\imath\omega)$, and if it is PDH then the example under consideration is stable along the pass.

Suppose now that $P(\imath\omega) = P^T(-\imath\omega)$ has been obtained as the solution of (4.3). Then it follows immediately that the PDH requirement on $P(\imath\omega)$ is equivalent to it having the so-called axis positivity property. In particular, the following result is an immediate consequence of Siljak's criteria [163] for axis positivity of $P(\imath\omega)$.

Lemma 4.1.1. *The linear repetitive process S generated by a differential unit memory linear repetitive process of the form defined by Example 1.2.3 is stable along the pass if, and only if,*

(a) the conditions listed under (a) of Theorem 4.1.1 hold; and
(b) the solution matrix $P(\imath\omega)$ of the 1D Lyapunov equation (4.3) satisfies

$$P(0) > 0$$

and

$$\det P(\imath\omega) > 0, \; \forall\, \omega \geq 0$$

The condition on $P(0)$ in this last result is easily tested, and the second is easily shown to be equivalent to the requirement that some real even order polynomial of the form

$$g(\omega^2) := \sum_{i=0}^{h} g_{2i}\omega^{2i}$$

has the so-called positive realness property

$$g(\omega^2) > 0, \; \forall\, \omega \geq 0 \tag{4.7}$$

This last condition is easily seen to be equivalent to the following requirement on the roots of $g(\omega^2)$.

Lemma 4.1.2. *The polynomial $g(\omega^2)$ has the property (4.7) if, and only if, it has no positive real roots and $g(\omega_o^2) > 0$ for some $\omega_o \geq 0$.*

Note that in the trivial case of $h = 0$, Lemma 4.1.2 reduces to $g_0 > 0$. Also $g(\omega_o^2) > 0$ for some $\omega_o \geq 0$ can be replaced by either $g_0 > 0$ or $g_{2h} > 0$, (i.e. $g(+\infty) > 0$) which are easy to test. In particular, (4.7) does not hold if $g_0 < 0$, and no further tests are required.

The classical Descartes rule of sign (see, for example, [91]) gives the following useful preliminary results on the positive realness of $g(\omega^2)$.

Lemma 4.1.3. *With $g_0 > 0$, a necessary condition for $g(\omega^2)$ to have no positive real roots is that there is an even number of changes of sign in the coefficients g_{2i} when arranged in descending order.*

If the number of sign changes is zero, i.e. all coefficients are positive, the result of Lemma 4.1.3 is sufficient and is a special case of the following result.

Lemma 4.1.4. *A sufficient condition for $g(\omega^2)$ to have no positive real roots is that the coefficients g_{2i} satisfy*

$$g_0 > 0, \ g_{2i} \geq 0, \ 1 \leq i \leq h$$

To test Lemma 4.1.2 in the general case requires a means of determining the sign of the real roots (if any) of the real polynomial $g(\omega^2)$. This is a well researched problem and numerous algorithms exist which avoid the need to compute the roots. For example, algorithms have been developed from the concept of a matrix inner [91] and the so-called modified Routh array [164]. Such tests are not considered further here since clearly their major use is in low order 'synthesis type', problems where some, or all, of the matrices defining the example under consideration contain design parameters.

Later in this section, it will be shown that the 1D Lyapunov equation condition for stability along the pass is equivalent to one expressed in terms of an eigenvalue problem for matrices with constant entries. As in the case of the discrete processes of the last chapter, this analysis uses the Kronecker product for matrices and, in effect, the basic starting point is the matrices which define the transfer-function matrix $G_1(s)$. Prior to this analysis, however, further special cases are analyzed.

Consider the SISO case when $|D_0| < 1$ (asymptotic stability). Suppose also that the matrix A has eigenvalue-eigenvector decomposition

$$\Lambda := T^{-1}AT = \text{diag}\{\lambda_i\}_{1 \leq i \leq n}$$

Then the conditions listed under (a) of Theorem 4.1.1 for stability along the pass hold if, and only if, $\text{Re}(\lambda_i) < 0, \ 1 \leq i \leq n$.

Now focus on the case when λ_i, $1 \leq i \leq n$, are real and introduce the so-called augmented plant matrix for a differential unit memory linear repetitive process of the form defined by Example 1.2.3 as (see also (3.4) for a discrete unit memory linear repetitive process of the form defined by Example 1.2.10)

$$\Phi := \begin{bmatrix} A & B_0 \\ C & D_0 \end{bmatrix}$$

Also define $\hat{T}$ as $\hat{T} := \text{diag}\{T, 1\}$ and transform Φ to

$$\hat{\Phi} := \begin{bmatrix} \Lambda & \hat{B}_0 \\ \hat{C} & D_0 \end{bmatrix}$$

where

$$\hat{B}_0 := T^{-1} B_0 = \begin{bmatrix} b_1 & \cdots & b_n \end{bmatrix}^T$$

and

$$\hat{C} := CT = \begin{bmatrix} c_1 & \cdots & c_n \end{bmatrix}$$

Consider also the case when all of the numbers

$$\delta_i := b_i c_i,\ 1 \leq i \leq n \tag{4.8}$$

have the same sign. Then the following result holds.

Lemma 4.1.5. *Consider a differential unit memory linear repetitive process of the form defined by Example 1.2.3 which is SISO and for which A is a diagonalizable matrix with real eigenvalues λ_i, $1 \leq i \leq n$. Suppose also that all the δ_i defined by (4.8) have the same sign. Then the linear repetitive process S generated in this case is stable along the pass if, and only if,*

(a) $|D_0| < 1$, $\lambda_i < 0$, $1 \leq i \leq n$; *and*
(b)

$$|D_0 - CA^{-1}B_0| < 1 \tag{4.9}$$

Proof. First note that the 1D Lyapunov equation for this case is

$$P(\imath\omega)(1 - |D_0 + C(\imath\omega I_n - A)^{-1}B_0|^2) = 1$$

Hence $P(\imath\omega) > 0$, $\forall\, \omega$ if, and only if,

$$|D_0 + C(\imath\omega I_n - A)^{-1}B_0| < 1, \forall\, \omega$$

and necessity is immediate.

To prove sufficiency, first note that

$$\begin{aligned} |\gamma(\imath\omega)| &:= |D_0 + C(\imath\omega I_n - A)^{-1}B_0| \\ &= |D_0 + \hat{C}\text{diag}\{\imath\omega - \lambda_i\}_{1\leq i\leq n}\hat{B}_0| \\ &= |D_0 + \text{sgn}(\delta_1)\sum_{i=1}^{n} \gamma_i(\imath\omega)| \end{aligned}$$

where

$$\gamma_i(\imath\omega) := \frac{|\delta_i|}{\imath\omega - \lambda_i}, \ 1 \leq i \leq n$$

and each of these functions maps the imaginary axis in the s-plane onto a circle centered on the real axis. This means that the maximum value can only occur when $\omega = 0$ and (4.9) follows immediately. □

In the case when the numbers δ_i of (4.8) have different signs, the following result gives a sufficient condition for stability along the pass.

Lemma 4.1.6. *Consider a differential unit memory linear repetitive process of the form defined by Example 1.2.3 for which A is diagonalizable matrix with real eigenvalues λ_i, $1 \leq i \leq n$. Suppose also that in the numbers δ_i, $1 \leq i \leq h$, of (4.8) have the same sign and so do the remainder δ_i, $h+1 \leq i \leq n$. Then the linear repetitive process S generated in this case is stable along the pass if*

(a) $|D_0| < 1$, $\lambda_i < 0$, $1 \leq i \leq n$; *and*
(b)

$$\max\{h_1 + h_2, \ h_3 + h_2\} < 1$$

where

$$h_1 := \left| D_0 + \sum_{i=1}^{h} \frac{|\delta_i|}{\lambda_i} \right|, \ h_2 := \left| \sum_{i=h+1}^{n} \frac{|\delta_i|}{-\lambda_i} \right|, \ h_3 := \left| D_0 - \sum_{i=1}^{h} \frac{|\delta_i|}{\lambda_i} \right|$$

Proof. In this case

$$\begin{aligned} |D_0 + C(sI_n - A)^{-1}B_0| &= \left| D_0 + \sum_{i=1}^{h} \frac{\delta_i}{s - \lambda_i} + \sum_{i=h+1}^{n} \frac{\delta_i}{s - \lambda_i} \right| \\ &\leq \max_{s=\imath\omega} \left| D_0 + \operatorname{sgn}(\delta_1) \sum_{i=1}^{h} \frac{|\delta_i|}{s - \lambda_i} \right| \\ &+ \max_{s=\imath\omega} \left| \sum_{i=h+1}^{n} \frac{|\delta_i|}{s - \lambda_i} \right| \end{aligned}$$

□

Consider now the general case when the conditions listed under (a) of Theorem 4.1.1 hold and the remaining task is to determine if (b) of this result – the 1D Lyapunov equation – holds and hence stability along the pass. The analysis which follows uses the matrix Kronecker product, again denoted by $\otimes$, to develop a solution to this problem expressed in terms of the locations of the eigenvalues of constant matrices. We mostly follow [157].

Using $\otimes$, the 1D Lyapunov equation (4.3) can be written as

$$(I_{m^2} - G_1^T(-\imath\omega) \otimes G_1^T(\imath\omega))S_{so}\left[P(\imath\omega)\right] = S_{so}\left[I_{m^2}\right]$$

where (as in the previous chapter) $S_{so}\left[\cdot\right]$ denotes the stacking operator. This now leads to the following set of necessary and sufficient conditions for stability along the pass of S generated by a differential unit memory linear repetitive process of the form defined by Example 1.2.3. These are then further developed to give stability tests which can be implemented using operations on matrices with constant entries.

Theorem 4.1.3. *Consider the linear repetitive process S generated by a differential unit memory linear repetitive process of the form defined by Example 1.2.3. Then S is stable along the pass if, and only if,*

(a) $r(D_0) < 1, \; \det(sI_n - A) \neq 0, \; Re(s) \geq 0$
(b) $P = P(\imath\omega_o)$, *the solution of*

$$G_1^T(-\imath\omega_o)PG_1(\imath\omega_o) - P = -I_m \tag{4.10}$$

is positive definite for some $\omega_o \geq 0$; *and*
(c)

$$\det(I_{m^2} - G_1^T(-\imath\omega) \otimes G_1^T(\imath\omega)) \neq 0, \; \forall\, \omega \tag{4.11}$$

Proof. It is clearly required to prove that (b) and (c) here are, together, equivalent to (b) of Theorem 4.1.1. First note, therefore, that (4.11) guarantees the existence of a unique solution $P(\imath\omega)$ of (4.3). Also $P(\imath\omega)$ is PDH if, and only if, its eigenvalues are positive $\forall\, \omega$. These are real continuous functions of ω and will always be positive if $P(\imath\omega_o) > 0$ for an arbitrary ω_o and (4.11) holds. Hence (4.10) and (4.11) are equivalent to (b) of Theorem 4.1.1 and the proof is complete. □

At this stage, the stability testing problem has been reduced to testing (or determining) the eigenvalue locations of three constant matrices and those of a second order matrix polynomial. Clearly therefore increased computational efficiency (relative to the necessary and sufficient conditions of Theorem 4.1.1 or equivalents) will only be achieved by (if possible) further refinement (i.e. reducing the computational complexity) of the condition of (4.11). The following result shows how it can be replaced by an equivalent one which is easily checked.

Theorem 4.1.4. *Consider the linear repetitive process S generated by a differential unit memory linear repetitive process of the form defined by Example 1.2.3. Then S is stable along the pass if, and only if,*

(a) the conditions listed under (a) and (b) of Theorem 4.1.3 hold; and

(b)

$$\det(\lambda^2 X_1 + \lambda X_2 + X_3) \neq 0,\ \lambda = \imath\omega,\ \forall\,\omega \tag{4.12}$$

where

$$\begin{aligned}
X_1 &:= \begin{bmatrix} 0 & 0 & 0 & 0 \\ 0 & 0 & 0 & 0 \\ 0 & 0 & 0 & 0 \\ 0 & 0 & 0 & -I_{n^2} \end{bmatrix} \\
X_2 &:= \begin{bmatrix} 0 & 0 & 0 & 0 \\ 0 & I_{nm} & 0 & 0 \\ 0 & 0 & -I_{nm} & 0 \\ 0 & 0 & 0 & I_n \otimes A^T - A^T \otimes I_n \end{bmatrix} \\
X_3 &:= \begin{bmatrix} I_{m^2} - D_0^T \otimes D_0^T & I_m \otimes B_0^T & B_0^T \otimes D_0^T & B_0^T \otimes B_0^T \\ D_0^T \otimes C^T & -I_m \otimes A^T & 0 & 0 \\ C^T \otimes I_m & 0 & -A^T \otimes I_m & 0 \\ C^T \otimes C^T & 0 & 0 & A^T \otimes A^T \end{bmatrix}
\end{aligned}$$

Proof. First set $\lambda = \imath\omega$ and introduce

$$f(\lambda) := \det(I_{m^2} - G_1^T(-\lambda) \otimes G_1^T(\lambda))$$

Then

$$\begin{aligned}
f(\lambda) = \det(&I_{m^2} - D_0^T \otimes D_0^T - (B_0^T \otimes D_0^T)((-\lambda I_n - A^T) \otimes I_m)^{-1}(C^T \otimes I_m) \\
&- (I_m \otimes B_0^T)(I_m \otimes (\lambda I_n - A^T)^{-1})(D_0^T \otimes C^T) \\
&- (B_0^T \otimes B_0^T)h(\lambda)^{-1}(C^T \otimes C^T))
\end{aligned} \tag{4.13}$$

where the matrix $h(\lambda)$ whose inverse appears in the last term of (4.13) is given by

$$h(\lambda) := -\lambda^2 I_{n^2} + \lambda(I_n \otimes A^T - A^T \otimes I_n) + A^T \otimes A^T$$

Hence

$$f(\lambda) = \det(I_{m^2} - D_0^T \otimes D_0^T - UV^{-1}W)$$

where

$$\begin{aligned}
U &:= \begin{bmatrix} (I_m \otimes B_0^T) & (B_0^T \otimes D_0^T) & (B_0^T \otimes B_0^T) \end{bmatrix} \\
V &:= \begin{bmatrix} \lambda I_{nm} - I_m \otimes A^T & 0 & 0 \\ 0 & -\lambda I_{nm} - A^T \otimes I_m & 0 \\ 0 & 0 & h(\lambda) \end{bmatrix}
\end{aligned}$$

and

$$W := \begin{bmatrix} D_0^T \otimes C^T \\ C^T \otimes I_m \\ C^T \otimes C^T \end{bmatrix}$$

Also it is easy to show (all eigenvalues of A have strictly negative real parts) that for $\lambda = \imath\omega$, $\forall\, \omega$

$$\det\left(\lambda I_{nm} - I_m \otimes A^T\right) \times \det\left(-\lambda I_{nm} - A^T \otimes I_m\right) \times \det h(\lambda) \neq 0 \quad (4.14)$$

At this stage, suppose that $f(\lambda)$ is pre-multiplied by the left-hand side of (4.14). In which case, it follows immediately that $f(\lambda) \neq 0$, $\lambda = \imath\omega$, $\forall\, \omega$ is equivalent to

$$\det \begin{pmatrix} I_{m^2} - D_0^T \otimes D_0^T & I_m \otimes B_0^T & B_0^T \otimes D_0^T & B_0^T \otimes B_0^T \\ D_0^T \otimes C^T & \lambda I_{nm} - I_m \otimes A^T & 0 & 0 \\ C^T \otimes I_m & 0 & -\lambda I_{nm} - A^T \otimes I_m & 0 \\ C^T \otimes C^T & 0 & 0 & h(\lambda) \end{pmatrix} \quad (4.15)$$

being non-zero for $\lambda = \imath\omega$, $\forall \omega$. Finally, it is easy to see that (4.12) and (4.15) are equivalent. □

At this stage, the stability testing problem has been reduced to testing (or determining) the eigenvalue locations of three constant matrices and those of a second order matrix polynomial. The difficulty with the latter condition is that the matrix X_1 here is singular and hence the required eigenvalues cannot be obtained directly using existing software packages. In the reminder of this section it will be shown that this stability condition can be further modified to result in one which is easily tested.

Theorem 4.1.5. *Consider the linear repetitive process S generated by a differential unit memory linear repetitive process of the form defined by Example 1.2.3. Then S is stable along the pass if, and only if,*

(a) the conditions listed under (a) of Theorem 4.1.4 hold; and
(b)

$$\det\left(\lambda I + L_{12} - L_{13} L_7^{-1} L_8\right) \neq 0$$

for $\lambda = \imath\omega$, $\forall\, \omega$, where

$$\begin{aligned} L_7 &:= I_{m^2} - D_0^T \otimes D_0^T \\ L_8 &:= \begin{bmatrix} I_m \otimes B_0^T & B_0^T \otimes D_0^T & B_0^T \otimes B_0^T & 0 \end{bmatrix} \\ L_{12} &:= \begin{bmatrix} -I_m \otimes A^T & 0 & 0 & 0 \\ 0 & A^T \otimes I_m & 0 & 0 \\ 0 & 0 & 0 & -I_{n^2} \\ 0 & 0 & -A^T \otimes A^T & A^T \otimes I_n - I_n \otimes A^T \end{bmatrix} \end{aligned}$$

and

$$L_{13} := \begin{bmatrix} D_0^T \otimes C^T \\ -C^T \otimes I_m \\ 0 \\ -C^T \otimes C^T \end{bmatrix}$$

Proof. The determinant of the matrix polynomial defining the condition of (4.12) can be rewritten as

$$\det\left(\lambda^2 X_1 + \lambda X_2 + X_3\right) = \det\begin{pmatrix} L_1\lambda + L_2 & L_3 \\ L_4 & -\lambda^2 I + L_5\lambda + L_6 \end{pmatrix}$$

where

$$\begin{aligned}
L_1 &:= \begin{bmatrix} 0 & 0 & 0 \\ 0 & I_{nm} & 0 \\ 0 & 0 & -I_{nm} \end{bmatrix} \\
L_2 &:= \begin{bmatrix} I_{m^2} - D_0^T \otimes D_0^T & I_m \otimes B_0^T & B_0^T \otimes D_0^T \\ D_0^T \otimes C^T & -I_m \otimes A^T & 0 \\ C^T \otimes I_m & 0 & -A^T \otimes I_m \end{bmatrix} \\
L_3 &:= \begin{bmatrix} B_0^T \otimes B_0^T \\ 0 \\ 0 \end{bmatrix}, \quad L_4 := \begin{bmatrix} C^T \otimes C^T & 0 & 0 \end{bmatrix} \\
L_5 &:= \left[I_n \otimes A^T - A^T \otimes I_n\right], \quad L_6 := \left[A^T \otimes A^T\right]
\end{aligned}$$

This implies that (4.12) is equivalent to

$$\det\left(-\lambda^2 I_{n^2} + \lambda L_5 + L_6\right) \times \det\left(\lambda L_1 + L_2 - L_3(-\lambda^2 I_{n^2} + \lambda L_5 + L_6)^{-1} L_4\right)$$

being non-zero for $\lambda = \imath\omega$, $\forall\, \omega$.

Routine algebraic manipulations applied to this last expression now shows that the condition of (4.12) is equivalent to

$$\det\begin{pmatrix} \lambda L_1 + L_2 & L_3 & 0 \\ 0 & \lambda I_{n^2} & -I_{n^2} \\ L_4 & L_6 & -\lambda I + L_5 \end{pmatrix} \neq 0$$

for $\lambda = \imath\omega$, for all ω. At this stage, the left-hand side of (4.12) has been transformed to one defined in terms of the determinant of a first order polynomial matrix which can be written in the form

$$\det\begin{pmatrix} L_7 & L_8 \\ L_9 & \lambda L_{10} + L_{11} \end{pmatrix} \neq 0$$

for $\lambda = \imath\omega$, $\forall\, \omega$, where

$$L_9 := \begin{bmatrix} D_0^T \otimes C^T \\ C^T \otimes I_m \\ 0 \\ C^T \otimes C^T \end{bmatrix}, \quad L_{10} := \begin{bmatrix} I_{nm} & 0 & 0 & 0 \\ 0 & -I_{nm} & 0 & 0 \\ 0 & 0 & I_{n^2} & 0 \\ 0 & 0 & 0 & -I_{n^2} \end{bmatrix}$$

and

$$L_{11} := \begin{bmatrix} -I_m \otimes A^T & 0 & 0 & 0 \\ 0 & -A^T \otimes I_m & 0 & 0 \\ 0 & 0 & 0 & -I_{n^2} \\ 0 & 0 & A^T \otimes A^T & I_n \otimes A^T - A^T \otimes I_n \end{bmatrix}$$

The result now follows immediately on applying the matrix block determinant formula. □

4.2 The 2D Lyapunov Equation

Suppose that $r(D_0) < 1$ (and all eigenvalues of the matrix A have strictly negative real parts). Then it follows immediately from the result of Theorem 2.2.7 that a differential unit memory linear repetitive process of the form defined by Example 1.2.3 is stable along the pass if, and only if,

$$\det\left(I_m - z_2 G_1(s)\right) \neq 0, \ |z_2| \leq 1, \ s = \imath\omega$$

(where $G_1(s)$ is given by setting $M = 1$ in (2.41)). Also we will require the following (see, for example, [1]) definition and result.

Definition 4.2.1. *Let $S(\gamma)$ be a square matrix of real rational functions in the complex variable γ. Then this matrix is said to be strictly bounded real (SBR) if, and only if,*

(a) all poles of $S(\gamma)$ have strictly negative real parts;
and
(b) $I - S^T(-\imath\omega)S(\imath\omega) > 0, \ \forall\, \omega$

Note 4.2.1. This last definition is the 1D differential linear systems counterpart of that of Definition 3.2.1 for 1D discrete linear systems and there is no general agreement in the literature as to how it is termed. Here we follow [81] and make no distinction as the context in which both definitions are applied is clear from the accompanying text.

The conditions under (a) and (b) in this last definition can be reduced to conditions on the matrices of a minimal state-space realization of $S(\gamma)$. In particular, suppose that $S(\gamma)$ has a minimal state-space realization given by the quadruple $\{F, G, H, J\}$ such that

$$S(\gamma) = H^T(\gamma I - F)^{-1}G + J \tag{4.16}$$

Then the following result is the so-called SBR lemma.

Lemma 4.2.1. *Suppose that $S(\gamma)$ is a square matrix of real rational functions in γ with minimal state-space realization defined by $\{F, G, H, J\}$ as per (4.16). Then this matrix is SBR if, and only if, $\exists$ a matrix $\widehat{P} > 0$ such that*

$$\tilde{Q} := \begin{bmatrix} F^T\widehat{P} + \widehat{P}F + HH^T & \widehat{P}G + HJ \\ G^T\widehat{P} + J^TH^T & J^TJ - I \end{bmatrix} < 0 \tag{4.17}$$

Suppose now that the conditions listed under (a) of Theorem 4.1.1 hold for the example under consideration. Then comparing the SBR definition applied to the transfer-function matrix $G_1(s)$ (which describes the contribution of the previous pass dynamics to those of the current pass profile) with that listed under (b) of this same theorem, it follows that (b) for the SBR property is equivalent to (4.3) admitting the constant solution $P = \tilde{T}^*\tilde{T}$ (where here $*$ denotes complex conjugate transpose) over $\mathbb{R}$. Hence it is clear that the requirement that $G_1(s)$ has the SBR property implies that (b) of Theorem 4.1.1 holds and hence stability along the pass.

In what follows, we use the SBR property to show that a solution of the following 2D Lyapunov equation is a sufficient condition for stability along the pass of a differential unit memory linear repetitive process of the form defined by Example 1.2.3

$$\Phi^T W^{1,0} + W^{1,0}\Phi + \Phi^T W^{0,1}\Phi - W^{0,1} = -Q \tag{4.18}$$

Here $W_1 > 0, W_2 > 0$ and $Q > 0$ are matrices of dimensions $n \times n, m \times m$, and $(n+m) \times (n+m)$ respectively, $W^{1,0} = W_1 \oplus 0_{m\times m}$, and $W_2 = 0_{n\times n} \oplus W_2$. Note also that, in effect, this 2D Lyapunov equation can be viewed as an appropriate combination of those for 1D differential and discrete linear systems respectively.

Theorem 4.2.1. *Consider the linear repetitive process S generated by a differential unit memory linear repetitive process of the form defined by Example 1.2.3. Then $\exists$ matrices $W_1 > 0, W_2 > 0$ and $Q > 0$ such that $W = W_1 \oplus W_2$ and Q satisfy the 2D Lyapunov equation (4.18) if $\exists$ an $m \times m$ nonsingular matrix T such that $\tilde{G}_1(s)$ defined by*

$$\tilde{G}_1(s) := TG_1(s)T^{-1} = T\left[C(sI_n - A)^{-1}B_0 + D_0\right]T^{-1}$$

is SBR.

Proof. To show that this condition is sufficient, first note that $\tilde{G}_1(s)$ has minimal realization $\{A, B_0T^{-1}, TC, TD_0T^{-1}\}$ and if it is SBR then by Lemma 4.2.1 $\exists$ a matrix $\widehat{P} > 0$ such that

$$Q_1 := \begin{bmatrix} A^T\widehat{P} + \widehat{P}A + C^T T^T TC & \widehat{P}B_0T^{-1} + C^T T^T TD_0T^{-1} \\ T^{-T}B_0^T\widehat{P} + T^{-T}D_0T^TTC & T^{-T}D_0^T T^T TD_0T^{-1} - I_m \end{bmatrix} < 0$$

Pre-multiplication by $\left(I_n \oplus T^T\right)$ and post-multiplication by $\left(I_n \oplus T\right)$ now gives the 2D Lyapunov equation (4.18) with

$$Q := -\left(I_n \oplus T^T\right)Q_1\left(I_n \oplus T\right),\ W := \left(\widehat{P} \oplus T^T T\right) > 0$$

To prove necessity, suppose that (4.18) holds. Then set $F - A, G = B_0T^{-1}, H^T = TC, J = TD_0T^{-1}$ and pre- and post-multiply (4.18) by $\left(I_n \oplus T^{-T}\right)$ and $\left(I_n \oplus T^{-1}\right)$ respectively to obtain (4.17) with $\widehat{P} = W_1$ and $\tilde{Q} = -\left(I_n \oplus T^{-T}\right)Q\left(I_n \oplus T^{-1}\right)$. This shows that $\tilde{G}_1(s)$ is SBR as required. □

We can now state the following corollary to this last theorem.

Corollary 4.2.1. *Consider the linear repetitive process S generated by a differential unit memory linear repetitive process of the form defined by Example 1.2.3. Then S is stable along the pass if $\exists$ matrices $W = W_1 \oplus W_2 > 0$ and $Q > 0$ which solve the 2D Lyapunov equation (4.18).*

In general, the reverse conclusion to this last corollary is not true. If, however, the example under consideration is SISO then it is true. Then stability along the pass is easily seen to hold [60] if, and only if,

$$||G_1^k(\imath\omega)|| \to 0,\ k \to \infty,\ 0 \le \omega \le \infty \tag{4.19}$$

where $||G_1(\imath\omega)|| := \sup_{0 \le \omega < \infty} |G_1(\imath\omega)|$, i.e. each frequency component of the initial pass profile is attenuated from pass-to-pass. Hence it is clear that stability along the pass is equivalent to $G_1(s)$ being SBR. Also since the characteristic polynomial (i.e. $\mathcal{C}(s, z_2)$ of (1.55)) in this case is first order in z_2, the converse also holds.

Solution algorithms for the 2D Lyapunov equation here can be developed as per those of the previous chapter for the 2D Lyapunov equation for discrete unit memory linear repetitive processes of the form defined by Example 1.2.10. Noting also the conclusions in the previous chapter, however, we do not detail such algorithms here but in Chap. 8 develop LMI based alternatives. It will be subsequently shown that this route alone leads to the use of the 2D Lyapunov equation in control systems design (both for design based on the assumption of a perfect process model and that under certain well defined uncertainty structures respectively). The major mechanism for this is, as in the discrete case of the previous chapter, through Lyapunov functions.

4.3 Differential Processes with Dynamic Boundary Conditions

In Chap. 2 it was shown that the 1D equivalent model can be used to develop computationally feasible tests for asymptotic stability of discrete unit memory linear repetitive processes in the presence of dynamic boundary conditions (recall Theorem 2.3.2). In this section, we show that, in one special case of practical interest, (1D) Lyapunov/Riccati equations can be used to develop computationally feasible tests for asymptotic stability of differential unit memory linear repetitive processes (for which no finite dimensional 1D equivalent model exists) in the presence of dynamic boundary conditions.

The state-space model of the sub-class of differential linear repetitive processes of the form defined by Example 1.2.5 considered in this section has the following form over $0 \le t \le \alpha$, $k \ge 0$,

$$\dot{y}_{k+1}(t) = Ay_{k+1}(t) + Bu_{k+1}(t) + B_0y_k(t)$$

with pass state initial vector sequence

$$y_{k+1}(0) = d_{k+1} + y_k(\alpha),\ k \geq 0$$

In this case, the current pass state vector is also the pass profile vector and in the case when this is an $n \times 1$ vector, asymptotic stability requires (see (2.26)) that all solutions of

$$\det\left(zI_n - e^{(A+z^{-1}B_0)\alpha}\right) = 0 \tag{4.20}$$

have modulus strictly less than unity $\forall\, \alpha \geq 0$. Now write $z = e^{s\alpha}$, and hence (4.20) reduces to the requirement that all solutions of

$$\det\left(sI_n - F(s)\right) = 0$$

have strictly negative real parts where

$$F(s) := A + B_0e^{-s\alpha}$$

Also it can be shown [97] that (4.20) reduces to the requirement that

$$\det\left(sI_n - F(e^{-\imath\omega\alpha})\right) \neq 0,\ \forall\, \mathrm{Re}(s) \geq 0,\ \forall\omega \in [0, 2\pi] \tag{4.21}$$

and we can establish the following result (see also [26] for the 1D delay differential systems case).

Theorem 4.3.1. *The condition of (4.21) holds if, and only if, for a given PDH matrix $Q(e^{\imath\omega})$, $\forall\, \omega \in [0, 2\pi]$, the solution $P(e^{\imath\omega})$ of the 1D matrix Lyapunov equation*

$$F^*(e^{\imath\omega})P(e^{\imath\omega}) + P(e^{\imath\omega})F(e^{\imath\omega}) = -Q(e^{\imath\omega}) \tag{4.22}$$

is PDH $\forall\, \omega \in [0, 2\pi]$.

Proof. To show sufficiency, first note that for any fixed $\omega \in [0, 2\pi]$, $F(e^{\imath\omega})$ is an $n \times n$ matrix with complex elements. Also let λ be an eigenvalue of this matrix and x the corresponding eigenvector. Now pre-multiply (4.22) by x^* (where $*$ denotes the complex conjugate transpose operation) and then post-multiply this same equation by x to yield

$$\begin{aligned} x^*Q(e^{\imath\omega})x &= -x^*(F^*(e^{\imath\omega})P(e^{\imath\omega}) + P(e^{\imath\omega})F(e^{\imath\omega}))x \\ &= -(\lambda^* + \lambda)x^*P(e^{\imath\omega})x \end{aligned}$$

Also if $P(e^{\imath\omega})$ and $Q(e^{\imath\omega})$ are PDH $\forall\, \omega \in [0, 2\pi]$, it follows that

$$\mathrm{Re}(\lambda) = \frac{1}{2}(\lambda^* + \lambda) = -\frac{1}{2}\left(\frac{x^*Q(e^{\imath\omega})x}{x^*P(e^{\imath\omega})x}\right) < 0$$

Hence (4.21) (i.e. asymptotic stability in this case) holds.

To show necessity, consider (4.22) with an arbitrary PDH matrix $Q(e^{\imath\omega})$ on $[0, 2\pi]$. Then if (4.21) holds, it can be shown that all eigenvalues of the matrix $F(e^{\imath\omega})$ have strictly negative real parts $\forall\, \omega \in [0, 2\pi]$. Now introduce

$$P(e^{\imath\omega}) := \int_0^\infty e^{F^*(e^{\imath\omega})t} Q(e^{\imath\omega}) e^{F(e^{\imath\omega})t}\, dt$$

which is well defined since the eigenvalues of $F(e^{\imath\omega})$ (and $F^*(e^{\imath\omega})$) are in the left half of the complex plane. Also $P^*(e^{\imath\omega}) = P(e^{\imath\omega})$, $\forall\, \omega \in [0, 2\pi]$ and

$$\begin{aligned}F^*(e^{\imath\omega})P(e^{\imath\omega})+P(e^{\imath\omega})F(e^{\imath\omega}) &= \int_0^\infty \Big(F^*(e^{\imath\omega})e^{F^*(e^{\imath\omega})t}Q(e^{\imath\omega})e^{F(e^{\imath\omega})t} \\ &\quad + e^{F^*(e^{\imath\omega})t}Q(e^{\imath\omega})e^{F(e^{\imath\omega})t}F(e^{\imath\omega})\Big)\, dt \\ &= \int_0^\infty \left(\left(\frac{d}{dt}e^{F^*(e^{\imath\omega})t}\right) Q(e^{\imath\omega})e^{F(e^{\imath\omega})t} \right.\\ &\quad \left. + e^{F^*(e^{\imath\omega})t}Q(e^{\imath\omega})\left(\frac{d}{dt}e^{F(e^{\imath\omega})t}\right)\right) dt \\ &= \int_0^\infty \frac{d}{dt}\left(e^{F^*(e^{\imath\omega})t}Q(e^{\imath\omega})e^{F(e^{\imath\omega})t}\right) dt \\ &= -\,Q(e^{\imath\omega})\end{aligned}$$

where this last equality follows from the facts that $e^{F(e^{\imath\omega})t} \to 0$, and $e^{F^*(e^{\imath\omega})t} \to 0$ as $t \to \infty$. To complete the proof, we now have that

$$F^*(e^{\imath\omega})P(e^{\imath\omega}) + P(e^{\imath\omega})F(e^{\imath\omega}) = -Q(e^{\imath\omega}),\ \forall\, \omega \in [0, 2\pi]$$

and $P^*(e^{\imath\omega}) = P(e^{\imath\omega})$, $\forall\, \omega \in [0, 2\pi]$. □

Consider now

$$F(z)|_{z=e^{\imath\omega}} = F_1(\omega) + \imath F_2(\omega)$$

where $F_1(\omega)$ and $F_2(\omega)$ are real $n \times n$ matrices. Also for a fixed $\omega_o \in [0, 2\pi]$, $F(e^{\imath\omega_o})$ is an $n \times n$ matrix with complex entries which can be written as

$$F(e^{\imath\omega_o}) = F_1(\omega_o) + \imath F_2(\omega_o)$$

The system $\dot{y} = F(e^{\imath\omega_o})y$ can be rewritten as

$$\dot{y}_r + \imath\dot{y}_i = (F_1(\omega_o) + \imath F_2(\omega_o))(y_r + \imath y_i) \tag{4.23}$$

where y_r and y_i denote the real and imaginary parts of y respectively. Separating (4.23) into real and imaginary parts now yields the equivalent expression

$$\begin{bmatrix} \dot{y}_r \\ \dot{y}_i \end{bmatrix} = \begin{bmatrix} F_1(\omega_o) & -F_2(\omega_o) \\ F_2(\omega_o) & F_1(\omega_o) \end{bmatrix} \begin{bmatrix} y_r \\ y_i \end{bmatrix}$$

Now consider the SISO case and write

$$\hat{F}(\omega) := \begin{bmatrix} f_1(\omega) & -f_2(\omega) \\ f_2(\omega) & f_1(\omega) \end{bmatrix}$$

Then in this case a necessary and sufficient condition for asymptotic stability is that $f_1(\omega) < 0$, $\forall\, \omega \in [0, 2\pi]$, i.e. 1D stability of the real part of $F(\imath\omega)$. This follows immediately from

$$\det(sI_2 - \hat{F}(\omega)) = s^2 - 2f_1(\omega)s + f_1^2(\omega) + f_2^2(\omega)$$

Example 4.3.1. Consider the following unforced process

$$\dot{y}_{k+1}(t) = \begin{bmatrix} 0 & 1 \\ -a & -b \end{bmatrix} y_{k+1}(t) + \begin{bmatrix} 0 & 0 \\ 0 & -c \end{bmatrix} y_k(t)$$

where a, b, and c are positive real numbers. Then in this case

$$F(z) = \begin{bmatrix} 0 & 1 \\ -a & -b - cz \end{bmatrix}, \; z = e^{\imath\omega}$$

The solution of the Lyapunov equation (4.22) with $Q = I_2$ is

$$P(z) = \frac{1}{ay} \begin{bmatrix} |b + cz|^2 + a(a+1) & b + cz^* \\ b + cz & a + 1 \end{bmatrix}, \; y = 2(b + c\cos\omega)$$

Then

$$\det P(z) = \frac{|b + cz|^2 + (a+1)^2}{ay^2}$$

and $P(e^{\imath\omega})$ is PDH $\forall\, \omega \in [0, 2\pi]$ if, and only if, $y > 0$, i.e.

$$b + c\cos\omega > 0, \; \forall\, \omega \in [0, 2\pi]$$

Hence asymptotic stability in this particular case holds if, and only if, $b > c$.

It is easy to generate examples which demonstrate that a generalization of these last results for the SISO case is not possible. One such process is that with

$$A = \begin{bmatrix} -1 & 0 \\ 0 & -1 \end{bmatrix}, \quad B_0 = \begin{bmatrix} -1 & -1 \\ 4 & -1 \end{bmatrix}$$

where both A and B_0 are stable but $A + \imath B_0$ does not satisfy the condition for asymptotic stability.

In what follows, we use results from the theory of transfer-function matrices with the SBR property to obtain alternative asymptotic stability tests to the one proposed above. This will lead to Riccati equation based conditions which can be checked numerically. Prior to this, however, it is useful to consider the development of equivalent conditions for asymptotic stability in terms of the Hermite matrix approach from 'classical' root clustering theory (see, for example, [91] and the relevant references therein).

First note that (4.21) is equivalent, on setting $z = e^{\imath\omega}$, to

$$\Delta(s,z) := \det(sI_n - F(z)) \neq 0, \ \mathrm{Re}(s) \geq 0, \ |z| = 1$$

or

$$\Delta(s, e^{\imath\omega}) \neq 0, \ \mathrm{Re}(s) \geq 0, \ \omega \in [0, 2\pi]$$

This is an equation with complex coefficients which are polynomials in $e^{\imath\omega}$ and it is required that all its roots should lie in the open-left half of the s-plane. Using 'classical' root clustering theory, the condition for this is that the Hermite matrix obtained from the coefficients in $\Delta(s, e^{\imath\omega})$ is positive definite or, alternatively, the inner-wise matrix obtained from the coefficients must be positive inner-wise.

Consider the complex polynomial

$$B(s) := \sum_{i=0}^{h} b_i s^i$$

Then the associated Hermite matrix, H, is obtained as follows

$$H := \{h_{q_1,q_2}\}, \ h_{q_1,q_2} = h_{q_2,q_1}$$

where

$$h(q_1, q_2) := 2(-1)^{\frac{q_1+q_2}{2}} \sum_{j=1}^{q_1} (-1)^j \mathrm{Re}(b_{h-j-1} b^*_{h-q_1-q_2+j})$$

for $q_1 + q_2 =$ even, $q_1 \leq q_2$, and

$$h(q_1, q_2) := 2(-1)^{\frac{q_1+q_2-1}{2}} \sum_{j=1}^{q_1} (-1)^j \mathrm{Im}(b_{h-j-1} b^*_{h-q_1-q_2+j})$$

for $q_1 + q_2 =$ odd, $q_1 \leq q_2$. Also (see the proof of Theorem 4.1.3) the requirement for $H(\imath\omega) > 0, \ \forall\, \omega \in [0, 2\pi]$ (or $|e^{\imath\omega}| \in [-1, 1]$) is equivalent to the following conditions

$$\begin{aligned} H(e^{\imath 0}) &= H(1) > 0 && (4.24) \\ \det H(e^{\imath\omega}) &> 0, \ \forall\, \omega \in [0, 2\pi] && (4.25) \end{aligned}$$

The checking of (4.24) is straightforward and to check the more difficult condition of (4.25) it is possible to use a positivity test. This is based on the fact that $\det H(e^{\imath\omega})$ is a function of $\cos\omega, \cos 2\omega, \cdots$ and on setting $x = \cos\omega$, $\det H(e^{\imath\omega})$ becomes a function of x and its powers. Hence (4.25) becomes

$$E(x) := \det H(e^{\imath\omega})|_{x=\cos\omega} > 0, \ x \in [-1, 1] \tag{4.26}$$

This last condition holds provided $E(x)$ has no real roots in the interval $[-1, 1]$. Also introduce the change of variable (a bilinear transform)

$$x = \frac{u-1}{u+1}$$

in $E(x)$ and let the result be denoted by $E_1(u)$. Then (4.26) is equivalent to

$$E_1(u) > 0, \ u \in [0, \infty]$$

and this condition can be checked computationally using any of the computational positivity tests (again see, for example, [91] and the relevant cited references therein).

In the remainder of this section we develop a computationally more feasible alternative to the approach just presented. The starting point is to note that the condition to be tested here can be expressed as the requirement that a two variable polynomial of the general form

$$a(s, z) := s^{q_1} + \sum_{j=0}^{q_1-1} \sum_{i=0}^{q_2} a_{ij} s^j z^i \tag{4.27}$$

should satisfy

$$a(s, z) \neq 0, \ \mathrm{Re}(s) \geq 0, \ |z| \leq 1 \tag{4.28}$$

Next we describe how to reduce (4.28) to a one-dimensional problem (see also [81] from where the following analysis is adapted) by showing that this condition is equivalent to positive realness of a certain 1D rational transfer-function matrix. This leads to a numerically efficient testing algorithm and requires, as background, the results given next relating to the SBR property and, in particular, starting with the following alternative to Lemma 4.2.1.

Lemma 4.3.1. *The condition of Lemma 4.2.1 is equivalent to the requirements that for any given real matrix $Q > 0$, $\exists$ a scalar $\epsilon > 0$ such that*

(a) $I - J^T J > 0$; and
(b) the algebraic Riccati equation

$$\begin{aligned} F^T \widehat{P} + \widehat{P} F \ &+ \ (\widehat{P} G + HJ)(I - J^T J)^{-1}(G^T \widehat{P} + J^T H^T) \\ &+ \ HH^T + \epsilon Q = 0 \end{aligned} \tag{4.29}$$

has a solution matrix $\widehat{P} > 0$.

Proof. This is well known in the literature but it is instructive to give it here. First consider a symmetric constant matrix of the form

$$\begin{bmatrix} X & Y \\ Y^T & Z \end{bmatrix}$$

Then this matrix is negative definite if, and only if, $Z < 0$ and $X - YZ^{-1}Y^T < 0$ since

$$\begin{bmatrix} X & Y \\ Y^T & Z \end{bmatrix} = \begin{bmatrix} I & Y \\ 0 & Z \end{bmatrix} \begin{bmatrix} X - YZ^{-1}Y^T & 0 \\ 0 & Z^{-1} \end{bmatrix} \begin{bmatrix} I & 0 \\ Y^T & Z \end{bmatrix}$$

Also it is easy to see that $X - YZ^{-1}Y^T < 0$ if, and only if, $\exists$ a matrix $Q > 0$ and a scalar $\epsilon > 0$ such that $X - YZ^{-1}Y^T + \epsilon Q = 0$. Then applying these facts to (4.17) of Lemma 4.2.1 with $X = F^T \widehat{P} + \widehat{P} F + HH^T$, $Y = \widehat{P} G + HJ$ and $Z = J^T J - I$ yields (4.29). □

The requirement of a minimal realization of $S(\gamma)$ in the last result is too restrictive for the application here and as an alternative we have the following.

Lemma 4.3.2. *Consider the transfer-function matrix $S(\gamma)$ of (4.16) in the case when (without loss of generality) $J = 0$ and the pair $\{F, G\}$ is controllable. Then this $S(\gamma)$ is SBR if, and only if, for any given matrix $Q > 0$, $\exists$ a scalar $\epsilon > 0$ such that the algebraic Riccati equation*

$$F^T \widehat{P} + \widehat{P} F + \widehat{P} G G^T \widehat{P} + HH^T + \epsilon Q = 0 \tag{4.30}$$

has solution matrix $\widehat{P} > 0$.

Proof. To show that (4.30) is sufficient, set $EE^T = HH^T + \epsilon_1 Q$, where $\epsilon_1 \in (0, \epsilon)$. Then, since such a matrix E is nonsingular, the pair $\{E^T, F\}$ is observable. Hence by Lemma 4.3.1, the transfer-function matrix $\hat{S}(\gamma) := E^T(\gamma I - F)^{-1} G$ is SBR and this implies that this property also holds for $S(\gamma)$ since

$$\hat{S}^T(-\imath\omega)\hat{S}(\imath\omega) \geq S^T(-\imath\omega)S(\imath\omega), \ \forall\, \omega \in \mathbb{R}$$

To prove the converse, suppose that $S(\gamma)$ is SBR and hence $\exists$ a scalar $\epsilon_2 > 0$ such that the matrix $(1-\epsilon_2)I - S^T(-\imath\omega)S(\imath\omega)$ is also PDH $\forall \omega \in [0, \infty]$, provided that $\epsilon_2 > 0$ is sufficiently small. Also since $S(\gamma)$ is SBR (and strictly proper)

$$\rho := \sup \left\{ \lambda_{\max} \left(G^T \left(-\imath\omega I - F^T \right)^{-1} Q \left(\imath\omega I - F \right)^{-1} G \right) : \omega \in \mathbb{R} \right\}$$

is a finite number for any matrix $Q > 0$, where $\lambda_{\max}(\cdot)$ denotes the eigenvalue of maximum modulus of the matrix concerned. Hence $\exists$ an $\epsilon_1 > 0$ such that the matrix

$$I - S^T(-\imath\omega)S(\imath\omega) - \epsilon_1 G^T \left(-\imath\omega I - F^T \right)^{-1} Q \left(\imath\omega I - F \right)^{-1} G$$

is PDH $\forall\, \omega \in [0, \infty]$ and it is clear that such a choice satisfies $\epsilon_1 \leq \frac{\epsilon_2}{\rho}$. This implies that $\hat{S}(\gamma)$ is SBR with $EE^T = HH^T + \epsilon_1 Q$ and it is clear that $\hat{S}(\gamma)$ is formed from a minimal realization. Hence by Lemma 4.3.1 the following algebraic Riccati equation has matrix solution $\widehat{P} > 0$

$$F^T \widehat{P} + \widehat{P} F + \widehat{P} G G^T \widehat{P} + E E^T + \epsilon' Q = 0 \tag{4.31}$$

provided $\epsilon' > 0$ is sufficiently small. The proof is now completed by setting $EE^T = HH^T + \epsilon_1 Q$ and $\epsilon = \epsilon' + \epsilon_1$ in (4.31) to obtain the algebraic Riccati equation (4.30). □

If (4.31) has solution matrix $\widehat{P} > 0$ for $\epsilon^* > 0$ then for any $\epsilon \in [0, \epsilon^*]$, this equation admits at least one positive definite solution. Note also that we do not need the realization to be observable since $EE^T = HH^T + \epsilon Q$ has full rank.

If $S(\gamma)$ is not strictly proper then we have the following result.

Lemma 4.3.3. *Let $\{F, G, H^T, J\}$ be a minimal realization of $S(\gamma)$. Then this transfer-function matrix is SBR if, and only if, $S_m(\gamma)$ is SBR where $\{F_m, G_m, H_m^T\}$ is a realization of $S_m(\gamma)$ with*

$$\begin{aligned} F_m &:= F + G(I - J^T J)^{-1} J^T H^T \\ G_m &:= G(I - J^T J)^{-\frac{1}{2}} \\ H_m^T &:= (I - J J^T)^{-\frac{1}{2}} H^T \end{aligned}$$

Proof. Follows immediately on noting that (after some routine manipulations which are omitted here) the algebraic Riccati equation of Lemma 4.3.1 in this case requires that $\exists$ a solution matrix $\widehat{P} > 0$ to

$$F_m^T \widehat{P} + \widehat{P} F_m + \widehat{P} G_m G_m^T \widehat{P} + H_m H_m^T + \epsilon' Q = 0 \tag{4.32}$$

for some $\epsilon' > 0$. □

The key point here is that if $S(\gamma)$ is SBR then stability of F (i.e. all eigenvalues have strictly negative real parts) implies that F_m is stable and also controllability of $\{F, G\}$ implies that $\{F_m, G_m\}$ is controllable.

To apply these results, we make use of the following well known result [97] (with a modification in [80]) whose proof is omitted here.

Lemma 4.3.4. *Consider the two variable polynomial $a(s, z)$ of (4.27) and suppose that $a(0, z) \neq 0$, $\forall\, |z| = 1$. Then (4.28) holds if, and only if,*

(a) $a(s, 0)$ is Hurwitz; and
(b)

$$a(s, z) \neq 0, \ \mathrm{Re}(s) = 0, \ |z| \leq 1 \tag{4.33}$$

Clearly it is (b) in this last result which is the most difficult to test. In what follows we develop a numerically efficient test based on treating $a(s, z)$ as a polynomial, denoted $a_s(z)$, in z with coefficients which are polynomials in s, with s taking values on the extended imaginary axis of the complex plane. In particular, we write

$$a(s,z) = a_s(z) = \sum_{i=0}^{q_2} a_i(s) z^i$$

with

$$a_0(s) := s^{q_1} + \sum_{j=0}^{q_1-1} a_{0j} s^j, \quad a_i(s) = \sum_{j=0}^{q_1-1} a_{ij} s^j,\ i = 1, 2, \cdots, q_2$$

The key point to note now is that (4.33) is true if, and only if, $a_s(z)$ has all its roots outside the unit circle for all s on the imaginary axis. Hence we can apply a 1D stability test to this condition using a point-wise approach, and here we use the Schur-Cohn test (again see, for example, [91] and the relevant references cited therein) expressed in the following well known form.

Lemma 4.3.5. *Let $a(z) = a_0 + a_1 z + \cdots + a_h z^h$, $a_0 \neq 0$, $a_h \neq 0$, be a polynomial with complex coefficients a_j, $j = 0, 1, \cdots, h$. Define also the triangular Toeplitz matrices*

$$\tilde{D} := \begin{bmatrix} a_0^* & a_1^* & \cdots & a_{h-2}^* & a_{h-1}^* \\ 0 & a_0^* & a_1^* & \cdots & a_{h-2}^* \\ \vdots & \vdots & \ddots & \ddots & \vdots \\ 0 & \cdots & \cdots & a_0^* & a_1^* \\ 0 & 0 & \cdots & 0 & a_0^* \end{bmatrix} \tag{4.34}$$

and

$$\tilde{N} := \begin{bmatrix} a_h & a_{h-1} & \cdots & a_2 & a_1 \\ 0 & a_h & a_{h-1} & \cdots & a_2 \\ \vdots & \vdots & \ddots & \ddots & \vdots \\ 0 & \cdots & \cdots & a_h & a_{h-1} \\ 0 & 0 & \cdots & \cdots & a_h \end{bmatrix} \tag{4.35}$$

Then $a(z) \neq 0,\ \forall\, |z| \leq 1$, if, and only if,

$$\Phi_{dn} := \tilde{D}^* \tilde{D} - \tilde{N}^* \tilde{N}$$

is PDH.

Note also that if $a_0 \neq 0$ then if Φ_{dn} is PDH this is equivalent to the matrix $G_{dn} := \tilde{N}\tilde{D}^{-1}$ being a strict contraction.

In the case under consideration here, the coefficient a_k is a polynomial in s, $s = \imath\omega$. Hence $a_i^*(s) = a_i(-s), i = 0, 1, \cdots, q_2$. Also the triangular Toeplitz matrices $\tilde{D}$ and $\tilde{N}$ in (4.34) and (4.35) respectively can be constructed for this case. Similarly, introduce

$$\Phi_{dn}(s) := \tilde{D}^T(-s)\tilde{D}(s) - \tilde{N}^T(-s)\tilde{N}(s)$$

and

$$G_{dn}(s) := \tilde{N}(s)\tilde{D}^{-1}(s)$$

Then a simple controllable realization for $G_{dn}(-s)$ is defined as follows

$$\hat{A} := \begin{bmatrix} -\hat{A}_1 & -\hat{A}_2 & -\hat{A}_3 & \cdots & -\hat{A}_{q_1} \\ I_{q_2} & 0 & 0 & \cdots & 0 \\ 0 & I_{q_2} & 0 & \cdots & 0 \\ \vdots & \vdots & \ddots & \ddots & \vdots \\ \cdots & \cdots & 0 & I_{q_2} & 0 \end{bmatrix}, \quad \hat{B} := \begin{bmatrix} I_{q_2} \\ 0 \\ \vdots \\ 0 \end{bmatrix}$$

$$\hat{C}^T := \begin{bmatrix} \hat{C}_1^T \\ \vdots \\ \hat{C}_{q_1}^T \end{bmatrix} \tag{4.36}$$

where

$$\hat{A}_{q_1-j} := \begin{bmatrix} a_{0j} & a_{1j} & a_{2j} & \cdots & a_{q_2-1j} \\ 0 & a_{0j} & a_{1j} & \cdots & a_{q_2-2j} \\ \vdots & \vdots & \ddots & \ddots & \vdots \\ 0 & \cdots & \cdots & a_{0j} & a_{1j} \\ 0 & 0 & \cdots & 0 & a_{0j} \end{bmatrix}$$

and

$$\hat{C}_{q_1-j} := (-1)^j \begin{bmatrix} a_{q_2j} & a_{q_2-1j} & a_{q_2-2j} & \cdots & a_{1j} \\ 0 & a_{q_2j} & a_{q_2-1j} & \cdots & a_{2j} \\ \vdots & \vdots & \ddots & \ddots & \vdots \\ 0 & \cdots & \cdots & a_{q_2j} & a_{q_2-1j} \\ 0 & 0 & \cdots & 0 & a_{q_2j} \end{bmatrix}$$

are upper triangular Toeplitz matrices with a_{ij} real as defined in (4.27).

The next stage is to show that the conditions of Lemma 4.3.4 are equivalent to $G_{dn}(-s)$ being SBR. To proceed, first note that by (a) of Lemma 4.3.4, $\det(sI - \hat{A}) = \det \tilde{D}(-s) = (a(s,0))^{q_2}$ is Hurwitz and hence $G_{dn}(-s)$ also has this property. Using (b) of Lemma 4.3.4 we now have that $\Phi_{dn}(\imath\omega)$ (or $\Phi_{dn}(-\imath\omega)$) is PDH $\forall \omega \in \mathbb{R}$ and this, in turn, is equivalent to $G_{dn}(-\imath\omega)$ being a strict contraction for each $\omega \in \mathbb{R}$. Hence $G_{dn}(-s)$ is SBR.

Suppose now that $G_{dn}(-s)$ is SBR. Then $\det(sI - \hat{A})$ is Hurwitz and hence (a) of Lemma 4.3.4 holds. Also if $G_{dn}(-\imath\omega)$ is a strict contraction for each ω then, by the Schur-Cohn test, (b) of Lemma 4.3.4 holds.

The arguments just given establish the following result.

Theorem 4.3.2. *Consider the two-variable polynomial $a(s,z)$ defined by (4.27) and $G_{dn}(-s)$ defined by the state-space matrices of (4.36). Suppose also that $a(0,z) \neq 0$, $\forall\, |z| = 1$. Then this polynomial satisfies (4.28) if, and only if, $G_{dn}(-s)$ is SBR.*

This leads immediately to the following algorithm for testing (4.28).

1. Input q_1, q_2, and a_{ij} as defined in (4.27).
2. Test if $a(s,0)$ is Hurwitz and if not then stop since (4.28) does not hold (and hence the example under consideration is not asymptotically stable (and hence unstable along the pass)).
3. Construct the matrices $\hat{A}$, $\hat{B}$, $\hat{C}^T$ and choose a $\hat{Q} > 0$ and a real scalar $\epsilon > 0$ to solve the algebraic Riccati equation (4.30) applied to this particular case, i.e.
$$\hat{A}^T\widehat{P} + \widehat{P}\hat{A} + \widehat{P}\hat{B}\hat{B}^T\widehat{P} + \hat{C}^T\hat{C} + \epsilon\hat{Q} = 0$$
If this equation has a solution then (4.28) holds. In this case proceed to test the other conditions for stability along the pass.

Note that the realization defined by (4.36) may not be minimal and hence there could be numerical problems in solving the algebraic Riccati equation if the product q_1q_2 is large. Hence an input normal realization should be used to obtain a minimal realization prior to testing $G_{dn}(-s)$ for the SBR property.

It is possible to avoid computing the solution of the algebraic Riccati equation here. This is based on the fact that since $G_{dn}(-s)$ is strictly proper, it is guaranteed to be strictly bounded real if $\det(I - G_{dn}^T(-s)G_{dn}(s)) \neq 0,\ \forall\ \mathrm{Re}(s) = 0$ or, equivalently, $\det \Phi_{dn}(s) \neq 0,\ \forall\ \mathrm{Re}(s) = 0$. Hence this transfer-function matrix is SBR if, and only if, the Hamiltonian matrix

$$H_a := \begin{bmatrix} \hat{A} & \hat{B}\hat{B}^T \\ -\hat{C}^T\hat{C} & -\hat{A}^T \end{bmatrix}$$

has no purely imaginary eigenvalues. Note that the dimensions of this matrix are $2q_1q_2 \times 2q_1q_2$ and hence if q_1q_2 is 'large' then the eigenvalue computation cannot be expected to produce 'high accuracy' results.

5 Robustness

Summary. In this chapter we study the stability robustness of discrete unit memory linear repetitive processes of the form defined by Example 1.2.10 and their differential counterparts of Example 1.2.3. In the former case, a range of techniques are considered which draw on results already known in 2D systems theory. For the differential case, it is only an LMI approach which can be applied. In subsequent chapters, it will be shown that (of those currently available) only the LMI approach can be extended to permit control law design, including the case when there is uncertainty associated with the process model.

5.1 Discrete Processes

5.1.1 Background

In common with 1D linear systems theory, the robustness problem can be studied from two starting points:

(a) structured, i.e. the perturbation model structure and the bounds on the individual elements of the perturbation matrix are known; and
(b) unstructured, where, at most, a spectral norm bound on the perturbation is known.

The analysis which follows in this section details three approaches to the robustness problem for discrete unit memory linear repetitive processes and then compares their relative performance, starting with the unstructured case. These approaches are the 2D Lyapunov equation, the theory of nonnegative matrices, and LMIs respectively. Prior to this, we introduce next the general problem to be considered.

Consider a discrete unit memory linear repetitive process of the form defined by Example 1.2.10 and suppose that it is stable along the pass where in this section we characterize this property as the requirement that

$$\rho(z_1^{-1}, z_2^{-1}) := \det \begin{bmatrix} z_1^{-1} I_n - A & -B_0 \\ -C & z_2^{-1} I_m - D_0 \end{bmatrix} \neq 0 \text{ in } \mathbb{U}_{oc}^2 \tag{5.1}$$

where $\mathbb{U}_{oc}^2 := \{(z_1^{-1}, z_2^{-1}) : |z_1^{-1}| \geq 1,\ |z_2^{-1}| \geq 1\}$ or, equivalently, using the augmented plant matrix Φ (see (3.4))

$$\rho(z_1^{-1}, z_2^{-1}) = \det\left(z_1^{-1} I_n \oplus z_2^{-1} I_m - \Phi\right) \neq 0, \ (z_1^{-1}, z_2^{-1}) \text{ in } \mathbb{U}_{oc}^2 \tag{5.2}$$

Given a stable along the pass discrete unit memory linear repetitive process of the form defined by Example 1.2.10, we define the set of unstructured, unstable perturbations as

$$S_u := \{\Delta\Phi : \Delta\Phi \in \mathbb{C}^{(n+m)\times(n+m)}, \ \Phi + \Delta\Phi \text{ unstable along the pass}\}$$

Also we define the exact bound for stable perturbations as

$$\nu := \inf_{\Delta\Phi \in S_u} ||\Delta\Phi||$$

where here $||\cdot||$ denotes the standard induced matrix 2-norm. Note that the perturbed process, i.e. $\Phi + \Delta\Phi$, will remain stable along the pass if $||\Delta\Phi|| < \nu$.

In what follows, we first seek efficient methods to evaluate ν and then, as an alternative, develop a 2D Lyapunov equation approach to obtaining good lower bounds on ν. This analysis will make use of the following notation

$$\begin{aligned} \Omega &:= \{(\omega_1, \omega_2) : -\pi \le \omega_1 \le \pi, \ -\pi \le \omega_2 \le \pi\} \\ T^2 &:= \left\{\left(z_1^{-1}, z_2^{-1}\right) : |z_1^{-1}| = 1, \ |z_2^{-1}| = 1\right\} \end{aligned}$$

where T^2 is the so-called distinguished boundary of $\mathbb{U}_{oc}^2$.

5.2 Methods for Exactly Calculating the Stable Perturbation Bound

To develop the first of these, follow [108] for 2D discrete linear systems described by the Roesser model and note that the stability along the pass condition of (5.1) or (5.2) can be replaced by the following.

Lemma 5.2.1. *Consider the linear repetitive process S generated by a discrete unit memory linear repetitive process of the form defined by Example 1.2.10. Then S is stable along the pass if, and only if,*

$$\rho(\tilde{h}, \tilde{h}) \neq 0, \ |\tilde{h}| \ge 1 \tag{5.3}$$

(where $\rho(\tilde{h}, \tilde{h})$ denotes $\rho(z_1^{-1}, z_2^{-1})$ evaluated for a common value of z_1^{-1} and z_2^{-1}) and

$$\rho(z_1^{-1}, z_2^{-1}) \neq 0, \ \ (z_1^{-1}, z_2^{-1}) \in T^2 \tag{5.4}$$

Proof. This follows from the well known DeCarlo et al. criterion [40] (for stability of 2D discrete linear systems described by the Roesser model). Hence the details are omitted here. □

The condition given by (5.4) can also be written as

$$\rho(e^{\imath\omega_1}, e^{\imath\omega_2}) \neq 0,\ (\omega_1, \omega_2) \in \Omega$$

Also write the singular value decomposition of a $q_1 \times q_2$ matrix, say $\widetilde{R}$, with complex entries as

$$\widetilde{R} = U\Sigma V^*$$

where U and V are $q_1 \times q_1$ and $q_2 \times q_2$ unitary matrices respectively (and the superscript $*$ denotes the complex conjugate transpose operation), and (assume $q_1 \leq q_2$)

$$\Sigma = \left[\begin{array}{ccccc|c} \sigma_1 & & & & & \\ & \sigma_2 & & & & \\ & & \ddots & & & 0 \\ & & & \ddots & & \\ & & & & \sigma_{q_1} & \end{array}\right]$$

contains the singular values ordered as $\sigma_1 \geq \ldots \geq \sigma_{q_1} \geq 0$. Now assume that $q_1 = q_2$ and write the matrices U and V as

$$U = \left[\begin{array}{cccc} u_1 & u_2 & \cdots & u_{q_1} \end{array}\right],\ V = \left[\begin{array}{cccc} v_1 & v_2 & \cdots & v_{q_1} \end{array}\right]$$

Then

$$\widetilde{R} = \sum_{i=1}^{q_1} \sigma_i u_i v_i^*$$

The following result is a special case of the well known Gastinel-Kahan theorem [94].

Lemma 5.2.2. *Suppose that $\widetilde{R} \in \mathbb{C}^{q_1 \times q_1}$ is nonsingular. Then*

$$\min_{\Delta\widetilde{R}} \{||\Delta\widetilde{R}|| : \widetilde{R} + \Delta\widetilde{R} \text{ singular}\} = \sigma_{q_1}$$

where σ_{q_1} is the smallest singular value of $\widetilde{R}$. Also this minimum can be achieved by the following perturbation

$$\Delta\widetilde{R} = -\sigma_{q_1} u_{q_1} v_{q_1}^*$$

In the case of discrete unit memory linear repetitive processes considered here, introduce $\Phi_d := \Phi + \Delta\Phi$ and

$$\Psi(z_1^{-1}, z_2^{-1}) := \left[I(z_1^{-1}, z_2^{-1}) - \Phi\right]^{-1}$$

and

$$\Psi_d(z_1^{-1}, z_2^{-1}) := \left[I(z_1^{-1}, z_2^{-1}) - \Phi_d\right]^{-1} = \left[I(z_1^{-1}, z_2^{-1}) - \Phi - \Delta\Phi\right]^{-1}$$

where (for notational convenience)

$$I(z_1^{-1}, z_2^{-1}) := z_1^{-1}I_n \oplus z_2^{-1}I_m = \begin{bmatrix} z_1^{-1}I_n & 0 \\ 0 & z_2^{-1}I_m \end{bmatrix}$$

Then the following result gives a sufficient condition for $\Delta\Phi$ to be a stable along the pass perturbation.

Lemma 5.2.3. *Consider the linear repetitive process S generated by a discrete unit memory linear repetitive process of the form defined by Example 1.2.10. Assume also that S is stable along the pass. Then any (additive) perturbation $\Delta\Phi$ to the augmented plant matrix Φ which satisfies the following condition does not destroy this property*

$$||\Delta\Phi|| < ||\Psi(e^{\imath\omega_1}, e^{\imath\omega_2})||^{-1}, \ \forall\, (\omega_1, \omega_2) \in \Omega \tag{5.5}$$

Proof. Assume that $\exists$ a $\Delta\Phi$ which satisfies (5.5) and $\Phi + \Delta\Phi$ arises from an unstable along the pass example. Then by the DeCarlo et al. criterion [40], either (5.3) or (5.4) (or both) does not hold for

$$\rho_d(z_1^{-1}, z_2^{-1}) := \det\left[I(z_1^{-1}, z_2^{-1}) - \Phi_d\right]$$

Also if $\rho_d(\tilde{h}, \tilde{h}) = 0$ for some $\tilde{h}$ satisfying $|\tilde{h}| \geq 1$ then $\Phi + \Delta\Phi$ is 1D unstable. Hence (using Lemma 2.1 from [114]) $\exists\, \delta \in (0, 1]$ and $\omega \in [-\pi, \pi]$ such that $\Psi^{-1}(e^{\imath\omega}, e^{\imath\omega}) - \delta\Delta\Phi$ is singular. By Lemma 5.2.2 this is equivalent to

$$||\delta\Delta\Phi|| \geq ||\Psi(e^{\imath\omega}, e^{\imath\omega})||^{-1} \text{ and } ||\Delta\Phi|| \geq ||\Psi(e^{\imath\omega}, e^{\imath\omega})||^{-1}$$

which contradicts (5.5).

Suppose now that $\rho_d(z_1^{-1}, z_2^{-1}) = 0$ for some $(z_1^{-1}, z_2^{-1}) \in T^2$. Then by Lemma 5.2.2 we conclude that for some $(\omega_1, \omega_2) \in \Omega$

$$||\Delta\Phi|| \geq ||\Psi(e^{\imath\omega_1}, e^{\imath\omega_2})||^{-1}$$

This again contradicts (5.5). □

By the definition of $\Psi(e^{\omega_1}, e^{\imath\omega_2})$, we can write

$$||\Psi(e^{\omega_1}, e^{\imath\omega_2})||^{-1} = \sigma_{n+m}\left[I(e^{\imath\omega_1}, e^{\imath\omega_2}) - \Phi\right]$$

and, since the singular values of a given matrix are continuous functions of its entries, $\sigma_{n+m}\left[I(e^{\imath\omega_1}, e^{\imath\omega_2}) - \Phi\right]$ achieves its minimum value on Ω. Hence an equivalent statement of (5.5) is

$$||\Delta\Phi|| < \nu$$

where

$$\nu := \min_{\Omega} \sigma_{n+m} \left[I(e^{\imath\omega_1}, e^{\imath\omega_2}) - \Phi \right] \tag{5.6}$$

and, as shown below, ν provides an appropriate measure of robustness.

Introduce

$$S_u := \{\Delta\Phi \; : \; \Phi + \Delta\Phi \text{ is unstable along the pass}\}$$

and

$$\mu := \min_{\Delta\Phi \in S_u} ||\Delta\Phi||$$

Then the following result shows that ν and μ are equal.

Theorem 5.2.1. *Suppose that the linear repetitive process S generated by a discrete unit memory linear repetitive process of the form defined by Example 1.2.10 is stable along the pass. Then μ and ν defined above are equal.*

Proof. From Lemma 5.2.3 it follows that $\mu \geq \nu$. Also $\exists \; (\omega_1^0, \omega_2^0) \in \Omega$ such that

$$\nu = \sigma_{n+m} \left[I(e^{\imath\omega_1^0}, e^{\omega_2^0}) - \Phi \right] \tag{5.7}$$

and introduce the singular value decomposition $I(e^{\imath\omega_1^0}, e^{\imath\omega_2^0}) - \Phi = U_0 \Sigma_0 V_0^*$. Then since the underlying process is stable along the pass, $I(e^{\imath\omega_1^0}, e^{\imath\omega_2^0}) - \Phi$ is nonsingular and Lemma 5.2.2 implies that the perturbation defined by $\Delta\Phi_0 = -\nu u_{n+m}^0 (v_{n+m}^0)^*$, where u_{n+m}^0 and v_{n+m}^0 are the last columns of U_0 and V_0 respectively, will make $I(e^{\imath\omega_1^0}, e^{\imath\omega_2^0}) - (\Phi + \Delta\Phi_0)$ singular. This means that $\Phi + \Delta\Phi_0$ is unstable along the pass. Hence $\Delta\Phi_0 \in S_u$ and $\mu \leq ||\Delta\Phi_0|| = || - \nu u_{n+m}^0 (v_{n+m}^0)^* || = \nu$. □

It follows that ν of (5.6) is the tightest upper bound for unstructured complex perturbations that will not cause a process to be unstable along the pass. In comparison with the result of [107], it is the least conservative and, as detailed below, is computationally more feasible.

To discuss the numerical computation of ν of (5.7), let $-\pi = \omega_{10} < \omega_{11} < \cdots < \omega_{1\mathcal{K}} = \pi$ and $-\pi = \omega_{20} < \omega_{21} < \cdots < \omega_{2\mathcal{L}} = \pi$ be two sets of points that equally partition the intervals $-\pi \leq \omega_1 \leq \pi$ and $-\pi \leq \omega_2 \leq \pi$ respectively and let $\Delta\theta_1 = \frac{\pi}{\mathcal{K}}$ and $\Delta\theta_2 = \frac{\pi}{\mathcal{L}}$. Also for each pair $(\omega_{1i}, \omega_{2j})$ define

$$d_{ij} := \sigma_{n+m} \left[I(e^{\imath\omega_{1i}}, e^{\imath\omega_{2j}}) - \Phi \right], \; 0 \leq i \leq \mathcal{K}, \; 0 \leq j \leq \mathcal{L} \tag{5.8}$$

and

$$\nu_{\mathcal{K}\mathcal{L}} := \min_{i,j} d_{ij} \tag{5.9}$$

Note also that for any $(\omega_1, \omega_2) \in \Omega$, $\exists \; (\omega_{1i}, \omega_{2j})$ such that $|\omega_1 - \omega_{1i}| \leq \Delta\theta_1$, $|\omega_2 - \omega_{2j}| \leq \Delta\theta_2$, and hence

$$\begin{aligned} |e^{\imath\omega_1} - e^{\imath\omega_{1i}}| &\leq \Delta\theta_1 \\ |e^{\imath\omega_2} - e^{\imath\omega_{2j}}| &\leq \Delta\theta_2 \end{aligned} \tag{5.10}$$

By (5.10) and a standard property of the singular value decomposition, we now have that

$$\begin{aligned}&|\sigma_{n+m}\left[I(e^{\imath\omega_1},e^{\imath\omega_2})-\Phi\right]-\sigma_{n+m}\left[I(e^{\imath\omega_{1i}},e^{\imath\omega_{2j}})-\Phi\right]|\\&\leq ||I(e^{\imath\omega_1},e^{\imath\omega_2})-I(e^{\imath\omega_{1i}},e^{\imath\omega_{2j}})||\\&\leq \max(\Delta\theta_1,\Delta\theta_2)\end{aligned} \tag{5.11}$$

and it now follows from (5.6), (5.9), and (5.11) that

$$\nu \leq \nu_{\mathcal{KL}} \leq \nu + \pi \max\left(\frac{1}{\mathcal{K}},\frac{1}{\mathcal{L}}\right) \tag{5.12}$$

Given (5.12), it follows that $\nu_{\mathcal{KL}}$ defined by (5.9) will yield a good approximation of ν if both $\mathcal{K}$ and $\mathcal{L}$ are large enough. The absolute and relative error of such an approximation can also be estimated based on (5.12) as

$$|v_{\mathcal{KL}} - \nu| \leq \pi \max\left(\frac{1}{\mathcal{K}},\frac{1}{\mathcal{L}}\right) \tag{5.13}$$

and

$$\left|\frac{\nu_{\mathcal{KL}}-\nu}{\nu}\right| \leq \frac{\pi\max\left(\frac{1}{\mathcal{K}},\frac{1}{\mathcal{L}}\right)}{\nu_{\mathcal{KL}}-\pi\max\left(\frac{1}{\mathcal{K}},\frac{1}{\mathcal{L}}\right)}$$

respectively. Also, since $\nu_{\mathcal{KL}} \to \nu$ as $\mathcal{K}$ and $\mathcal{L} \to \infty$ and $\nu > 0$, we conclude that both the absolute and relative errors approach zero as $\mathcal{K},\mathcal{L} \to \infty$. This leads to the following computational algorithm.

Algorithm 1.

Step 1. Set $\mathcal{L} = \mathcal{K}$ and determine the smallest positive integer $\mathcal{K}$ such that

$$\frac{\pi}{\mathcal{K}} \leq \epsilon$$

where ϵ is the prescribed tolerance on ν.

Step 2. For $0 \leq i \leq \mathcal{K}$, $0 \leq j \leq \mathcal{L}$, compute d_{ij} defined by (5.8).

Step 3. Compute $\nu_{\mathcal{KL}}$ defined by (5.9).

Step 4. Set $\nu = \nu_{KL}$.

Using (5.13), it follows that the robustness measure computed by the above algorithm differs from the true ν by at most ϵ. Also it is easy to implement numerically and we next give an example to illustrate its application.

Example 5.2.1. Consider the following special case of Example 1.2.10

$$\Phi = \begin{bmatrix} -0.5 & -1 \\ 0.395 & -0.01 \end{bmatrix}$$

Then applying the above algorithm yields $\nu = 0.070694$ which occurs at $(\omega_1,\omega_2) = (\pi,0)$. This compares with $\nu = 0.0483972$ given by an alternative algorithm ((4.4) of [107]).

In practice, the above algorithm requires the computation of the infimum of the minimum singular values of a two-variable complex matrix of dimension $(n+m)\times(n+m)$ over the so-called base-band Ω and hence it could be very expensive computationally. To develop more efficient methods for computing ν, we follow [111] for 2D discrete linear systems described by the Roesser model.

To begin, recall the definitions of $\widehat{A}_1$ and $\widehat{A}_2$ from (2.66). Also introduce

$$F(\gamma) := \widehat{A}_1 + e^{\imath\gamma}\widehat{A}_2$$

Then we have the following result, where $\underline{\sigma}(\cdot)$ denotes the minimum singular value of its matrix argument.

Lemma 5.2.4.

$$\nu = \inf_{0\le\gamma\le\pi}\ \inf_{-\pi\le\omega\le\pi} \underline{\sigma}\left[e^{\imath\omega}I_{n+m} - F(\gamma)\right]$$

Proof. Write

$$\begin{aligned} I(e^{\imath\omega_1},e^{\imath\omega_2}) - \Phi &= e^{\imath\omega_1}\begin{bmatrix} I_n & 0 \\ 0 & e^{\imath(\omega_2-\omega_1)}I_m \end{bmatrix} - \widehat{A}_1 - \widehat{A}_2 \\ &= \begin{bmatrix} I_n & 0 \\ 0 & e^{\imath(\omega_2-\omega_1)}I_m \end{bmatrix}\left[e^{\imath\omega_1}I_{n+m} - (\widehat{A}_1 + e^{\imath(\omega_1-\omega_2)}\widehat{A}_2)\right] \end{aligned}$$

Then, since the matrix

$$\begin{bmatrix} I_n & 0 \\ 0 & e^{\imath(\omega_2-\omega_1)}I_m \end{bmatrix}$$

is unitary, we have that

$$\underline{\sigma}\left[I(e^{\imath\omega_1},e^{\imath\omega_2}) - \Phi\right] = \underline{\sigma}\left[e^{\imath\omega_1}I_{n+m} - F(\gamma)\right]$$

where $\gamma := \omega_1 - \omega_2$. Hence

$$\begin{aligned} \nu &= \inf_{-\pi\le\gamma\le\pi}\ \inf_{-\pi\le\omega\le\pi} \underline{\sigma}\left[e^{\imath\omega}I_{n+m} - F(\gamma)\right] \\ &= \inf_{0\le\gamma\le\pi}\ \inf_{-\pi\le\omega\le\pi} \underline{\sigma}\left[e^{\imath\omega}I_{n+m} - F(\gamma)\right] \end{aligned}$$

where the last equality is due to the fact that

$$f(\gamma) := \inf_{-\pi\le\omega\le\pi} \underline{\sigma}\left[e^{\imath\omega}I_{n+m} - F(\gamma)\right] \tag{5.14}$$

is an even function of γ. □

Given this last result, we can now write

$$\nu = \inf_{0 \leq \gamma \leq \pi} f(\gamma)$$

where $f(\gamma)$ is as defined by (5.14) and this relationship suggests that ν can be found by solving a scalar minimization problem. A key feature of this minimization problem is the fact that, for a fixed γ, $f(\gamma)$ is the exact stable perturbation bound for the 1D discrete linear system whose state matrix is $F(\gamma)$. Also such an $f(\gamma)$ can be evaluated using, for example, the well known bisection method as discussed next.

The bisection method for evaluating $f(\gamma)$ is based on the following result [29].

Lemma 5.2.5. *For a real scalar γ, let $F(\gamma)$ be an $(n+m) \times (n+m)$ matrix with complex entries. Then there exists a real number $\beta > f(\gamma)$ such that for $\beta \geq \mu \geq f(\gamma)$ the $2(n+m) \times 2(n+m)$ matrix pencil*

$$H(\mu) - \beta H_1(\mu) := \begin{bmatrix} -\mu I_{n+m} & F(\gamma) \\ I_{n+m} & 0 \end{bmatrix} - \lambda \begin{bmatrix} 0 & I_{n+m} \\ F^*(\gamma) & -\mu I_{n+m} \end{bmatrix} \quad (5.15)$$

has a generalized eigenvalue λ with modulus one. Also if $\mu < f(\gamma)$ then (5.15) has no generalized eigenvalue with modulus equal to one.

This last result now leads to the following algorithm [29] which estimates $f(\gamma)$ to within a positive integer factor, say 10, or indicates that $f(\gamma)$ is less than a specified tolerance, say ϵ, where the input is $F(\gamma)$ with associated tolerance $\epsilon > 0$, and the output is $\zeta \in \mathbb{R}$ and $\delta \in \mathbb{R}$ such that

$$\frac{\delta}{10} \leq \zeta \leq F(\gamma) \leq \delta \text{ or } 0 = \zeta \leq F(\gamma) \leq \delta \leq 10\epsilon$$

Algorithm 2.

Step 1. $\zeta := 0$.
Step 2. $\delta := \underline{\sigma}\left(I_{n+m} - F(\gamma)\right)$.
Step 3. While $\delta > 10 \max\,(\epsilon, \zeta)$, $\mu := \sqrt{\delta \times (\max(\epsilon, \delta)}$, and if (5.15) has a generalized eigenvalue of modulus one, then $\delta := \mu$, otherwise $\zeta := \mu$.
Step 4. End.

This algorithm first evaluates $F(\gamma)$ for a given γ and hence δ as defined in this algorithm is just the smallest singular value of the constant matrix $I_{n+m} - F(\gamma)$. Also the generalized eigenvalue for a given pair of matrices is a well defined concept and numerous efficient and reliable algorithms for computing it exist (see, for example, [76]).

In terms of the algorithm itself, note that when ζ and δ are specified, an interval $[\zeta, \delta]$ containing the value of $f(\gamma)$ is identified, and the algorithm reduces the length of this interval to either

$$[\zeta, \sqrt{\delta \times \max(\delta, \epsilon)}]$$

or

$$[\sqrt{\delta \times \max(\delta, \epsilon)}, \delta]$$

based on whether (5.15) has a generalized eigenvalue of modulus equal to one. In execution, the reduced interval is again denoted by $[\zeta, \delta]$ and the algorithm continues until the length of the interval is reduced to within a factor of 10 or it indicates that $f(\gamma)$ is less than the tolerance level 10ϵ. If the factor of 10 is not accurate enough, the algorithm is easily modified (the details are omitted here) to give higher accuracy (clearly the computational cost will increase 3with the accuracy demanded).

It is possible to use a singular value perturbation analysis to show that $f(\gamma)$ is continuous with respect to μ. This fact then leads to another computational algorithm for the problem under consideration here.

Lemma 5.2.6. *The function $f(\gamma)$ defined by (5.14) is continuous over $[0, \pi]$.*

Proof. Let $\gamma \in [0, \pi]$ be fixed and let $\Delta\gamma$ be a real perturbation. Also define the two so-called auxiliary functions $g_\gamma(\omega)$ and $g_{\gamma+\Delta\gamma}(\omega)$ as

$$g_\gamma(\omega) := \underline{\sigma}\,[e^{\imath\omega} I_{n+m} - F(\gamma)]$$

and

$$g_{\gamma+\Delta\gamma}(\omega) := \underline{\sigma}\,[e^{\imath\omega} I_{n+m} - F(\gamma + \Delta\gamma)]$$

respectively. Then by a standard result in matrix analysis

$$|g_{\gamma+\Delta\gamma}(\omega) - g_\gamma(\omega)| \leq ||F(\gamma + \Delta\gamma) - F(\gamma)|| \leq |\Delta\gamma|\; ||\widehat{A}_2|| \tag{5.16}$$

or, equivalently, as functions of ω the two curves $g_{\gamma+\Delta\gamma}(\omega)$ and $g_\gamma(\omega)$ are uniformly close to each other over the entire interval $[0, \pi]$ with maximum deviation no greater than $|\Delta\gamma|\; ||\widehat{A}_2||$. Now let $\hat{\omega}_1$ and $\hat{\omega}_2$ be the points at which $g_\gamma(\omega)$ and $g_{\gamma+\Delta\gamma}(\omega)$ achieve their minimum values respectively, i.e.

$$f(\gamma) = g_\gamma(\hat{\omega}_1), \;\; f(\gamma + \Delta\gamma) = g_{\gamma+\Delta\gamma}(\hat{\omega}_2) \tag{5.17}$$

Then there are two cases we need to consider.

The first of these is when $f(\gamma) > f(\gamma + \Delta\gamma)$. In this case, we conclude from (5.16) and (5.17) that at the point $\hat{\omega}_2$

$$|f(\gamma + \Delta\gamma) - g_\gamma(\hat{\omega}_2)| \leq |\Delta\gamma|\; ||\widehat{A}_2|| \tag{5.18}$$

By (5.14) and $f(\gamma) > f(\gamma + \Delta\gamma)$ we have that

$$g_\gamma(\hat{\omega}_2) \geq f(\gamma) > f(\gamma + \Delta\gamma) \tag{5.19}$$

Also (5.18) and (5.19) imply that

$$|f(\gamma) - f(\gamma + \Delta\gamma)| \leq |\Delta\gamma|| \, ||\widehat{A}_2|| \tag{5.20}$$

The second case is when

$$f(\gamma) \leq f(\gamma + \Delta\gamma) \tag{5.21}$$

Here at $\hat{\omega}_1$, (5.16) and (5.17) yield

$$|g_{\gamma+\Delta\gamma}(\hat{\omega}_1) - f(\gamma)| \leq |\Delta\gamma| \, ||\widehat{A}_2|| \tag{5.22}$$

Using (5.14) and (5.21) we obtain

$$f(\gamma) \leq f(\gamma + \Delta\gamma) \leq g_{\gamma+\Delta\gamma}(\hat{\omega}_1)$$

which, together with (5.22), again leads to (5.20). Hence the continuity of $f(\gamma)$ is established and the proof is complete. □

Consider now the search for the minimum point of a continuous scalar function such as $f(\gamma)$ over a finite interval. Then a standard way of proceeding is to first identify all the brackets over the interval, where a bracket is defined as a sub-interval over which the function has a unique local minimum. This can be routinely implemented by a search of $f(\gamma)$ over a fairly coarse sampling of points in the interval. In each bracket, a modified quadratic curve-fitting method (see, for example, [59] for the details) and, in particular, the three-point pattern method can be used to find the minimum point in the bracket with guaranteed global convergence. Finally, the global minimum of $f(\gamma)$ over $[0, \pi]$ can be found by comparing the values of this function over the local minimum points just determined.

An algorithm for implementing the procedure just outlined can now be stated as follows.

Algorithm 3.

Step 1. Identify over $[0, \pi]$ all brackets $\hat{B}_j$, $j = 1, 2, \cdots, \mathcal{J}$, where evaluation of $f(\gamma)$ at each sampling point is done using Algorithm 2.

Step 2. In each bracket $\hat{B}_j$ find the minimum point γ_j of $f(\gamma)$ using the three-point pattern method ([59], Sect. 7.3)

Step 3. Obtain the global minimum point of $f(\gamma)$ by comparing $f(\gamma_j)$, $j = 1, 2, \cdots, \mathcal{J}$.

An alternative set of algorithms arises from employing a direct method for computing ν using a numerical minimization method with the objective function given by

$$f(\omega_1, \omega_2) := \underline{\sigma}\left[I(e^{\imath\omega_1}, e^{\imath\omega_2}) - \Phi\right]$$

Then, since $f(\omega_1, \omega_2)$ is well defined over $\mathbb{R}^2$, it is an unconstrained optimization problem which has to be solved and many well-established solution techniques are available. The following analysis based on singular-value-perturbation theory (see, for example, [165]) also shows that $f(\omega_1, \omega_2)$ as a two-variable function is continuous on $\mathbb{R}^2$.

First we have that

$$\begin{aligned} |f(\omega_1 + \Delta\omega_1, \omega_2) - f(\omega_1, \omega_2)| &= |\underline{\sigma}[I(e^{\imath(\omega_1+\Delta\omega_1)}, e^{\imath\omega_2}) - \Phi] \\ &- \underline{\sigma}\,[I(e^{\imath\omega_1}, e^{\imath\omega_2}) - \Phi]\,| \\ &\leq ||I(e^{\imath(\omega_1+\Delta\omega_1)}, e^{\imath\omega_2}) - I(e^{\imath\omega_1}, e^{\imath\omega_2})|| \\ &\leq |\Delta\omega_1| \end{aligned} \tag{5.23}$$

and in a similar manner

$$|f(\omega_1, \omega_2 + \Delta\omega_2) - f(\omega_1, \omega_2)| \leq |\Delta\omega_2| \tag{5.24}$$

Also for numerical computation of the minimum of $f(\omega_1, \omega_2)$, it is reasonable to assume that this function is differentiable over Ω. If this is not the case, it can be approximated by a smooth function with any desired degree of accuracy over the entire region Ω and the minimum value obtained via this approximation can be regarded as the minimum value of $f(\omega_1, \omega_2)$ within a known tolerance.

With this smoothness assumption, (5.23) and (5.24) imply that the gradient of $f(\omega_1, \omega_2)$ satisfies

$$||\nabla f(\omega_1, \omega_2)|| \leq \sqrt{2} \tag{5.25}$$

Also (5.23)–(5.25) can be used to establish that a reliable approximation of $\nabla f(\omega_1, \omega_2)$ is given by the numerical differentiation expression

$$\begin{aligned} \nabla f(\omega_1, \omega_2) &= \begin{bmatrix} \frac{\partial f(\omega_1,\omega_2)}{\partial \omega_1} \\ \frac{\partial f(\omega_1,\omega_2)}{\partial \omega_2} \end{bmatrix} \\ &= \frac{1}{\delta} \begin{bmatrix} f(\omega_1 + \delta, \omega_2) - f(\omega_1, \omega_2) \\ f(\omega_1, \omega_2 + \delta) - f(\omega_1, \omega_2) \end{bmatrix} \end{aligned}$$

Among the many available minimization methods are quasi-Newton algorithms. These have a descent structure and provide fast convergence without the need to evaluate the Hessian matrix of the objective function. They are well known in the literature (see, for example, [59] or [112]) and are hence not detailed here.

Next we aim to develop lower bounds for ν using a Lyapunov approach. To begin, write the dynamics of the perturbed process with no current pass input terms in the form

$$\begin{bmatrix} x_{k+1}(p+1) \\ y_{k+1}(p) \end{bmatrix} = (\Phi + \Delta\Phi) \begin{bmatrix} x_{k+1}(p) \\ y_k(p) \end{bmatrix}$$

Suppose now that the 2D Lyapunov equation developed in Chap. 3 (in particular, Sect. 3.2.1) holds, i.e. $\exists$ matrices $W = W_1 \oplus W_2 > 0$ and $Q > 0$ such that

$$\Phi^T W \Phi - W = -Q \tag{5.26}$$

Also use $W_1 > 0$ and $W_2 > 0$ to construct the Lyapunov function

$$V(k,p) := V_1(k,p) + V_2(k,p) = x_{k+1}^T(p) W_1 x_{k+1}(p) + y_k^T(p) W_2 y_k(p) \tag{5.27}$$

or, equivalently,

$$V(k,p) = X_k^T(p) W X_k(p) \tag{5.28}$$

where $X_k(p) := \begin{bmatrix} x_{k+1}^T(p) & y_k^T(p) \end{bmatrix}^T$.

Now introduce the so-called increment

$$\Delta V(k,p) := V_1(k,p+1) - V_1(k,p) + V_2(k+1,p) - V_2(k,p) \tag{5.29}$$

and hence (using (5.26) and (5.27)–(5.29))

$$\Delta V(k,p) \leq - \left[\underline{\sigma}(Q) - 2||\Phi^T W^{\frac{1}{2}}|| \, \overline{\sigma}^{\frac{1}{2}}(W) \, ||\Delta\Phi|| - \overline{\sigma}(W)||\Delta\Phi||^2 \right] ||X_k(p)||^2$$

where $\overline{\sigma}(\cdot)$ denotes the maximum singular value of its matrix argument. Also from the 2D Lyapunov equation (5.26) we have that

$$\Phi^T W^{\frac{1}{2}} W^{\frac{1}{2}} \Phi = W - Q$$

and hence

$$||\Phi^T W^{\frac{1}{2}}|| \leq \overline{\sigma}^{\frac{1}{2}}(W - Q)$$

Consequently we have the following estimate for $\Delta V(k,p)$

$$\Delta V(k,p) \leq - \left[\underline{\sigma}(Q) - 2\overline{\sigma}^{\frac{1}{2}}(W - Q)\overline{\sigma}^{\frac{1}{2}}(W) \, ||\Delta\Phi|| - \overline{\sigma}(W)||\Delta\Phi||^2 \right] ||X_k(p)||^2$$

It now follows that if

$$||\Delta\Phi|| < \frac{[\overline{\sigma}(W - Q) + \underline{\sigma}(Q)]^{\frac{1}{2}} - \overline{\sigma}^{\frac{1}{2}}(W - Q)}{\overline{\sigma}^{\frac{1}{2}}(W)} \tag{5.30}$$

then $\Delta V(k,p) < 0$, $\forall \, 0 \leq p \leq \alpha$, $k \geq 0$, and a routine argument can now be used to show that the perturbed process is stable along the pass.

An alternative Lyapunov function can be defined as

$$\widetilde{V}(k,p) := \sqrt{V(k,p)}$$

Then by repeating the steps in the above analysis, we eventually arrive at the conclusion that the perturbed process is stable along the pass if

$$||\Delta\Phi|| < \frac{\underline{\sigma}(Q)}{\overline{\sigma}(W) + \overline{\sigma}^{\frac{1}{2}}(W) \, \overline{\sigma}^{\frac{1}{2}}(W - Q)} \tag{5.31}$$

then $\Delta V_k(p) < 0, \; \forall\, 0 \leq p \leq \alpha, k \geq 0$, and the perturbed process is stable along the pass.

At this stage, we have developed two stable (along the pass) perturbation bounds, i.e.

$$\nu_{p1} := \frac{[\overline{\sigma}(W-Q)+\underline{\sigma}(Q)]^{\frac{1}{2}} - \overline{\sigma}^{\frac{1}{2}}(W-Q)}{\overline{\sigma}^{\frac{1}{2}}(W)} \tag{5.32}$$

and

$$\nu_{p2} := \frac{\underline{\sigma}(Q)}{\overline{\sigma}(W)+\overline{\sigma}^{\frac{1}{2}}(W)\,\overline{\sigma}^{\frac{1}{2}}(W-Q)} \tag{5.33}$$

Clearly $\nu_{p1} \leq \nu$ and $\nu_{p2} \leq \nu$, i.e. they are lower bounds on ν, and to compare them first note that

$$\overline{\sigma}(W-Q) \leq \overline{\sigma}(W) - \underline{\sigma}(Q) \tag{5.34}$$

This leads to the following result.

Lemma 5.2.7. *The following relationship holds*

$$\nu_{p2} \leq \nu_{p1}$$

Proof. For ease of notation, set $a = \underline{\sigma}(Q)$, $b = \overline{\sigma}(W)$ and $c = \overline{\sigma}(W-Q)$. Then $a > 0$, $b > 0$, $c \geq 0$ and (5.34) yields $a + c \leq b$. Hence

$$a[(a+c)^{\frac{1}{2}} + c^{\frac{1}{2}}] \leq a(b^{\frac{1}{2}} + c^{\frac{1}{2}})$$

and therefore

$$\frac{a}{b + b^{\frac{1}{2}}c^{\frac{1}{2}}} \leq \frac{(a+c)^{\frac{1}{2}} - c^{\frac{1}{2}}}{b^{\frac{1}{2}}}$$

and the proof is complete. □

The major conclusion here is that on comparing (5.30) and (5.31) we see that the former produces a better stable perturbation bound. Also it is of interest now to consider the special case when $||\Phi|| < 1$.

First note again that, in general, to compute ν_{p1} and ν_{p2} we need to solve the 2D Lyapunov equation, but if $||\Phi|| < 1$ then

$$Q := I_{n+m} - \Phi^T\Phi > 0 \tag{5.35}$$

This, in turn, implies that the 2D Lyapunov equation (5.26) has a positive definite solution with $W = I_{n+m}$ and Q as in (5.35). Then since $\overline{\sigma}(\Phi^T\Phi) = ||\Phi||^2$ and $\underline{\sigma}(I_{n+m} - \Phi^T\Phi) = 1 - ||\Phi||^2$, (5.32) and (5.33) become

$$\begin{aligned}\nu_{p1} &= \left[\overline{\sigma}(\Phi^T\Phi) + \underline{\sigma}(I_{n+m} - \Phi^T\Phi)\right]^{\frac{1}{2}} - \overline{\sigma}^{\frac{1}{2}}(\Phi^T\Phi)\\ &= 1 - ||\Phi||\end{aligned}$$

and

$$\begin{aligned}\nu_{p2} &= \frac{\underline{\sigma}(I_{n+m} - \Phi^T \Phi)}{1 + \overline{\sigma}^{\frac{1}{2}}(\Phi^T \Phi)} = \frac{1 - ||\Phi||^2}{1 + ||\Phi||} \\ &= 1 - ||\Phi||\end{aligned}$$

respectively. Hence the two bounds are identical and equal to $1 - ||\Phi||$ and an immediate consequence is that among all equivalent augmented plant matrices $T\Phi T^{-1}$ we should seek a similarity transformation matrix $T = T_1 \oplus T_2$ such that $||T\Phi T^{-1}||$ is minimized, since this gives the largest possible stability robustness bound ν_{p1} (and ν_{p2}). We now discuss this point in more detail by developing another (numerically feasible) solution of the 2D Lyapunov equation (5.26) as an alternative to the algorithms of Sect. 3.2.3.

Lemma 5.2.8. *Solution matrices $W = W_1 \oplus W_2 > 0$ and $Q > 0$ exist for the 2D Lyapunov equation (5.26) if, and only if, $\exists$ nonsingular $T = T_1 \oplus T_2$ such that*

$$||T\Phi T^{-1}|| < 1 \tag{5.36}$$

(where $\oplus$ again denotes the matrix direct sum).

Proof. If (5.26) holds for some $W = W_1 \oplus W_2 > 0$ and $Q > 0$, then T can be obtained via the decomposition $W = T^T T$, and this fact and (5.26) now lead to

$$(T\Phi T^{-1})^T (T\Phi T^{-1}) - I_{n+m} = -T^{-T} Q T^{-1} < 0$$

which implies (5.36).

Conversely, the existence of $T = T_1 \oplus T_2$ satisfying (5.26) implies that

$$\hat{Q} := I_{n+m} - (T\Phi T^{-1})^T (T\Phi T^{-1}) > 0$$

and on re-writing this last equation as

$$T^T \hat{Q} T = T^T T - \Phi^T (T^T T) \Phi$$

we get (5.26) with $W := T^T T = T_1^T T_1 \oplus T_2^T T_2 > 0$ and $Q := T^T \hat{Q} T > 0$. □

The key fact from this last result is that solving the 2D Lyapunov equation (5.26) is equivalent to solving the minimization problem

$$\tau := \inf_{T = T_1 \oplus T_2} ||T\Phi T^{-1}|| \tag{5.37}$$

and solution matrices $W > 0$ and $Q > 0$ exist if, and only if, $\tau < 1$. Also using a QR decomposition argument, it is straightforward to show that it is sufficient to implement the optimal search problem for (5.37) over the set

$$S_T := \{T : T = T_1 \oplus T_2,\ T_1, T_2 \text{ upper triangular}\}$$

This reduces the number of parameters in the optimization from $n^2 + m^2$ to $\frac{[n(n+1)+m(m+1)]}{2}$. Also quasi-Newton optimization methods (see, for example, [59]) can be employed here where the gradient of $||T\Phi T^{-1}||$ can be evaluated using numerical differentiation.

Example 5.2.2. Consider the following special case of Example 1.2.10 when

$$\begin{aligned} x_{k+1}(p+1) &= \begin{bmatrix} 0.95440 & -0.18730 \\ 0.10790 & 0.93460 \end{bmatrix} x_{k+1}(p) + Bu_{k+1}(p) \\ &+ \begin{bmatrix} -0.24430 & 0.20880 \\ 0.19230 & -0.16440 \end{bmatrix} y_k(p) \\ y_{k+1}(p) &= \begin{bmatrix} 0.05670 & -0.03700 \\ -0.00340 & 0.03750 \end{bmatrix} x_{k+1}(p) + Du_{k+1}(p) \\ &+ \begin{bmatrix} 0.94450 & 0.19240 \\ -0.10460 & 0.94450 \end{bmatrix} y_k(p) \end{aligned}$$

which is known to be stable along the pass.

As detailed earlier in this section, one method of obtaining the exact (complex) stable perturbation bound ν is to approximate $\underline{\sigma}\left[e^{\imath\omega} I_{n+m} - F(\gamma)\right]$ over a sufficiently fine grid in the region $\{0 \leq \gamma \leq \pi,\ -\pi \leq \omega \leq \pi\}$, and then find the minimum value over these points. The result of this evaluation is a large matrix with each row representing a discrete version of $\underline{\sigma}\left[e^{\imath\omega} I_{n+m} - F(\gamma)\right]$ for $-\pi \leq \omega \leq \pi$ and a fixed $\gamma \in [0, \pi]$. Here it turns out that in order to obtain an estimate, say $\tilde{\nu}$, with $|\tilde{\nu} - \nu| < 10^{-4}$ requires a grid to be as fine as $\frac{\pi}{300}$ (for γ) by $\frac{\pi}{300}$ (for ω). The algorithm here gives $\tilde{\nu} = 0.01108$.

In this case $||\Phi|| = 1.6125$ and hence (5.26) must be solved to find the lower bound ν_{p1}. After QR decomposition and scaling the transformation matrix T in (5.37) can always be assumed to have the form

$$T = \begin{bmatrix} 1 & t_4 & 0 & 0 \\ 0 & t_1 & 0 & 0 \\ 0 & 0 & t_2 & t_5 \\ 0 & 0 & 0 & t_3 \end{bmatrix}$$

Hence (5.37) in this case is a minimization problem involving a parameter vector, say x, of dimension 5 where $x = \begin{bmatrix} t_1 & t_2 & t_3 & t_4 & t_5 \end{bmatrix}^T$. The objective function is

$$f(x) := ||T\Phi T^{-1}||$$

Starting from $T = I$, i.e. $x = \begin{bmatrix} 1 & 1 & 1 & 0 & 0 \end{bmatrix}^T$, use of an appropriate algorithm leads to a local minimum point at

$$x_{\min} = \begin{bmatrix} 1.17486 & 2.40857 & 3.66415 & -0.12800 & -0.05464 \end{bmatrix}^T$$

with $f(x^*) = 0.98565 < 1$. Hence a solution of the 2D Lyapunov equation (5.26) in this case is given by

$$W = T^T T = \begin{bmatrix} 1 & -0.128 & 0 & 0 \\ -0.128 & 1.39667 & 0 & 0 \\ 0 & 0 & 0.80121 & -0.13160 \\ 0 & 0 & -0.13160 & 13.42896 \end{bmatrix} > 0$$

and

$$Q = \begin{bmatrix} 0.08037 & 0.03568 & -0.09236 & -0.20490 \\ 0.03568 & 0.06963 & -0.07003 & -0.15537 \\ -0.09236 & -0.07003 & 0.32978 & 0.36110 \\ -0.20490 & -0.15537 & 0.36110 & 1.9220 \end{bmatrix} > 0$$

This yields $\nu_{p1} = 0.00144$ and $\nu_{p2} = 0.00141$ respectively.

5.3 Nonnegative Matrix Theory Approach

This approach starts from the following necessary and sufficient stability along the pass condition arising from a simple rearrangement of the characteristic polynomial condition for stability along the pass of Theorem 2.2.8 written in the form

$$\det\left(I_{n+m} - z_1\widehat{A}_1 - z_2\widehat{A}_2\right) \neq 0 \text{ in } \mathbb{U}_c^2 \tag{5.38}$$

where we assume that $r(D_0) < 1$, $r(A) < 1$ (necessary conditions for the above condition) and hence $r(\widehat{A}_1) < 1$ and $r(\widehat{A}_2) < 1$. Also we will require the following result (recall from Chap. 2 that $[\cdot]_p$ denotes the result of replacing each element in a matrix by its absolute value).

Lemma 5.3.1. *Let*

$$\widehat{G}_1(z_1) := \left(I_{n+m} - z_1\widehat{A}_1\right)^{-1}$$

and write its Maclaurin series expansion as

$$\widehat{G}_1(z_1) = \left(I_{n+m} - z_1\widehat{A}_1\right)^{-1} = \sum_{i=0}^{\infty} \widehat{A}_1^i z_1^i$$

Then

$$\begin{aligned} \left[\widehat{G}_1(z_1)z_2\widehat{A}_2\right]_p &= \left[\sum_{i=0}^{\infty} \widehat{A}_1^i \widehat{A}_2 z_1^i z_2\right]_p \\ &\leq \sum_{i=0}^{\infty} \left[\widehat{A}_1^i \widehat{A}_2\right]_p =: H_1 \text{ in } \mathbb{U}_c^2 \end{aligned} \tag{5.39}$$

Proof. A straightforward application of elementary properties of $[\cdot]_p$ and hence the details are omitted here. □

Using this last result, the following is another sufficient condition for stability along the pass. (See also [171] for a similar approach in the case of 2D discrete linear systems described by the Fornasini-Marchesini model.)

Theorem 5.3.1. *Consider the linear repetitive process S generated by a discrete unit memory linear repetitive process of the form defined by Example 1.2.10. Then S is stable along the pass if*

$$r(H_1) < 1 \tag{5.40}$$

where H_1 is defined by (5.39).

Proof. Using (5.38) we have that

$$\begin{aligned}\det\left(I_{n+m} - z_1\widehat{A}_1 - z_2\widehat{A}_2\right) &= \det\left(I_{n+m} - z_1\widehat{A}_1\right)\\ &\times \det\left(I_{n+m} - \left(I_{n+m} - z_1\widehat{A}_1\right)^{-1} z_2\widehat{A}_2\right)\\ &\neq 0 \text{ in } \mathbb{U}_c^2\end{aligned}$$

Since $r(\widehat{A}_1) < 1$ by assumption, $\det\left(I_{n+m} - z_1\widehat{A}_1\right) \neq 0$ for $|z_1| \leq 1$. Hence to complete the proof of this result, it remains to show that

$$\det\left(I_{n+m} - \left(I_{n+m} - z_1\widehat{A}_1\right)^{-1} z_2\widehat{A}_2\right) \neq 0 \text{ in } \mathbb{U}_c^2 \tag{5.41}$$

A routine argument now shows that this last condition holds provided

$$r\left(\left(I_{n+m} - z_1\widehat{A}_1\right)^{-1} z_2\widehat{A}_2\right) < 1 \text{ in } \mathbb{U}_c^2$$

Use elementary properties of nonnegative matrices now yields

$$\begin{aligned}r\left(\left(I_{n+m} - z_1\widehat{A}_1\right)^{-1} z_2\widehat{A}_2\right) &= r\left(\widehat{G}(z_1)z_2\widehat{A}_2\right)\\ &\leq r\left(\left[\widehat{G}(z_1)z_2\widehat{A}_2\right]_p\right)\\ &\leq r(H_1) \text{ in } \mathbb{U}_c^2\end{aligned}$$

Hence (5.40) implies (5.41) and the proof is complete. □

Corollary 5.3.1. *The linear repetitive process S generated by a discrete unit memory linear repetitive process of the form defined by Example 1.2.10 is stable along the pass if*

$$r(H_2) < 1$$

where

$$H_2 := \sum_{i=0}^{\infty} \left[\widehat{A}_2^i \widehat{A}_1\right]_p \tag{5.42}$$

Note that since it has been assumed that $r(\widehat{A}_1) < 1$ and $r(\widehat{A}_2) < 1$, the convergence of the series in (5.39) and (5.42) is ensured. Hence they are easily computed. Also in the case when all elements in the matrices $\widehat{A}_1$ and $\widehat{A}_2$ are positive, it is easy to show that the matrices H_1 and H_2 are given by

$$H_1 = (I_{n+m} - \widehat{A}_1)^{-1}\widehat{A}_2$$

and

$$H_2 = (I_{n+m} - \widehat{A}_2)^{-1}\widehat{A}_1$$

respectively.

Consider now the case of structured perturbations of the form $[\Delta\hat{A}_i]_p \leq \hat{a}_i F_i$ to the matrices $\widehat{A}_i$, $i = 1, 2$. Here $\hat{a}_i$ is a positive real number, F_i is a nonnegative matrix and together they represent highly structured information relating to the perturbations of the entries in $\widehat{A}_i$.

The following result will be required in the proof of the next theorem.

Lemma 5.3.2. *Suppose that* $\left(I_{n+m} - z_1\widehat{A}_1 - z_2\widehat{A}_2\right)^{-1}$ *is expressed as*

$$\begin{aligned} \widehat{G}(z_1, z_2) &:= \left(I_{n+m} - z_1\widehat{A}_1 - z_2\widehat{A}_2\right)^{-1} \\ &= \sum_{i=0}^{\infty}\left(z_1\widehat{A}_1 + z_2\widehat{A}_2\right)^i \end{aligned}$$

Then

$$[\widehat{G}(z_1, z_2)]_p \leq I_{n+m} + \widehat{S}_1 + \widehat{S}_2 =: \widehat{S} \text{ in } \mathbb{U}_c^2 \tag{5.43}$$

where

$$\begin{aligned} \widehat{S}_1 &:= \left(I_{n+m} - \widehat{T}_1\widehat{T}_2\right)^{-1}\left(\widehat{T}_1 + \widehat{T}_1\widehat{T}_2\right) \\ \widehat{S}_2 &:= \left(I_{n+m} - \widehat{T}_2\widehat{T}_1\right)^{-1}\left(\widehat{T}_2 + \widehat{T}_2\widehat{T}_1\right) \end{aligned} \tag{5.44}$$

and

$$\widehat{T}_1 := \sum_{i=1}^{\infty}[\widehat{A}_1^i]_p, \quad \widehat{T}_2 := \sum_{i=1}^{\infty}[\widehat{A}_2^i]_p \tag{5.45}$$

Proof. After some routine analysis which is omitted here, we can write the Maclaurin expansion for $\widehat{G}(z_1, z_2)$ in the form

$$\widehat{G}(z_1, z_2) = I_{n+m} + \mathcal{S}_1 + \mathcal{S}_2$$

where

$$\mathcal{S}_1 := \sum_{i=0}^{\infty}\widehat{A}_1^i z_1^i(I_{n+m} + \mathcal{S}_2), \quad \mathcal{S}_2 := \sum_{i=0}^{\infty}\widehat{A}_2^i z_2^i(I_{n+m} + \mathcal{S}_1)$$

respectively. Also from these last two equations in turn it follows that for $(z_1, z_2) \in \mathbb{U}_c^2$, we have the following upper bounds for $\mathcal{S}_1$ and $\mathcal{S}_2$

$$[\mathcal{S}_1]_p \leq \widehat{T}_1([I_{n+m} + \mathcal{S}_2]_p) \leq \widehat{T}_1 + \widehat{T}_1[\mathcal{S}_2]_p$$

and

$$[\mathcal{S}_2]_p \leq \widehat{T}_2([I_{n+m} + \mathcal{S}_1]_p) \leq \widehat{T}_2 + \widehat{T}_2[\mathcal{S}_1]_p$$

respectively. Solving these last two inequalities immediately gives the expressions of (5.44) and completes the proof. □

Note also that since all elements in the matrix $\widehat{\mathcal{S}}$ of (5.43) are required to be positive it follows from (5.44) and (5.45) that $\widehat{\mathcal{S}}$ exists provided

$$r(\widehat{T}_1\widehat{T}_2) = r(\widehat{T}_2\widehat{T}_1) < 1$$

This last condition may also be regarded as another sufficient condition for stability along the pass.

Consider now the case when $[\widehat{A}_i]_p = \widehat{A}_i$, $i = 1, 2$, and $r(\widehat{A}_1 + \widehat{A}_2) < 1$. Then it is routine to show that

$$\widehat{\mathcal{S}} = \left(I_{n+m} - \widehat{A}_1 - \widehat{A}_2\right)^{-1}$$

Also when $[\widehat{A}_i]_p \neq \widehat{A}_i$, $i = 1, 2$, but $r\left([\widehat{A}_1]_p + [\widehat{A}_2]_p\right) < 1$ then $\widehat{\mathcal{S}}$ satisfies the upper bound

$$\widehat{\mathcal{S}} \leq \left(I_{n+m} - [\widehat{A}_1]_p - [\widehat{A}_2]_p\right)^{-1}$$

Now we have the following result, where we assume that the nominal process (i.e. no perturbations) is stable along the pass.

Theorem 5.3.2. *Consider the linear repetitive process S generated by a stable along the pass discrete unit memory linear repetitive process of the form defined by Example 1.2.10. Suppose also that the matrices $\widehat{A}_i$, associated with this case are subject to additive perturbations $\Delta\widehat{A}_i$, $i = 1, 2$, which satisfy $[\Delta\widehat{A}_i]_p \leq \hat{a}_i F_i$, where $\hat{a}_i$ is a positive real number and F_i is a nonnegative matrix. Then the following are sufficient conditions for stability along the pass of S generated by the perturbed process*

$$r\left(H_1 + (I_{n+m} + \widehat{T}_1)\sum_{i=1}^{2} \hat{a}_i F_i\right) < 1 \tag{5.46}$$

or

$$r\left(H_2 + (I_{n+m} + \widehat{T}_2)\sum_{i=1}^{2} \hat{a}_i F_i\right) < 1$$

Proof. The proofs of these two conditions follow the same basic steps and hence only the proof of (5.46) is given here.

First note that the perturbed process in this case is stable along the pass if, and only if,

$$\det\left(I_{n+m} - z_1(\widehat{A}_1 + \Delta\widehat{A}_1) - z_2(\widehat{A}_2 + \Delta\widehat{A}_2)\right) \neq 0 \text{ in } \mathbb{U}_c^2$$

or, since $r(\widehat{A}_1) < 1$,

$$\begin{aligned}\det(I_{n+m} &- (I_{n+m} - z_1\widehat{A})^{-1} z_2\widehat{A}_2 - \\ (I_{n+m} &- z_1\widehat{A}_1)^{-1}(z_1\Delta\widehat{A}_1 + z_2\Delta\widehat{A}_2)) \neq 0 \text{ in } \mathbb{U}_c^2\end{aligned}$$

This last condition holds if

$$r((I_{n+m} - z_1\widehat{A}_1)^{-1} z_2\widehat{A}_2 + (I_{n+m} - z_1\widehat{A}_1)^{-1}(z_1\Delta\widehat{A}_1 + z_2\Delta\widehat{A}_2)) < 1 \quad (5.47)$$

in $\mathbb{U}_c^2$, and, in turn, (5.47) holds if

$$\begin{aligned} r(\widehat{G}_1(z_1)z_2\widehat{A}_2 &+ \widehat{G}_1(z_1)(z_1\Delta\widehat{A}_1 + z_2\Delta\widehat{A}_2)) \leq r(H_1 + [\tilde{\Delta}]_p) \\ &\leq r\left(H_1 + (I_{n+m} + \widehat{T}_1)\sum_{i=1}^{2}\hat{a}_i F_i\right) \text{ in } \mathbb{U}_c^2 \end{aligned}$$

where

$$[\tilde{\Delta}]_p := [\widehat{G}_1(z_1)]_p\left([\Delta\widehat{A}_1]_p + [\Delta\widehat{A}_2]_p\right)$$

and the proof is complete. □

The following result gives sufficient conditions for stability along the pass of a discrete unit memory linear repetitive process of the form defined by Example 1.2.10 under unstructured perturbations of the form $||A_i|| \leq f_i$, $i = 1, 2$, where $f_i > 0$. Again we assume that the unperturbed process is stable along the pass.

Theorem 5.3.3. *Consider the linear repetitive process S generated by a stable along the pass discrete unit memory linear repetitive process of the form defined by Example 1.2.10. Suppose also that the matrices $\widehat{A}_i$ associated with this case are subject to unstructured perturbations of the form $||\Delta\widehat{A}_i|| \leq f_i$, $f_i > 0$, $i = 1, 2$. Then the following are sufficient conditions for stability along the pass of S generated by the perturbed process*

$$f_1 + f_2 < \frac{1 - \overline{\sigma}(H_1)}{\overline{\sigma}(I_{n+m} + \widehat{T}_1)} \quad (5.48)$$

or

$$f_1 + f_2 < \frac{1 - \overline{\sigma}(H_2)}{\overline{\sigma}(I_{n+m} + \widehat{T}_2)} \quad (5.49)$$

Proof. Again, the proofs for these two results follow identical steps and hence only that for (5.48) is given here. This relies on the fact that for a square matrix, say X, $r(X) \leq \overline{\sigma}(X)$. Hence we have the inequality

$$r(H_1 + (I_{n+m} + \widehat{T}_1)([\Delta\widehat{A}_1]_p + [\Delta\widehat{A}_2]_p)) \leq (f_1 + f_2)\overline{\sigma}(I_{n+m} + \widehat{T}_1) + \overline{\sigma}(H_1)$$

and the result follows immediately. □

The following sufficient conditions for stability along the pass apply to either type of perturbation (again assuming that the unperturbed process is stable along the pass).

Theorem 5.3.4. *Consider the linear repetitive process S generated by a stable along the pass discrete unit memory linear repetitive process of the form defined by Example 1.2.10. Suppose also that the matrices $\widehat{A}_i$, $i = 1, 2$ associated with this case are subject to either type of perturbation considered here. Then the following are sufficient conditions for stability along the pass of S generated by the perturbed process*

$$r\left(\widehat{\mathcal{S}}\sum_{i=1}^{2}\hat{a}_i F_i\right) < 1$$

for the structured perturbations and

$$f_1 + f_2 < \frac{1}{\overline{\sigma}(\widehat{\mathcal{S}})} \tag{5.50}$$

for the unstructured perturbations, where $\widehat{\mathcal{S}}$ is defined by (5.43).

Proof. If the nominal process is stable along the pass, it follows that the perturbed process will also be stable along the pass if, and only if,

$$\det(I_{n+m} - (I_{n+m} - z_1\widehat{A}_1 - z_2\widehat{A}_2)^{-1}(z_1\Delta\widehat{A}_1 + z_2\Delta\widehat{A}_2)) \neq 0 \text{ in } \mathbb{U}_c^2$$

in $\mathbb{U}_c^2$. The result now follows on using (5.43) and (5.44) and the proofs of the previous two theorems as appropriate. □

Note here that the conditions of (5.48) and (5.49) require that $\overline{\sigma}(H_1) < 1$ and $\overline{\sigma}(H_2) < 1$ respectively. This restriction is not required by (5.50) and hence it is less conservative than these two conditions. Also upper bounds for $\hat{a}_i$, $i = 1, 2$, can be computed from

$$\hat{a}_1\overline{\sigma}(\widehat{\mathcal{S}}F_1) + \hat{a}_2\overline{\sigma}(\widehat{\mathcal{S}}F_2) < 1$$

In computation terms, the conditions of Theorems 5.3.1–5.3.3 are easy to execute. For Theorem 5.3.1, the series (5.39) converges and hence the matrix

H_1 can be obtained by just summing a finite number of terms. With regard to Theorems 5.3.2 and 5.3.3 and the matrix $\widehat{\mathcal{S}}$, of Lemma 5.3.2, it is required to compute the matrices $\widehat{T}_1$ and $\widehat{T}_2$ which are summations of infinite series. Since it is assumed that $r(\widehat{A}_1) < 1$ and $r(\widehat{A}_2) < 1$, the convergence of the defining series is again assured, i.e. $\widehat{T}_1$ and $\widehat{T}_2$ can be obtained by summing a finite number of terms. Moreover, the rate of convergence of the infinite series here is related to $r(\widehat{A}_i)$, $i = 1, 2$ – the smaller these numbers the faster the corresponding convergence.

It is possible to give an alternative upper bound for the elements of $\widehat{G}(z_1, z_2)$ to that of (5.43) as shown next.

Lemma 5.3.3. *$\widehat{G}(z_1, z_2)$ satisfies the following inequality in $\mathbb{U}_c^2$*

$$[\widehat{G}(z_1, z_2)]_p \leq \widetilde{\mathcal{S}} := I_{n+m} + \widehat{T}_1 + \widehat{T}_2 + \widetilde{T}_1 + \widetilde{T}_2$$

where

$$\widetilde{T}_1 := \widetilde{B}^{-1}\widetilde{C}, \ \widetilde{T}_2 := \widetilde{D} + \widetilde{E}\widetilde{B}^{-1}\widetilde{C}$$

and

$$\begin{aligned}
\widetilde{B} &:= I_{n+m} - \overline{H}_1 - \overline{H}_1(I_{n+m} - \overline{H}_2)^{-1}\overline{H}_2 \\
\widetilde{C} &:= \overline{H}_1(I_{n+m} + \widehat{T}_1 + \widehat{T}_2) + \overline{H}_1(I_{n+m} - \overline{H}_2)^{-1} \\
&\times \overline{H}_2(I_{n+m} + \widehat{T}_1 + \widehat{T}_2) \\
\widetilde{D} &:= (I_{n+m} - \overline{H}_2)^{-1}\overline{H}_2(I_{n+m} + \widehat{T}_1 + \widehat{T}_2) \\
\widetilde{E} &:= (I_{n+m} - \overline{H}_2)^{-1}\overline{H}_2 \qquad (5.51) \\
\overline{H}_1 &:= \sum_{i=1}^{\infty}[\widehat{A}_1^i\widehat{A}_2]_p \\
\overline{H}_2 &:= \sum_{i=1}^{\infty}[\widehat{A}_2^i\widehat{A}_1]_p
\end{aligned}$$

and $\widehat{T}_1$ and $\widehat{T}_2$ are defined in (5.45).

Proof. The Maclaurin series for $\widehat{G}(z_1, z_2)$ can be written as follows

$$\widehat{G}(z_1, z_2) = I_{n+m} + \sum_{i=1}^{\infty}\widehat{A}_1^i z_1^i + \sum_{i=1}^{\infty}\widehat{A}_2^i z_2^i + J_1 + J_2$$

where (after some routine manipulations)

$$J_1 := \sum_{i=1}^{\infty}\widehat{A}_1^i\widehat{A}_2 z_1^i z_2(I_{n+m} + \widehat{\mathcal{S}}_1 + \widehat{\mathcal{S}}_2) \qquad (5.52)$$

and

$$J_2 := \sum_{i=1}^{\infty} \widehat{A}_2^i \widehat{A}_1 z_1 z_2^i (I_{n+m} + \widehat{S}_1 + \widehat{S}_2) \tag{5.53}$$

Hence

$$\widehat{S}_1 = \sum_{i=1}^{\infty} \widehat{A}_1^i z_1^i + J_1 \tag{5.54}$$

$$\widehat{S}_2 = \sum_{i=1}^{\infty} \widehat{A}_2^i z_2^i + J_2 \tag{5.55}$$

Substituting (5.54) and (5.55) into (5.52) and (5.53) now gives

$$J_1 = \sum_{i=1}^{\infty} \widehat{A}_1^i \widehat{A}_2 z_1^i z_2 \left(I_{n+m} + J_1 + J_2 + \sum_{i=1}^{\infty} \widehat{A}_1^i z_1^i + \sum_{i=1}^{\infty} \widehat{A}_2^i z_2^i \right)$$

and

$$J_2 = \sum_{i=1}^{\infty} \widehat{A}_2^i \widehat{A}_1 z_1 z_2^i \left(I_{n+m} + J_1 + J_2 + \sum_{i=1}^{\infty} \widehat{A}_1^i z_1^i + \sum_{i=1}^{\infty} \widehat{A}_2^i z_2^i \right)$$

Hence, as in the proof of Lemma 5.3.2, the following upper bounds for J_1 and J_2 can be obtained

$$[J_1]_p \le \overline{H}_1 \left(I_{n+m} + \widehat{T}_1 + \widehat{T}_2 + [J_1]_p + [J_2]_p \right) \tag{5.56}$$

and

$$[J_2]_p \le \overline{H}_2 \left(I_{n+m} + \widehat{T}_1 + \widehat{T}_2 + [J_1]_p + [J_2]_p \right) \tag{5.57}$$

where $\widehat{T}_i$, $i = 1, 2$, are defined in (5.45) and $\overline{H}_i$, $i = 1, 2$, in (5.51). Solving the inequalities (5.56) and (5.57) gives

$$[J_1]_p \le \widetilde{B}^{-1} \widetilde{C} =: \widetilde{T}_1$$

and

$$[J_2]_p \le \widetilde{D} + \widetilde{E} \widetilde{B}^{-1} \widetilde{C} =: \widetilde{T}_2$$

where the matrices on the right-hand sides of these last two equations are defined in (5.51) and the proof is complete. □

The matrix $\widetilde{S}$ exists provided

$$r \left(\overline{H}_1 + \overline{T}_2 \left(I_{n+m} - \overline{T}_2 \right)^{-1} \overline{T}_2 \right) < 1 \tag{5.58}$$

and (5.58) can be regarded as another sufficient condition for stability along the pass.

5.4 LMI Based Approaches

In this section, the following results, whose proofs are well known, are extensively used.

Lemma 5.4.1. *Given constant matrices W, L, V of appropriate dimensions where $W = W^T$ and $V = V^T > 0$, then*

$$W + L^T V L < 0$$

if, and only if,

$$\begin{bmatrix} W & L^T \\ L & -V^{-1} \end{bmatrix} < 0 \tag{5.59}$$

or, equivalently,

$$\begin{bmatrix} -V^{-1} & L \\ L^T & W \end{bmatrix} < 0 \tag{5.60}$$

The matrix (5.59) or (5.60) is known as the Schur's complement (see, for example, [25]) of V.

Lemma 5.4.2. *[98] For any matrix $\mathcal{F}$ such that $\mathcal{F}^T\mathcal{F} \leq I$ and scalar $\epsilon > 0$ the following holds*

$$\Sigma_1 \mathcal{F} \Sigma_2 + \Sigma_2^T \mathcal{F}^T \Sigma_1^T \leq \epsilon^{-1} \Sigma_1 \Sigma_1^T + \epsilon \Sigma_2^T \Sigma_2$$

5.4.1 Discrete Processes

Consider a process of the form defined by Example 1.2.10 and recall again the definitions of the matrices $\widehat{A}_1$ and $\widehat{A}_2$ of (2.66). Then we have the following sufficient condition for stability along the pass of such a process.

Theorem 5.4.1. *Consider the linear repetitive process S generated by a discrete unit memory linear repetitive process of the form defined by Example 1.2.10. Then S is stable along the pass if $\exists$ matrices $P > 0$ and $Q > 0$ satisfying the following LMI*

$$\begin{bmatrix} \widehat{A}_1^T P \widehat{A}_1 + Q - P & \widehat{A}_1^T P \widehat{A}_2 \\ \widehat{A}_2^T P \widehat{A}_1 & \widehat{A}_2^T P \widehat{A}_2 - Q \end{bmatrix} < 0 \tag{5.61}$$

Proof. First pre-multiply (5.61) by $\begin{bmatrix} z_1^* I_n & z_2^* I_m \end{bmatrix}$ and post-multiply it by $\begin{bmatrix} z_1 I_n & z_2 I_m \end{bmatrix}^T$ to yield

$$\left(z_1\widehat{A}_1 + z_2\widehat{A}_2\right)^* P \left(z_1\widehat{A}_1 + z_2\widehat{A}_2\right) + z_1^* Q z_1 - z_1^* P z_1 - z_2^* Q z_2 < 0$$

Also for this last equation to hold we must have that $Q - P < 0$ and hence in $\mathbb{U}_c^2$

$$\begin{aligned} z_1^* Q z_1 - z_1^* P z_1 - z_2^* Q z_2 &= z_1^* (Q - P) z_1 - z_2^* Q z_2 \\ &= -\left[z_1^* (P - Q) z_1 + z_2^* Q z_2 \right] \\ &\geq -[P - Q + Q] = -P \end{aligned}$$

Hence we require that $r(z_1 \widehat{A}_1 + z_2 \widehat{A}_2) < 1$ in $\mathbb{U}_c^2$, and this last condition in turn implies that

$$\det \left(I_{n+m} - z_1 \widehat{A}_1 - z_2 \widehat{A}_2 \right) \neq 0 \text{ in } \mathbb{U}_c^2$$

which is the condition for stability along the pass and the proof is complete. □

The following corollary to Theorem 5.4.1 links this LMI approach to a 2D Lyapunov inequality interpretation of stability along the pass, where the inequality condition arises directly from the 2D Lyapunov equation developed in Chap. 3 (Sect. 3.2.1) (and Φ is the augmented plant matrix of (3.4)).

Corollary 5.4.1. *Consider the linear repetitive process S generated by a discrete unit memory linear repetitive process of the form defined by Example 1.2.10. Then S is stable along the pass if $\exists$ matrices $W_1 > 0,\ W_2 > 0$ such that the 2D Lyapunov equation*

$$\Phi^T W \Phi - W < 0 \tag{5.62}$$

holds, where $W = W_1 \oplus W_2$.

Proof. If (5.62) holds then $\exists$ a sufficiently small scalar $\epsilon > 0$ such that

$$\Phi^T W \Phi - W + \epsilon I < 0$$

or, on substituting for Φ in terms of $\widehat{A}_1$ and $\widehat{A}_2$,

$$\widehat{A}_1^T W \widehat{A}_1 + \widehat{A}_2^T W \widehat{A}_2 - W + \epsilon I < 0$$

Now select Q and P as $Q = \widehat{A}_2^T W \widehat{A}_2 + \epsilon I$, and $P = W$ respectively. Then this last expression can be rewritten in the form

$$\begin{bmatrix} \widehat{A}_1^T P \widehat{A}_1 + Q - P & \widehat{A}_1^T P \widehat{A}_2 \\ \widehat{A}_2^T P \widehat{A}_1 & \widehat{A}_2^T P \widehat{A}_2 - Q \end{bmatrix} < 0$$

where we have also used the fact that $\widehat{A}_1^T W \widehat{A}_2 = 0$. □

Next, we use this last result to develop an LMI approach to stability analysis in the presence of uncertainty in the process definition.

Here we treat the case when the augmented plant matrix is subject to additive perturbations (as in Sect. 6.2) defined as follows

$$\Phi_d := \Phi + \Delta\Phi \tag{5.63}$$

where

$$\Delta\Phi := \begin{bmatrix} \Delta A & \Delta B_0 \\ \Delta C & \Delta D_0 \end{bmatrix} \tag{5.64}$$

Also we assume that the uncertainties considered here have the following typical norm bounded structure

$$\Delta\Phi = \begin{bmatrix} H_1 \\ H_2 \end{bmatrix} \mathcal{F} \begin{bmatrix} E_1 & E_2 \end{bmatrix} =: \begin{bmatrix} H_1 \\ H_2 \end{bmatrix} \mathcal{F} E \tag{5.65}$$

where the matrices on the right-hand side are of compatible dimensions and also $\mathcal{F}^T\mathcal{F} \leq I$.

Now introduce the following matrices

$$\widehat{\Delta A}_1 := \begin{bmatrix} \Delta A & \Delta B_0 \\ 0 & 0 \end{bmatrix}, \quad \widehat{\Delta A}_2 := \begin{bmatrix} 0 & 0 \\ \Delta C & \Delta D_0 \end{bmatrix}$$

Then we can write $\Delta\Phi$ in the form

$$\Delta\Phi = \widehat{\Delta A}_1 + \widehat{\Delta A}_2 = \widehat{H}_1 \mathcal{F} E + \widehat{H}_2 \mathcal{F} E$$

where

$$\widehat{H}_1 := \begin{bmatrix} H_1 \\ 0 \end{bmatrix}, \quad \widehat{H}_2 := \begin{bmatrix} 0 \\ H_2 \end{bmatrix}$$

The LMI sufficient condition for stability along the pass of Theorem 5.4.1 is equivalent to the existence of matrices $P > 0$ and $Q > 0$ such that

$$\widehat{A}^T P \widehat{A} + \widehat{Q} < 0$$

where

$$\widehat{A} := \begin{bmatrix} \widehat{A}_1 & \widehat{A}_2 \end{bmatrix}, \quad \widehat{Q} := \begin{bmatrix} Q - P & 0 \\ 0 & -Q \end{bmatrix}$$

Applying this theorem to the case when the uncertainty is of the form detailed above gives the following result.

Theorem 5.4.2. *Consider the linear repetitive process S generated by a discrete unit memory linear repetitive process of the form defined by Example 1.2.10 in the presence of an uncertainty structure satisfying (5.63)–(5.65). Then S is stable along the pass if $\exists$ matrices $P > 0$ and $Q > 0$ such that*

$$\left(\widehat{A} + \widehat{H}\widehat{\mathcal{F}}\widehat{E}\right)^T P \left(\widehat{A} + \widehat{H}\widehat{\mathcal{F}}\widehat{E}\right) + \widehat{Q} < 0 \tag{5.66}$$

where

$$\widehat{H} = \begin{bmatrix} \widehat{H}_1 & \widehat{H}_2 \end{bmatrix}, \quad \widehat{\mathcal{F}} := I_2 \otimes \mathcal{F}, \ \ \widehat{E} := I_2 \otimes E$$

(where as before $\otimes$ denotes the matrix Kronecker product).

Also it can be shown that, for any choice of Q, $\exists\ P > 0$ such that (5.66) holds if, and only if, $\exists$ a scalar $\epsilon > 0$ such that

$$\begin{bmatrix} -P^{-1} + \epsilon \widehat{H}\widehat{H}^T & \widehat{A} \\ \widehat{A}^T & \epsilon^{-1}\widehat{E}^T\widehat{E} + \widehat{Q} \end{bmatrix} < 0 \tag{5.67}$$

Now we have the following result which gives an LMI based sufficient condition for stability along the pass of a discrete linear repetitive process of the form considered here under the uncertainty structure defined above.

Theorem 5.4.3. *Consider the linear repetitive process S generated by a discrete unit memory linear repetitive process of the form defined by Example 1.2.10 in the presence of an uncertainty structure satisfying (5.63)–(5.65). Then S is stable along the pass if $\exists$ matrices $Y > 0$ and $Z > 0$ such that the following LMI holds*

$$\begin{bmatrix} -Y & \widehat{A}_1 Y & \widehat{A}_2 Y & \varepsilon\widehat{H}_1 & \varepsilon\widehat{H}_2 & 0 & 0 \\ Y\widehat{A}_1^T & Z - Y & 0 & 0 & 0 & YE^T & 0 \\ Y\widehat{A}_2^T & 0 & -Z & 0 & 0 & 0 & YE^T \\ \varepsilon\widehat{H}_1^T & 0 & 0 & -\varepsilon I & 0 & 0 & 0 \\ \varepsilon\widehat{H}_2^T & 0 & 0 & 0 & -\varepsilon I & 0 & 0 \\ 0 & EY & 0 & 0 & 0 & -\varepsilon I & 0 \\ 0 & 0 & EY & 0 & 0 & 0 & -\varepsilon I \end{bmatrix} < 0$$

Proof. First apply the Schur's complement formula (5.59) to (5.67) with

$$W = \begin{bmatrix} -P^{-1} & \widehat{A} \\ \widehat{A}^T & \widehat{Q} \end{bmatrix}, \quad V = \begin{bmatrix} \epsilon I & 0 \\ 0 & \epsilon^{-1} I \end{bmatrix}, \quad L = \begin{bmatrix} \widehat{H}^T & 0 \\ 0 & \widehat{E} \end{bmatrix}$$

Then the required result follows on pre- and post-multiplying the result of this last step by $\mathrm{diag}\{I, P^{-1}, P^{-1}, \epsilon I, \epsilon I, I, I\}$, and use of the substitutions $Y = P^{-1},\ \ Z = P^{-1}QP^{-1} > 0$. □

Next we give an example to illustrate the results just obtained.

Example 5.4.1. Consider the special case of Example 1.2.10 when

$$A = \begin{bmatrix} -0.1 & 0.2 & 0.1 \\ 0.1 & -0.2 & 0.2 \\ 0.2 & -0.3 & -0.2 \end{bmatrix}, \quad B_0 = \begin{bmatrix} -0.1 & -0.5 & -0.4 & 0.1 \\ -0.7 & 0.2 & -0.5 & 0.5 \\ -0.5 & 0.4 & -0.5 & -0.5 \end{bmatrix}$$

$$C = \begin{bmatrix} -0.05 & -0.05 & 0.2 \\ -0.3 & -0.2 & -0.2 \\ 0.2 & -0.1 & 0.2 \\ -0.1 & 0.1 & -0.1 \end{bmatrix}, \quad D_0 = \begin{bmatrix} -0.1 & 0.2 & 0.2 & 0.1 \\ 0.3 & 0.2 & -0.1 & 0.2 \\ 0.2 & 0.2 & -0.2 & 0.1 \\ 0.2 & 0 & -0.2 & -0.1 \end{bmatrix}$$

with uncertainty structure defined by

$$H_1 = \begin{bmatrix} 0.1 & 0.1 & 0.1 \\ 0.1 & 0.1 & 0.1 \\ 0.1 & 0.1 & 0.1 \end{bmatrix}, \quad H_2 = \begin{bmatrix} 0.2 & 0.2 & 0.2 \\ 0.2 & 0.2 & 0.2 \\ 0.2 & 0.2 & 0.2 \\ 0.2 & 0.2 & 0.2 \end{bmatrix}$$

and

$$E = \begin{bmatrix} -0.1 & -0.1 & -0.1 & -0.1 & -0.1 & -0.1 & -0.1 \\ -0.1 & -0.1 & -0.1 & -0.1 & -0.1 & -0.1 & -0.1 \\ -0.1 & -0.1 & -0.1 & -0.1 & -0.1 & -0.1 & -0.1 \end{bmatrix}$$

The LMI condition of Theorem 5.4.3 holds with matrices Z and Y having the following eigenvalues

$$1.5240, 1.4939, 1.7366, 1.8542, 2.1623, 0.7657, 0.7018$$

and

$$1.1935, 1.2172, 0.9060, 0.9650, 0.6988, 0.2722, 0.2147$$

respectively, and $\epsilon = 1.1703$. Hence the process is stable along the pass for all admissible uncertainties.

5.4.2 Differential Processes

Here we show that, unlike the other two considered in this chapter for discrete processes, the LMI approach extends naturally to the case of a differential unit memory linear repetitive process of the form defined by Example 1.2.3. The starting point here is the analysis of Chap. 4 which showed that a sufficient condition for stability along the pass is that the 2D Lyapunov inequality (obtained from the 2D Lyapunov equation of (4.18))

$$\Phi^T W^{1,0} + W^{1,0}\Phi + \Phi^T W^{0,1}\Phi - W^{0,1} = -Q < 0 \tag{5.68}$$

holds, where $W^{1,0} := W_1 \oplus 0_{m\times m} = \mathrm{diag}\{W_1, 0_{m\times m}\}$, $W^{0,1} := 0_{n\times n} \oplus W_2 = \mathrm{diag}\{0_{n\times n}, W_2\}$, $W_1 > 0$, $W_2 > 0$, and Φ is defined exactly as for discrete processes of the form defined by Example 1.2.10 (see (3.4)). To write this last condition in terms of an LMI, first note that (5.68) can be rewritten in the form

$$\widehat{A}_2^T \widehat{W}_2 \widehat{A}_2 - W^{0,1} + \widehat{A}_1^T W^{1,0} + W^{1,0}\widehat{A}_1 < 0 \tag{5.69}$$

where $\widehat{W}_2 := W_3 \oplus W_2$, the matrix $W_3 > 0$ is arbitrary, and the matrices $\widehat{A}_1$ and $\widehat{A}_2$ are defined exactly as in the case of discrete processes of the form defined in Example 1.2.10 (see (2.66)).

Now apply the Schur's complement formula (5.59) to (5.69) with $W = -W^{0,1} + \widehat{A}_1^T W^{1,0} + W^{1,0}\widehat{A}_1$, $L = \widehat{A}_2$, and $V = \widehat{W}_2$, and then pre- and post-multiply the result by $(I \oplus \widehat{W}_2)$ (which clearly does not affect the definiteness property of this matrix since by assumption $\widehat{W}_2 > 0$) to obtain the following equivalent condition in terms of an LMI

$$\begin{bmatrix} -W^{0,1} + \widehat{A}_1^T W^{1,0} + W^{1,0}\widehat{A}_1 & \widehat{A}_2^T \widehat{W}_2 \\ \widehat{W}_2 \widehat{A}_2 & -\widehat{W}_2 \end{bmatrix} < 0 \tag{5.70}$$

and hence we have established the following result.

Theorem 5.4.4. *Consider the linear repetitive process S generated by a differential unit memory linear repetitive process of the form defined by Example 1.2.3. Then S is stable along the pass if the LMI of (5.70) holds.*

Before giving a numerical example, it is of interest to note that (5.68) is a combination of the Lyapunov equations for the 1D differential case (i.e. $Q_1 := \widehat{A}_1^T W^{1,0} + W^{1,0}\widehat{A}_1$) and the 1D discrete case (i.e. $Q_2 := \widehat{A}_2^T \widehat{W}_2 \widehat{A}_2 - W^{0,1}$) but these cannot be split into two independent LMI conditions, i.e. $Q_1 + Q_2 < 0$ cannot be replaced by the two conditions $Q_1 \leq 0$ and $Q_2 < 0$. To see why $Q_1 < 0$ is not possible first note that

$$Q_1 = \widehat{A}_1^T W^{1,0} + W^{1,0}\widehat{A}_1 = \begin{bmatrix} A^T W_1 + W_1 A & W_1 B_0 \\ B_0^T W_1 & 0 \end{bmatrix}$$

and the zero block entry in the lower right corner means that $Q_1 < 0$ only holds under the assumption that $W_1 B_0 = 0$ which is clearly never going to be a feasible assumption for the underlying state-space model of the process dynamics. Also Q_2 can be rewritten as

$$Q_2 = \widehat{A}_2^T \widehat{W}_2 \widehat{A}_2 - W^{0,1} = \begin{bmatrix} C^T W_2 C & C^T W_2 D_0 \\ D_0^T W_2 C & D_0^T W_2 D_0 - W_2 \end{bmatrix} < 0$$

which can clearly never hold since $C^T W_2 C < 0$ is not possible with $W_2 > 0$. Hence, we can only work with LMI condition (5.70).

To illustrate the LMI solution of the 2D Lyapunov equation, we now give the following example.

Example 5.4.2. Consider the following special case of Example 1.2.3

$$\begin{aligned} \dot{x}_{k+1}(t) &= \begin{bmatrix} -0.2816 & -0.1082 & -0.0899 \\ -0.4585 & -0.4178 & -0.3160 \\ -0.1234 & -0.2776 & -0.3721 \end{bmatrix} x_{k+1}(t) \\ &+ \begin{bmatrix} -0.4281 & -0.3296 & -0.0872 \\ -0.2266 & -0.1015 & -0.1499 \\ -0.1668 & -0.3199 & -0.4697 \end{bmatrix} y_k(t) \end{aligned}$$

$$\begin{aligned} y_{k+1}(t) &= \begin{bmatrix} -0.0650 & -0.3013 & -0.3679 \\ -0.1946 & -0.1315 & -0.3798 \\ -0.3382 & -0.2288 & -0.1596 \end{bmatrix} x_{k+1}(t) \\ &+ \begin{bmatrix} -0.1096 & -0.4937 & -0.2752 \\ -0.2489 & -0.0506 & -0.3407 \\ -0.2941 & -0.1331 & -0.4178 \end{bmatrix} y_k(t) \end{aligned}$$

In this case the LMI of (5.70) holds when

$$W_1 = \begin{bmatrix} 22.3223 & 0.9343 & -1.1712 \\ 0.9343 & 10.1877 & 6.9433 \\ -1.1712 & 6.9433 & 15.9771 \end{bmatrix}, \quad W_2 = \begin{bmatrix} 18.3606 & 7.6396 & 1.4437 \\ 7.6396 & 16.6431 & 3.2674 \\ 1.4437 & 3.2674 & 20.8302 \end{bmatrix}$$

Next we use the LMI setting to study robust stability along the pass of the differential unit memory linear repetitive processes under consideration here. In particular, we consider two classes of uncertainty structure which are introduced in turn below.

(a) Here we assume that the uncertainty in the differential part (i.e. the along the pass dynamics) is of a polytopic form. In particular, we assume that all possible choices for the matrices which define the current pass state dynamics in the model of Example 1.2.3 can be expressed as

$$\begin{bmatrix} A & B_0 \end{bmatrix} \in \mathrm{Co} \begin{bmatrix} A^i & B_0^i \end{bmatrix}$$

where $i = 1, 2, \ldots, h$, h denotes the number of polytope vertices, and

$$\mathrm{Co} \begin{bmatrix} A^i & B_0^i \end{bmatrix} := \left\{ X : X = \sum_{i=1}^{h} \alpha_i \begin{bmatrix} A^i, & B_0^i \end{bmatrix}, \; \alpha_i \geq 0, \; \sum_{i=1}^{h} \alpha_i = 1 \right\}$$

For the current pass profile updating equation we assume a standard (in the 1D systems literature) norm bound on the perturbations, i.e.

$$y_{k+1}(t) = (C + \Delta C)x_{k+1}(t) + (D_0 + \Delta D_0)y_k(t)$$

where

$$\begin{bmatrix} \Delta C & \Delta D_0 \end{bmatrix} = H\mathcal{F} \begin{bmatrix} E_1 & E_2 \end{bmatrix}$$

and $\mathcal{F}$ satisfies the same norm bounded condition as for the discrete case, i.e. $\mathcal{F}^T\mathcal{F} \leq I$.

(b) Here we assume a norm bounded perturbation in both the current pass state and pass profile updating equations. In particular, the state vector updating equation assumes the form

$$\dot{x}_{k+1}(t) = (A + \Delta A)x_{k+1}(t) + (B_0 + \Delta B_0)y_k(t)$$

where

$$\begin{bmatrix} \Delta A & \Delta B_0 \\ \Delta C & \Delta D_0 \end{bmatrix} = \begin{bmatrix} H_1 \\ H_2 \end{bmatrix} \mathcal{F} \begin{bmatrix} E_1 & E_2 \end{bmatrix}$$

and again $\mathcal{F}^T\mathcal{F} \leq I$.

Uncertainties satisfying either of the above conditions are said to be admissible.

Consider now the case when uncertainty type (a) is present. Then a sufficient condition of the form (5.69) for stability along the pass is given by

$$\widehat{A}_1^{iT} W^{1,0} + W^{1,0} \widehat{A}_1^i - W^{0,1} + (\widehat{A}_2 + \widetilde{H}\widehat{F}\widetilde{E})^T \widehat{W}_2 (\widehat{A}_2 + \widetilde{H}\widehat{F}\widetilde{E}) < 0 \quad (5.71)$$

$\forall i = 1, 2, \ldots, h$, and

$$\widehat{A}_1^i := \begin{bmatrix} A^i & B_0^i \\ 0 & 0 \end{bmatrix}, \; \widetilde{H} := \begin{bmatrix} 0 & 0 \\ H & H \end{bmatrix}$$
$$\widehat{F} := \begin{bmatrix} \mathcal{F} & 0 \\ 0 & \mathcal{F} \end{bmatrix}, \; \widetilde{E} := \begin{bmatrix} E_1 & 0 \\ 0 & E_2 \end{bmatrix}$$

This follows immediately from a standard result which states that for the polytopic uncertainty considered here, a stability property holds provided it also holds for each 'vertex' model A^i, B_0^i, C, D_0, (For the background here see, for example, [75].)

This result is still difficult to apply since the underlying matrix inequality still contains an uncertainty matrix $\mathcal{F}$ and is not in exact LMI form. To avoid these problems we use the modified version of the equivalence between inequalities (5.71) given in the next result.

Lemma 5.4.3. *The conditions of (5.71) hold if, and only if, $\exists$ a real scalar $\epsilon > 0$ and matrices $W_1 > 0$, $W_2 > 0$ such that $\forall i = 1, 2, \ldots, h$,*

$$\begin{bmatrix} -\widehat{W}_2^{-1} + \epsilon \widetilde{H}\widetilde{H}^T & \widehat{A}_2 \\ \widehat{A}_2^T & \epsilon^{-1}\widetilde{E}^T\widetilde{E} + \widehat{A}_1^{iT} W^{1,0} + W^{1,0}\widehat{A}_1^i - W^{0,1} \end{bmatrix} < 0 \quad (5.72)$$

Now we have the following result which is in the desired LMI form.

Theorem 5.4.5. *Consider the linear repetitive process S generated by a differential unit memory linear repetitive process of the form defined by Example 1.2.3 in the presence of an uncertainty structure of the form of (a) above. Then S is stable along the pass if $\exists$ matrices $W_1 > 0$, $W_2 > 0$, a real scalar $\mu > 0$ such that the following LMI holds $\forall i = 1, 2, \ldots, h$*

$$\begin{bmatrix} -\widehat{W}_2 & \widehat{W}_2\widehat{A}_2 & \widehat{W}_2\widetilde{H} & 0 \\ \widehat{A}_2^T\widehat{W}_2 & \widehat{A}_1^{iT}W^{1,0} + W^{1,0}\widehat{A}_1^i - W^{0,1} & 0 & \mu\widetilde{E}^T \\ \widetilde{H}^T\widehat{W}_2 & 0 & -\mu I & 0 \\ 0 & \mu\widetilde{E} & 0 & -\mu I \end{bmatrix} < 0 \quad (5.73)$$

Proof. First apply the Schur's complement formula (5.59) to (5.72) with

$$W = \begin{bmatrix} -\widehat{W}_2^{-1} & \widehat{A}_2 \\ \widehat{A}_2^T & \widehat{A}_1^{iT}W^{1,0} + W^{1,0}\widehat{A}_1^i - W^{0,1} \end{bmatrix}$$
$$V = \begin{bmatrix} \epsilon I & 0 \\ 0 & \epsilon^{-1} I \end{bmatrix}, \; L = \begin{bmatrix} \widetilde{H}^T & 0 \\ 0 & \widetilde{E} \end{bmatrix}$$

to obtain

$$\begin{bmatrix} -\widehat{W}_2^{-1} & \widehat{A}_2 & \widetilde{H} & 0 \\ \widehat{A}_2^T & \widehat{A}_1^{iT} W^{1,0} + W^{1,0} \widehat{A}_1^i - W^{0,1} & 0 & \widetilde{E}^T \\ \widetilde{H}^T & 0 & -\epsilon^{-1} I & 0 \\ 0 & \widetilde{E} & 0 & -\epsilon I \end{bmatrix} < 0 \qquad (5.74)$$

Finally, pre- and post-multiply (5.74) by $\mathrm{diag}(\widehat{W_2}, I, I, \epsilon^{-1} I)$ and then introduce the change of variable $\epsilon^{-1} = \mu$ to obtain (5.73). □

To illustrate this last result, we now give the following example.

Example 5.4.3. Consider the following special case of Example 1.2.3

$$C = \begin{bmatrix} -0.34 & 0 \\ 0.1 & 0.8 \end{bmatrix}, \quad D_0 = \begin{bmatrix} 0.1 & 0 \\ 0 & -0.4 \end{bmatrix}$$

Suppose also that the uncertainty in the current pass state vector is defined by

$$S_i := \left[\begin{array}{c|c} A^i & B_0^i \end{array}\right], \; i = 1, 2, 3$$

with

$$\begin{aligned} S_1 &= \left[\begin{array}{cc|cc} -0.6543 & -0.2965 & -0.0727 & -0.0893 \\ -0.7161 & -0.9360 & 0.0357 & -0.2052 \end{array}\right] \\ S_2 &= \left[\begin{array}{cc|cc} -1.0591 & -0.3174 & 0.0324 & 0.1106 \\ -0.7423 & -0.7382 & 0.0573 & -0.1613 \end{array}\right] \\ S_3 &= \left[\begin{array}{cc|cc} -1.1930 & -0.8477 & 0.0318 & -0.1677 \\ -0.6706 & -0.7636 & 0.0473 & -0.1069 \end{array}\right] \end{aligned}$$

where S_1, S_2, and S_3 respectively denote the 3 vertices of the polytope determined by the triple $\left[\begin{array}{c|c} A^i & B_0^i \end{array}\right]$. Also

$$\begin{aligned} H &= \begin{bmatrix} 0.1471 \\ 0.0576 \end{bmatrix}, \quad E_1 = \begin{bmatrix} 0.0634 & 0.7228 \end{bmatrix} \\ E_2 &= \begin{bmatrix} 0.0337 & 0.6957 \end{bmatrix} \end{aligned}$$

For this data set, the LMI (5.73) holds with $\mu = 0.0068$ and

$$W_1 = \begin{bmatrix} 345.3073 & -214.3344 \\ -214.3344 & 518.7656 \end{bmatrix}, \quad W_2 = \begin{bmatrix} 241.8881 & -29.6818 \\ -29.6818 & 254.0108 \end{bmatrix}$$

To consider uncertainty type (b), we first assume that we can write

$$\widehat{\Delta A_1} = \begin{bmatrix} \Delta A & \Delta B_0 \\ 0 & 0 \end{bmatrix} = \begin{bmatrix} H_1 \\ 0 \end{bmatrix} \mathcal{F} \begin{bmatrix} E_1 & E_2 \end{bmatrix} =: \widehat{H}_1 \mathcal{F} E \qquad (5.75)$$

and

$$\widehat{\Delta A}_2 = \begin{bmatrix} 0 & 0 \\ \Delta C & \Delta D_0 \end{bmatrix} = \begin{bmatrix} 0 \\ H_2 \end{bmatrix} \mathcal{F} \begin{bmatrix} E_1 & E_2 \end{bmatrix} =: \widehat{H}_2 \mathcal{F} E \tag{5.76}$$

Then we have the following result.

Theorem 5.4.6. *Consider the linear repetitive process S generated by a differential unit memory linear repetitive process of the form defined by Example 1.2.3 in the presence of an uncertainty structure of the form of (b) above. Then S is stable along the pass if $\exists$ matrices $W_1 > 0,\ W_2 > 0,\ W_3 > 0$ a real scalar $\epsilon > 0$ such that the following LMI holds*

$$\begin{bmatrix} -W^{0,1} + \widehat{A}_1^T W^{1,0} + W^{1,0}\widehat{A}_1 + \epsilon E^T E & \widehat{A}_2^T \widehat{W}_2 & W^{1,0}\widehat{H}_1 \\ \widehat{W}_2 \widehat{A}_2 & -\widehat{W}_2 & \widehat{W}_2 \widehat{H}_2 \\ \widehat{H}_1^T W^{1,0} & \widehat{H}_2^T \widehat{W}_2 & -\epsilon I \end{bmatrix} < 0$$

Proof. It follows immediately from Theorem 5.4.4 that stability along the pass holds in this case if

$$\begin{bmatrix} -W^{0,1} + \widehat{A}_1^T W^{1,0} + W^{1,0}\widehat{A}_1 & \widehat{A}_2^T \widehat{W}_2 \\ \widehat{W}_2 \widehat{A}_2 & -\widehat{W}_2 \end{bmatrix} + \begin{bmatrix} \widehat{\Delta A}_1^T W^{1,0} + W^{1,0}\widehat{\Delta A}_1 & \widehat{\Delta A}_2^T \widehat{W}_2 \\ \widehat{W}_2 \widehat{\Delta A}_2 & 0 \end{bmatrix} < 0$$

Also rewrite the second term in this last expression as

$$\begin{bmatrix} \widehat{\Delta A}_1^T W^{1,0} + W^{1,0}\widehat{\Delta A}_1 & \widehat{\Delta A}_2^T \widehat{W}_2 \\ \widehat{W}_2 \widehat{\Delta A}_2 & 0 \end{bmatrix} = \begin{bmatrix} \widehat{\Delta A}_1^T W^{1,0} & \widehat{\Delta A}_2^T \widehat{W}_2 \\ 0 & 0 \end{bmatrix} + \begin{bmatrix} W^{1,0}\widehat{\Delta A}_1 & 0 \\ \widehat{W}_2 \widehat{\Delta A}_2 & 0 \end{bmatrix}$$

and hence

$$\begin{aligned} &\begin{bmatrix} \widehat{\Delta A}_1^T W^{1,0} + W^{1,0}\widehat{\Delta A}_1 & \widehat{\Delta A}_2^T \widehat{W}_2 \\ \widehat{W}_2 \widehat{\Delta A}_2 & 0 \end{bmatrix} \\ &= \begin{bmatrix} E^T \mathcal{F}^T \widehat{H}_1^T W^{1,0} & E^T \mathcal{F}^T \widehat{H}_2^T \widehat{W}_2 \\ 0 & 0 \end{bmatrix} + \begin{bmatrix} W^{1,0}\widehat{H}_1 \mathcal{F} E & 0 \\ \widehat{W}_2 \widehat{H}_2 \mathcal{F} E & 0 \end{bmatrix} \\ &= \begin{bmatrix} E^T \\ 0 \end{bmatrix} \mathcal{F}^T \begin{bmatrix} \widehat{H}_1^T W^{1,0} & \widehat{H}_2^T \widehat{W}_2 \end{bmatrix} + \begin{bmatrix} W^{1,0}\widehat{H}_1 \\ \widehat{W}_2 \widehat{H}_2 \end{bmatrix} \mathcal{F} \begin{bmatrix} E & 0 \end{bmatrix} \end{aligned}$$

Now introduce

$$\Sigma_1 := \begin{bmatrix} W^{1,0}\widehat{H}_1 \\ \widehat{W}_2 \widehat{H}_2 \end{bmatrix}, \Sigma_2 := \begin{bmatrix} E & 0 \end{bmatrix}$$

and hence

$$\begin{bmatrix} \widehat{\Delta A}_1^T W^{1,0} + W^{1,0}\widehat{\Delta A}_1 & \widehat{\Delta A}_2^T \widehat{W}_2 \\ \widehat{W}_2 \widehat{\Delta A}_2 & 0 \end{bmatrix} = \Sigma_1 \mathcal{F} \Sigma_2 + \Sigma_2^T \mathcal{F}^T \Sigma_1^T$$

Application of the result of Lemma 5.4.2 now yields

$$\begin{bmatrix} \widehat{\Delta A}_1^T W^{1,0} + W^{1,0}\widehat{\Delta A}_1 & \widehat{\Delta A}_2^T \widehat{W}_2 \\ \widehat{W}_2 \widehat{\Delta A}_2 & 0 \end{bmatrix} \leq \epsilon^{-1}\Sigma_1 \Sigma_1^T + \epsilon \Sigma_2^T \Sigma_2$$

where

$$\epsilon^{-1}\Sigma_1\Sigma_1^T = \epsilon^{-1}\begin{bmatrix} W^{1,0}\widehat{H}_1 \\ \widehat{W}_2 \widehat{H}_2 \end{bmatrix} \begin{bmatrix} \widehat{H}_1^T W^{1,0} & \widehat{H}_2^T \widehat{W}_2 \end{bmatrix}$$

and

$$\epsilon \Sigma_2^T \Sigma_2 = \epsilon \begin{bmatrix} E^T \\ 0 \end{bmatrix} \begin{bmatrix} E & 0 \end{bmatrix} = \begin{bmatrix} \epsilon E^T E & 0 \\ 0 & 0 \end{bmatrix}$$

Hence

$$\begin{bmatrix} \widehat{\Delta A}_1^T W^{1,0} + W^{1,0}\widehat{\Delta A}_1 & \widehat{\Delta A}_2^T \widehat{W}_2 \\ \widehat{W}_2 \widehat{\Delta A}_2 & 0 \end{bmatrix} \leq \epsilon^{-1}\begin{bmatrix} W^{1,0}\widehat{H}_1 \\ \widehat{W}_2 \widehat{H}_2 \end{bmatrix}$$
$$\times \begin{bmatrix} \widehat{H}_1^T W^{1,0} & \widehat{H}_2^T \widehat{W}_2 \end{bmatrix} + \begin{bmatrix} \epsilon E^T E & 0 \\ 0 & 0 \end{bmatrix}$$

and therefore

$$\begin{bmatrix} -W^{0,1} + \widehat{A}_1^T W^{1,0} + W^{1,0}\widehat{A}_1 & \widehat{A}_2^T \widehat{W}_2 \\ \widehat{W}_2 \widehat{A}_2 & -\widehat{W}_2 \end{bmatrix}$$
$$+ \begin{bmatrix} \widehat{\Delta A}_1^T W^{1,0} + W^{1,0}\widehat{\Delta A}_1 & \widehat{\Delta A}_2^T \widehat{W}_2 \\ \widehat{W}_2 \widehat{\Delta A}_2 & 0 \end{bmatrix}$$
$$\leq \begin{bmatrix} -W^{0,1} + \widehat{A}_1^T W^{1,0} + W^{1,0}\widehat{A}_1 + \epsilon E^T E & \widehat{A}_2^T \widehat{W}_2 \\ \widehat{W}_2 \widehat{A}_2 & -\widehat{W}_2 \end{bmatrix}$$
$$+\epsilon^{-1}\begin{bmatrix} W^{1,0}\widehat{H}_1 \\ \widehat{W}_2 \widehat{H}_2 \end{bmatrix} \begin{bmatrix} \widehat{H}_1^T W^{1,0} & \widehat{H}_2^T \widehat{W}_2 \end{bmatrix}$$

Stability along the pass now holds if

$$\begin{bmatrix} -W^{0,1} + \widehat{A}_1^T W^{1,0} + W^{1,0}\widehat{A}_1 + \epsilon E^T E & \widehat{A}_2^T \widehat{W}_2 \\ \widehat{W}_2 \widehat{A}_2 & -\widehat{W}_2 \end{bmatrix}$$
$$+\epsilon^{-1}\begin{bmatrix} W^{1,0}\widehat{H}_1 \\ \widehat{W}_2 \widehat{H}_2 \end{bmatrix} \begin{bmatrix} \widehat{H}_1^T W^{1,0} & \widehat{H}_2^T \widehat{W}_2 \end{bmatrix} < 0$$

An obvious application of the Schur's complement formula now yields the result. □

To illustrate this last result, we now give the following example.

Example 5.4.4. Consider the following special case of Example 1.2.3

$$A = \begin{bmatrix} -0.0920 & -0.0024 \\ -0.0021 & -0.0910 \end{bmatrix}, \quad B_0 = \begin{bmatrix} 0.0164 & 0.0800 \\ 0.0831 & 0.0144 \end{bmatrix}$$

$$C = \begin{bmatrix} -0.0100 & 0 \\ 0 & 0.2000 \end{bmatrix}, \quad D_0 = \begin{bmatrix} -0.1000 & 0 \\ 0 & -0.1000 \end{bmatrix}$$

when the uncertainty structure (5.75)–(5.76) is specified by

$$H_1 = \begin{bmatrix} 0.2779 \\ 0.3722 \end{bmatrix}, \quad H_2 = \begin{bmatrix} 0.1705 \\ 0.3644 \end{bmatrix}$$
$$E_1 = \begin{bmatrix} 0.0332 & 0.0404 \end{bmatrix}, \quad E_2 = \begin{bmatrix} 0.0397 & 0.0944 \end{bmatrix}$$

One solution of the LMI of Theorem 5.4.6 is

$$W_1 = \begin{bmatrix} 1.1780 & -0.4271 \\ -0.4271 & 0.9405 \end{bmatrix}, \quad W_2 = \begin{bmatrix} 0.9435 & -0.0322 \\ -0.0322 & 0.8396 \end{bmatrix}$$

$$W_3 = \begin{bmatrix} 0.9598 & 0 \\ 0 & 0.9598 \end{bmatrix}, \quad \epsilon_1 = 1.1253, \; \epsilon_2 = 0.9958$$

and hence this example is stable along the pass for all admissible uncertainties.

6 Controllability, Observability, Poles and Zeros

Summary. In this chapter, we first investigate controllability and observability of discrete and differential linear repetitive processes. As with the 2D linear systems case, it will be shown that more than one physically relevant controllability property can be defined in the discrete case and also that clear differences exist with 2D discrete linear systems. Observability is only addressed for differential processes since for the discrete case this property is almost identical to that for 2D discrete linear systems described by the Roesser or Fornasini-Marchesini state-space models. The next topic covered in this chapter is the extension of the 1D theory of system equivalence to linear repetitive processes using the 2D transfer-function matrix descriptions of Chap. 1. Finally, a behavioral approach is used to define poles and zeros for these processes which, in contrast to those in the 2D/nD systems literature, have interpretations in terms of the existence of exponential trajectories in the dynamics, i.e. a natural generalization of the well known 1D linear systems case.

6.1 Controllability For Discrete Processes

6.1.1 2D Discrete Linear Systems

It is instructive to first briefly consider the 2D discrete linear systems case before proceeding with the analysis of this section. Here we mostly follow [103] (and the relevant cited references in this paper) and note first that the (most) general Fornasini-Marchesini state-space model has the form

$$\begin{aligned} Ex(i+1,j+1) &= Fx(i+1,j)+Gx(i,j+1) \\ &+ Hx(i,j)+B_1u(i+1,j) \\ &+ B_2u(i,j+1)+B_0u(i,j) \\ y(i,j) &= Cx(i,j)+Du(i,j) \end{aligned} \tag{6.1}$$

where $u(i,j)$ is the input vector, $y(i,j)$ is the output vector and $x(i,j)$ is the $\widetilde{n}\times 1$ local semi-state vector. The corresponding Roesser model is

$$\begin{aligned} \begin{bmatrix} E_1 & E_2 \\ E_3 & E_4 \end{bmatrix} \begin{bmatrix} x_h(i+1,j) \\ x_v(i,j+1) \end{bmatrix} &= \begin{bmatrix} A_1 & A_2 \\ A_3 & A_4 \end{bmatrix} \begin{bmatrix} x_h(i,j) \\ x_v(i,j) \end{bmatrix} \\ &+ \begin{bmatrix} B_1 \\ B_2 \end{bmatrix} u(i,j) \\ y(i,j) &= \begin{bmatrix} C_1 & C_2 \end{bmatrix} \begin{bmatrix} x_h(i,j) \\ x_v(i,j) \end{bmatrix} + Du(i,j) \end{aligned} \tag{6.2}$$

or, more compactly,

$$\begin{aligned} E\tilde{x}(i,j) &= Ax(i,j) + Bu(i,j) \\ y(i,j) &= Cx(i,j) + Du(i,j) \end{aligned} \tag{6.3}$$

where

$$\tilde{x}(i,j) := \begin{bmatrix} x_h(i+1,j) \\ x_v(i,j+1) \end{bmatrix}, \quad x(i,j) := \begin{bmatrix} x_h(i,j) \\ x_v(i,j) \end{bmatrix}$$

Here the $n_1 \times 1$ vector $x_h(i,j)$ is termed the horizontal semi-state, the $n_2 \times 1$ vector $x_v(i,j)$ the vertical semi-state, and the $\widetilde{n} \times 1$ vector $\tilde{x}(i,j)$ with $\widetilde{n} := n_1 + n_2$ the local semi-state.

If $E \neq I$, the above models are termed implicit 2D systems and here we term (6.1) the implicit Fornasini-Marchesini model and (6.2) (or (6.3)) the implicit Roesser model. Also if E is non-square or $\det E = 0$, we term these models singular 2D linear systems. Moreover, there are several ways of showing that the implicit Roesser and Fornasini-Marchesini models are equivalent. For example, in the case of the Roesser model define

$$\tilde{E}_1 := \begin{bmatrix} E_1 & 0 \\ E_2 & 0 \end{bmatrix}, \quad \tilde{E}_2 := \begin{bmatrix} 0 & E_3 \\ 0 & E_4 \end{bmatrix}$$

Then (6.2) can be written as

$$\tilde{E}_1 x(i+1,j) + \tilde{E}_2 x(i,j+1) = Ax(i,j) + Bu(i,j)$$

i.e. a special case of (6.1).

There are several ways to specify the boundary conditions and the region of interest for models such as (6.1) and (6.2). Here, however, we need only consider the case when the boundary conditions are specified along the i and j axes, where for the Fornasini-Marchesini model this means that $x(i,0)$, $i \geq 0$ and $x(0,j)$, $j \geq 0$ are known, and for the Roesser model that $x_h(i,0)$, $i \geq 0$, and $x_v(0,j)$, $j \geq 0$ are known. Then, if the region of interest is the rectangle $[0, N_1] \times [0, N_2]$ in the (i,j)-plane, we are concerned with finding the so-called 'forward solution'. If, however, the boundary conditions are specified along the upper and right-hand sides of the rectangle, i.e. along (i, N_2) and (N_1, j), then the so-called 'backward solution' on $[0, N_1] \times [0, N_2]$ can be constructed. Note also that if $\det E = 0$ then, as for 1D singular linear systems, not all boundary conditions are allowed – for the existence of a solution to the 2D implicit model they must lie in an admissible subspace (see also the discussion after (6.6) below).

Note 6.1.1. In the remainder of this sub-section we mostly follow [103] (and the relevant references cited in this paper) and we use the notation of this paper for the 2D z-transform, despite the fact that it clashes with that for discrete linear repetitive processes used in this monograph.

Suppose now, for simplicity, that $D = 0$, $B_1 = 0$, $B_2 = 0$, and $B_0 = B$ in (6.1). Suppose also that the boundary conditions are given along the i and j axes and define the 2D z-transform in terms of negative powers of z_i, e.g.

$$X(z_1, z_2) := \sum_{i=0}^{\infty}\sum_{j=0}^{\infty} x(i,j) z_1^{-i} z_2^{-j}$$

Then applying this transform to (6.1) yields

$$\begin{aligned} P_{FM}(z_1,z_2)X(z_1,z_2) &= BU(z_1,z_2) + z_2(z_1E - G)X(z_1,0) \\ &+ z_1(z_2E - F)X(0,z_2) - z_1z_2EX(0,0) \\ Y(z_1,z_2) &= CX(z_1,z_2) \end{aligned} \tag{6.4}$$

where

$$P_{FM}(z_1,z_2) := (z_1z_2E - z_1F - z_2G - H)$$

and to (6.3)

$$\begin{aligned} (EZ - A)X(z_1,z_2) &= BU(z_1,z_2) + z_2\tilde{E}^1X_v(z_1,0) \\ &+ z_1\tilde{E}^2X_h(0,z_2) \\ Y(z_1,z_2) &= CX(z_1,z_2) \end{aligned} \tag{6.5}$$

where

$$\begin{aligned} X_v(z_1,0) &:= \sum_{i=0}^{\infty} x(i,0)z_1^{-i} \\ X_h(0,z_2) &:= \sum_{j=0}^{\infty} x(0,j)z_2^{-j} \end{aligned}$$

and

$$\tilde{E}^1 := \begin{bmatrix} E_1 \\ E_2 \end{bmatrix}, \ \tilde{E}^2 := \begin{bmatrix} E_3 \\ E_4 \end{bmatrix}, \ Z := \begin{bmatrix} z_1I_{n_1} & 0 \\ 0 & z_2I_{n_2} \end{bmatrix}$$

The so-called characteristic matrix or pencil for the FM model considered here is $P_{FM}(z_1,z_2)$ as defined above and the corresponding matrix or pencil for the Roesser model is

$$P_R(z_1, z_2) := EZ - A$$

For simplicity, assume now that the boundary conditions in each case are zero. Also denote the so-called generic 2D pencil as $P(z_1, z_2)$. Then with zero boundary conditions (6.4) and (6.5) can be written generically as

$$\begin{aligned} P(z_1, z_2)X(z_1, z_2) &= BU(z_1, z_2) \\ Y(z_1, z_2) &= CX(z_1, z_2) \end{aligned} \tag{6.6}$$

Also we say that there exists a solution (with zero boundary conditions) to the implicit Fornasini-Marchesini or Roesser model if (6.6), with the corresponding $P(z_1, z_2)$, has a solution $X(z_1, z_2)$ for some $U(z_1, z_2)$.

Suppose now that E is singular. Then (as noted above) not all inputs are allowed since (6.6) implies that some relationships exist between the boundary conditions of $X(z_1, z_2)$ and the coefficients of $U(z_1, z_2)$. For example, consider the Fornasini-Marchesini model with $\widetilde{n} = 1$ and $E = F = H = B_1 = B_2 = 0$ and $G = B_0 = 1$, and take the region of interest to be $[0, 1] \times [0, 1]$. Then by (6.6) (corresponding to zero boundary conditions) it is required that $u(0, 0) = -x(0, 1) = 0$. Those inputs which satisfy (6.6) for some semi-state under zero boundary conditions are termed admissible.

Use of standard manipulation tools over the field of rational $\widetilde{n} \times 1$ vector functions in two indeterminates immediately yields the following result where the concept of rank used is the usual (normal) rank over the rational functions.

Theorem 6.1.1. *A solution to the 2D implicit model (for either model with the appropriate choice of $P(z_1, z_2)$) exists $\forall$ admissible $u(i, j)$ if ,and only if,*

$$\text{rank}\begin{bmatrix} P(z_1, z_2) \\ C \end{bmatrix} = \text{rank } P(z_1, z_2)$$

Also if a solution $x(i, j)$ exists then it is unique if, and only if,

$$\text{rank} P(z_1, z_2) = \widetilde{n}$$

These results are not contingent on E being a square matrix but if this is true then in all cases the conditions in this last theorem are guaranteed by the regularity condition

$$\Delta(z_1, z_2) := \det P(z_1, z_2) \neq 0$$

for some (z_1, z_2). In the case of the Fornasini-Marchesini model (which is used subsequently here), the regularity condition is

$$\det(z_1 z_2 E - z_1 F - z_2 G - H) \neq 0$$

and numerical algorithms for testing a given example for regularity are available (when $P(z_1, z_2)$ is regular) – see, for example, [20].

The Laurent expansion about infinity of the resolvent matrix may be written as

$$P^{-1}(z_1, z_2) = z_1^{-1} z_2^{-1} \sum_{i=-\mu_1}^{\infty} \sum_{j=-\mu_2}^{\infty} T_{i,j} z_1^{-i} z_2^{-j} \tag{6.7}$$

In the case when $\det E \neq 0$, $\mu_1 = 0$ and $\mu_2 = 0$ and hence $T_{i,j}$ reduces to the usual 2D transition matrix. Suppose, however, that

$$E = \begin{bmatrix} 1 & 1 \\ 0 & 0 \end{bmatrix}, \quad A = \begin{bmatrix} 0 & 0 \\ -1 & -1 \end{bmatrix}$$

then

$$(EZ - A)^{-1} = \frac{1}{(z_1 - z_2)} \begin{bmatrix} 1 & -z_2 \\ -1 & z_1 \end{bmatrix}$$

and this Roesser model example shows that, in general, μ_1 and μ_2 are not finite in (6.7).

It can be shown that a necessary and sufficient condition for the existence of (6.7) with finite μ_1 and μ_2 is that the characteristic polynomial $\Delta(z_1, z_2)$ satisfies

$$\deg \Delta(z_1, z_2) = \deg_{z_1} \Delta(z_1, z_2) + \deg_{z_2} \Delta(z_1, z_2) \tag{6.8}$$

where the degree of $\Delta(z_1, z_2)$ is defined as the degree in z of $\Delta(z, z)$ and $\deg_{z_i}(z_1, z_2)$ denotes the degree in z_i, $i = 1, 2$. This property requires that the highest power in z_1 and z_2 occurs in the same term.

Singular Roesser/Fornasini-Marchesini models for 2D linear systems which have this last property are termed principal. Suppose also that in the (i, j)-plane we let w at (i, j) denote the occurrence in $\Delta(z_1, z_2)$ of the term $z_1^i z_2^j$. Also we use the term region of support to denote the collection of all such w for $\Delta(z_1, z_2)$. Then (6.8) is equivalent to the requirement that the region of support should have a corner on the line $i + j = \deg\Delta(z_1, z_2)$ of maximum degree in both variables z_1 and z_2. Roesser models satisfying (6.8) are also termed principal.

If (6.7) holds for finite μ_1, μ_2, then we define the minimum value of (μ_1, μ_2) for which it holds as the 2D index of the system. In this case $\{T_{i,j}\}$ is unique and is termed the fundamental matrix sequence, or transition matrix sequence, for the 2D implicit system and recursive formulas for computing this matrix are known for both models. Next, we obtain the transition matrix sequence for a discrete unit memory linear repetitive process of the form defined by Example 1.2.10 and then proceed to define and characterize so-called local reachability and controllability.

6.1.2 The Transition Matrix Sequence for Discrete Linear Repetitive Processes

Following [68], consider again a discrete unit memory linear repetitive process of the form defined by Example 1.2.10 expressed in terms of its singular Fornasini-Marchesini model interpretation (1.38), i.e.

$$E\mathcal{X}(k+1,p+1) = A_8\mathcal{X}(k+1,p) + A_9\mathcal{X}(k,p) + B_5u(k+1,p) \tag{6.9}$$

where $\mathcal{X}(k,p)$ is the augmented state vector, i.e.

$$\mathcal{X}(k,p) := \begin{bmatrix} x_k^T(p) & y_k^T(p) \end{bmatrix}^T$$

and the matrices E, A_8, A_9, and B_5 are defined by (1.39). Then using available results for the singular 2D Fornasini-Marchesini model of the previous subsection, we have that the elements of $\{T_{i,j}\}$ are given by

$$ET_{i,j} = \begin{cases} A_9T_{-1,-1} + A_8T_{0,-1} + I_{n+m}, & i = j = 0 \\ A_9T_{i-1,j-1} + A_8T_{i,j-1}, & i \neq 0 \text{ and/or } j \neq 0 \end{cases} \tag{6.10}$$

$i \geq -\mu_1$, $j \geq -\mu_2$. Also, exploiting the structure of A_8 and A_9 in this particular case, it is appropriate to partition $T_{i,j}$ as follows

$$T_{i,j} = \begin{bmatrix} T_{i,j}^{11} & T_{i,j}^{12} \\ T_{i,j}^{21} & T_{i,j}^{22} \end{bmatrix}$$

where $T_{i,j}^{11}$ has dimension $n \times n$. Then the block form of (6.10) for $i = j = 0$ is

$$\left.\begin{aligned} T_{0,0}^{11} &= AT_{0,-1}^{11} + B_0T_{-1,-1}^{21} + I_n \\ T_{0,0}^{12} &= AT_{0,-1}^{12} + B_0T_{-1,-1}^{22} \\ 0 &= CT_{0,-1}^{11} - T_{0,-1}^{21} + D_0T_{-1,-1}^{21} \\ 0 &= CT_{0,-1}^{12} - T_{0,-1}^{22} + D_0T_{-1,-1}^{22} + I_m \end{aligned}\right\} \tag{6.11}$$

and for $i \neq 0$ and/or $j \neq 0$

$$\left.\begin{aligned} T_{i,j}^{1h} &= B_0T_{i-1,j-1}^{2h} + AT_{i,j-1}^{1h} \\ 0 &= D_0T_{i-1,j-1}^{2h} + CT_{i,j-1}^{1h} - T_{i,j-1}^{2h} \end{aligned}\right\}, \; h = 1, 2 \tag{6.12}$$

Note that in each set of four equations defined by (6.11) and (6.12), only two give an explicit recurrence, i.e. an explicit formula for $T_{i,j}$. The other two in each set are implicit and link block entries in $T_{i,j-1}$ with those for $i-1$ and/or $j-1$. Hence when dealing with $(i,j) = (0,0)$, for example, it will also be necessary to consider $(i,j) = (0,-1), (-1,0)$ and $(-1,-1)$ individually.

Now we must compute the index (μ_1, μ_2) for this case, where this is equivalent to finding the integers $\mu_1 \geq 0$, $\mu_2 \geq 0$ such that $T_{i,j} = 0$ when $i < -\mu_1$ and/or $j < -\mu_2$, and $T_{i,j} \neq 0$ otherwise. These indices are finite as a consequence of the fact that the underlying 2D systems equivalent model representation is principal. Next, we show that

$$\mu_1 = 0, \; \mu_2 = 1 \tag{6.13}$$

and also

$$T_{0,-1} = \begin{bmatrix} 0 & 0 \\ 0 & I_m \end{bmatrix} \tag{6.14}$$

and

$$\left.\begin{array}{l} T_{-\hat{\delta},0} = 0, \ \hat{\delta} \geq 1 \\ T_{-\hat{\delta},-\gamma} = 0, \ \hat{\delta}, \ \gamma \geq 1 \\ T_{-\hat{\delta},\gamma} = 0, \ \hat{\delta} \geq 1, \ \gamma > 0 \\ T_{\hat{\delta},-\gamma} = 0, \ \gamma > 1, \ \hat{\delta} > 0 \end{array}\right\} \quad (6.15)$$

To prove (6.13)–(6.15), the route is to proceed by an induction argument and show that these expressions solve the defining equations for all feasible choices of i and j in (6.11) and (6.12). To begin, consider these expressions for $(i,j) = (0,0), \ (0,-1), \ (-1,0)$ and $(-1,-1)$. Then since $T^{1h}_{0,-1} = 0, \ T^{2h}_{-1,-1} = 0, \ h = 1,2$, we have that

$$\left.\begin{array}{l} T^{11}_{0,0} = I_n \\ T^{12}_{0,0} = 0 \\ T^{21}_{0,-1} = 0 \\ T^{22}_{0,-1} = I_m \end{array}\right\} \quad (6.16)$$

Also, since $T^{1h}_{i,j} = 0, \ h = 1,2, \ i \neq 0$ and/or $j \neq 0, \ T_{-1,-1} = 0$. Moreover, $T_{0,-2} = T_{-1,-2}, \ T_{-2,-1} = T_{-2,-2} = 0$.

Use of the second pair of equations for $T_{-1,1}$ now yields $T_{-1,0} = T_{-2,0} = 0$ and a simple inductive argument, based on the second pair of equations for $T_{-h,1}, \ h \geq 1$, yields $T_{-h,0} = 0, \ h \geq 1$. Employing the same general argument for $T_{i,j}$ with i and/or $j < -1$ now establishes the four terms given in (6.15). Hence the index in this case is given by (6.13) and the matrix $T_{0,-1}$ defined by (6.14) can be viewed as the initial matrix for the general recurrence of (6.12).

Using the second pair of equations for $T_{i,0}, \ i \geq 0$, immediately yields

$$T_{i,-1} = \begin{bmatrix} 0 & 0 \\ 0 & D_0^i \end{bmatrix}$$

Also, before obtaining the formula for the general element $T_{i,j}, \ i,j \geq 1$, it remains to construct $T_{0,0}, \ T_{i,0}, \ i > 0$ and $T_{0,j}, \ j > 0$. In the case of $T_{0,0}$ it remains, noting (6.16), to construct $T^{2h}_{0,0}, \ h = 1,2$. These follow immediately from the second pair of equations for $T_{0,1}$ as $T^{21}_{0,0} = CT^{11}_{0,0}, \ T^{22}_{0,0} = 0$, i.e.

$$T_{0,0} = \begin{bmatrix} I_n & 0 \\ C & 0 \end{bmatrix}$$

In the case of $T_{i,0}, \ i \geq 1$, the first pair of equations for $T_{i,0}$ and the second pair of equations for $T_{i,j}$ immediatcly yield

$$\left.\begin{array}{l} T^{11}_{i,0} = 0 \\ T^{12}_{i,0} = B_0 D_0^{i-1} \\ T^{21}_{i,0} = D_0 T^{21}_{i-1,0} = D_0^i C \\ T^{22}_{i,0} = CT^{12}_{i,0} + D_0 T^{22}_{i-1,0} = CB_0 D_0^{i-1} + D_0 T^{22}_{i-1,0} \end{array}\right\}$$

By way of illustration, consider the cases when $i = 1, 2$ and 3 respectively. Then

$$\left.\begin{aligned} T^{22}_{1,0} &= CB_0 \\ T^{22}_{2,0} &= CB_0D_0 + D_0CB_0 \\ T^{22}_{3,0} &= CB_0D_0^2 + D_0CB_0D_0 + D_0^2CB_0 \end{aligned}\right\}$$

To obtain $T_{0,j}$, $j \geq 1$, the route is to consider the first pair of equations for $T_{0,j}$ and the second pair of equations for $T_{0,j+1}$. Hence

$$T_{0,j} = \begin{bmatrix} C & 0 \\ CA^j & 0 \end{bmatrix}$$

At this stage, it is possible to compute the general element $T_{i,j}$, $i, j \geq 1$. This involves solving the first pair of equations for $T_{i,j}$ and the second pair of equations for $T_{i,j+1}$. The resulting recursive formulae are

$$\left.\begin{aligned} T^{1h}_{i,j} &= AT^{1h}_{i,j-1} + B_0T^{2h}_{i-1,j-1} \\ T^{2h}_{i,j} &= CT^{1h}_{i,j} + D_0T^{2h}_{i-1,j} \end{aligned}\right\}$$

At this stage, consider again the general singular 2D Fornasini-Marchesini state-space model (6.1). Then the key difference between the transition matrix sequence for this case and that for a discrete unit memory linear repetitive process of the form defined by Example 1.2.10 is that in the latter the support of the defining relations must be extended by adding the value -1 to the index j (the along the pass direction). Next we use this transition matrix sequence to develop a general response formula for the repetitive process case as an essential step before beginning local reachability/controllability analysis.

6.1.3 The General Response Formula

Again following [68], suppose that a discrete unit memory linear repetitive process of the form defined by Example 1.2.10 has a known input sequence applied to it and the boundary conditions are also known. Equivalently, consider the singular 2D Fornasini-Marchesini state-space model representation (6.9) under these conditions. Then the following result uses this latter representation to develop a general response formula for this case, i.e. an explicit formula for computing the resulting pass profile sequence.

Theorem 6.1.2. *Consider a discrete unit memory linear repetitive process of the form defined by Example 1.2.10 when both the control input sequence and the boundary conditions are known. Then the general response formula in this case is given by*

$$
\begin{aligned}
\mathcal{X}(k,p) &= \sum_{i=0}^{k-1}\sum_{j=-1}^{p-1} T_{i,j}B_5u(k-i,p-j-1)+\sum_{i=0}^{k-1}T_{i,p-1}A_8\mathcal{X}(k-i,0)\\
&+ \sum_{i=0}^{k-2}T_{i,p-1}A_9\mathcal{X}(k-i-1,0)+\sum_{j=-1}^{p-2}T_{k-1,j}A_9\mathcal{X}(0,p-j-1)\\
&+ T_{k-1,p-1}A_9\mathcal{X}(0,0)
\end{aligned}
\tag{6.17}
$$

Proof. This is by induction, based on the assumption that (6.17) is valid for (k,p), $(k,p+1)$ and $(k+1,p)$ and then showing that this implies it is also true for $(k+1,p+1)$. We begin by computing the augmented state vector at $(k+1,p+1)$, i.e. $\mathcal{X}(k+1,p+1)$, under these assumptions, where we have

$$
\begin{aligned}
x_{k+1}(p+1) &= Ax_{k+1}(p)+Bu_{k+1}(p)+B_0y_k(p)\\
y_{k+1}(p+1) &= Cx_{k+1}(p+1)+Du_{k+1}(p+1)+D_0y_k(p+1)\\
&= CAx_{k+1}(p)+CBu_{k+1}(p)+Du_{k+1}(p+1)\\
&+ CB_0y_k(p)+D_0y_k(p+1)
\end{aligned}
$$

Hence

$$
\begin{aligned}
\mathcal{X}(k+1,p+1) &= \begin{bmatrix} A & 0\\ CA & 0\end{bmatrix}\mathcal{X}(k+1,p)+\begin{bmatrix} 0 & 0\\ 0 & D_0\end{bmatrix}\mathcal{X}(k,p+1)\\
&+ \begin{bmatrix} 0 & B_0\\ 0 & CB_0\end{bmatrix}\mathcal{X}(k,p)+\begin{bmatrix} B\\ CB\end{bmatrix}u(k+1,p)\\
&+ \begin{bmatrix} 0\\ D\end{bmatrix}u(k+1,p+1)
\end{aligned}
$$

(where $u(k+1,p) := u_{k+1}(p)$ and $u(k+1,p+1) := u_{k+1}(p+1)$). Evaluating the general response formula for (k,p), $(k+1,p)$ and $(k,p+1)$ respectively and substituting the results into this last expression now gives

$$
\begin{aligned}
\mathcal{X}(k+1,p+1) &= \begin{bmatrix} A & 0\\ CA & 0\end{bmatrix}\\
&\times \left\{\sum_{i=0}^{k}\sum_{j=-1}^{p-1} T_{i,j}B_5u(k+1-i,p-j-1)\right.\\
&+ \sum_{i=0}^{k}T_{i,p-1}A_8\mathcal{X}(k+1-i,0)+\sum_{i=0}^{k-1}T_{i,p-1}A_9\mathcal{X}(k-i,0)\\
&+ \left.\sum_{j=-1}^{p-1}T_{k,j}A_9\mathcal{X}(0,p-j-1)\right\}
\end{aligned}
$$

$$
\begin{aligned}
&+ \begin{bmatrix} 0 & 0 \\ 0 & D_0 \end{bmatrix} \times \left\{ \sum_{i=0}^{k-1} \sum_{j=-1}^{p} T_{i,j} B_5 u(k-i, p-j) \right. \\
&+ \sum_{i=0}^{k-1} T_{i,p} A_8 \mathcal{X}(k-i, 0) + \sum_{i=0}^{k-2} T_{i,p} A_9 \mathcal{X}(k-i-1, 0) \\
&+ \left. \sum_{j=-1}^{p} T_{k-1,j} A_9 \mathcal{X}(0, p-j) \right\} \\
&+ \begin{bmatrix} 0 & B_0 \\ 0 & CB_0 \end{bmatrix} \left\{ \sum_{i=0}^{k-1} \sum_{j=-1}^{p-1} T_{i,j} B_5 u(k-i, p-j-1) \right. \\
&+ \sum_{i=0}^{k-1} T_{i,p-1} A_8 \mathcal{X}(k-i, 0) + \sum_{i=0}^{k-2} T_{i,p-1} A_9 \mathcal{X}(k-i-1, 0) \\
&+ \left. \sum_{j=-1}^{p-1} T_{k-1,j} A_9 \mathcal{X}(0, p-j-1) \right\} \\
&+ \begin{bmatrix} B \\ CB \end{bmatrix} u(k+1, p) + \begin{bmatrix} 0 \\ D \end{bmatrix} u(k+1, p+1)
\end{aligned} \tag{6.18}
$$

Let $W(k+1,p+1)$ denote the right hand side of (6.18). Then extensive, but routine and hence omitted here, computations show that

$$
\begin{aligned}
W(k+1,p+1) &= \sum_{i=0}^{k} \sum_{j=-1}^{p} T_{i,j} B_5 u(k+1-i, p-j) \\
&+ \sum_{i=0}^{k} T_{i,p} A_9 \mathcal{X}(k+1-i, 0)
\end{aligned}
$$

or, equivalently,

$$
\begin{aligned}
W(k+1,p+1) &= \sum_{i=0}^{k-1} T_{i,p} A_9 \mathcal{X}(k-i, 0) + \sum_{i=0}^{p-1} T_{k,p} A_9 \mathcal{X}(0, p-j) \\
&+ T_{k,p} A_9 \mathcal{X}(0,0)
\end{aligned}
$$

i.e. $W(k+1,p+1) = \mathcal{X}(k+1,p+1)$ as required. □

The following example illustrates the use of Theorem 6.1.2.

Example 6.1.1. Suppose that it is required to compute $\mathcal{X}(2,2)$ for a given process. Then use of the formula (6.17) gives

$$
\begin{aligned}
\mathcal{X}(2,2) &= T_{0,-1}B_5u(2,2) + T_{0,0}B_5u(2,1) + T_{0,1}B_5u(2,0) \\
&+ T_{1,-1}B_5u(1,2) + T_{1,0}B_5u(1,1) + T_{1,1}B_5u(1,0) \\
&+ T_{0,1}A_8\mathcal{X}(2,0) + T_{1,1}A_8\mathcal{X}(1,0) + T_{0,1}A_9\mathcal{X}(1,0) \\
&+ T_{1,-1}A_9\mathcal{X}(0,2) + T_{1,0}A_9\mathcal{X}(0,1) + T_{1,1}A_9\mathcal{X}(0,0)
\end{aligned}
$$

Note 6.1.2. In the proof of Theorem 6.1.2, we have, in effect, transformed the singular model of (6.1) to a particular case of the nonsingular general 2D Fornasini-Marchesini model in the form

$$
\begin{aligned}
\mathcal{X}(k+1,p+1) &= A_{10}\mathcal{X}(k+1,p) + A_{11}\mathcal{X}(k,p+1) \\
&+ A_{12}\mathcal{X}(k,p) + B_6u(k+1,p+1) + B_7u(k+1,p)
\end{aligned}
$$

where

$$
\begin{aligned}
A_{10} &:= \begin{bmatrix} A & 0 \\ CA & 0 \end{bmatrix}, \ A_{11} := \begin{bmatrix} 0 & 0 \\ 0 & D_0 \end{bmatrix} \\
A_{12} &:= \begin{bmatrix} 0 & B_0 \\ 0 & CB_0 \end{bmatrix}, \ B_6 := \begin{bmatrix} 0 \\ D \end{bmatrix}, \ B_7 := \begin{bmatrix} B \\ CB \end{bmatrix}
\end{aligned}
$$

This analysis clearly demonstrates that the discrete unit memory linear repetitive process state-space model used in singular form above is principal and can be transformed to nonsingular form.

6.1.4 Local Reachability/Controllability

Here we use the transition matrix sequence (and Theorem 6.1.2) to define and characterize so-called local reachability and controllability of the dynamics of a discrete unit memory linear repetitive process of the form defined by Example 1.2.10. In the analysis which follows, the following partial ordering of two-tuple integers will be used

$$
\left.
\begin{array}{l}
(i,j) \le (k,p), \text{ if } i \le k \text{ and } j \le p \\
(i,j) = (k,p), \text{ if } i = k, \ j = p \\
(i,j) < (k,p), \text{ if } (i,j) \le (k,p) \text{ and } (i,j) \ne (k,p)
\end{array}
\right\}
$$

The dynamics of the processes considered here evolve over

$$
D_e := \{(k,p) : k \ge 0, \ 0 \le p \le \alpha\}
$$

but in practice only a finite number of passes, say K, will actually be completed. Hence a natural way to define reachability/controllability properties for them is to ask if it is possible to 'reach' all possible vectors in the rectangle whose boundary in the pass-to-pass direction is defined by $0 \le k \le K$ and in the along-the-pass direction by $0 \le p \le \alpha$. The analysis which follows formalizes this intuitive idea in terms of so-called local reachability and controllability properties, where, for $(a, b) < (c, d)$, the rectangle $[(a, b), (c, d)]$ is defined as follows

$$[(a, b), (c, d)] := \{(a, b) \leq (i, j) \leq (c, d)\}$$

In general, local reachability and controllability are quite distinct concepts and are defined as follows.

Definition 6.1.1. *A discrete unit memory linear repetitive process of the form defined by Example 1.2.10 is said to be locally reachable in the rectangle*

$$[(0, 0), (f, h)],\ 0 \leq f \leq K,\ 0 \leq h \leq \alpha$$

if $\forall z_r \in \mathbb{R}^{n+m} \exists$ *a sequence of control input vectors* $u(i, j)$ *on* $(0, 0) \leq (i, j) \leq (f, h)$ *such that* $\mathcal{X}(0, 0) = 0$ *and* $\mathcal{X}(f, h) = z_r$ *in the singular Fornasini-Marchesini state-space model interpretation of the dynamics defined by (6.9).*

Definition 6.1.2. *A discrete unit memory linear repetitive process of the form defined by Example 1.2.10 is said to be locally controllable in the rectangle*

$$[(0, 0), (f, h)],\ 0 \leq f \leq K,\ 0 \leq h \leq \alpha$$

if $\forall z_c \in \mathbb{R}^{n+m} \exists$ *a sequence of control vectors* $u(i, j)$ *on* $(0, 0) \leq (i, j) \leq (f, h)$ *such that* $\mathcal{X}(0, 0) = z_c$ *and* $\mathcal{X}(f, h) = 0$ *in the singular Fornasini-Marchesini state-space model interpretation of the dynamics defined by (6.9).*

The following results now characterize local reachability and controllability respectively in terms of conditions on the rank of a matrix with constant entries.

Theorem 6.1.3. *A discrete unit memory linear repetitive process of the form defined by Example 1.2.10 is locally reachable in* $[(0, 0), (f, h)]$ *if, and only if,*

$$\text{rank } R_{fh} = n + m$$

where

$$\begin{aligned} R_{fh} := & \left[\begin{matrix} M_{0,-1} & M_{0,0} & \cdots & M_{0,h-1} & M_{1,-1} & M_{1,0} \end{matrix} \right. \\ & \left. \begin{matrix} \cdots & M_{1,h-1} & \cdots & \cdots & M_{f-1,-1} & M_{f-1,0} & \cdots & M_{f-1,h-1} \end{matrix} \right] \end{aligned} \tag{6.19}$$

and

$$M_{ij} := T_{ij} B_5,\ 0 \leq i \leq f - 1,\ -1 \leq j \leq h - 1$$

Proof. Since $\mathcal{X}(f, h) = z_r$, we can write (using the result of Theorem 6.1.2)

$$\begin{aligned} z_r &- \sum_{i=0}^{k-1} T_{i,p-1}A_8\mathcal{X}(k-i,0) - \sum_{i=0}^{k-2} T_{i,p-1}A_9\mathcal{X}(k-i-1,0) \\ &- \sum_{j=-1}^{p-2} T_{k-1,j}A_9\mathcal{X}(0,p-j-1) - T_{k-1,p-1}A_9\mathcal{X}(0,0) \\ &= R_{fh}U_{fh} \end{aligned} \tag{6.20}$$

where

$$U_{fh} := \begin{bmatrix} u^T(f,h) & \cdots, & u^T(f,0) & u^T(f-1,h) \\ \cdots\ \cdots & u^T(f-1,0) & \cdots & u^T(1,h) & \cdots & u^T(1,0) \end{bmatrix}^T$$

Using (6.19), the result follows immediately. □

Theorem 6.1.4. *A discrete unit memory linear repetitive process of the form defined by Example 1.2.10 is locally controllable in* $[(0,0),(f,h)]$ *if, and only if,*

$$\text{rank } R_{fh} = \text{rank} \begin{bmatrix} R_{fh} & P_{fh} \end{bmatrix}$$

where R_{fh} *is again defined by (6.19) and*

$$P_{fh} := \begin{bmatrix} \widetilde{P}_0 & \widetilde{P}_{1,1} & \cdots & \widetilde{P}_{1,f-1} & \cdots & \widetilde{P}_{2,-1} & \widetilde{P}_{2,0} & \cdots & \widetilde{P}_{2,h-1} \end{bmatrix}$$

with

$$\begin{aligned} \widetilde{P}_0 &:= T_{0,h-1}A_8 \\ \widetilde{P}_{1,i} &:= T_{i,h-1}A_8 + T_{i-1,h-1}A_9,\ i = 1,2,\cdots,f-1 \\ \widetilde{P}_{2,j} &:= T_{f-1,j}A_9,\ j = -1,0,\cdots,h-1 \end{aligned}$$

Proof. Using (6.19) and (6.20) and setting $\mathcal{X}(f,h) = 0$ yields

$$R_{fh} = -P_{fh}X_{fh}$$

where

$$X_{fh} := \begin{bmatrix} \mathcal{X}^T(f,0) & \cdots & \mathcal{X}^T(1,0) & \mathcal{X}^T(0,h) & \cdots & \cdots & \mathcal{X}^T(0,0) \end{bmatrix}^T$$

□

It is clear from these last two results that local reachability implies locally controllability, but the converse is not necessarily true. Hence it is of interest to establish exactly when local controllability implies local reachability. This problem has been considered elsewhere [93] for the Fornasini-Marchesini state-space model and the results obtained generalize in a natural manner. Hence the details are omitted here.

In common with other classes of 2D linear systems, at least one other distinct definition of reachability/controllability for a discrete unit memory linear repetitive process of the form defined by Example 1.2.10 can be formulated (see also the next section). For example, other work has highlighted the need for so-called simultaneous, or pass, reachability/controllability – termed pass profile controllability here. The definition is as follows.

Definition 6.1.3. *A discrete unit memory linear repetitive process of the form defined by Example 1.2.10 is said to be pass profile controllable if* $\exists$ *a pass number* K^* *and control input vectors* $u_{k+1}(p)$, $0 \leq p \leq \alpha$, $0 \leq k \leq K^*$, *which will drive the process from any pre-specified initial pass profile to any pre-specified pass profile on pass* K^*.

Consider now the problem of characterizing this property using the equivalent 2D linear systems state-space models. Then, in fact, it is not possible to completely characterize this property using this general approach. The reason for this is explained next.

Pass controllability is easily seen to play the same role for discrete unit memory linear repetitive processes as so-called global controllability for 2D linear systems described by the Roesser or Fornasini-Marchesini state-space models. In particular, global controllability in these cases is expressed in terms of the global state vector with entries along the entire separation set, but, in the case of discrete unit (and non-unit) memory linear repetitive processes, this collection of local state vector entries is finite by definition. As shown next, however, the approach used above can be used to completely characterize the weaker property of so-called simultaneous local reachability.

Definition 6.1.4. *A discrete unit memory linear repetitive process of the form defined by Example 1.2.10 is said to be simultaneously locally reachable at all points along a given pass* $\widetilde{K} > 0$ *where* $\widetilde{K}$ *is an arbitrarily chosen pass number, if the process is locally reachable in every rectangle of the set*

$$\Sigma := \left\{ \; [(0,0),(\widetilde{K},1)] \quad [(0,0),(\widetilde{K},2)] \quad \cdots \quad [(0,0),(\widetilde{K},\alpha)] \; \right\}$$

It is easy to see that for any valid index (i,j) the matrix R_{ij} is a sub-matrix of R_{ij+1} and this fact establishes the following result.

Theorem 6.1.5. *Suppose a discrete unit memory linear repetitive process of the form defined by Example 1.2.10 is locally reachable in the rectangle*

$$[(0,0),(i,j)], \; 0 \leq i \leq \widetilde{K}-1, \; 0 \leq j \leq \alpha - 1$$

Then this property also holds for the rectangles $[(0,0),(i,j+1)]$ *and* $[(0,0),(i+1,j)]$.

Corollary 6.1.1. *A discrete unit memory linear repetitive process of the form defined by Example 1.2.10 is locally reachable in every rectangle of the set*

$$\{\ [(0,0),(\widetilde{K},1)]\quad [(0,0),(\widetilde{K},2)]\quad \cdots\quad [(0,0),(\widetilde{K},\alpha)]\ \}$$

if, and only if, it is locally reachable in the rectangle $[(0,0),(\widetilde{K},1)]$.

Alternatively, use of the transition matrix sequence yields the following result.

Theorem 6.1.6. *A discrete unit memory linear repetitive process of the form defined by Example 1.2.10 is locally reachable in all the rectangles of pass* $\widetilde{K}$ *if, and only if,*

$$\text{rank} \begin{bmatrix} D_0^{\widetilde{K}-1}D & \cdots & D \end{bmatrix} = m$$

and

$$\text{rank} \begin{bmatrix} B_0 D_0^{\widetilde{K}-2}D & \cdots & B_0 D_0 & B \end{bmatrix} = n$$

Proof. This follows immediately since $R_{\widetilde{K}1}$ has the structure

$$R_{\widetilde{K}1} = \begin{bmatrix} R_{\widetilde{K}0} & Y \end{bmatrix}$$

where

$$Y := \begin{bmatrix} B_0 D_0^{\widetilde{K}-2}D & \cdots & B_0 D & B \\ * & \cdots & * & * \end{bmatrix}$$

$$R_{\widetilde{K}0} := \begin{bmatrix} 0 & 0 & \cdots & 0 \\ D_0^{\widetilde{K}-1}D & D_0^{\widetilde{K}-2}D & \cdots & D \end{bmatrix}$$

and $*$ denotes a compatibly dimensioned sub-matrix whose structure is not relevant here.

6.1.5 Controllability of Discrete Processes with Dynamic Boundary Conditions

In this sub-section we extend the pass profile controllability analysis to the case of a discrete unit memory linear repetitive process with dynamic boundary conditions of the form defined by Example 1.2.13, where we follow [156].

The starting point is the 1D equivalent model of the dynamics of such a process when the pass state initial vector sequence is of the most general form given in (1.35), i.e.

$$x_{k+1}(0) = d_{k+1} + \sum_{j=0}^{\alpha-1} J_j y_k(j),\ k \geq 0 \tag{6.21}$$

Here the matrix $\mathfrak{B}_7$ in the 1D equivalent model (1.47)–(1.49) has a somewhat more complex structure than its counterpart $\mathfrak{B}_1$ in the case of Example 1.2.10. In fact, however, it is easy to see that pass profile controllability here is equivalent to controllability of the 1D equivalent model, i.e. to controllability of the pair $\{\mathfrak{B}_7, \mathfrak{B}_2\}$ in (1.47)–(1.49). This is stated formally as the following result.

Theorem 6.1.7. *A discrete unit memory linear repetitive process of the form defined by the state-space model of Example 1.2.13 and pass state initial vector sequence (6.21) is pass profile controllable if, and only if,*

$$\text{rank} \begin{bmatrix} zI - \mathfrak{B}_7 & \mathfrak{B}_2 \end{bmatrix} = m\alpha, \ \forall\, z \in \mathbb{C} \tag{6.22}$$

The condition of (6.22) can be decomposed into a simpler structure equivalent form by introducing the following matrices

$$\begin{aligned}
\Omega_0 &:= \begin{bmatrix} CJ_1 & CJ_2 & \cdots & CJ_{\alpha-1} & D \end{bmatrix} \\
\Omega_1 &:= \begin{bmatrix} C(B_0 + AJ_0) & CAJ_2 & \cdots & CAJ_{\alpha-1} & CB & D \end{bmatrix} \\
&\ \ \vdots \quad \vdots \\
\Omega_{\alpha-1} &:= \begin{bmatrix} CA^{\alpha-2}(B_0 + AJ_0) & \cdots & C(B_0 + A^{\alpha-2}J_{\alpha-2}) \\ CA^{\alpha-2}B \quad CA^{\alpha-3}B \quad CA^{\alpha-4}B & \cdots & CAB \quad CB \quad D \end{bmatrix}
\end{aligned}$$

Then we have the following result.

Corollary 6.1.2. *A discrete unit memory linear repetitive process of the form defined by the state-space model of Example 1.2.13 and pass state initial vector sequence (6.21) is pass profile controllable if, and only if,* $\forall\, j = 0, 1, \cdots, \alpha - 1$

$$\text{rank} \begin{bmatrix} \zeta I_m - (D_0 + CA^j J_j) & \Omega_j \end{bmatrix} = m, \ \forall\, \zeta \in \mathbb{C}$$

Proof. This follows immediately from an obvious decomposition of (6.22). □

Now we have the following result.

Theorem 6.1.8. *A discrete unit memory linear repetitive process of the form defined by the state-space model of Example 1.2.13 and pass state initial vector sequence (6.21) is pass profile controllable if the following conditions hold* $\forall\, \zeta \in \mathbb{C}$

$$\begin{aligned}
\text{rank} \begin{bmatrix} \zeta I_m - (D_0 + CK_0) & D \end{bmatrix} &= m \\
\text{rank} \begin{bmatrix} \zeta I_m - (D_0 + CAK_1) & D \end{bmatrix} &= m \\
&\ \ \vdots \\
\text{rank} \begin{bmatrix} \zeta I_m - (D_0 + CA^{\alpha-1}K_{\alpha-1}) & D \end{bmatrix} &= m
\end{aligned}$$

Proof. Simply note that the matrix $\mathfrak{B}_7$ in the 1D equivalent model has a lower block triangular structure in this case. □

From the above analysis, we immediately see the essential difference between pass profile controllability of processes with boundary conditions of the form of Examples 1.2.10 and 1.2.13 respectively. In the former case, it follows immediately that the pass profile controllability horizon, i.e. the number of passes $\widetilde{K}$ required to attain the prescribed pass profile, can be equal to m – the dimension of the pass profile vector. This is due to the block triangular structure of the matrices which define this property for these processes. In the latter case, however, the pass profile controllability horizon must generally be assumed to be equal to the full (state) dimension of the 1D equivalent model, i.e. $m\alpha$.

Next we give a simple example to highlight the critical role of the boundary conditions on the pass profile controllability properties of discrete linear repetitive processes. (This again (recall the stability analysis of Chap. 2) highlights the critical influence of the structure of the pass state initial vector sequence on the systems theoretic and applications properties of linear repetitive process.)

Example 6.1.2. Consider the case when $\alpha = 2$ and

$$C = \begin{bmatrix} 1 & 1 \end{bmatrix}, \quad D = \begin{bmatrix} 1 \\ 2 \end{bmatrix}, \quad D_0 = \begin{bmatrix} 0 & 1 \\ 1 & 0 \end{bmatrix}$$

This particular case is pass profile controllable when the boundary conditions are of the simple form of Example 1.2.10, but if they are of the form of Example 1.2.13, i.e. $J = \begin{bmatrix} J_0 & J_1 \end{bmatrix}$ then it is easy to see that pass profile controllability does not hold when $J_0 + 2J_1 = -3$.

In common with the 1D case, transformations of both the state and/or pass profile vector play an important role in the analysis of discrete unit (and non-unit) memory linear repetitive processes. The most basic form of these is defined next where T_1 and T_2 are nonsingular matrices of dimensions $n \times n$ and $m \times m$ respectively

$$\begin{aligned} x_{k+1}(t) &= T_1 \hat{x}_{k+1}(t) \\ y_{k+1}(t) &= T_2 \hat{y}_{k+1}(t) \end{aligned} \tag{6.23}$$

These are termed similarity transformations.

Theorem 6.1.9. *Pass profile controllability of a discrete unit memory linear repetitive process of the form defined by the state-space model of Example 1.2.13 and pass state initial vector sequence (6.21) is invariant under the similarity transformations defined by (6.23).*

Proof. Applying (6.23) to the process state-space model yields

$$\begin{aligned} \hat{x}_{k+1}(p+1) &= T_1^{-1} A T_1 \hat{x}_{k+1}(p) + T_1^{-1} B u_{k+1}(t) + T_1^{-1} B_0 T_2 \hat{y}_k(p) \\ \hat{y}_{k+1}(p) &= T_2^{-1} C T_1 \hat{x}_{k+1}(p) + T_2^{-1} D u_{k+1}(p) + T_2^{-1} D_0 T_2 \hat{y}_k(p) \\ \hat{x}_{k+1}(0) &= T_1^{-1} d_{k+1} + \sum_{j=0}^{\alpha-1} T_1^{-1} J_j T_2 \hat{y}_k(j) \end{aligned}$$

Hence in the 1D model the matrices $\mathfrak{B}_7$ and $\mathfrak{B}_2$ are transformed to

$$\mathfrak{B}_7^* := \widehat{T}_2^{-1}\mathfrak{B}_7\widehat{T}_2, \text{ and } \mathfrak{B}_2^* := \widehat{T}_2^{-1}\mathfrak{B}_2$$

respectively, where $\widehat{T}_2 = \text{diag}\{T_2, T_2, \cdots, T_2\}$, and the proof is complete. □

The property of pass state controllability is defined as follows.

Definition 6.1.5. *A discrete unit memory linear repetitive process with dynamics of the form defined by the state-space model of Example 1.2.13 and pass state initial vector sequence (6.21) is said to be pass state controllable if $\exists$ a pass number N and control input vectors defined on the rectangle $[(0,0),(N,\alpha-1)]$ which will drive the process to an arbitrarily specified state vector sequence on pass N.*

In order to study this property, consider first the equivalent 2D Roesser state-space model representation defined by (1.42). Then in this model, the current pass state vector of the repetitive process is treated as a local horizontal state sub-vector and the pass profile vector as a local vertical state sub-vector. Previously, after introducing the 1D state-space model we have investigated only pass controllability relative to the global pass profile vector as it represents the global state of the process. The subject of what follows is the pass controllability properties of the state super-vector $X(l)$ of the 1D equivalent model. Note that it is never possible to simultaneously achieve arbitrarily specified pass profile and state vector sequences along a given pass, i.e. the processes considered here can never be simultaneously pass and pass state controllable.

It is clear from (1.47)–(1.49) that $X(l)$ depends statically on the global pass profile vector $Y(l)$, since its dynamics have been 'absorbed' into the matrices which define the 1D equivalent model. Hence, $X(l)$ alone cannot be the subject of the controllability investigations. However, in applications we may only be interested in reaching $X(l)$ instead of $Y(l)$. The conditions under which this can be achieved are established next.

The only way of obtaining a pre-defined global state vector along some pass number, say $N+1$, is to first obtain the corresponding pass profile on pass number $N+1$, say Y^*,

$$Y^* := Y(N+1) = \mathfrak{B}_1 Y(N) + \mathfrak{B}_2 U(N) + \mathfrak{B}_3 x_N(0)$$

such that there exists on the next pass $N+1$ a control input vector $U(N+1)$ which gives the required X^*, i.e.

$$X^* := X(N+1) = \mathfrak{B}_4 Y(N+1) + \mathfrak{B}_5 U(N+1) + \mathfrak{B}_6 x_{N+1}(0) \qquad (6.24)$$

Also $X(N+1,0)$ is pre-specified on each pass, i.e. not the subject of any control action. Consequently introduce

$$\tilde{X} := X^* - \mathfrak{B}_6 x_{N+1}(0)$$

to yield from (6.24)

$$\tilde{X} = \mathfrak{B}_4 Y(N+1) + \mathfrak{B}_6 U(N+1) \tag{6.25}$$

The problem now is to find an admissible pass profile trajectory $Y(N+1)$ and input $U(N+1)$ such that (6.25) holds for a prescribed $\tilde{X}$.

The necessary and sufficient condition for a solution to this problem clearly is

$$\text{rank} \begin{bmatrix} \mathfrak{B}_4 & \mathfrak{B}_6 \end{bmatrix} = \text{rank} \begin{bmatrix} \mathfrak{B}_4 & \mathfrak{B}_6 & \tilde{X} \end{bmatrix} \tag{6.26}$$

Since it is not possible to use $U(N+1)$ (this would change the final pass profile, i.e. on pass $N+1$), set this vector equal to zero and then (6.25) and (6.26) become

$$\tilde{X} = \mathfrak{B}_4 Y(N+1)$$

and

$$\text{rank}\, \mathfrak{B}_4 = \text{rank} \begin{bmatrix} \mathfrak{B}_4 & \tilde{X} \end{bmatrix}$$

respectively. Hence to obtain a given Y^* we must have pass profile controllability for the special case when the boundary conditions are of the form defined by Example 1.2.10, i.e. $\{D_0, D\}$ must be a controllable pair.

6.2 Controllability and Observability of Differential Processes

6.2.1 Controllability

The state-space model considered is a special case of that of Example 1.2.3 where the pass profile vector is equal to the state vector, i.e. the dynamics are described by

$$\dot{x}_k(t) = Ax_k(t) + Bu_k(t) + B_0 x_{k-1}(t) \tag{6.27}$$

over $0 \leq t \leq \alpha$, $k \geq 1$. The boundary conditions are as in Example 1.2.3 specialized to this case, i.e.

$$\begin{aligned} x_k(0) &= d_k, \ \ k \geq 1 \\ x_0(t) &= f(t), \ \ 0 \leq t \leq \alpha \end{aligned} \tag{6.28}$$

where the $n \times 1$ vector d_k has known constant entries and the entries in the $n \times 1$ vector $f(t)$ are known functions of t for $0 \leq t \leq \alpha$. The analysis which follows is taken from [34] and we follow the notation in this paper and write $x_k(t)$ for the pass profile on pass k.

We can obtain a representation of the pass profile vector $x_k(t)$ on pass k in terms of a set of transition matrices $\mathcal{K}_i(t)$, $i = 1, 2, \cdots, k$ as

$$\begin{aligned} x_k(t) &= \sum_{i=1}^{k} \mathcal{K}_i(t) d_{k+1-i} + \int_0^t \mathcal{K}_k(t-\tau) B_0 f(\tau)\, d\tau \\ &+ \sum_{i=1}^{k} \int_0^t \mathcal{K}_i(t-\tau) B u_{k+1-i}(\tau)\, d\tau \end{aligned} \tag{6.29}$$

over $0 \leq t \leq \alpha$, $k = 1, 2, \cdots$. The $n \times n$ matrices $\mathcal{K}_k(t)$ are the transition matrices of a differential unit memory linear repetitive process of the form considered here, where for $k = 1$, $\mathcal{K}_1(t) = e^{At}$, the matrix exponential for A (see, for example, [27]), and for $k \geq 2$

$$\dot{\mathcal{K}}_k(t) = A\mathcal{K}_k(t) + B_0\mathcal{K}_{k-1}(t),\ \mathcal{K}_k(0) = 0 \tag{6.30}$$

Another representation for these transition matrices is the integral form

$$\mathcal{K}_k(t-\sigma) = \int_\sigma^t \mathcal{K}_{k-i}(t-\tau) B_0 \mathcal{K}_i(\tau - \sigma)\, d\tau$$

for any $i = 1, 2, \cdots, k-1$, $k = 2, 3, \cdots$. They can also be represented in series form as

$$\mathcal{K}_k(t-\sigma) = \sum_{i=1}^{k} \mathcal{K}_i(t-\tau) \mathcal{K}_{k+1-i}(\tau - \sigma),\ k = 1, 2, \cdots$$

which can be regarded as a generalization of the familiar transition property of the matrix exponential ([27]).

It is also possible to represent the transition matrices in series form as

$$\mathcal{K}_k(t) = \sum_{i=k-1}^{\infty} \frac{1}{i!} t^i P_k^{i+1},\ k = 1, 2, \cdots \tag{6.31}$$

where the $n \times n$ matrices P_i^k are given by

$$\begin{aligned} P_1^1 &= I_n,\ P_i^k = 0, \text{ when } i > k \text{ or } i = 0 \\ P_i^{k+1} &= AP_i^k + B_0 P_{i-1}^k, \text{ when } i = 1, 2, \cdots, k;\ k = 1, 2, \cdots \end{aligned}$$

Hence we have that

$$P_1^k = A^{k-1},\ P_k^k = B_0^{k-1},\ k = 1, 2, \cdots$$

and it is possible to show that

$$P_i^{k+1} = P_i^k A + P_{i-1}^k B_0,\ i = 1, 2, \cdots, k;\ k = 1, 2, \cdots$$

Note also that these matrices arise in connection with classes of delay differential systems [172].

Cayley-Hamilton type relations for these matrices follow from the Cayley-Hamilton theorem for the matrix $A+\gamma B_0$ where γ is an arbitrary scalar. In particular,

$$(A+\gamma B_0)^n = \sum_{i=1}^{n} \nu_i(\gamma)(A+\gamma B_0)^{n-i} \tag{6.32}$$

where the functions $\nu_i(\gamma)$, $i = 1, 2, \cdots, n$, are polynomials of degree i in γ and are given by

$$\nu_i(\gamma) = \sum_{j=0}^{i} b_{j+1}^{(i)} \gamma^j, \; i = 1, 2, \cdots, n$$

where $b_1^{(0)} = 0$.

For $h \geq 1$ it is possible to show that

$$(A+\gamma B_0)^h = \sum_{k=0}^{h} \gamma^k P_{k+1}^{h+1}$$

Substitution in (6.32) and equating powers in γ on both sides of the result yields for $n \geq h$

$$P_{h+1}^{n+1} = \sum_{i=0}^{h} \sum_{k=i}^{n+i-h} b_{h+1-i}^{(n-k)} P_{i+1}^{k+1} \tag{6.33}$$

We can now obtain a generalized Cayley-Hamilton relation for the matrix functions $\mathcal{K}_k(t)$. In particular, suppose that the differential equations (6.30) are extended to the interval $[0, \infty)$. Then the Laplace transforms, $L_i(s)$, of the matrices $\mathcal{K}_i(t)$ are given by

$$L_i(s) = \left[(sI_n - A)^{-1} B_0\right]^{i-1} (sI_n - A)^{-1}, \; i = 1, 2, \cdots$$

or, alternatively, as

$$L_i(s) = \sum_{h=0}^{\infty} s^{-(i+h)} P_i^{i+h} \tag{6.34}$$

Note now that $(sI - A)^{-1} B_0$ is an $n \times n$ matrix and hence there exists, by the Cayley-Hamilton theorem, scalars $a_i(s)$, $i = 1, 2, \cdots, n$, such that

$$\left[(sI_n - A)^{-1} B_0\right]^n = \sum_{i=1}^{n} a_{n+1-i}(s) \left[(sI - A)^{-1} B_0\right]^{i-1}$$

and hence from (6.34) we have that

$$L_{n+1}(s) = \sum_{i=1}^{n} a_{n+1-i}(s) L_i(s)$$

Also taking the inverse Laplace transform of this last equation gives

$$\mathcal{K}_{n+1}(t) = \sum_{i=1}^{n} \int_0^t \beta_{n+1-i}(t-\tau)\mathcal{K}_i(\tau)\, d\tau \tag{6.35}$$

where the scalar functions $\beta_i(t)$ are the inverse Laplace transforms of the functions $a_i(s)$.

Using the Leverrier algorithm (see, for example, [27]) for the coefficients $a_i(s)$ gives the following formulas for the functions $\beta_i(t)$

$$\begin{aligned}\beta_1(t) &= \text{trace } \mathcal{K}_1(t)B_0 \\ \beta_k(t) &= \frac{1}{k}\int_0^t \text{trace } [\mathcal{K}_1(t-\tau)B_0\eta_{k-1}(\tau)]\, d\tau,\ k = 2, 3, \cdots, n\end{aligned}$$

where the $n \times n$ matrix functions $\eta_k(t)$ are given by

$$\begin{aligned}\eta_1(t) &= \mathcal{K}_1(t)B_0 - \beta_1(t)I_n \\ \eta_k(t) &= \mathcal{K}_k(t)B_0 - \sum_{i=1}^{k-1}\int_0^t \beta_i(t-\tau)\mathcal{K}_{k-i}(\tau)\, B_0\, d\tau \\ &- \beta_k(t)I_n,\ k = 2, 3, \cdots, n-1\end{aligned}$$

Next we give an illustrative example consisting of two cases.

Example 6.2.1. Consider the case when

$$A = \begin{bmatrix} 2 & 0 \\ 0 & 1 \end{bmatrix},\quad B_0 = \begin{bmatrix} 1 & 2 \\ 0 & 1 \end{bmatrix}$$

Then $\beta_1(t) = e^t + e^{2t}$ and $\beta_2(t) = e^t - e^{2t}$.

As a second case, take

$$A = \begin{bmatrix} 1 & 0 \\ 0 & 0 \end{bmatrix},\quad B_0 = \begin{bmatrix} 1 & 0 \\ 0 & 1 \end{bmatrix}$$

Then $\beta_1(t) = 1 + e^t$ and $\beta_2(t) = 1 - e^t$.

An alternative form of the Cayley-Hamilton relation is obtained by differentiating (6.35) with respect to t and using the differential equation (6.30) for $\mathcal{K}_{n+1}(t)$ to yield

$$\mathcal{K}_n(t)B_0 = \sum_{i=1}^{n-1}\int_0^t \beta_{n-i}(t-\tau)\mathcal{K}_i(\tau)\, B_0 d\tau + \beta_n(t)I_n$$

Also an alternative formula to that of (6.29) for the pass profile vector $x_k(t)$ on pass k can be obtained which corresponds to eigenfunction representations of the pass profile vector for the processes considered here. In particular, introduce the modified pass profile vector $w_k(t)$ as

$$w_k(t) := x_k(t) - \sum_{i=1}^{k} \mathcal{K}_i(t) d_{k+1-i}, \; k = 1, 2, \cdots$$
$$w_0(t) := f(t) - f(0)$$

Then $w_k(0) = 0$ and also $w_k(t)$ can be shown to satisfy the equation

$$w_k(t) = \int_0^t e^{A(t-\tau)} B_0 w_{k-1}(\tau)\, d\tau + \int_0^t e^{A(t-\tau)} B u_k(\tau)\, d\tau, k = 1, 2, \cdots \quad (6.36)$$

obtained from (6.27). Suppose also that the process is uncontrolled, i.e. $u_k = 0, k \geq 1$. Then (6.36) can be considered as a difference equation on the Hilbert resolution space (H, Γ) where H is $L^2([0, \alpha]; \mathbb{R}^n)$ and Γ is a resolution of the identity on H (see, for example, [56]). Also this equation can be written in the form

$$w_k(t) = \mathbb{A} w_{k-1}(t), \;\; k = 1, 2, \cdots$$

where $\mathbb{A}$ is the linear causal operator on H

$$w_k(t) = \int_0^t e^{A(t-\tau)} B_0 w_{k-1}(\tau)\, d\tau \quad (6.37)$$

Noting that $\mathbb{A}$ is also a convolution operator (see [54]), we can formulate a convolution eigenvalue problem for it as the determination of a scalar function μ and corresponding $n \times 1$ vector functions g which satisfy

$$\int_0^t e^{A(t-\tau)} B_0 g(\tau)\, d\tau = \int_0^t \mu(t-\tau) g(\tau)\, d\tau \quad (6.38)$$

This problem is defined on the interval $[0, \alpha]$ but the causal nature of the operators in (6.38) enables us to consider the same problem on the half-line $[0, \infty)$ provided we extend μ and g from $[0, \alpha]$ to $[0, \infty)$ to have at most exponential growth. Then we can take the Laplace transform of (6.38) to obtain an eigenvalue problem in the transform space for the matrix $[sI_n - A]^{-1} B_0$ as a function of the transform variable s.

This last matrix has n eigenvalues and corresponding eigenvectors which are functions of s. Here we assume that these eigenvalues are distinct and hence there exists n distinct scalar functions μ_i and corresponding vectors g_i, $i = 1, 2, \cdots, n$, which satisfy (6.38). Since the operators in (6.38) are causal, these functions and vectors restricted to the interval $[0, \alpha]$ are the convolution eigenfunctions and eigenvectors of the operator $\mathbb{A}$.

In order to obtain a characteristic equation satisfied by the function μ, first define $\mu^{(k)}$ as the convolution of μ with itself k times, i.e.

$$\mu^{(k)}(t) := \int_0^t \mu(t-\tau) \mu^{(k-1)}(\tau)\, d\tau, \; t \geq 0, \; k \geq 2 \quad (6.39)$$

and $\mu^{(1)}(t) = \mu(t)$. Applying $\mathbb{A}$ repeatedly to both sides of (6.39) and using properties of convolution operators [54] we have that

$$\int_0^t \mathcal{K}_h(t-\tau)B_0 g(\tau)\, d\tau = \int_0^t \mu^{(h)}(t-\tau)g(\tau)\, d\tau,\ t \geq 0,\ h \geq 1 \tag{6.40}$$

Now using (6.35) and (6.40) we have that for $t \geq 0$

$$\int_0^t \left[\mu^{(n+1)}(t-\tau) - \sum_{i=1}^{n} \int_\sigma^t \beta_{n+1-i}(t-\sigma)\mu^{(i)}(\sigma-\tau)\, d\sigma \right] g(\tau)\, d\tau = 0$$

This integral equation either has the solution $g(t) = 0$ or for $t \geq 0$

$$\mu^{(n+1)}(t) - \sum_{i=1}^{n} \int_0^t \beta_{n+1-i}(t-\tau)\mu^{(i)}(\tau)\, d\tau = 0 \tag{6.41}$$

and hence (since g is not identically zero) we obtain (6.41), the convolution characteristic equation of the function μ for which (6.35) is the corresponding Cayley-Hamilton equation. Note that (6.41) can also be derived from the characteristic equation of the matrix $[sI_n - A]^{-1}B_0$ in the transform space.

The $n \times 1$ vector functions g_i satisfy

$$\int_0^t e^{A(t-\tau)}B_0 g_i(\tau)\, d\tau = \int_0^t \mu_i(t-\tau)g_i(\tau)\, d\tau,\ i = 1, 2, \cdots, n$$

and we now show that these functions are linearly independent in the sense that the vector integral equation for the scalar functions γ_i

$$\sum_{i=1}^{n} \int_0^t \gamma_i(t-\tau)g_i(\tau)\, d\tau = 0,\ t \geq 0 \tag{6.42}$$

has only the solution $\gamma_i = 0,\ i = 1, 2, \cdots, n$.

Using the notation

$$\gamma \star g := \int_0^t \gamma(t-\tau)g(\tau)\, d\tau,\ t \geq 0$$

we write (6.42) as

$$\sum_{i=1}^{n} \gamma_i \star g_i = 0$$

Now apply to this last equation the convolution operator $\mathbb{A} - \mu_1\Psi$, where $\mu_1\Psi$ is the operator which maps the $n \times\ 1$ vector g into the $n \times 1$ vector $\mu_1 \star g$. Using the properties of convolution, we obtain

$$\sum_{i=2}^{n} \gamma_i \star (\mu_i - \mu_1) \star g_i = 0$$

Repeating this with the operators $\mathbb{A} - \mu_h \Psi$ $(h = 2, \cdots, n-1)$ in turn now gives

$$\sum_{i=h}^{n} \gamma_i \star (\mu_i - \mu_1) \star \cdots \star (\mu_i - \mu_{h-1}) \star g_i = 0, \ h = 2, 3, \cdots, n \tag{6.43}$$

and, in particular, for $h = n$, we have that

$$\gamma_n \star (\mu_n - \mu_1) \star \cdots \star (\mu_n - \mu_{n-1}) \star g_n = 0$$

It now follows from this last integral equation that either $\gamma_n = 0$ or $\mu_n = \mu_i$ for some $(i,\ 1 \leq i < n)$, or $g_n = 0$. Only the first case is possible and hence $\gamma_n = 0$. This then means, using (6.43), that $\gamma_i = 0,\ i = 1, \cdots, n$. Hence the vector functions g_i are linearly independent in the sense defined.

Example 6.2.2. Return now to the processes considered in the previous example. Then in the first of these cases, $\mu_1(t) = e^{2t}$, $\mu_2(t) = e^t$, and

$$g_1(t) = \begin{bmatrix} 1 \\ 0 \end{bmatrix}, \ g_2(t) = \begin{bmatrix} 0 \\ 1 \end{bmatrix}$$

In the second case, $\mu_1(t) = e^t,\ \ \mu_2(t) = 1$, and

$$g_1(t) = \begin{bmatrix} 1 \\ 0 \end{bmatrix}, \ g_2(t) = \begin{bmatrix} 0 \\ 1 \end{bmatrix}$$

In the case when $B_0 = I_n$, and A has distinct eigenvalues $\lambda_i,\ 1 \leq i \leq n$, with corresponding eigenvectors e_i, then

$$\mu_i(t) = e^{\lambda_i t}, \ g_i(t) = e_i, \ i = 1, 2, \cdots, n$$

To obtain an expression for the modified pass profile vector $w_k(t)$ in terms of the convolution eigenfunctions, we assume that the modified initial pass profile is differentiable and we extend $w_0(t)$ from $[0, \alpha]$ to $[0, \infty)$ in some appropriate way. The transforms of the vector functions g_i are linearly independent and give a basis for the transform space. Expressing the transform of $w_0(t)$ as a sum over the transforms of the g_i and inverting gives the unique expression

$$w_0(t) = \sum_{i=1}^{n} \int_0^t \sigma_i(t - \tau) g_i(\tau)\, d\tau, \quad 0 \leq t \leq \alpha \tag{6.44}$$

where the σ_i are scalar functions. Also since the operators in this last equation are causal, they are independent of the way in which the w_0 and g_i are extended.

Using (6.37) and (6.44), we have that

$$w_k(t) = \sum_{i=1}^{n} \int_0^t \sigma_{k,i}(t-\tau) g_i(\tau)\, d\tau,\ k \geq 1$$

where

$$\sigma_{k,i}(t) := \int_0^t \sigma_i(t-\tau)\mu_i^{(k)}(\tau)\, d\tau,\ i = 1, 2, \cdots, n,\ k \geq 1$$

Hence

$$\sigma_{k,i}(t) = \int_0^t \mu_i(t-\tau)\sigma_{k-1,i}(\tau)\, d\tau,\ i = 1, 2, \cdots, n,\ k \geq 1$$

where $\sigma_{0,i}(\tau) = \sigma_i(\tau)$. In effect, we can treat this last expression as a diagonalized form of the operator equation (6.37) and hence it is a special canonical form for these processes.

6.2.2 Point Controllability

The formal definition of this property is as follows.

Definition 6.2.1. *The differential linear repetitive process (6.27) is said to be point controllable if for any $N \geq 1\ \exists$ a sequence, denoted by $(u_k)_{k=1}^N$, of control vectors with $u_k \in L^2([0,\alpha];\mathbb{R}^r)$ for each $k = 1, 2, \cdots, N$, such that the pass profile vector x_N on pass N can be steered to $x_N(t_1) = t_v$, where t_v is an arbitrarily chosen $n \times 1$ vector with real entries, at an arbitrary position $t = t_1$, $0 < t_1 \leq \alpha$ independent of the initial pass profile $x_0(t) = f(t)$ and pass profile initial vectors d_k, $k = 1, 2, \cdots, N$.*

We now proceed to obtain conditions on the process matrices A, B, and B_0 such that this property holds.

On pass $N \geq 1$ use of (6.29) gives that

$$z_N(t_1) = \sum_{i=1}^{N} \int_0^{t_1} \mathcal{K}_i(t_1-\tau) B u_{N+1-i}(\tau)\, d\tau \tag{6.45}$$

where

$$z_N(t_1) := x_N(t_1) - \sum_{i=1}^{N} \mathcal{K}_i(t_1) d_{N+1-i} - \int_0^{t_1} \mathcal{K}_N(t_1-\tau) B_0 f(\tau)\, d\tau$$

and since $x_N(t_1) = t_v$ is given, $z_N(t_1)$ is given. It now follows from (6.45) that we need only consider control vectors in the interval $[0, t_1]$ and hence this equation defines an operator

$$G : [L^2([0,t_1];\mathbb{R}^r)]^N \to \mathbb{R}^n$$

which maps a vector $(u_k)_{k=1}^N$ (or (u_k) for short) to a vector $z_N(t_1)$. Also the process is point controllable if the range of G is $\mathbb{R}^n$.

Existing theory for distributed parameter systems (see, for example, [35]) can now be invoked to show that a necessary and sufficient condition for G to have this property is that the kernel of the adjoint operator $G^* : \mathbb{R}^n \rightarrow [L^2([0,t_1];\mathbb{R}^r)]^N$ comprises only the zero vector. Since the inner product on $[L^2([0,t_1],\mathbb{R}^r)]^N$ is

$$([v_k],[u_k]) := \sum_{k=1}^N \int_0^{t_1} v_k^T(t)u_k(t)\,dt$$

we have that for $y \in \mathbb{R}^n$

$$\begin{aligned} y^T G([u_k]) &= \sum_{i=1}^N \int_0^{t_1} y^T \mathcal{K}_i(t_1-\tau)Bu_{N+1-i}(\tau)\,d\tau \\ &= \sum_{i=1}^N \int_0^{t_1} u_i^T(\tau)B^T\mathcal{K}_{N+1-i}^T(t_1-\tau)y\,d\tau \end{aligned}$$

and hence for $0 \leq \tau \leq t_1$

$$G^*y = \left[\begin{array}{cccc} B^T\mathcal{K}_N^T(t_1-\tau)y & B^T\mathcal{K}_{N-1}^T(t_1-\tau)y & \cdots & B^T\mathcal{K}_1^T(t_1-\tau)y \end{array}\right]$$

Now we proceed to obtain conditions on A, B, and B_0 which ensure that $G^*y = 0$ implies that $y = 0$. Hence we consider the equations

$$B^T\mathcal{K}_i^T(t_1-\tau)y = 0,\ i = 1,2,\cdots,N,\ 0 \leq \tau \leq t_1$$

Using the Cayley-Hamilton relation, we can replace this last expression by

$$B^T\mathcal{K}_i^T(t_1-\tau)y = 0,\ i = 1,2,\cdots,\min(N,n),\ 0 \leq \tau \leq t_1$$

Repeated differentiation of this last equation with respect to τ and setting $\tau = t_1$ each time gives, on use of (6.31),

$$B^T(P_i^k)^Ty = 0,\ k \geq i,\ i = 1,2,\cdots,\min\,(N,n)$$

or, using the series representation of (6.33) for P_{h+1}^{n+1},

$$B^T(P_i^k)^Ty = 0,\ k = i,\cdots,n,\ i = 1,2,\cdots,\min\,(N,n)$$

Hence provided

$$\text{rank}\,[P_i^kB]_{k=i,\cdots,n;i=1,2,\cdots,\min\,(N,n)} = n$$

it follows that $y = 0$, and therefore this process is point controllable when this last condition holds. Moreover, this condition can also be shown to be necessary for this property.

It is also interesting to note that this condition is also that for point controllability (see, for example, [172, 174]) of the delay-differential equation (see also Example 1.2.6)

$$\dot{x}(t) = Ax(t) + Bu(t) + B_0 x(t - \tau)$$

Finally, we say that the process is point controllable over $N \geq 1$ passes if $\exists$ a sequence of controls $(u_k)_{k=1}^N$ such that the pass profile vectors x_k, $k = 1, 2, \cdots, N$ can be steered to arbitrary values at the points $t = t_1$, i.e. $x_k(t_1) = t_v^{(k)}$, $k = 1, 2, \cdots, N, t_v^{(k)}$ given, independent of the boundary conditions. Using a similar method to that developed here for point controllability, it is possible to show in the case when the control input is a scalar that the process is point controllable over N passes if, and only if,

$$\text{rank} \begin{bmatrix} B & AB & A^2B & \cdots & A^{n-1}B \end{bmatrix} = n$$

6.2.3 Sufficient Conditions for Approximate Reachability and Controllability

Here we are dealing with processes which have discrete dynamics in the k variable and continuous dynamics in the t variable. Hence we need to distinguish between reachability and controllability for these processes as for 1D discrete linear systems (see, for example, [96]) (recall also the analysis earlier in this chapter for discrete unit memory linear repetitive processes of the form defined by Example 1.2.10). Also because of the smoothing properties of the solutions [33], we need to distinguish between approximate and exact reachability and controllability (see, for example, [44, 160]).

Approximate reachability and controllability are formally defined as follows.

Definition 6.2.2. *For the differential linear repetitive process (6.27), a pass profile vector $g \in L^2([0, \alpha]; \mathbb{R}^n)$ is said to be approximately reachable in N passes if given any $\epsilon > 0$ $\exists$ an admissible sequence $(u_k)_{k=1}^N$ of control input vectors and a sequence, denoted by $(d_k)_{k=1}^M$, of constant pass initial vectors such that if the initial pass profile is identically zero and (6.28) is satisfied $\forall\, t \in [0, \alpha]$ then*

$$\left[\int_0^\alpha ||x_N(t) - g(t)||^2 \, dt \right]^{\frac{1}{2}} < \epsilon$$

where $||x|| := (x^T x)^{\frac{1}{2}}$. The sequence $(u_k)_{k=1}^N$ is said to be admissible if $u_k \in L^2([0, \alpha]; \mathbb{R}^r)$, $k = 1, \cdots, N$. The process (6.27) is approximately reachable in N passes if all pass profile vectors $g \in L^2([0, \alpha]; \mathbb{R}^n)$ are approximately reachable in N passes.

Definition 6.2.3. *A pass profile vector $f \in L^2([0, \alpha]; \mathbb{R}^n)$ is approximately controllable to zero in N passes if given any $\epsilon > 0$ $\exists$ a pass profile vector $f_e \in L^2([0, \alpha]; \mathbb{R}^n)$ controllable to zero in N passes with*

$$\left[\int_0^\alpha ||f(t) - f_e(t)||^2 \, dt \right]^{\frac{1}{2}} < \epsilon$$

The repetitive process is approximately controllable to zero in N passes if all states $f \in L^2([0, \alpha]; \mathbb{R}^n)$ are approximately controllable to zero in N passes.

We now establish sufficient conditions on the matrices A, B and B_0 for approximate reachability in n passes when the control input is a scalar, i.e. $r = 1$, and the vectors $d_k = 0,\ k = 1, 2, \cdots, n$.

For any N and r when $f = 0$ and $d_k = 0,\ k = 1, 2, \cdots, N$, (6.29) shows that the pass profile vectors x_N reachable on pass N are given by

$$x_N(t) = \sum_{i=1}^{N} \int_0^t \mathcal{K}_{N+1-i}(t - \tau) u_i(\tau) \, d\tau, \ 0 \le t \le \alpha \tag{6.46}$$

for all admissible sequences of control inputs $(u_k)_{k=1}^N$. Note also that $x_N(0) = 0$ and there are extra conditions satisfied by the derivatives of x_N at $t = 0$. Here (6.46) also shows that x_N has a first derivative in $L^2([0, \alpha]; \mathbb{R}^n)$. Hence we can differentiate (6.46) and form

$$\begin{aligned} \Phi_N(t) &:= \dot{x}_N(t) - A x_N(t) \\ &= B u_N(t) + \sum_{i=1}^{N-1} \int_0^t B_0 \mathcal{K}_{N-i}(t - \tau) B u_i(\tau) \, d\tau \end{aligned} \tag{6.47}$$

where $\Phi_N \in L^2([0, \alpha]; \mathbb{R}^n)$.

Now let X denote the space $L^2([0, \alpha]; \mathbb{R}^n)$ and let Z_N denote the product Hilbert space $Z_N := L^2([0, \alpha]; \mathbb{R}^r)^N$ so that

$$U_N := \begin{pmatrix} u_1 & u_2 & \cdots & u_N \end{pmatrix}^T \in Z_N$$

In this setup, (6.46) defines a causal operator $Q : Z_N \to X$ such that

$$(QU_N)(t) := \sum_{i=1}^{N} \int_0^t \mathcal{K}_{N+1-i}(t - \tau) B u_i(\tau) \, d\tau$$

and (6.47) a causal operator $Q_{pc} : Z_N \to X$ such that

$$(Q_{pc} U_N)(t) := B u_N(t) + \sum_{i=1}^{N-1} \int_0^t B_0 \mathcal{K}_{N-i}(t - \tau) B u_i(\tau) \, d\tau$$

The process is approximately reachable if

$$\overline{\mathcal{R}(Q)} = X$$

where $\mathcal{R}(Q)$ denotes the range of Q and the bar closure. It has been shown previously [33] that if $\overline{\mathcal{R}(Q_{pc})} = X$ then $\overline{\mathcal{R}(Q)} = X$ and hence it is sufficient to establish conditions under which $\mathcal{R}(Q_{pc})$ is dense in X.

The range of Q_{pc} is dense in X if the null space $\mathcal{N}(Q_{pc}^*)$ of the adjoint operator $Q_{pc}^* : X \to Z_N$ of Q_{pc} contains only the zero function, i.e. $\mathcal{N}(Q_{pc}^*) = \{0\}$ (see, for example, [35]). Also since the inner product on $X = L^2([0,\alpha];\mathbb{R}^n)$ is

$$(\psi, \phi)_X := \int_0^\alpha \psi^T(t)\phi(t)\, dt$$

we have that $Q_{pc}^* : X \to Z_N$ is the anti-causal operator

$$(Q_{pc}^*\psi)(t) = \left(B^T\psi(t), \int_t^\alpha B^T K_{N-i}^T(\tau - t) B_0^T \psi(\tau)\, d\tau \right)^T_{i=1,\cdots,N-1}$$

Hence we need to establish conditions on A, B, and B_0 such that

$$B^T\psi(t) = 0 \tag{6.48}$$

and

$$\int_0^\alpha B^T \mathcal{K}_{N-i}^T(t-\tau) B_0^T \psi(\tau)\, d\tau = 0,\ i = 1, \cdots, N-1,\ 0 \le t \le \alpha \tag{6.49}$$

implies

$$\psi(t) = 0,\ 0 \le t \le \alpha$$

When $N = n$ and $r = 1$, sufficient conditions for this property are reported elsewhere [33]. In particular, it is clearly sufficient that the vectors

$$B, B_0B, \cdots, B_0^{n-1}B$$

are linearly independent, but alternative sets of conditions can also be derived. Since $\psi = 0$ is the only solution of (6.48) and (6.49), it follows that $\mathcal{N}(Q_{pc}^*) = \{0\}$, and hence $\mathcal{R}(Q_{pc})$ is dense in X and this, in turn, means that $\mathcal{R}(Q)$ is dense in X.

We now consider approximate controllability and again assume that $d_k = 0$, $k = 1, \cdots, N$. When the pass profile vector x_N on pass N is zero, (6.29) gives

$$-\int_0^t \mathcal{K}_N(t-\tau)B_0 f(\tau)\, d\tau = \sum_{i=1}^N \int_0^t \mathcal{K}_{N+1-i}(t-\tau)Bu_i(\tau)\, d\tau$$

and define the operator $\mathcal{S} : X \to X$ by

$$(\mathcal{S}f)(t) := -\int_0^t \mathcal{K}_N(t-\tau)B_0 f(\tau)\, d\tau$$

Then the process is approximately controllable if

$$\mathcal{R}(\mathcal{S}) \subseteq \overline{\mathcal{R}(Q)} \tag{6.50}$$

Since the process is approximately reachable when $\overline{\mathcal{R}(Q)} = X$, approximate reachability implies approximate controllability and hence a sufficient condition for approximate controllability when $r = 1$ and $N = n$ is that the vectors $B, B_0B, \cdots, B_0^{n-1}B$ are linearly independent.

The condition (6.50) is satisfied if

$$\mathcal{N}(Q^*) \subseteq \mathcal{N}(\mathcal{S}^*)$$

where the adjoint operator $\mathcal{S}^* : X \to X$ of $\mathcal{S}$ is defined by

$$(\mathcal{S}^*\psi)(t) := -\int_t^\alpha B_0^T \mathcal{K}_N^T(\tau - t)\psi(\tau)\, d\tau$$

Hence the process is approximately controllable in N passes if

$$\int_t^\alpha B^T \mathcal{K}_{N+1-k}^T(\tau - t)\psi(\tau)\, d\tau = 0,\ 0 \le t \le \alpha,\ k = 1, 2, \cdots, N \tag{6.51}$$

implies that

$$\int_t^\alpha B_0^T \mathcal{K}_N^T(\tau - t)\psi(\tau)\, d\tau = 0,\ 0 \le t \le \alpha$$

When B_0 is invertible, this last expression gives

$$\int_t^\alpha \mathcal{K}_N^T(\tau - t)\psi(\tau)\, d\tau = 0,\ 0 \le t \le \alpha$$

Differentiating and using (6.30) gives

$$\int_t^\alpha B_0^T \mathcal{K}_{N-1}^T(\tau - t)\psi(\tau)\, d\tau = 0,\ 0 \le t \le \alpha$$

Repeating this, finally gives $\psi(t) = 0,\ 0 \le t \le \alpha$, so that, when B_0 is invertible, the process is approximately controllable in N passes if (6.51) implies that $\psi(t) = 0$. Since then $\mathcal{N}(Q^*) = 0$, it follows that $\overline{\mathcal{R}(Q)} = X$, and hence the process is also approximately reachable. Hence when B_0 is invertible, approximate controllability implies approximate reachability, paralleling the (well known) result (see, for example, [96]) for 1D discrete linear systems.

6.2.4 Observability and Control Canonical Forms

Observability for the differential linear repetitive processes considered here is defined in the case when one scalar function is observed on each pass. Hence with the model of (6.27) and (6.28) we associate the observations (i.e. a state-space model of the form defined by Example 1.2.3 with $D = 0$ and $D_0 = 0$.)

$$y_k(t) = Cx_k(t),\ k = 0, 1, 2, \cdots$$

where C is a $1 \times n$ row vector. The process is observable over n passes in the case when f in (6.28) is not the zero function in $L^2([0, \alpha]; \mathbb{R}^n)$ if at least one of the observations, denoted by $[y_k]_{k=0}^{n-1}$, is non-zero. It follows that the process is observable if $y_k = 0$ implies that f is the zero function.

Since

$$y_k(t) = C \int_0^t \mathcal{K}_k(t-\tau)B_0 f(\tau)\, d\tau$$

we have that if $f \in L^2([0, \alpha]; \mathbb{R}^n)$, then $y_k \in H^k([0, \alpha]; \mathbb{R})$ and hence

$$y_k^{(k)}(t) = CB_0^k f(t) + \int_0^t \sum_{i=1}^{k} P_i^{k+1} \mathcal{K}_{k+1-i}(t-\tau)B_0 f(\tau)\, d\tau$$

for $k = 1, \cdots, n-1$. Here $H^k([0, \alpha]; \mathbb{R})$ is the Sobolev space of functions which with their first k derivatives belong to $L^2([0, \alpha]; \mathbb{R}^n)$ (see, for example, [14]).

Hence for the process to be observable, we require that the set of n integral equations

$$Cf(t) = 0,\ 0 \le t \le \alpha$$

$$CB_0^k f(t) + \int_0^t \sum_{i=1}^{k} P_i^{k+1} \mathcal{K}_{k+1-i}(t-\tau)B_0 f(\tau)\, d\tau = 0$$

$0 \le t \le \alpha$, $k = 1, \cdots, N-1$ only has the zero solution. Making the change of variable $t = \alpha - \eta$ and in the integrals $\tau = \alpha - \sigma$ and comparing the resulting equations with those for controllability ((6.48) and (6.49)) we find that they are the same if we replace A by A^T, B_0 by B_0^T, and B by C^T.

Sufficient conditions for observability now follow from the results of the previous sub-section. Also we have the duality result that the process under consideration here is approximately controllable if, and only if, the process

$$\begin{aligned} \dot{x}_k(t) &= -A^T x_k(t) - B_0^T x_{k-1}(t) \\ y_k(t) &= B^T x_k(t),\ 0 \le t \le \alpha,\ k = 0, 1, 2, \cdots \end{aligned}$$

is observable.

For a 1D SISO differential linear system $\dot{x} = Ax + Bu$, controllability implies that the system can be transformed to the so-called controllable companion form. Using a method first due to [161], it is possible to extend this analysis to the differential unit memory linear repetitive processes considered here. The details are in [34] which also includes the derivation of a second control canonical form and a deadbeat controller.

6.3 System Equivalence

It is instructive to first develop some results for the general nD case. In particular, suppose that $G(\underline{z})$ is a $h \times q$ matrix ($\underline{z} = (z_1, z_2, \cdots, z_n)$) – termed a rational polynomial matrix. Then $G(\underline{z})$ may be written as

$$\begin{aligned} G(\underline{z}) &= N_1(\underline{z})D_1^{-1}(\underline{z}) \\ &= D_2^{-1}(\underline{z})N_2(\underline{z}) \end{aligned} \tag{6.52}$$

where $N_1(\underline{z}), N_2(\underline{z})$ are $h \times q$ polynomial matrices, $D_1(\underline{z})$ is a $q \times q$ polynomial matrix and $D_2(\underline{z})$ is a $h \times h$ polynomial matrix. The pair $N_1(\underline{z}), D_1(\underline{z})$ defines a right matrix fraction description of $G(\underline{z})$ and the pair $D_2(\underline{z}), N_2(\underline{z})$ a left matrix fraction description. If these pairs, i.e., $N_1(\underline{z}), D_1(\underline{z})$ and $D_2(\underline{z}), N_2(\underline{z})$ do not have any common matrix divisors then the corresponding matrix fraction descriptions can be viewed as irreducible in some sense. In this case, the following definitions are taken as defining nonessential singularities.

Definition 6.3.1. *The singularities of the irreducible right matrix fraction in (6.52) are said to be*

(a) nonessential singularities of the first kind if $\det D_1(\underline{z}) = 0$ *and the matrix*

$$\tilde{W}(\underline{z}) := \begin{bmatrix} D_1(\underline{z}) \\ N_1(\underline{z}) \end{bmatrix}$$

has full rank; and

(b) nonessential singularities of the second kind if $\det D_1(\underline{z}) = 0$ *and* $\tilde{W}(\underline{z})$ *does not have full rank.*

Note also that similar statements can be made about a left matrix fraction description $D_2(\underline{z}), N_2(\underline{z})$.

Consider now input/output (I/O) description for the 2D case (which is the most relevant here). Then the Rosenbrock system matrix for 2D systems can be defined in an analogous manner to that of the 1D case as

$$P(z_1, z_2) := \begin{bmatrix} T(z_1, z_2) & U(z_1, z_2) \\ -V(z_1, z_2) & W(z_1, z_2) \end{bmatrix} \tag{6.53}$$

Here $T(z_1, z_2)$, $U(z_1, z_2)$, $V(z_1, z_2)$, and $W(z_1, z_2)$ are of dimensions $n \times n$, $n \times r$, $m \times n$, and $m \times r$ respectively. These matrices may be rational but the most physically important case is when they are polynomial matrices arising from a 2D transfer-function matrix

$$G(z_1, z_2) := V(z_1, z_2)T^{-1}(z_1, z_2)U(z_1, z_2) + W(z_1, z_2) \tag{6.54}$$

The concept of equivalent I/O response (or behavior) for 2D systems with this structure can now be defined as follows.

Definition 6.3.2. *Two system matrices $P_1(z_1, z_2)$ and $P_2(z_1, z_2)$ of dimensions $(n_1+m)\times(n_1+r)$ and $(n_2+m)\times(n_2+r)$ are said to be I/O equivalent if*

$$G_1(z_1, z_2) = G_2(z_1, z_2)$$

i.e. provided their 2D transfer-function matrices, defined by (6.54), are equal.

By analogy with the 1D case, the transformations of strict system equivalence and system equivalence can be defined, but now the transforming matrices are polynomial in two indeterminates. The first point to consider is the existence or otherwise of necessary and sufficient conditions for I/O equivalence, where in the 1D case an exact characterization in a closed matrix form is known. The following result extends this characterization to the 2D linear systems case.

Theorem 6.3.1. *The system matrices*

$$P_i(z_1, z_2) = \begin{bmatrix} T_i(z_1, z_2) & U_i(z_1, z_2) \\ -V_i(z_1, z_2) & W_i(z_1, z_2) \end{bmatrix}, \ i = 1, 2$$

are realizations of the same 2D transfer-function matrix $G(z_1, z_2)$ if, and only if, $\exists$ rational matrices $Q_1(z_1, z_2)$, $Q_2(z_1, z_2)$, $R_1(z_1, z_2)$, and $R_2(z_1, z_2)$ such that

$$\begin{bmatrix} Q_1(z_1, z_2) & 0 \\ R_1(z_1, z_2) & I_m \end{bmatrix} \begin{bmatrix} T_2(z_1, z_2) & U_2(z_1, z_2) \\ -V_2(z_1, z_2) & W_2(z_1, z_2) \end{bmatrix}$$

$$= \begin{bmatrix} T_1(z_1, z_2) & U_1(z_1, z_2) \\ -V_1(z_1, z_2) & W_1(z_1, z_2) \end{bmatrix} \begin{bmatrix} Q_2(z_1, z_2) & R_2(z_1, z_2) \\ 0 & I_r \end{bmatrix} \tag{6.55}$$

Proof. Suppose that $\exists$ rational matrices $Q_1(z_1, z_2)$, $Q_2(z_1, z_2)$, $R_1(z_1, z_2)$, and $R_2(z_1, z_2)$ such that (6.55) holds. Then the constituent equations are

$$\begin{aligned} Q_1 T_2 &= T_1 Q_2 \\ Q_1 U_2 &= T_1 R_2 + U_1 \\ R_1 T_2 - V_2 &= -V_1 Q_2 \\ R_1 U_2 + W_2 &= -V_1 R_2 + W_1 \end{aligned}$$

Then it follows by direct substitution that

$$\begin{aligned} G_2(z_1, z_2) &= V_2 T_2^{-1} U_2 + W_2 \\ &= V_1 T_1^{-1} U_1 + W_1 = G_1(z_1, z_2) \end{aligned}$$

and hence $P_1(z_1, z_2)$ and $P_2(z_1, z_2)$ are I/O equivalent.

Conversely, suppose that $P_1(z_1, z_2)$ and $P_2(z_1, z_2)$ are I/O equivalent; then $G_1(z_1, z_2) = G_2(z_1, z_2) \Rightarrow$

$$V_1 T_1^{-1} U_1 + W_1 = V_2 T_2^{-1} U_2 + W_2$$

Hence for an arbitrary $n_2 \times n_1$ matrix $Q_1(z_1, z_2)$

$$\begin{bmatrix} Q_1 & 0 & I_{n_2} & 0 \\ 0 & I_m & 0 & V_1T_1^{-1}U_1 + W_1 \end{bmatrix} \begin{bmatrix} I_{n_1} & 0 \\ 0 & V_2T_2^{-1}U_2 + W_2 \\ -Q_1 & 0 \\ 0 & -I_r \end{bmatrix} = 0 \quad (6.56)$$

Also

$$H := \begin{bmatrix} T_2^{-1} & 0 & 0 & 0 \\ V_2T_2^{-1} & I_m & 0 & 0 \\ 0 & 0 & T_1 & U_1 \\ 0 & 0 & 0 & I_r \end{bmatrix}$$

is a rational invertible matrix with inverse

$$H^{-1} = \begin{bmatrix} T_2 & 0 & 0 & 0 \\ -V_2 & I_m & 0 & 0 \\ 0 & 0 & T_1^{-1} & -T_1^{-1}U_1 \\ 0 & 0 & 0 & I_r \end{bmatrix}$$

Post-multiplying the first matrix on the left-hand side of (6.56) by H and pre-multiplying the second by H^{-1} (i.e. insert HH^{-1} between the two matrices on the left-hand side) yields

$$\begin{bmatrix} Q_1T_2^{-1} & 0 & T_1 & U_1 \\ V_2T_2^{-1} & I_m & 0 & V_1T_1^{-1}U_1 + W_1 \end{bmatrix} \begin{bmatrix} T_2 & 0 \\ -V_2 & V_2T_2^{-1}U_2 + W_2 \\ -T_1^{-1}Q_1 & T_1^{-1}U_1 \\ 0 & -I_r \end{bmatrix} = 0$$

Finally, pre-multiplying this last equation by

$$H_1 := \begin{bmatrix} I_m & 0 \\ -V_1T_1^{-1} & I_r \end{bmatrix}$$

and post-multiplying it by

$$H_2 := \begin{bmatrix} I_m & T_2^{-1}U_2 \\ 0 & I_r \end{bmatrix}$$

yields, after some manipulations,

$$\begin{bmatrix} Q_1T_2^{-1} & 0 \\ V_2T_2^{-1} - V_1T_1^{-1}Q_1T_2^{-1} & I_m \end{bmatrix} \begin{bmatrix} T_2 & U_2 \\ -V_2 & W_2 \end{bmatrix} =: H_3$$

where

$$H_3 = \begin{bmatrix} T_1 & U_1 \\ -V_1 & W_1 \end{bmatrix} \begin{bmatrix} T_1^{-1}Q_1 & T_1^{-1}Q_1T_1^{-1}U_1 - T_2^{-1}U_1 \\ 0 & I_r \end{bmatrix}$$

which is of the form (6.55) and the proof is complete. □

Note 6.3.1. The transformation of (6.55) is based on the use of rational transforming matrices and an open research problem is to determine what extra conditions are required to ensure that (6.55) can be applied using polynomial matrices. Given that this problem is still open for the 1D case, it can be conjectured that if it has a solution in the 2D case then the conditions could be much more restrictive.

In 1D linear systems theory, a major result is that two least-order realizations of a given transfer-function matrix are related by strict system equivalence. Considering the Rosenbrock system matrix in the 1D differential case (s again denotes the Laplace transform variable)

$$P(s) := \begin{bmatrix} T(s) & U(s) \\ -V(s) & W(s) \end{bmatrix}$$

least-order is defined as the absence of input and output decoupling zeros. Equivalently, the pair $T(s), U(s)$ are relatively (left) prime and the pair $T(s), V(s)$ are relatively (right) prime.

In 2D discrete linear systems theory the term least-order can, by analogy with the 1D case, be expressed in terms of (6.53) as the requirement that $T(z_1, z_2)$ and $U(z_1, z_2)$ are minor (left) coprime and $T(z_1, z_2)$ and $V(z_1, z_2)$ are minor (right) coprime. As shown below, however, this is not a satisfactory definition of least-order in terms of having similar properties to the 1D case, i.e. a necessary and sufficient condition for a polynomial equivalence relation to exist between two system matrices of the same system. Later in this section an alternative definition of least-order will be introduced and to simplify notation the obvious definition above expressed in terms of (6.53) will be termed (minor) least-order.

It is clearly not appropriate to define least-order entirely using zero coprimeness as this would exclude a large class of system matrices that contain nonessential singularities of the second kind. A simple example here is $G(z_1, z_2) = \frac{z_1}{z_2}$, which always has a nonessential singularity of the second kind at $(0, 0)$, and any system matrix realization would clearly need to reflect this given the central role (see, for example, [77]) of these nonessential singularities in 2D/nD systems theory (and also they have no 1D systems counterparts).

All systems have a (minor) least-order realization. To detail this point, consider (6.53) as the system matrix of the 2D transfer-function matrix $G(z_1, z_2)$. Then, using the greatest common divisor algorithm [89], compute matrices $Q_1(z_1, z_2)$ and $Q_2(z_1, z_2)$ such that

$$\begin{aligned} P(z_1, z_2) &= \begin{bmatrix} Q_1(z_1, z_2) & 0 \\ 0 & I_m \end{bmatrix} \begin{bmatrix} \overline{T}(z_1, z_2) & \overline{U}(z_1, z_2) \\ -\overline{V}(z_1, z_2) & \overline{W}(z_1, z_2) \end{bmatrix} \\ &\times \begin{bmatrix} Q_2(z_1, z_2) & 0 \\ 0 & I_r \end{bmatrix} \end{aligned}$$

where $\overline{T}(z_1,z_2), \overline{U}(z_1,z_2)$ are minor (left) coprime and $\overline{T}(z_1,z_2), \overline{V}(z_1,z_2)$ are minor (right) coprime and hence a (minor) least-order realization has been constructed.

Example 6.3.1. Consider the following system matrices.

$$\begin{aligned} P_1(z_1,z_2) &= \begin{bmatrix} z_2^2 & z_1 \\ -(z_1-1) & 0 \end{bmatrix} \\ P_2(z_1,z_2) &= \begin{bmatrix} z_2^2 & z_1-1 \\ -z_1 & 0 \end{bmatrix} \end{aligned} \tag{6.57}$$

These are both (minor) least-order system matrices corresponding to

$$G(z_1,z_2) = \frac{z_1(z_1-1)}{z_2^2}$$

Hence by Theorem 6.3.1 the matrices of (6.57) can be related through rational transforming matrices which in this case can be expressed as

$$\begin{aligned} &\begin{bmatrix} q_1(z_1,z_2) & 0 \\ r_1(z_1,z_2) & 1 \end{bmatrix} \begin{bmatrix} z_2^2 & z_1 \\ -(z_1-1) & 0 \end{bmatrix} \\ &= \begin{bmatrix} z_2^2 & z_1-1 \\ -z_1 & 0 \end{bmatrix} \begin{bmatrix} q_2(z_1,z_2) & r_2(z_1,z_2) \\ 0 & 1 \end{bmatrix} \end{aligned} \tag{6.58}$$

for rational functions $q_i(z_1,z_2)$ and $r_i(z_1,z_2)$, $i=1,2$.

Equating the entries in row one, column two of the matrices on the left and right-hand sides of (6.58) now gives

$$q_1(z_1,z_2)z_1 - r_2(z_1,z_2)z_2^2 = z_1 - 1$$

and it follows from Hilbert's Nullstellensatz result (Lemma 1.4.1) that this last equation does not have a polynomial solution for $q_1(z_1,z_2)$ and $r_1(z_1,z_2)$. Consequently a polynomial connection of the form (6.55) does not exist for the system matrices of (6.57).

Interchanging the roles of $P_1(z_1,z_2)$ and $P_2(z_1,z_2)$ in (6.58) yields the equation

$$q_1(z_1-1) = z_2^2 r_2 + z_1$$

on equating the entries in row one, column two as before and again there does not exist a polynomial transformation of the required type. Hence, despite the fact that these system matrices are both (minor) least-order realizations of the same 2D transfer-function, there is no polynomial-based transformation connecting them. In particular, there do not exist q_1, r_2 such that a polynomial transformation of the form required for strict system equivalence based on unimodular matrices can be formed.

This example demonstrates that the well known 1D result that least-order realizations are strictly system equivalent does not generalize to 2D (minor) least-order systems. Also a polynomial transformation of the form in Theorem 6.3.1 may not even exist. If it does exist, however, the following result shows that the nature of this transformation can be established.

Theorem 6.3.2. *Consider the case when two (minor) least-order system matrices $P_i(z_1, z_2)$, $i = 1, 2$, are related by a polynomial form of the relationship*

$$\begin{bmatrix} Q_1(z_1,z_2) & 0 \\ R_1(z_1,z_2) & I_m \end{bmatrix} \begin{bmatrix} T_2(z_1,z_2) & U_2(z_1,z_2) \\ -V_2(z_1,z_2) & W_2(z_1,z_2) \end{bmatrix}$$

$$= \begin{bmatrix} T_1(z_1,z_2) & U_1(z_1,z_2) \\ -V_1(z_1,z_2) & W_1(z_1,z_2) \end{bmatrix} \begin{bmatrix} Q_2(z_1,z_2) & R_2(z_1,z_2) \\ 0 & I_r \end{bmatrix} \tag{6.59}$$

Then $Q_1(z_1, z_2)$ and $T_1(z_1, z_2)$ are (minor) left coprime and $Q_2(z_1, z_2)$ and $T_2(z_1, z_2)$ are (minor) right coprime.

Proof. Write (6.59) in the form

$$\begin{bmatrix} Q_1 & 0 & T_1 & U_1 \\ R_1 & I_m & -V_1 & W_1 \end{bmatrix} \begin{bmatrix} T_2 & U_2 \\ -V_2 & W_2 \\ -Q_2 & -R_2 \\ 0 & -I_r \end{bmatrix} = 0 \tag{6.60}$$

Then the block entry in row one, column two of the left-hand side here yields

$$U_1 = Q_1 U_2 - T_1 R_2 \tag{6.61}$$

in which T_1 and U_1 are minor (left) coprime. Hence by the Bezout identities of Theorem 1.4.2 (part (ii)) $\exists\ \overline{X}_i, \overline{Y}_i,\ i = 1, 2$, such that

$$T_1 \overline{X}_i + U_1 \overline{Y}_i = \psi_i(\underline{z}_i^c) I_m,\ i = 1, 2$$

for scalar polynomials $\psi_i(\underline{z}_i^c)$. This identity and (6.61) now give

$$T_1(\overline{X}_i - R_2 \overline{Y}_i) + Q_1(U_2 \overline{Y}_i) = \psi_i(\underline{z}_i^c) I_m,\ i = 1, 2$$

and hence T_1, Q_1 are minor (left) coprime.

The block entry in row two, column one of (6.60) yields

$$V_2 = R_1 T_2 + V_1 Q_2 \tag{6.62}$$

where V_2 and T_2 are minor (right) coprime and hence by the Bezout identities of Theorem 1.4.2 (again part (ii)) $\exists$ polynomial matrices X_i, Y_i, $i = 1, 2$, such that

$$X_i T_2 + Y_i V_2 = \hat{\psi}_i(\underline{z}_i^c) I_r,\ i = 1, 2 \tag{6.63}$$

Pre-multiplying (6.62) by Y_i and substituting for $Y_i V_2$ from (6.63) yields

$$\hat{\psi}_i(\underline{z}_i^c)I_r - X_iT_2 = Y_iR_1T_2 + Y_iV_1Q_2$$

which implies that

$$(Y_1R_1 + X_i)T_2 + Y_iV_1Q_2 = \hat{\psi}_i(\underline{z}_i^c)I_r,\ i = 1, 2$$

i.e. T_2 and Q_2 are minor (right) coprime and the proof is complete. □

Note 6.3.2. Some conditions of (minor) least-order are redundant in the proof of the last result. These are the minor (right) coprime conditions on T_1 and V_1 and the minor (left) coprime condition on T_2 and U_2.

The following definition introduces an alternative form of least-order by imposing a stronger form of coprimeness on the T, V part of the system matrix. It is then shown that this form is sufficiently general to enable every 2D transfer-function matrix to be realized in this way.

Definition 6.3.3. *A 2D polynomial system matrix $P(z_1, z_2)$ which is of the form (6.53) is said to be least-order if $T(z_1, z_2)$ and $U(z_1, z_2)$ are minor (left) coprime and $T(z_1, z_2)$ and $V(z_1, z_2)$ are zero (right) coprime.*

This definition aims to realize the nonessential singularities of the second kind of the transfer-function matrix in such a way that they always occur in the $[T(z_1, z_2), U(z_1, z_2)]$ part of the system matrix. Note also that $G(z_1, z_2)$ always has least-order realizations of this type since any (left (necessarily minor)) coprime matrix fraction description gives rise to a system matrix that satisfies the last definition here.

Note 6.3.3. It is, of course, not necessary to treat the nonessential singularities of the second kind in this manner and they could equally well be treated in the $\left[T^T(z_1, z_2), -V^T(z_1, z_2)\right]^T$ part of the system matrix. The important point is that it is necessary to agree at the outset exactly how these singularities of $G(z_1, z_2)$ are to be realized in the system matrix.

In spite of these facts, it is still not possible to establish that any two least-order realizations of $G(z_1, z_2)$ are related by a polynomial form of (6.55). It is, however, possible to determine the nature of the polynomial connection when it exists.

Theorem 6.3.3. *Consider two least-order system matrices $P_i(z_1, z_2)$, $i = 1, 2$, which are related by a polynomial form of the relationship*

$$\begin{bmatrix} Q_1(z_1, z_2) & 0 \\ R_1(z_1, z_2) & I_m \end{bmatrix} \begin{bmatrix} T_2(z_1, z_2) & U_2(z_1, z_2) \\ -V_2(z_1, z_2) & W_2(z_1, z_2) \end{bmatrix}$$
$$= \begin{bmatrix} T_1(z_1, z_2) & U_1(z_1, z_2) \\ -V_1(z_1, z_2) & W_1(z_1, z_2) \end{bmatrix} \begin{bmatrix} Q_2(z_1, z_2) & R_2(z_1, z_2) \\ 0 & I_r \end{bmatrix}$$

Then $Q_1(z_1, z_2)$ and $T_1(z_1, z_2)$ are zero (left) coprime and $Q_2(z_1, z_2)$ and $T_2(z_1, z_2)$ are zero (right) coprime.

Proof. Consider the case of Q_2 and T_2. Then following arguments similar to those used in establishing Theorem 6.3.2 shows that Q_2 and T_2 are zero (right) coprime and since $T_1^{-1}Q_1 = Q_2T_2^{-1}$ are 2D irreducible matrix fraction descriptions they must be of the same type. Hence since Q_2 and T_2 are zero (right) coprime it follows that Q_1 and T_1 are zero (left) coprime. □

At this stage, it has been shown that any two system matrix realizations of the same system are related by a closed-form matrix relationship as defined by (6.55) in which the transforming matrices are, in general, rational. This has been shown to be a necessary and sufficient condition for the system matrices to be I/O equivalent. Also if this relationship were polynomial and the system matrices were least-order (as defined here) the T blocks have been shown to possess a zero coprimeness relationship to the Q blocks.

This relationship can be developed further for system matrices that are not necessarily least-order. A major result here is that, under certain coprimeness conditions, the polynomial form of the relationship (6.59) is an equivalence relation. Also this analysis can be extended to the nD, $n > 2$, case.

One of the major considerations for the equivalence of system matrices is the invariance of the zeros possessed by the system. In the remainder of this section, a possible definition of a zero for a differential linear repetitive process of the form defined by Example 1.2.3 is introduced and characterized. Also the feasibility of constructing a least-order system matrix in the form of Definition 6.3.2 is investigated. (The extension of these results to the discrete case is routine and hence not detailed here.)

Return now to the transfer-function matrices $G_0(s)$ and $G_1(s)$ of Chap. 1 which, in effect, define the 2D transfer-function matrix $G(s, z_2)$ for a differential unit memory linear repetitive process of the form defined by Example 1.2.3 and, in particular, (see (1.56))

$$\begin{aligned} G_0(s) &= D^{-1}(s)N(s) \in \mathbb{R}^{m\times r}(s) \\ G_1(s) &= \overline{N}(s)\overline{D}^{-1}(s) \in \mathbb{R}^{m\times m}(s) \end{aligned}$$

where $N(s)$ and $D(s)$ are (zero) left coprime and $\overline{N}(s)$ and $\overline{D}(s)$ are (zero) right coprime. Also the coprimeness type is zero because all matrices involved are polynomial in one indeterminate and consequently all three definitions of 2D coprimeness are equivalent. Hence the 2D transfer-function matrix $G(s, z)$ can (see (1.57)) be written as

$$\begin{aligned} G(s, z_2) &= \overline{D}(s)\left[\overline{D}(s) - z_2\overline{N}(s)\right]^{-1} D^{-1}(s)N(s) \\ &= \overline{D}(s)\left[D(s)(\overline{D}(s) - z_2\overline{N}(s)\right]^{-1} N(s) \end{aligned} \tag{6.64}$$

and a Rosenbrock system matrix (see (1.58)) over $\mathbb{R}[s, z_2]$ giving rise to $G(s, z_2)$ is the following $2m \times (m + r)$ polynomial matrix

$$P(s,z_2)=\begin{bmatrix} D(s)\left[\overline{D}(s)-z_2\overline{N}(s)\right] & N(s) \\ -\overline{D}(s) & 0 \end{bmatrix} \tag{6.65}$$

Below, following [90], it is shown that the zeros in this case can be completely characterized for the output matrix pair $D(s)\left[\overline{D}(s)-z_2\overline{N}(s)\right], \overline{D}(s)$ and partially for the input matrix pair $D(s)\left[\overline{D}(s)-z_2\overline{N}(s)\right], N(s)$. Also conditions are derived such that $P(s,z_2)$ is least-order in the sense of Definition 6.3.3.

Consider first the output zeros which are formally defined as follows.

Definition 6.3.4. *The output zeros of a differential unit memory linear repetitive process of the form defined by Example 1.2.3 are those values of* $(s,z_2)\in\mathbb{C}\times\mathbb{C}$ *such that*

$$Z_{od}(s,z_2):=\begin{bmatrix} D(s)[\overline{D}(s)-z_2\overline{N}(s)] \\ -\overline{D}(s) \end{bmatrix} \tag{6.66}$$

does not have full rank.

The following result can now be established.

Lemma 6.3.1. *The matrix of (6.66) does not have full rank whenever the matrix*

$$\tilde{Z}_{od}(s,z_2):=\begin{bmatrix} -z_2 D(s)\overline{N}(s) \\ \overline{D}(s) \end{bmatrix} \tag{6.67}$$

does not have full rank.

Proof. Simply consider pre-multiplication of $\tilde{Z}_{od}(s,z_2)$ by a unimodular matrix, i.e.

$$\begin{bmatrix} I_m & D(s) \\ 0 & I_m \end{bmatrix}\tilde{Z}_{od}(s,z_2)=\begin{bmatrix} -z_2 D(s)\overline{N}(s) \\ -\overline{D}(s) \end{bmatrix}$$

and let $Z(s)$ be the greatest common right divisor of $D(s)\overline{N}(s), \overline{D}(s)$. Then

$$\begin{bmatrix} D(s)\overline{N}(s) \\ \overline{D}(s) \end{bmatrix}=\begin{bmatrix} \widetilde{N}(s) \\ \widetilde{D}(s) \end{bmatrix}Z(s) \tag{6.68}$$

where $\widetilde{N}(s)$ and $\widetilde{D}(s)$ are $m\times m$ polynomial matrices, i.e. the zeros of (6.67) correspond to the values of $s:\det Z(s)=0$. □

The zeros of the last result are removable and are attributable to the realization and not the transfer-function matrix. Also the only other type of (output) zeros of (6.65) are those such that $z_2=0$. Hence the non-removable output zeros for $G(s,z_2)$ written in the form (6.64) are given by the set

$$\Omega_1:=\{(s_{\mathrm{rz}},0):\det\widetilde{D}(s_{\mathrm{rz}})=0\}$$

These are nonessential singularities of the second kind of (6.65) since, at these points, the system matrix does not have full rank and

$$\det D(s_{rz}) \ \det(\widetilde{D}(s_{rz})Z(s_{rz}) - z_2\widetilde{N}(s_{rz})Z(s_{rz})) = 0$$

The following result now gives a condition for a differential unit memory linear repetitive process of the form defined by Example 1.2.3 to have a least-order realization in the sense of Definition 6.3.3.

Theorem 6.3.4. *A least-order realization of*

$$G(s, z_2) = (I_m - G_1(s)z_2)^{-1}G_0(s)$$

can be derived from $P(s, z_2)$ of (6.65) using, in essence, 1D linear systems techniques if one of the following hold

(a) $\widetilde{D}(s)$ of (6.68) is a unimodular matrix,

(b) $\overline{D}(s)$ is a right divisor of $D(s)\overline{N}(s)$,

(c) $D(s)\overline{N}(s)\overline{D}^{-1}(s)$ is a polynomial matrix.

Proof. Here we need only to show (a). In this case, note that a system matrix realization is least-order if the input pair are minor (right) coprime and the output pair are zero (left) coprime. A system matrix realization in which both the input and output pairs are minor coprime is achieved by extracting the greatest common divisor of the input and output pair, i.e. let $Z_r(s)$ be a greatest common (right) divisor of the output pair such that

$$P(s, z_2) = \begin{bmatrix} \tilde{D}(s) - z_2\tilde{N}(s) & N(s) \\ -\tilde{D}(s) & 0 \end{bmatrix} \begin{bmatrix} Z_r(s) & 0 \\ 0 & 0 \end{bmatrix}$$

and let $Z_l(s)$ be a greatest common left divisor of $\tilde{D}(s) - z_2\tilde{N}(s), \tilde{N}(s)$, i.e.

$$\begin{bmatrix} \tilde{D}(s) - z_2\tilde{N}(s) & N(s) \end{bmatrix} = Z_l(s) \begin{bmatrix} \tilde{N}'(s, z_2) & N'(s) \end{bmatrix}$$

Then

$$P(s, z_2) = \begin{bmatrix} R(s) & 0 \\ 0 & 0 \end{bmatrix} P'(s, z_2) \begin{bmatrix} Z_r(s) & 0 \\ 0 & 0 \end{bmatrix}$$

where

$$P'(s, z_2) := \begin{bmatrix} \tilde{N}'(s, z_2) & N'(s) \\ -\tilde{D}(s) & 0 \end{bmatrix}$$

If $\tilde{D}(s)$ is a unimodular matrix the output pair possess a constant high-order minor therefore $\tilde{D}(s), \tilde{N}'(s, z)$ are zero (right) coprime and the proof is complete. □

Note 6.3.4. If $\widetilde{D}(s)$ is not unimodular, it may be possible to apply constant output feedback control around $G_1(s)$ to remove these nonessential singularities of the second kind, i.e. use feedback control around $G_1(s)$ to create a unimodular $\tilde{D}(s)$ such that the new system matrix is least-order.

As a consequence of the structure of the system matrix, it is not possible to explicitly characterize the input zeros of a differential unit memory linear repetitive process of the form defined by Example 1.2.3. If, however, $m \leq r$ then it is possible to give a set of values of which these zeros are a subset and hence narrow the possible candidates to a finite set. Note also that this assumption is not unrealistic in practical terms since it means that there are not more outputs than inputs. (If there are more outputs than inputs then it can be shown that some of the outputs are dependent on the others.)

Definition 6.3.5. *The input zeros of the system matrix of Definition 6.3.3 are defined to be the values* $(s, z_2) \in \mathbb{C} \times \mathbb{C}$ *such that*

$$Z_{ic}(s, z_2) := \begin{bmatrix} T(s, z_2) & U(s, z_2) \end{bmatrix}$$

does not have full row rank.

Clearly a necessary condition for (6.65) not to have any zeros is that $N(s)$ has full rank $\forall s \in \mathbb{C}$. Also a necessary condition for (s, z_2) to be an input zero of (6.65) is that it is simultaneously a zero of the two constituent matrices, i.e. $D(s)\left[\overline{D}(s) - z_2\overline{N}(s)\right]$ and $N(s)$, but these conditions do not guarantee that it is also a zero of the compound matrix of (6.65). A set of candidate input zeros for (6.65), denoted by Ω_{input} can now be derived by characterizing the set of values of (s, z_2) that are simultaneous zeros of $D(s)\left[\overline{D}(s) - z_2\overline{N}(s)\right]$ and $N(s)$.

Theorem 6.3.5. *Suppose that* $m \leq r$ *and let* Ω_{input} *denote the set of input zeros of a differential unit memory linear repetitive process of the form defined by Example 1.2.3. Then*

$$\Omega_{\text{input}} \subseteq \cup_{s_i \in A_N} \Omega_{s_i}$$

where

$$\begin{aligned} \Omega_{s_i} &:= \{(s_i, \tilde{z}_2) : \det(\overline{D}(s_i) - \tilde{z}_2\overline{N}(s_i)) = 0\} \\ A_N &:= \{s_{\text{iz}} \in \mathbb{C} \; : \; N(s_{\text{iz}}) \text{ has less than full rank}\} \end{aligned}$$

Proof. By definition, the input zeros in this case are defined to be the values $(s_{\text{iz}}, \tilde{z}_2) \in \mathbb{C} \times \mathbb{C}$ such that $\begin{bmatrix} D(s_{\text{iz}})(\overline{D}(s_{\text{iz}}) - \tilde{z}_2\overline{N}(s_{\text{iz}})) & N(s_{\text{iz}}) \end{bmatrix}$ does not have full rank. Suppose, therefore, that this matrix does not have full (row) rank. Then $N(s_{\text{iz}})$ must lose (row) rank, i.e. s_{iz} is a zero of the previous pass transfer-function matrix. Also

$$\begin{bmatrix} D(s_{\text{iz}})(\overline{D}(s_{\text{iz}})\tilde{z}_2 - \overline{N}(s_{\text{iz}})) & N(s_{\text{iz}}) \end{bmatrix} \begin{bmatrix} \overline{D}(s_{\text{iz}}) - \tilde{z}_2 N(s_{\text{iz}}) & 0 \\ 0 & I_r \end{bmatrix} =$$

$$\begin{bmatrix} D(s_{\text{iz}}) & N(s_{\text{iz}}) \end{bmatrix} \begin{bmatrix} \overline{D}(s_{\text{iz}}) - \tilde{z}_2 N(s_{\text{iz}}) & 0 \\ 0 & I_r \end{bmatrix}$$

and it follows that the only way for $\left[\begin{array}{cc} D(s_{\text{iz}})(\overline{D}(s_{\text{iz}})\tilde{z}_2 - \overline{N}(s_{\text{iz}})) & N(s_{\text{iz}}) \end{array}\right]$ to lose rank is for

$$\det(\overline{D}(s_{\text{iz}}) - \tilde{z}_2\overline{N}(s_{\text{iz}})) = 0$$

As s_{iz} is fixed, this last equation determines the appropriate values of $\tilde{z}_2$. □

6.4 Poles and Zeros – A Behavioral Approach

There has long been interest in developing physically meaningful definitions of poles and zeros for 2D linear systems and repetitive processes. This is also reflected in the analysis of the previous section on zeros for a differential unit memory linear repetitive process of the form defined by Example 1.2.3. Some of the previous work (for 2D linear systems) has produced definitions (and characterizations) that are purely algebraic in nature. In particular, these results do not preserve the exponential trajectory interpretation of poles and zeros for 1D differential and discrete linear systems. Recently, the behavioral setting (for an introduction to the behavioral approach (for the 1D case) see, for example, [135]) has been used to develop a theory of poles for general nD linear systems. Here we apply this to a discrete unit memory linear repetitive process of the form defined by Example 1.2.10 and relate the results to stability. Also it is shown how this approach can be extended define and characterize zeros. Next we give the relevant background.

6.4.1 Behavioral Theory – Background

Consider again a discrete unit memory linear repetitive process of the form defined by Example 1.2.10. Then, since the state on pass 0 plays no role in stability analysis (see also Sect. 2.4 relating to the development of the 2D transfer-function matrix description), it is convenient to re-label the state vector as $x_{k+1}(p) \mapsto x_k(p)$ to obtain

$$\begin{aligned} x_k(p+1) &= Ax_k(p) + Bu_{k+1}(p) + B_0y_k(p) \\ y_{k+1}(p) &= Cx_k(p) + Du_{k+1}(p) + D_0y_k(p) \end{aligned} \qquad (6.69)$$

We characterize stability along the pass in what follows as the requirement that

$$\rho(\sigma_1, \sigma_2) := \det\left(\left[\begin{array}{cc} \sigma_1 I_n - A & -B_0 \\ -C & \sigma_2 I_m - D_0 \end{array}\right]\right) \neq 0 \text{ in } \mathbb{U}_b^2 \qquad (6.70)$$

where $\mathbb{U}_b^2 := \{(\sigma_1, \sigma_2) : |\sigma_1| \geq 1,\ |\sigma_2| \geq 1\}$. (This is the condition of (5.1) where, for ease of notation we have set $\sigma_i = z_i^{-1}, i = 1, 2.$)

In behavioral theory, the system (or process here) is viewed (following the notation in, for example, [177]) as a triple $(\mathcal{A}, q, \mathcal{B})$, where the signal space $\mathcal{A}$ is a vector space over the field $k = \mathbb{C}$ (or $\mathbb{R}$) or more generally a $\mathbb{K}[\sigma]$-module of n-dimensional mappings, where here $n = 2$. The signal space $\mathcal{A}$ here is taken as the discrete space $k^{\mathbb{N}}$, q is the number of system variables, and the behavior $\mathcal{B} \subseteq \mathcal{A}^q$ is the solution space of the finite set of 2-dimensional difference equations describing the process dynamics. In particular, the behavior for the processes considered here is given by

$$\mathcal{B}_{x,u,y} = \left\{ \begin{pmatrix} x \\ u \\ y \end{pmatrix} \in \mathcal{A}^{\bullet} \;\middle|\; \begin{pmatrix} x \\ u \\ y \end{pmatrix} \text{ satisfy (6.69)} \right\}$$

Note also that $\mathcal{A}^{\bullet}$ denotes the appropriate number of copies of the signal space $\mathcal{A}$.

For the polynomial ring $\mathbb{K}[\sigma]$, the ring action $\mathbb{K}[\sigma] \times \mathcal{A} \longrightarrow \mathcal{A}$ for discrete processes is defined as the shift operator $\sigma_i, i = 1, 2$. Using this notation we can write any repetitive process described by (6.69) in the form of a behavior, and similarly we can write any sub-process (such as when the output pass profile vector is zero – see below) in terms of sub-behaviors. For example, we can write (6.69) in a behavioral kernel representation [176] as

$$\mathcal{B}_{x,u,y} = Ker_{\mathcal{A}} \begin{pmatrix} \sigma_1 I_n - A & -\sigma_2 B & -B_0 \\ -C & -\sigma_2 D & \sigma_2 I_m - D_0 \end{pmatrix} \subseteq \mathcal{A}^q$$

where we use the suffix notation $Ker_{\mathcal{A}}$ to denote the kernel of the ring action of the matrix. An important sub-behavior is the one containing all outputs that are zero, i.e. the sub-behavior

$$\mathcal{B}_{x,u,0} := \left\{ \begin{pmatrix} x \\ u \\ y \end{pmatrix} \in \mathcal{B}_{x,u,y} \mid y = 0 \right\}$$

which we see is given by the kernel representation

$$\mathcal{B}_{x,u,0} \cong \mathrm{Ker}_{\mathcal{A}} \begin{pmatrix} \sigma_1 I_n - A & -\sigma_2 B \\ -C & -\sigma_2 D \end{pmatrix} \tag{6.71}$$

and also the zero input sub-behavior

$$\mathcal{B}_{x,0,y} \cong \mathrm{Ker}_{\mathcal{A}} \begin{pmatrix} \sigma_1 I_n - A & -B_0 \\ -C & \sigma_2 I_m - D_0 \end{pmatrix}$$

The first of the sub-behavior will be very important when considering invariant zeros and the second when considering poles. For example, the invariant zeros in the 1D case are given by the set of points in $\mathbb{C}$ where the Rosenbrock system matrix loses rank (see, for example [113]). We will show below that the definition introduced for the invariant zeros of a discrete unit memory linear repetitive process of the form defined by Example 1.2.10 are the varieties in $\mathbb{C}^2$ such that a 2D polynomial matrix loses rank, and hence our analysis will have to consider the rank loss points of such a matrix. We next outline some preliminary results concerning behaviors.

For any matrix $E \in \mathcal{R}^{g,q}$, where $\mathcal{R}$ is some ring, define the modules

$$
\begin{aligned}
\mathrm{Ker}_{\mathcal{R}} E &:= \{v \in \mathcal{R}^{1,g} \,|vE = 0\} \\
\mathrm{Im}_{\mathcal{R}} E &:= \{v \in \mathcal{R}^{1,q} \,\big|v = xE \text{ for some } x \in \mathcal{R}^{1,g}\} \\
\mathrm{Coker}_{\mathcal{R}} E &:= \mathcal{R}^{1,q}/\mathrm{Im}_{\mathcal{R}} E \\
\mathrm{Im}_{\mathcal{A}} E &:= \{w \in \mathcal{A}^{g} \mid w = Ef \text{ for some f} \in \mathcal{A}^{\mathrm{q}}\} \\
\mathrm{Ker}_{\mathcal{A}} E &:= \{w \in \mathcal{A}^{q} \mid Ew = 0\}
\end{aligned}
$$

Note the different subscripts used to denote different ring actions. The modules $\mathrm{Ker}_{\mathcal{R}} E, \mathrm{Im}_{\mathcal{R}} E$, and $\mathrm{Coker}_{\mathcal{R}} E$ are defined with respect to a left action on E, whereas $\mathrm{Ker}_{\mathcal{A}} E$ and $\mathrm{Im}_{\mathcal{A}} E$ are defined with respect to a right action.

Let M be a finitely generated $\mathcal{R}$-module. The dual of M with respect to $\mathcal{A}$, denoted $D(M)$, is defined by

$$D(M) := \mathrm{Hom}_{\mathcal{R}}(M, \mathcal{A})$$

If $\psi : M \to N$ is a morphism of finitely generated $\mathcal{R}$-modules, then the dual map $D(\psi) : D(N) \to D(M)$ is given by $\forall\, v \in D(N), (D(\psi))(v) := v \circ \psi$. The next result (which also holds for behaviors defined by the partial differential equation operator $\frac{\partial}{\partial t_i}$ and for the nD, $n > 2$ case but here we only state the relevant case) tells us precisely what the module M is.

Theorem 6.4.1. *[118] Discrete behaviors are precisely the dual modules of finitely generated $\mathcal{R}$-modules. Specifically, if $\mathcal{B} = \mathrm{Ker}_{\mathcal{A}} E$ then $\mathcal{B} = D(M)$, where M is the finitely generated module* $\mathrm{Coker}_{\mathcal{R}} E$.

For each signal space $\mathcal{A}$ (an injective cogenerator) we have the important property that given a complex of modules

$$\dots \longrightarrow M_{i+2} \xrightarrow{\phi_{i+1}} M_{i+1} \xrightarrow{\phi_i} M_i \longrightarrow \dots \tag{6.72}$$

and its dual complex

$$\dots \longrightarrow D(M_i) \xrightarrow{D(\phi_i)} D(M_{i+1}) \xrightarrow{D(\phi_{i+1})} D(M_{i+2}) \longrightarrow \dots \tag{6.73}$$

then (6.72) is exact if, and only if, (6.73) is exact. We define the submodule $\mathcal{B}^{\perp} \subset \mathcal{R}^{1,q}$ called the orthogonal module as

$$\mathcal{B}^{\perp} := \{v \in \mathcal{R}^{1,q} \,|vw = 0 \text{ for all } w \in \mathcal{B}\}$$

Consequently if $\mathcal{B} = \mathrm{Ker}_{\mathcal{A}} E$ then $\mathcal{B}^{\perp} = \mathrm{Im}_{\mathcal{R}} E$ and therefore M is the finitely generated module $\mathrm{Coker}_{\mathcal{R}} E = \mathcal{R}^{1,q}/\mathcal{B}^{\perp}$.

The set of variables $\{w_i \mid i \in \mathcal{G}\}$ for some subset $\mathcal{G}$ of $\{1, \dots, q\}$ is said to be free if the mapping $\rho : \mathcal{A}^q \longrightarrow \mathcal{A}^{\mathcal{G}}$, which projects a trajectory onto the components of $\mathcal{G}$, is epic when restricted to $\mathcal{B}$. The maximum cardinality of such a set $\mathcal{G}$ is an invariant of the behavior, is denoted here by $m(\mathcal{B})$, and is given by

$$m(\mathcal{B}) = q - \text{rank}(E)$$

where $\mathcal{B} = \text{Ker}_{\mathcal{A}} E$. The number of free variables is an additive property, i.e. given the sub-behavior $\mathcal{B}' \subset \mathcal{B}$ and the short exact sequence

$$0 \longrightarrow \mathcal{B}' \longrightarrow \mathcal{B} \longrightarrow \mathcal{B}/\mathcal{B}' \longrightarrow 0$$

where $\mathcal{B}/\mathcal{B}'$ has the structural properties of a behavior [177], then $m(\mathcal{B}) = m(\mathcal{B}') + m(\mathcal{B}/\mathcal{B}')$.

Consider the generators $\{e_1, \ldots, e_q\}$ of the free module $\mathcal{R}^{1,q}$. The module M is generated by the set of generators $e_1 + \mathcal{B}^\perp, \ldots, e_q + \mathcal{B}^\perp$, and for a maximal set of free variables indexed by $\mathcal{G} \subset \{1, \ldots, q\}$, the corresponding set of elements $\{e_i + \mathcal{B}^\perp \mid i \in \mathcal{G}\}$ form a set of $m(\mathcal{B})$ linearly independent elements of M, i.e. they generate a free sub-module, Λ, of M. The system variables are assumed to be partitioned into inputs u, which are free and outputs x, y, which contain no free elements (once the inputs are fixed).A partitioning of the form described above is called an input/output structure of the behavior and we have the following construction on M. Suppose $\mathcal{B}$ has h latent variables x (which are to be treated as outputs – in the 1D case these form the state variables), r input variables u and w output variables y. Moreover, if we let $\mathcal{G} = \{h+1, \ldots, h+r\}$, then

$$\Lambda = \langle e_{h+1} + \mathcal{B}^\perp, \ldots, e_{h+w} + \mathcal{B}^\perp \rangle$$

Similarly, let

$$\begin{aligned} \Omega_1 &:= \langle e_1 + \mathcal{B}^\perp, \ldots, e_h + \mathcal{B}^\perp \rangle \\ \Omega_2 &:= \langle e_{h+w+1} + \mathcal{B}^\perp, \ldots, e_q + \mathcal{B}^\perp \rangle \end{aligned}$$

where Ω_1 and Ω_2 correspond to x and y respectively. Then $M = \Omega_1 + \Lambda + \Omega_2$ and

$$\mathcal{B}_{x,u,y} = D(M) = D(\Omega_1 + \Lambda + \Omega_2)$$

We define the annihilator of a behavior $\mathcal{B}$ as

$$\text{ann}\, \mathcal{B} = \{s \in \mathcal{R} \mid s\widetilde{w} = 0 \,\forall\, \widetilde{w} \in \mathcal{B}\}$$

From [176] we have ann $\mathcal{B}$ = ann M. A behavior containing no free variables is an autonomous behavior and is precisely one which has a non-zero annihilator. In the discrete case, we say that a behavior [177] $\mathcal{B}$ with signal domain $T = \mathbb{N}^n$ is controllable if there exits a number $\rho > 0$ such that for any sets $T_1, T_2 \subseteq T$ with $d(T_1, T_2) > \rho$, for any $\underline{b}_1, \underline{b}_2 \in T$, and any two trajectories $w^{(1)}, w^{(2)} \in \mathcal{B}$, there exists $w \in \mathcal{B}$ such that

$$w(\underline{t}) = \begin{cases} w^{(1)}(\underline{t} - \underline{b}_1) & \text{if } \underline{t} \in T_1 \text{ and } \underline{t} - \underline{b}_1 \in T \\ w^{(2)}(\underline{t} - \underline{b}_2) & \text{if } \underline{t} \in T_2 \text{ and } \underline{t} - \underline{b}_2 \in T \end{cases}$$

where we can take $\underline{b}_1 = 0$ without loss of generality. Controllability in this setting expresses the idea of being able to join with a system trajectory any two system trajectories defined on regions which are sufficiently far apart. Characterizations for this property are well known in the literature and in the case of the discrete linear repetitive processes considered here further progress is possible [166].

For a given behavior, we define the controllable part as the unique maximal controllable sub-behavior, and we denote this, the controllable part of $\mathcal{B}$, by $\mathcal{B}^c$. It is well known that for $\mathcal{B} = D(M)$,

$$\mathcal{B}^c = D(M/tM) \text{ and } \mathcal{B}/\mathcal{B}^c = D(tM)$$

where tM is the torsion submodule of M. For any behavior we can write its corresponding dual module representation using the fact that $M = \Omega_1 + \Lambda + \Omega_2$. As one possible example, we see that

$$\mathcal{B}_{x,u,0} = D(M/\Omega_2)$$

and

$$\mathcal{B}^c_{x,u,0} := \mathcal{B}^c_{x,u,y} \cap \mathcal{B}_{x,u,0} = D(M/tM) \cap D(M/\Omega_2) = D(M/(tM + \Omega_2))$$

6.4.2 Characteristic Varieties

Linear systems with constant coefficients are entirely characterized by the exponential trajectories contained in their behavior. We now give the definition of such trajectories, again for the case of $n = 2$ which is required here.

Definition 6.4.1. *Let $w(t) = w(t_1, t_2) \in \mathcal{A}^q$. Then w is said to be an exponential trajectory of frequency $(a_1, a_2) \in \mathbb{C}^2$ if it is of the form*

$$w(t) = v_0 a_1^{t_1} a_2^{t_2}$$

where $v_0 \in \mathcal{C}^q$. Also w is said to be a polynomial exponential trajectory of pure frequency (a_1, a_2) if it is of the form

$$w(t) = p(t) a_1^{t_1} a_2^{t_2}$$

where $p(t) = p(t_1, t_2) \in \mathbb{C}[t_1, t_2]^q$. A polynomial exponential trajectory is any trajectory which is a finite sum of polynomial trajectories of pure frequencies.

Let $J \subseteq \mathcal{R}$, be an ideal where $k = \mathbb{R}$ or $\mathbb{C}$. Define the variety $V(J)$ as

$$V(J) := \{a \in \mathbb{C}^2 \mid p(a) = 0 \, \forall \, p \in J\}$$

Note that $V(J)$ is defined as a subset of $\mathbb{C}^2$ even when $k = \mathbb{R}$.

Definition 6.4.2. *[176] The characteristic variety of a behavior $\mathcal{B} = \mathrm{Ker}_{\mathcal{A}} R$ is the set $\mathcal{V}(\mathcal{B})$ of all points $(a_1, a_2) \in \mathbb{C}^2$ such that the following equivalent conditions hold*

(a) $(a_1, a_2) \in \mathcal{V}(\mathrm{ann}\mathcal{B})$.
(b) $R(a_1, a_2)$ *has less than full column rank.*
(c) $\mathcal{B}$ *contains a non-zero exponential trajectory of frequency* (a_1, a_2).

The points in $\mathcal{V}(\mathcal{B})$ are called the characteristic points of $\mathcal{B}$.

Note that if $\mathcal{B}$ contains a non-zero polynomial exponential trajectory of pure frequency (a_1, a_2) then by repeated differentiation it also contains a non-zero exponential trajectory of the same frequency.

The next result provides a characterization of the characteristic variety of a factor behavior $\mathcal{B}/\mathcal{B}'$.

Theorem 6.4.2. *[181] Let $\mathcal{B}' \subseteq \mathcal{B}$ be behaviors and $(a_1, a_2) \in \mathbb{C}^2$. Then the following are equivalent;*

(a) (a_1, a_2) *is a characteristic point of* $\mathcal{B}/\mathcal{B}'$.
(b) There exists a polynomial vector x such that $x\tilde{w}' = 0$ for all $\tilde{w}' \in \mathcal{B}'$ but $x\tilde{w}$ equals a non-zero exponential trajectory of frequency (a_1, a_2) for some $\tilde{w} \in \mathcal{B}$.
(c) There exists a polynomial exponential trajectory of pure frequency (a_1, a_2) in $\mathcal{B}\backslash\mathcal{B}'$.

6.4.3 Generalized Characteristic Varieties

In order to define the zeros of a behavior, it is necessary to consider the rank loss points of the representation matrix of $\mathcal{B}$. By rank loss points we mean those values of $(a_1, a_2) \in \mathbb{C}^2$ such that the representation matrix of the behavior loses rank. Consider then the matrix $R \in \mathcal{R}^{g,q}$ of rank $b \leq q$, and let $I_b(R) \subset \mathcal{R}$ denote the ideal generated by the order b minors of R. The rank loss points of R are given by the elements of the variety of the ideal $I_b(R)$, i.e. by $V(I_b(R))$.

Definition 6.4.3. *[53] Let $M = \mathrm{Coker}_{\mathcal{R}} R$. Then for any positive integer i the ith Fitting invariant of M denoted by $\mathrm{Fitt}_i\, M$, is defined as*

$$\mathrm{Fitt}_i\, M = I_{q-i}(R)$$

where q is the number of columns of R.

The following theorem, the first part of which is a well known result, enables us to work exclusively with the module M when considering the rank loss points of the representation matrix R.

Theorem 6.4.3. *[28] For any finitely generated module M over a commutative domain,* $\mathrm{rad}(\mathrm{ann}M) = \mathrm{rad}(Fitt_0 M)$. *Furthermore for any $n \geq 0$ then*

$$\mathrm{rad}(\mathrm{Fitt}_n M) = \mathrm{rad}(\mathrm{ann}(\bigwedge^{n+1})M)$$

Since ideals with the same radical have the same variety, Definition 6.4.3 and Theorem 6.4.3 immediately give

$$V(I_b(R)) = V(\mathrm{Fitt}_{q-b}(M)) = V(\mathrm{ann}\,(\bigwedge^{q-b+1})M) \tag{6.74}$$

where $m(\mathcal{B}) = q - b$, and $\bigwedge$ denotes the wedge product. Since the rank loss points of R are given by (6.74), these points are in fact independent of the choice of representation matrix R. We therefore speak of the rank loss points of $\mathcal{B}$ or M.

The following definition generalizes the concept of the characteristic variety.

Definition 6.4.4. *Let $\mathcal{B} = D(M)$ be an nD behavior and define the generalized characteristic variety of $\mathcal{B}$ to be the variety*

$$\mathcal{V}(\mathcal{B}) := \mathrm{ann}(\bigwedge^{m(\mathcal{B})+1})M)$$

and the generalized characteristic points to be the elements of $\mathcal{V}(\mathcal{B})$.

Now we have the following corollary to this last result.

Corollary 6.4.1. *The rank loss points of any kernel representation of $\mathcal{B}$ are precisely the generalized characteristic points of $\mathcal{B}$. Moreover, if $\mathcal{B}$ is autonomous (i.e., M is a torsion module) then $\mathcal{V}(\mathcal{B}) = V(\mathrm{ann}\, M)$.*

The above corollary therefore states that for an autonomous system the rank loss points of $\mathcal{B}$ are precisely the characteristic points of $\mathcal{B}$. The following non-trivial theorem, first established in [180], is central to the development of zeros here (and for nD systems).

Theorem 6.4.4. *For 2D difference behaviors $\mathcal{B}' \subseteq \mathcal{B}$ such that the sequence $0 \longrightarrow \mathcal{B}' \longrightarrow \mathcal{B} \longrightarrow \mathcal{B}/\mathcal{B}' \longrightarrow 0$ is exact, we have the following*

(a) in general $\mathcal{V}(\mathcal{B}) \subseteq \mathcal{V}(\mathcal{B}') \cup \mathcal{V}(\mathcal{B}/\mathcal{B}')$
(b) specifically in the case that $\mathcal{B}/\mathcal{B}'$ is autonomous we have that $\mathcal{V}(\mathcal{B}) = \mathcal{V}(\mathcal{B}') \cup \mathcal{V}(\mathcal{B}/\mathcal{B}')$.

As with the characteristic variety, we have an interpretation of the generalized characteristic points in terms of exponential trajectories and rank loss points.

Theorem 6.4.5. *[181] The following conditions are equivalent for a behavior $\mathcal{B} = \mathrm{Ker}_{\mathcal{A}} R$ and $(a_1, a_2) \in \mathbb{C}^2$*

(a) $(a_1, a_2) \in \mathcal{V}(\mathcal{B})$
(b) the rank of $R(a_1, a_2)$ is less than the rank of $R(z_1, z_2)$
(c) for any of up to $m(\mathcal{B})$ variables ω_i, $\mathcal{B}$ contains a non-zero exponential trajectory of frequency (a_1, a_2) which is zero in the specified components

6.4.4 Poles and Zeros in the Behavioral Setting

Consider the case of poles for a discrete unit memory linear process of the form defined by Example 1.2.10, where it is appropriate to first recall the 1D case. In particular, point a is a pole of the system (defined by the standard state-space model with output vector $y(t)$, state vector $x(t)$, control input vector $u(t)$) if, when zero input $u(t)$ is fed to the system, there exists a non-zero initial condition $x(0)$ such that the resulting state trajectory has the form $x(t) = x(0)a^t$. Note that as $y(t)$ is determined linearly by $x(t)$ ($u(t)$ being zero), $y(t)$ it also be of the form $y(0)a^t$. The analysis which follows is from [158].

In the case considered here, the solutions of the defining equations can be considered as functions from $\mathbb{N}^2$ to $\mathbb{R}$, though for purposes of interpretation they are cut off in one dimension at the pass length α. Also the poles of this process are defined as the characteristic points of the zero-input behavior $\mathcal{B}_{x,0,y}$, i.e. the set of all trajectories which can arise when the input vanishes. The zero-input behavior is given (to within trivial isomorphism) by

$$\begin{pmatrix} \sigma_1 I_n - A & -B_0 \\ -C & \sigma_2 I_m - D_0 \end{pmatrix} \begin{pmatrix} x \\ y \end{pmatrix} = 0 \tag{6.75}$$

Applying Thm/Defn 4.4 of [176], we can define the poles as follows.

Definition 6.4.5. *The poles of a discrete unit memory linear repetitive process whose state-space model can be written in the form (6.69) are the component-wise non-zero points in 2D complex space where the matrix on the left-hand-side of (6.75) fails to have full rank, i.e. they are given by the set*

$$\mathcal{V}(\mathcal{B}_{x,0,y}) \quad := \quad \{(a_1, a_2) \in \mathbb{C}^2 \mid \rho(a_1, a_2) = 0\} \tag{6.76}$$

where $\rho(\sigma_1, \sigma_2)$ is defined by (6.70). The set $\mathcal{V}$ is called the pole variety of the process.

Since in this case the pole variety is given by the vanishing of a single 2D non-unit polynomial, it is guaranteed to be a one-dimensional geometric set in

2D complex space, i.e. a curve. In particular, the pole variety cannot be zero-dimensional (i.e. finite). This corresponds to the fact that proper principal ideals in the ring $\mathbb{C}[\sigma_1, \sigma_2]$ have codimension 1. Note also that the pole variety is a complex variety, even though the entries of the matrices A, B_0, C and D_0 are real. This is essential in order to capture the full exponential-type dynamics of the process.

Poles can be interpreted in terms of exponential trajectories [176], which in the case considered here have a clear physical interpretation. Assume therefore that $(a_1, a_2) \in \mathbb{C}^2$ is such that $\rho(a_1, a_2) = 0$, and write $a_1 = r_1 e^{\imath\theta_1}, a_2 = r_2 e^{\imath\theta_2}$ (with $r_1 = 0$ for $a_1 = 0$ and $r_2 = 0$ for $a_2 = 0$). The existence of such a 'zero' guarantees [176] (and, in particular, Thm/Defn 4.4 of this paper) the existence of an 'exponential trajectory' in the process having the form

$$\begin{aligned} x'_k(p) &= x^1_{00} r_1^p r_2^k \cos(\theta_1 p + \theta_2 k) \\ &+ x^2_{00} r_1^p r_2^k \sin(\theta_1 p + \theta_2 k) \end{aligned} \tag{6.77}$$

$$\begin{aligned} y'_k(p) &= y^1_{00} r_1^p r_2^k \cos(\theta_1 p + \theta_2 k) \\ &+ y^2_{00} r_1^p r_2^k \sin(\theta_1 p + \theta_2 k) \end{aligned} \tag{6.78}$$

$$u'_k(p) = 0 \tag{6.79}$$

where $x^1_{00}, x^2_{00} \in \mathbb{R}^n$, $y^1_{00}, y^2_{00} \in \mathbb{R}^m$, and at least one of these four is non-zero. This form of exponential trajectory has been characterized algebraically in [119]. Conversely, the existence of such a trajectory implies that $\rho(r_1 e^{\imath\theta_1}, r_2 e^{\imath\theta_2}) = 0$, i.e. the 'frequency' $(r_1 e^{\imath\theta_1}, r_2 e^{\imath\theta})$ is a pole of the repetitive process.

In the case where $(a_1, a_2) \in \mathbb{R}^2$, it is straightforward to construct such trajectories from the 'zeroes'. Take a_1 and a_2 to be real numbers satisfying $\rho(a_1, a_2) = 0$. There must then exist a non-zero vector $(x_{00}, y_{00}) \in \mathbb{R}^{n+m}$ satisfying

$$\begin{pmatrix} a_1 I_n - A & -B_0 \\ -C & a_2 I_m - D_0 \end{pmatrix} \begin{pmatrix} x_{00} \\ y_{00} \end{pmatrix} = 0$$

Now extend (x_{00}, y_{00}) to a process trajectory by

$$x'_k(p) = x_{00} a_1^p a_2^k \tag{6.80}$$

$$y'_k(p) = y_{00} a_1^p a_2^k \tag{6.81}$$

$$u'_k(p) = 0 \tag{6.82}$$

A quick computation yields

$$\begin{aligned} x'_k(p+1) &= a_1 x_{00} a_1^p a_2^k \\ &= A x_{00} a_1^p a_2^k + B_0 y_{00} a_1^p a_2^k \\ &= A x'_k(p) + B u'_{k+1}(p) + B_0 y'_k(p) \\ y'_{k+1}(p) &= a_2 y_{00} a_1^p a_2^k \\ &= D_0 y_{00} a_1^p a_2^k + C x_{00} a_1^p a_2^k \\ &= D_0 y'_k(p) + C x'_k(p) \end{aligned}$$

proving that (6.80)–(6.82) indeed describes a solution of the process. Returning to the general case (6.77)–(6.79), we see that if $|a_2| = r \geq 1$ then we have a non-zero exponential (or sinusoidal) state-output trajectory in the process, which tends towards infinity as the pass number increases (but may remain stable along any given pass). Conversely, if $|a_2| = r < 1$ for all poles (a_1, a_2), then no trajectory tends to infinity for a given value of p as the pass number increases, but there may be trajectories tending to infinity along the pass. Thus we again encounter the distinction between asymptotic stability and stability along the pass. In order to avoid having trajectories of the form (6.77)–(6.79) which are unstable either along the pass or in the k-direction, we also need to avoid poles (a_1, a_2) with $|a_1| \geq 1$. Equivalently, we need that the characteristic variety (6.76) of the zero-input behavior lies inside

$$\overline{\mathcal{P}}_1 \ := \ \{(a_1, a_2) \in \mathbb{C}^2 \mid |a_1| \leq 1, |a| \leq 1\}$$

The necessary and sufficient condition of (6.70) states that stability along the pass is equivalent to all poles being inside $\overline{\mathcal{P}}_1$ (recall the 1D case!). Equivalently, with zero input there should be no exponential/ sinusoidal state-output trajectories which tend to infinity either in the pass-to-pass direction or along the pass. (Similar results for the more general setting of stability of 2D behaviors over the lattice $\mathbb{I}^2$ can be found in [173]).

Note finally that poles of the discrete unit memory linear repetitive processes considered here can be decomposed into controllable and uncontrollable, observable and unobservable poles (again recall the 1D case!), as described in [176]. The only one of these sets which can be easily described is the set of unobservable poles, which give the (2D) frequencies which can occur in the state when both input and output vanish. These are given by the rank-loss points of the matrix

$$\begin{pmatrix} \sigma_1 I_n - A \\ -C \end{pmatrix}$$

and so indeed describe the defect of observability.

In the case of zeros for the discrete linear repetitive processes considered here (where again the analysis (and proofs of the results given) is from [158]), first recall the 1D case and, in particular, the system defined over some field $\mathbb{K}$, by the polynomial matrix description

$$\begin{aligned} A(z)x &= B(z)u \\ y &= C(z)x + D(z)u \end{aligned}$$

where z denotes the shift operator, A, B, C, D are polynomial matrices over $\mathbb{K}[z]$ and x, u, y are state variables, input and output variables respectively. Then the invariant zeros of a 1D discrete linear system are the rank loss points of its Rosenbrock system matrix

$$P(z) := \begin{bmatrix} A(z) & -B(z) \\ C(z) & D(z) \end{bmatrix}$$

and it is easy to see that this corresponds to the behavior $\mathcal{B}_{x,u,0}$ and therefore the invariant zeros are given by the generalized characteristic points of the behavior $\mathcal{B}_{x,u,0}$, i.e.

$$\{\text{invariant zeros}\} = \mathcal{V}(\mathcal{B}_{x,u,0})$$

where we term $\mathcal{V}(\mathcal{B}_{x,u,0})$, the invariant zero variety. We can easily generalize this to nD systems and, in particular, for the behavior $\mathcal{B}_{x,u,0}$ of (6.71) here we can define the invariant zero points to be the elements of the variety $\mathcal{V}(\mathcal{B}_{x,u,0})$. As expected we can extend this concept very easily to define controllable and uncontrollable invariant zeros etc to develop a zero structure – the structure $\mathcal{B}_{x,u,y}$ itself provides a map for this via the following pair of exact commutative diagrams

$$\begin{array}{ccccccccc}
 & & 0 & & 0 & & 0 & & \\
 & & \downarrow & & \downarrow & & \downarrow & & \\
0 & \longrightarrow & \mathcal{B}^c_{x,u,0} & \longrightarrow & \mathcal{B}^c_{x,u,y} & \longrightarrow & \mathcal{B}^c_{y} & \longrightarrow & 0 \\
 & & \downarrow & & \downarrow & & \downarrow & & \\
0 & \longrightarrow & \mathcal{B}_{x,u,0} & \longrightarrow & \mathcal{B}_{x,u,y} & \longrightarrow & \mathcal{B}_{y} & \longrightarrow & 0 \\
 & & \downarrow & & \downarrow & & \downarrow & & \\
0 & \longrightarrow & \frac{\mathcal{B}_{x,u,0}}{\mathcal{B}^c_{x,u,0}} & \longrightarrow & \frac{\mathcal{B}_{x,u,y}}{\mathcal{B}^c_{x,u,y}} & \longrightarrow & \frac{\mathcal{B}_{y}}{\mathcal{B}^c_{y}} & \longrightarrow & 0 \\
 & & \downarrow & & \downarrow & & \downarrow & & \\
 & & 0 & & 0 & & 0 & &
\end{array}$$

$$\begin{array}{ccccccccc}
 & & 0 & & 0 & & 0 & & \\
 & & \downarrow & & \downarrow & & \downarrow & & \\
0 & \longrightarrow & \mathcal{B}^c_{x,0,0} & \longrightarrow & \mathcal{B}^c_{x,u,0} & \longrightarrow & \mathcal{B}^c_{u,0} & \longrightarrow & 0 \\
 & & \downarrow & & \downarrow & & \downarrow & & \\
0 & \longrightarrow & \mathcal{B}_{x,0,0} & \longrightarrow & \mathcal{B}_{x,u,0} & \longrightarrow & \mathcal{B}_{u,0} & \longrightarrow & 0 \\
 & & \downarrow & & \downarrow & & \downarrow & & \\
0 & \longrightarrow & \frac{\mathcal{B}_{x,0,0}}{\mathcal{B}^c_{x,0,0}} & \longrightarrow & \frac{\mathcal{B}_{x,u,0}}{\mathcal{B}^c_{x,u,0}} & \longrightarrow & \frac{\mathcal{B}_{u,0}}{\mathcal{B}^c_{u,0}} & \longrightarrow & 0 \\
 & & \downarrow & & \downarrow & & \downarrow & & \\
 & & 0 & & 0 & & 0 & &
\end{array} \tag{6.83}$$

where

$$\mathcal{B}_y := \left\{ y \in \mathcal{A}^m \;\middle|\; \exists \begin{pmatrix} x \\ u \end{pmatrix} \in \mathcal{A}^{n+r};\; \begin{pmatrix} x \\ u \\ y \end{pmatrix} \in \mathcal{B}_{x,u,y} \right\}$$

$$\mathcal{B}_{x,0,0} := \left\{ \begin{pmatrix} x \\ u \\ y \end{pmatrix} \in \mathcal{B}_{x,u,y} \mid u = y = 0 \right\}$$

$$\mathcal{B}_{u,0} := \left\{ \begin{pmatrix} u \\ 0 \end{pmatrix} \in \mathcal{A}^{r+m} \;\middle|\; \exists\, x \in \mathcal{A}^n ; \begin{pmatrix} x \\ u \\ 0 \end{pmatrix} \in \mathcal{B}_{x,u,y} \right\}$$

We make the following definitions.

Definition 6.4.6. *For the behavior $\mathcal{B}_{x,u,y}$ we have the following*

(a) The invariant [invariant controllable] zero variety is defined to be $\mathcal{V}(\mathcal{B}_{x,u,0})$ [$\mathcal{V}(\mathcal{B}^c_{x,u,0})$] and the invariant [invariant controllable] zero points are taken as the elements of $\mathcal{V}(\mathcal{B}_{x,u,0})$ [$\mathcal{V}(\mathcal{B}^c_{x,u,0})$].

(b) The invariant uncontrollable zero variety is defined to be $\mathcal{V}(\mathcal{B}_{x,u,0}/\mathcal{B}^c_{x,u,0})$ and the invariant uncontrollable zero points are taken as the elements of $\mathcal{V}(\mathcal{B}_{x,u,0}/\mathcal{B}^c_{x,u,0})$.

(c) The observable [observable controllable] zero variety is defined to be $\mathcal{V}(\mathcal{B}_{u,0})[\mathcal{V}(\mathcal{B}^c_{u,0})]$ and the observable [observable controllable] zero points are taken as the elements of $\mathcal{V}(\mathcal{B}_{u,0})[\mathcal{V}(\mathcal{B}^c_{u,0})]$.

(d) The observable uncontrollable zero variety is defined to be $\mathcal{V}(\mathcal{B}_{u,0}/\mathcal{B}^c_{u,0})$ and the observable uncontrollable zero points are taken as the elements of $\mathcal{V}(\mathcal{B}_{u,0}/\mathcal{B}^c_{u,0})$.

From [176] the uncontrollable pole points are defined to be the elements of the variety $\mathcal{V}(\mathcal{B}_{x,u,y}/\mathcal{B}^c_{x,u,y})$.

Now we have the following result – for a proof see [180].

Theorem 6.4.6. *For the behavior $\mathcal{B}_{x,u,y}$ we have*

(a) The invariant zero points are precisely the union of the invariant controllable and invariant uncontrollable zero points, i.e.

$$\mathcal{V}(\mathcal{B}_{x,u,0}) = \mathcal{V}(\mathcal{B}^c_{x,u,0}) \cup \mathcal{V}(\mathcal{B}_{x,u,0}/\mathcal{B}^c_{x,u,0})$$

(b) The observable zero points are precisely the union of the observable controllable and observable uncontrollable zero points, i.e.

$$\mathcal{V}(\mathcal{B}_{u,0}) = \mathcal{V}(\mathcal{B}^c_{u,0}) \cup \mathcal{V}(\mathcal{B}_{u,0}/\mathcal{B}^c_{u,0})$$

(c) The invariant uncontrollable zero points are precisely the union of the unobservable uncontrollable and observable uncontrollable zero points, i.e.

$$\mathcal{V}(\mathcal{B}_{x,u,0}/\mathcal{B}^c_{x,u,0}) = \mathcal{V}(\mathcal{B}_{x,0,0}/\mathcal{B}^c_{x,0,0}) \cup \mathcal{V}(\mathcal{B}_{u,0}/\mathcal{B}^c_{u,0})$$

(d) *The invariant uncontrollable zero points are contained in the uncontrollable pole points. In general we have*

$$\mathcal{V}(\mathcal{B}_{x,u,0}/\mathcal{B}^c_{x,u,0}) = \mathcal{V}(\mathcal{B}_{x,u,y}/\mathcal{B}^c_{x,u,y}) \cup \mathcal{V}(\mathcal{B}_y/\mathcal{B}^c_y)$$

(e) *The invariant [controllable invariant] zero points are contained in the union of the unobservable [controllable unobservable] and observable [controllable observable] zeros, i.e.*

$$\begin{aligned} \mathcal{V}(\mathcal{B}_{x,u,0}) &\subset \mathcal{V}(\mathcal{B}_{x,0,0}) \cup \mathcal{V}(\mathcal{B}_{u,0}) \\ \mathcal{V}(\mathcal{B}^c_{x,u,0}) &\subset \mathcal{V}(\mathcal{B}^c_{x,0,0}) \cup \mathcal{V}(\mathcal{B}^c_{u,0}) \end{aligned}$$

In fact we can now show that the transmission zeros are not only contained in the invariant zeros but in the controllable invariant zeros – a subset of the invariant zeros. We need the following result from [180]

Lemma 6.4.1. *For any 1D differential/difference behavior $\mathcal{B} = D(M)$, and any submodule $L \subset M$, we have for $\mathcal{B}' = D(L)$, that $\mathcal{V}(\mathcal{B}') \subset \mathcal{V}(\mathcal{B})$.*

As we have already noted the invariant zeros in the classical framework correspond to the invariant zeros in the behavioral framework. Similarly the transmission zeros correspond to the observable controllable zeros. Therefore applying Lemma 6.4.1 to the exact commutative diagram (6.83), we have the following results for the 1D case

(a) The observable zero variety is contained in the invariant zero variety, i.e.

$$\mathcal{V}(\mathcal{B}_{u,0}) \subset \mathcal{V}(\mathcal{B}_{x,u,0})$$

(b) The observable controllable zero variety is contained in the invariant controllable zero variety, i.e.

$$\mathcal{V}(\mathcal{B}^c_{u,0}) \subset \mathcal{V}(\mathcal{B}^c_{x,u,0})$$

From (b) of this last result we see in the 1D case that the transmission zeros (observable controllable zeros) are certainly contained in the invariant zeros (since the invariant zeros are the union of the invariant controllable and invariant uncontrollable zeros). More precisely, they are, in fact, contained in the invariant controllable zeros. We have the following physical characterization of invariant zeros in terms of exponential and polynomial exponential trajectories.

Lemma 6.4.2. *[180] Let $\mathcal{B}^c_{x,u,0} \subset \mathcal{B}_{x,u,0}$ where always $m' = m(\mathcal{B}^c_{x,u,0}) = m(\mathcal{B}_{x,u,0})$. Then we have the following*

(a) The point $\zeta \in \mathbb{C}^n$ is an invariant [resp. controllable] zero point of $\mathcal{B}$ if, and only if, for any choice of up to m' free (input) variables, $\exists$ a non-zero exponential trajectory of frequency ζ contained in $\mathcal{B}_{x,u,0}$ [resp. $\mathcal{B}^c_{x,u,0}$] with given choice of variables set to zero.

(b) The point $\zeta \in \mathbb{C}^n$ is an invariant uncontrollable zero point of $\mathcal{B}$ if, and only if, there exists a non-zero polynomial exponential trajectory of frequency ζ contained in $\mathcal{B}_{x,u,0}$ but not in $\mathcal{B}^c_{x,u,0}$.

We now consider the case when the behavior $\mathcal{B}_{x,u,y}$ is a 2D behavior associated with a discrete linear repetitive process of the form considered here. We shall refer to the unobservable pole points as output decoupling zeros, and the unobservable uncontrollable pole points as the input-output decoupling zeros, and the observable controllable zero points as the transmission zeros. Now we have the following result.

Lemma 6.4.3. *Let $\mathcal{B}_{x,u,y}$ be an 2D behavior as described above for the discrete linear repetitive processes considered here. Then we have the following.*

(a) The uncontrollable observable zeros and input output decoupling zeros are always contained in the invariant zeros, i.e.

$$\{i.o.d.z\} \cup \{\text{unc. obs. zeros}\} \subset \{\text{invariant zeros}\}$$

(b) If $\mathcal{B}_{x,u,y}$ is such that the outputs y are free, then the input decoupling zeros are contained in the invariant zeros, i.e.

$$\{\text{input dec. zeros}\} \subset \{\text{invariant zeros}\}$$

(c) If $\mathcal{B}_{x,u,0}$ is autonomous then the transmission zeros and output decoupling zeros are contained in the invariant zeros, i.e.

$$\begin{aligned} \{\text{output dec. zeros}\} &\subset \{\text{invariant zeros}\} \\ \{\text{transmission zeros}\} &\subset \{\text{invariant zeros}\} \end{aligned}$$

At this stage return to the kernel representation of $\mathcal{B}_{x,u,y}$. Then given this representation, we can find the kernel representation of the controllable part by first supposing that

$$\mathcal{B}^c_{x,u,y} = \text{Ker}_{\mathcal{A}} \begin{pmatrix} R^c_x & R^c_u & R^c_y \end{pmatrix} \subseteq \mathcal{A}^q$$

Then we have

$$\begin{aligned} \mathcal{B}_{x,0,0} &= \text{Ker}_{\mathcal{A}} \begin{pmatrix} A \\ C \end{pmatrix} \\ \mathcal{B}^c_{x,0,0} &= \text{Ker}_{\mathcal{A}} (R^c_x) \end{aligned}$$

Similarly, by eliminating the latent variables x we get

$$\begin{aligned}
\mathcal{B}_{u,y} &= \operatorname{Ker}_{\mathcal{A}} \begin{pmatrix} R_u^m & R_y^m \end{pmatrix} \\
\mathcal{B}_{u,y}^c &= \operatorname{Ker}_{\mathcal{A}} \begin{pmatrix} R_u^{mc} & R_y^{mc} \end{pmatrix} \\
\mathcal{B}_{0,y}^c &= \operatorname{Ker}_{\mathcal{A}} (R_y^{mc}) \\
\mathcal{B}_{u,0}^c &= \operatorname{Ker}_{\mathcal{A}} (R_u^{mc})
\end{aligned}$$

Now, the output decoupling zeros are the rank loss points of $\begin{pmatrix} A \\ C \end{pmatrix}$, the transmission zeros are the rank loss points of R_u^{mc} and the invariant zeros are the rank loss points of $\begin{pmatrix} A & -B \\ C & D \end{pmatrix}$. For the case of input decoupling zeros and input-output decoupling zeros, first find a kernel representation of the corresponding behaviors. Then the rank loss points of the kernel representations give the corresponding zeros.

7 Feedback and Optimal Control

Summary. In this chapter a substantial volume of results on the control of differential and discrete linear repetitive processes are developed and illustrated, where applicable, by numerical examples. The control laws (or controllers) used make use of both current and previous pass (or passes in the non-unit memory case) and the resulting design algorithms fall, in the main, into two general classes. The first of these execute the design using Linear Matrix Inequalities and the second by minimizing a suitably formulated cost function.

7.1 Control Objectives and Structures

7.1.1 Control Objectives

By analogy with the 1D linear systems approach, consider the case of a differential non-unit memory linear repetitive process of the form defined by Example 1.2.2, i.e.

$$\begin{aligned}\dot{x}_{k+1}(t) &= Ax_{k+1}(t) + Bu_{k+1}(t) + \sum_{j=1}^{M} B_{j-1}y_{k+1-j}(t) \\ y_{k+1}(t) &= Cx_{k+1}(t) + Du_{k+1}(t) + \sum_{j=1}^{M} D_{j-1}y_{k+1-j}(t) \qquad (7.1)\end{aligned}$$

Then a study of physical examples, such as bench mining systems, leads to the following three general control objectives. Note also that these extend in a natural manner to the corresponding discrete non-unit memory linear repetitive processes and hence the details are omitted here.

Stability along the Pass – This is an obvious necessary item in the general case. Note, however, that there will be cases where the weaker property of asymptotic stability suffices or is all that can be achieved. (In the optimal control example, see Example 1.2.7, asymptotic stability is all that can ever be achieved.)

The Limit Profile – Suppose that the particular example of (7.1) under consideration is asymptotically stable, i.e. the condition of Theorem 2.1.4 holds. Then if a strongly convergent control input sequence $\{u_k\}_{k\geq 1}$ is applied the corresponding limit profile can be obtained by replacing all variables in the defining state-space model by their strong limits. The result over $0 \leq t \leq \alpha$ (with $D = 0$ for simplicity) is

$$\begin{aligned} \dot{x}_\infty(t) &= Ax_\infty(t) + Bu_\infty(t) + \hat{B}y_\infty(t) \\ y_\infty(t) &= Cx_\infty(t) + \hat{D}y_\infty(t), \ x_\infty(0) = d_\infty \end{aligned}$$

where (see (2.21))

$$\hat{B} = \sum_{j=1}^{M} B_{j-1}, \qquad \hat{D} = \sum_{j=1}^{M} D_{j-1}$$

and d_∞ is the strong limit of the sequence $\{d_k\}_{k\geq 1}$. Alternatively, since asymptotic stability ensures that $\det(I_m - \hat{D}) \neq 0$,

$$\begin{aligned} \dot{x}_\infty(t) =&(A + \hat{B}(I_m - \hat{D})^{-1}C)x_\infty(t) + Bu_\infty(t) \\ y_\infty(t) =&(I_m - \hat{D})^{-1}Cx_\infty(t) \end{aligned} \tag{7.2}$$

Hence, in effect, the repetitive dynamics in this case can, after a 'sufficiently large' number of passes, be described by a 1D linear systems state-space model. Also, stability along the pass implies that (7.2) is stable in the 1D sense, i.e. all eigenvalues of $A + \hat{B}(I_m - \hat{D})^{-1}C$ have strictly negative real parts.

Given stability along the pass and (7.2), it follows immediately that the 'transient' along the pass response of a differential non-unit memory linear repetitive process of the form considered here can, after a 'sufficiently large' number of passes have elapsed, be classified in terms of well known 1D linear systems criteria (such as the damping ratio and time constant from SISO 1D differential linear systems). Also there is clearly a need to consider performance from pass-to-pass, i.e. in the k direction. Combining these requirements leads to the control objective stated next, where the quotation marks indicate that the precise interpretation of the terms within them are a matter for judgement based on detailed knowledge of the particular application under consideration.

(i) The process must be stable along the pass and hence the existence of a stable limit profile is guaranteed. Also the limit profile dynamics along the pass should satisfy such other additional 1D differential linear systems performance criteria as appropriate. For example, the interaction effects in response to a unit step demand should be within 'acceptable limits' and also the steady-state error should be within an 'acceptable bound'.

(ii) The output pass profile sequence produced, $\{y_k\}_{k\geq 1}$, must be within a specified bound, or band, of y_∞ after a prescribed number of passes, say k^*, and remain within it for all successive passes $k > k^*$.

(iii) The error $y_k - y_\infty$, $0 \leq k \leq k^*$ must be 'acceptable'.

Several variations of (i)–(iii) above exist and clearly much further work remains to be done on developing rules for refining the terms in quotation marks into design criteria which, where appropriate, should (ideally) display similar characteristics to existing well used 1D linear systems counterparts. Recall also that the simulation based stability tests of Sect. 2.3 yield, at no extra cost, computable information concerning (ii) and (iii) above in one special case of major practical interest. We will return to this point later in this chapter.

Repetitive Process Disturbance Decoupling – This problem has more than one aspect, the first of which is based on regarding the previous pass profiles as disturbances acting on the current one. Its requirements can be summarized as the following control objective and note that this obviously implies stability along the pass.

(i) The pass profiles $y_{k-j}(t)$, $0 \leq t \leq \alpha$, $1 \leq j \leq M$, make no contribution to the pass profile $y_k(t)$, $0 \leq t \leq \alpha$ for all passes $k \geq k^* \geq 1$ with an optimum choice of $k^* = 1$.
(ii) If (i) above holds, then for all passes $k \geq k^* \geq 1$, the process dynamics are, in effect, described by the 1D differential linear system defined by the state-space quadruple $\{A, B, C, D\}$. In effect, therefore, the limit profile is reached exactly on pass k^* and is completely described by the 1D differential linear system defined by the state-space quadruple $\{A, B, C, D\}$. Also the dynamics of this limit profile along the pass should satisfy such other additional 1D differential linear systems performance criteria as appropriate.

Next we begin to consider the structure of control laws (or schemes) for differential (and discrete) linear repetitive processes.

7.1.2 Control Laws

By analogy with the 1D linear systems case, one intuitively obvious approach to the control of differential and discrete linear repetitive processes is to use appropriately defined feedback control laws. In which context note that the design of control laws for 2D discrete linear systems described by the Roesser and Fornasini-Marchesini state-space models has received considerable attention in the literature over the years. A valid criticism of such work, however, is that the structure of the control algorithms are not well founded physically. In particular, for a 1D linear system with state vector $x(t)$ and input vector $u(t)$, a state feedback control law with zero tracking vector is of the form $u(t) = Fx(t)$ but in the 2D case the natural generalization could replace $x(t)$ with either the local or global state vectors.

The difficulty above does not arise with either differential or discrete linear repetitive processes. For example, it is physically meaningful to define the current pass error as the difference, at each point along the pass, between a specified reference trajectory for that pass, which in most cases will be the same on each, and the actual pass profile produced. Then it is possible to define a so-called current pass error actuated control law which uses the generated error vector to construct the current pass control input vector. Such control laws are termed current pass, or memoryless, to denote the fact that they do not explicitly use information from previous passes in computing the control input. This term also applies to control laws which only make use of the current pass state vector to compute the current pass control input vector (where if all entries in this vector are not directly available then they would have to be estimated with a suitable observer structure).

For a repetitive process executing pass k, all previous pass profiles have been completed. Hence the pass profile vectors are causal information and are therefore available for use as control input signals on the current pass. Clearly, however, such control laws will be more complex than current pass schemes and hence if they are to be used then we should aim to make the minimum possible use of previous pass data. Next we introduce the control laws considered in this chapter.

A current pass state feedback control law for a differential non-unit memory linear repetitive process described by (7.1) has the form

$$u_{k+1}(t) = Fx_{k+1}(t) + Gr_{k+1}(t) \tag{7.3}$$

where F and G are $r \times n$ and $r \times m$ matrices respectively to be selected and $r_{k+1}(t)$ is an $m \times 1$ new external reference vector taken to represent desired behavior on pass $k+1$, $k \geq 0$. In effect, this control law is its well known 1D linear systems counterpart applied on the current pass. The resulting controlled process (with $D = 0$ for simplicity) is described by

$$\begin{aligned} \dot{x}_{k+1}(t) &= (A + BF)x_{k+1}(t) + BGr_{k+1}(t) + \sum_{j=1}^{M} B_{j-1}y_{k+1-j}(t) \\ y_{k+1}(t) &= Cx_{k+1}(t) + \sum_{j=1}^{M} D_{j-1}y_{k+1-j}(t) \end{aligned}$$

This state-space model has an identical structure to (7.1) and therefore necessary and sufficient conditions for stability along the pass immediately result on interpreting any of the (necessary and sufficient) conditions given in Chap. 2 (Sect. 2.2). An alternative is to use the tests of Sect. 2.3 based on the total variation (see, for example, Lemma 2.3.1) which are, in general, sufficient but not necessary, but if they hold release the performance information for the important case when the reference vector $r_{k+1}(t)$ is constant from pass-to-pass (i.e. has the same entries along each pass).

In terms of the controlled process, the first fact to note is that asymptotic stability, and hence stability along the pass, is invariant under current pass state feedback. If, however, the current pass 'direct feedthrough' term (i.e. $Du_{k+1}(t)$) is restored to the uncontrolled (or open-loop) process state-space model then this fact is no longer true. Also even if this term is absent, it can be argued that asymptotic stability will be present in many cases of practical interest due to the stabilizing influence of resetting the pass initial conditions before the start of each new pass (recall from the discussion of Chap. 1 that concerns about the influence of these conditions was one of the main reasons behind the development of the abstract model based stability theory).

To discuss this last point further, consider for simplicity the unit memory version of (7.1) with $u_{k+1}(t) = 0,\ 0 \leq t \leq \alpha,\ k \geq 0$, and zero state initial vector on each pass. Then the initial output on each pass is given by

$$y_k(0) = D_0^k y_0(0),\ k \geq 0$$

where for asymptotic stability we require that $r(D_0) < 1$. Also in physical terms, this property requires (see Chap. 2) that the initial output on each pass does not become unbounded as $k \to +\infty$. This will most certainly be the case in industrially oriented examples where the initial conditions on each pass are always finite. An example here is the bench mining systems referred to in Chap. 1.

Under this control law, the state matrix A is mapped to $A+BF$ and hence a necessary condition for stability along the pass of the controlled process is that an F exists such that all eigenvalues of $A+BF$ have strictly negative real parts. Equivalently, solvability of the pole placement problem for the 1D differential linear system defined by the state-space triple (state, input and output respectively) $\{A, B, C\}$ under the feedback control law $u(t) = Fx(t)$ is a necessary condition for stability along the pass in this case. The question then is whether or not there is sufficient freedom left in the choice of F to ensure that the controlled process is stable along the pass.

If the complete current pass state vector is not available for measurement then either a suitably structured observer must be included in the control scheme or else output, i.e. pass profile, information only used. Again by analogy with the 1D case, the current pass error is defined as

$$e_{k+1}(t) := r_{k+1}(t) - y_{k+1}(t),\ 0 \leq t \leq \alpha,\ k \geq 0 \tag{7.4}$$

and a so-called memoryless dynamic unity negative feedback control law for (7.1) is defined as

$$\begin{aligned} \dot{x}_{k+1}^c(t) &= A^c x_{k+1}^c(t) + B^c e_{k+1}(t) \\ u_{k+1}(t) &= C^c x_{k+1}^c(t) + D^c e_{k+1}(t) \end{aligned} \tag{7.5}$$

where the $n_1 \times 1$ vector $x_{k+1}^c(t)$ denotes the internal state of the controller. It is also clear that (7.4) and (7.5) describe a memoryless dynamic unity negative feedback control scheme for (7.1), which also includes proportional control as a special case (set $A^c = 0,\ B^c = 0,\ C^c = 0$).

Introduce

$$X_{k+1}(t) := \begin{bmatrix} x_{k+1}^T(t) & (x^c)_{k+1}^T(t) \end{bmatrix}^T \in \mathbb{R}^{n_c},\ n_c = n + n_1$$

Then the state-space model of the controlled process is given by

$$\begin{aligned} \dot{X}_{k+1}(t) &= (\tilde{A} - \tilde{B}\tilde{C})X_{k+1}(t) + \tilde{B}r_{k+1}(t) + \sum_{j=1}^{M}(\tilde{B}_{j-1} - \tilde{B}D_{j-1})y_{k+1-j}(t) \\ y_{k+1}(t) &= \tilde{C}X_{k+1}(t) + \sum_{j=1}^{M} D_{j-1}y_{k+1-j}(t) \end{aligned} \tag{7.6}$$

where

$$\tilde{A} := \begin{bmatrix} A & BC^c \\ 0 & A^c \end{bmatrix},\ \tilde{B} := \begin{bmatrix} B \\ 0 \end{bmatrix},\ \tilde{B}_{j-1} := \begin{bmatrix} B_{j-1} \\ 0 \end{bmatrix},\ \tilde{C} := \begin{bmatrix} C & 0 \end{bmatrix}$$

The 2D transfer-function matrix block diagram (specialized to the differential case) of Fig. 1.10 or Fig. 1.11 generalizes in a natural manner to this case. In particular, let $R(s, z_2)$ and (as in Chap. 1) $Y(s, z_2)$ denote the results of applying the s and z_2 transforms to the reference and pass profile vectors respectively and denote the 2D transfer-function matrix of the controlled process by $G_c(s, z_2)$. Then

$$Y(s, z_2) = G_c(s, z_2)R(s, z_2)$$

where it is also easy to show that

$$G_c(s, z_2) = (I_m + Q(s, z_2))^{-1}Q(s, z_2)$$

Also

$$Q(s, z_2) = G(s, z_2)K(s, z_2)$$

where (see Sect. 1.4) $G(s, z_2)$ is given by

$$\begin{aligned} G(s, z_2) &= \left(I_m - \sum_{j=1}^{M} G_j(s)z_2^j\right)^{-1} G_0(s) \\ G_0(s) &= C(sI_n - A)^{-1}B \\ G_j(s) &= C(sI_n - A)^{-1}B_{j-1} + D_{j-1},\ 1 \le j \le M \\ K(s, z_2) &= K(s) = C^c(sI_{n_1} - A^c)^{-1}B^c + D^c \end{aligned}$$

As with current pass state feedback, this scheme is essentially its 1D differential linear systems counterpart applied on the current pass, and necessary and sufficient conditions for stability along the pass of the controlled process can be obtained by interpreting any of the conditions for this property given in Sect. 2.2 in terms of its state-space model or 2D transfer-function matrix as appropriate. Also asymptotic stability, and hence stability along the pass, is invariant unless the direct feedthrough term is present in the pass profile updating equation. With the usual assumptions, a necessary condition for stability along the pass of the controlled process is that $K(s)$ stabilizes $G_0(s)$ under unity negative feedback control. The question then is whether or not there is sufficient freedom left in the choice of $K(s)$ to ensure that the controlled process is stable along the pass. In this case, the remaining condition can be given a Nyquist diagram interpretation as discussed next.

Consider, for simplicity, the SISO unit memory case. Suppose also that $K(s)$ has been designed to stabilize $G_0(s)$ under unity negative feedback control. Then (by applying (b) of Corollary 2.2.3) stability along the pass holds in this case if, and only if, the Nyquist locus generated by

$$G_c(s) := \frac{G_1(s)}{1 + G_0(s)K(s)} \tag{7.7}$$

lies entirely within the unit circle in the complex plane.

It is easy to see that there exist cases when the memoryless schemes given above cannot ensure stability along the pass (of the controlled process). One option in such cases is to consider the use of control laws with memory, i.e. where some, or all, of the control actuation explicitly uses information from some, or all, of the previous pass profiles (or even states). There are clearly a very large number of possibilities here but in implementation terms it is immediate that if control laws with memory are required then they should make the minimum possible explicit use of information from previous passes. Here we will consider a number of such control laws.

To introduce the first of these, consider the unit memory version of (7.1). Then one form of control law which explicitly uses previous pass information is

$$u_{k+1}(t) = K_1 x_{k+1}(t) + K_2 y_k(t) =: \widetilde{K} \begin{bmatrix} x_{k+1}(t) \\ y_k(t) \end{bmatrix} \tag{7.8}$$

where K_1 and K_2 are appropriately dimensioned matrices to be designed. In effect, this control law uses feedback of the current state vector (which is assumed to be available for use) and 'feedforward' of the previous pass profile vector. (Note that here the term 'feedforward' is used to describe the case where state or pass profile information from the previous pass (or passes) is used as (part of) the input to a control law applied on the current pass, i.e. to information which is propagated in the pass-to-pass (k) direction.) We give conditions for stability along the pass of the controlled process in context later in this chapter. Also the counterpart of (7.8) for a discrete unit memory linear repetitive process of the form defined by Example 1.2.10 is

$$u_{k+1}(p) = K_1 x_{k+1}(p) + K_2 y_k(p) =: K \begin{bmatrix} x_{k+1}(p) \\ y_k(p) \end{bmatrix} \tag{7.9}$$

In many cases the state vector $x_{k+1}(t)$ (or $x_{k+1}(p)$) may not be available or, at best, only some of its entries are. Hence, we now consider the use of control laws to achieve stability along the pass which are only actuated by pass profile information. The first law considered has the following form over $0 \le t \le \alpha,\ k \ge 0$, for the differential case

$$u_{k+1}(t) = \widetilde{K}_1 y_{k+1}(t) + \widetilde{K}_2 y_k(t) =: \widetilde{K}_{do} \begin{bmatrix} y_{k+1}(t) \\ y_k(t) \end{bmatrix} \tag{7.10}$$

This control law is, in general, weaker than that of (7.8) and examples are easily given where stability along the pass can be achieved using (7.8) but not (7.10). It is important to note here that by definition the pass profile vector produced on each pass is available for control purposes before the start of each new pass. As such, this control law (and extensions) assumes storage of the required previous pass profiles and that they are not corrupted by noise etc. The counterpart of this last control law for a discrete unit memory linear repetitive process of the form defined by Example 1.2.10 is

$$u_{k+1}(p) = \widetilde{K}_1 y_{k+1}(p) + \widetilde{K}_2 y_k(p) =: K_{dd} \begin{bmatrix} y_{k+1}(p) \\ y_k(p) \end{bmatrix} \tag{7.11}$$

In the differential non-unit memory case, a natural extension of (7.8) is given over $0 \le t \le \alpha,\ k \ge 0$ by

$$u_{k+1}(t) = F x_{k+1}(t) + G r_{k+1}(t) - \sum_{j=1}^{M} K_j y_{k+1-j}(t)$$

where the first two terms are as in (7.3) and the extra control action results from the introduction of the $r \times m$ matrices $K_j,\ 1 \le j \le M$, which describe the (explicit) contribution of the previous M pass profiles to the control input vector on the current pass.

Substitution into (7.1) gives the following state-space model for the controlled process

$$\begin{aligned} \dot{x}_{k+1}(t) &= (A + BF)x_{k+1}(t) + BGr_{k+1}(t) + \sum_{j=1}^{M}(B_{j-1} - BK_j)y_{k+1-j}(t) \\ y_{k+1}(t) &= Cx_{k+1}(t) + \sum_{j=1}^{M} D_{j-1} y_{k+1-j}(t) \end{aligned}$$

This state-space model again has an identical structure to (7.1) and hence the same comments as for memoryless state feedback control hold regarding conditions for stability along the pass etc. Also if full state feedback is not possible, then one option is to use a so-called memoryless dynamic unity negative feedback control law with proportional repetitive minor loop compensation which constructs the input $u_{k+1}(t),\ k \ge 0$, as

$$u_{k+1}(t) = y^c_{k+1}(t) - \sum_{j=1}^{M} K_j y_{k+1-j}(t),\ 0 \le t \le \alpha,\ k \ge 0$$

where $K_j,\ 1 \le j \le M$, are $r \times m$ matrices and $y^c_{k+1}(t)$ is (compare with (7.5)) generated as

$$\begin{aligned} \dot{x}^c_{k+1}(t) &= A^c x^c_{k+1}(t) + B^c e_{k+1}(t) \\ y^c_{k+1}(t) &= C^c x^c_{k+1}(t) + D^c e_{k+1}(t) \end{aligned}$$

The state-space model of the controlled process only differs from the memoryless case ($K_j = 0,\ 1 \le j \le M$) in that the matrix corresponding to $\tilde{B}_{j-1}$ (see (7.6)) is given by

$$\tilde{B}_{j-1} = B_{j-1} - BK_j,\ 1 \le j \le M$$

This change is, however, very significant in that it augments the memoryless scheme by providing extra design parameters with which (to attempt) to satisfy the conditions for stability along the pass. For example, suppose that $m = r$ and B is a nonsingular matrix (a generic assumption). Then the choice of $K_j = B^{-1}B_{j-1},\ 1 \le j \le M$, completely decouples the dynamics of the previous passes pass profiles from those of the current pass state vector. In the case when also $D_j = 0,\ 1 \le j \le M$, this is a complete solution to the problem of completely decoupling the effects of the previous pass vector(s) from the updating of the current pass dynamics. (See the repetitive process disturbance decoupling control objective discussed earlier in this chapter.)

Recall also the 1D equivalent model for a discrete unit memory linear repetitive process of the form defined by Example 1.2.10 which is given by (1.45) with associated boundary conditions (1.46). Then this representation can be used to design a control law of the form

$$U(l) = K_{1D} Y(l) \tag{7.12}$$

This control law is actuated solely by the previous pass profile and as we will see later can be used to effect when asymptotic stability alone is required. Note also that a direct extension to processes described by Example 1.2.13 is possible.

Note 7.1.1. The control laws (7.8)–(7.12) are termed static since they do not involve any dynamic updating.

When we consider H_∞ based control (and related approaches such as H_2 based design and guaranteed cost control) in the next chapter, it will be shown that a so-called dynamic pass profile (or output) controller is particularly powerful. The state-space model of such a controller for a discrete unit memory linear repetitive process of the form defined by Example 1.2.10 is

$$\begin{aligned}\begin{bmatrix} x_{k+1}^c(p+1) \\ y_{k+1}^c(p) \end{bmatrix} &= \begin{bmatrix} A_{c11} & A_{c12} \\ A_{c21} & A_{c22} \end{bmatrix} \begin{bmatrix} x_{k+1}^c(p) \\ y_k^c(p) \end{bmatrix} + \begin{bmatrix} B_{c1} \\ B_{c2} \end{bmatrix} y_{k+1}(p) \\ u_{k+1}(p) &= \begin{bmatrix} C_{c1} & C_{c2} \end{bmatrix} \begin{bmatrix} x_{k+1}^c(p) \\ y_k^c(p) \end{bmatrix} + D_c y_{k+1}(p) \end{aligned} \tag{7.13}$$

where $x_k^c(p)$ and $y_k^c(p)$ are state vectors for the controller. In effect, this controller is a dynamic 2D linear system described by a Roesser model whose input and output vectors are formed by the process current pass profile and input vectors respectively.

To obtain the state-space model of the resulting controlled process, introduce

$$\begin{aligned} A_c &:= \begin{bmatrix} A_{c11} & A_{c12} \\ A_{c21} & A_{c22} \end{bmatrix}, \quad B_2 := \begin{bmatrix} B \\ D \end{bmatrix} \\ B_c &:= \begin{bmatrix} B_{c1} \\ B_{c2} \end{bmatrix}, \quad C_c := \begin{bmatrix} C_{c1} & C_{c2} \end{bmatrix} \end{aligned}$$

Also define the so-called augmented state and pass profile vectors respectively for the controlled process as

$$\overline{x}_{k+1}(p) := \begin{bmatrix} x_{k+1}(p) \\ x_{k+1}^c(p) \end{bmatrix}, \quad \overline{y}_k(p) := \begin{bmatrix} y_k(p) \\ y_k^c(p) \end{bmatrix}$$

Then the resulting model is

$$\begin{aligned} \begin{bmatrix} \overline{x}_{k+1}(p+1) \\ \overline{y}_{k+1}(p) \end{bmatrix} &= \overline{A} \begin{bmatrix} \overline{x}_{k+1}(p) \\ \overline{y}_k(p) \end{bmatrix} \\ y_{k+1}(p) &= \overline{C} \begin{bmatrix} \overline{x}_{k+1}(p) \\ \overline{y}_{k+1}(p) \end{bmatrix} \end{aligned}$$

where (noting again the definition of the augmented plant matrix Φ given in (3.4))

$$\overline{A} := \Pi \begin{bmatrix} \Phi + B_2 D_c C_2 & B_2 C_c \\ B_c C_2 & A_c \end{bmatrix} \Pi, \quad \overline{B} := \Pi \begin{bmatrix} B_1 + B_2 D_c \\ B_c \end{bmatrix}$$

$$\overline{C} := \begin{bmatrix} C_2 & 0 \end{bmatrix} \Pi, \quad C_2 = \begin{bmatrix} 0 & I \end{bmatrix}, \quad \Pi := \begin{bmatrix} I & 0 & 0 & 0 \\ 0 & 0 & I & 0 \\ 0 & I & 0 & 0 \\ 0 & 0 & 0 & I \end{bmatrix}$$

The version of this controller for the differential case will be introduced in context in the next chapter. Note also that the control laws considered in this monograph by no means exhaust the possibilities either in terms of the information required to actuate them or the basic control law (or controller) structure.

7.2 Design of Memoryless Control Laws

First we consider two special cases for differential processes. The first of these is when the example under consideration is SISO, unit memory, and the vectors B and B_0 are related through $B_0 = \gamma B$, where γ is a positive real scalar. Then in this case $G_0(s)$ and $G_1(s)$ from the 2D transfer-function matrix description have the same zeros which we further assume to be real, negative and distinct, i.e. a special case of the minimum phase property. Suppose also that $CB \neq 0$, or, equivalently, these transfer-functions have n poles and $q = n - 1$ zeros. Then if the pair $\{A, B\}$ is controllable any one of numerous well documented algorithms can be used to design the row vector F in the current pass state feedback control law (7.3) such that all eigenvalues of $A + BF$ are assigned to a pre-specified set of locations λ_i, $1 \leq i \leq n$, in the open left-half of the complex plane.

Stability along the pass of the controlled process now requires that the Nyquist locus generated by $G_c(s)$ of (7.7) lies entirely with the unit circle in the complex plane. In this particular case

$$G_c(s) = \gamma \frac{\prod_{i=1}^{q}(s - h_i)}{\prod_{i=1}^{n}(s - \lambda_i)}$$

where the h_i are the zeros of $G_0(s)$ which by assumption are real, distinct and negative. Suppose also that the λ_i are chosen to be real, distinct, negative and such that they satisfy the interlacing property

$$\lambda_1 < h_1 < \lambda_2 < \cdots < h_{n-1} < \lambda_n$$

Then applying the results of (2.64) and (2.65) from Chap. 2, it follows immediately that the controlled process is stable along the pass if, and only if,

$$\gamma \left| \prod_{i=1}^{q} h_i \right| < \left| \prod_{i=1}^{n} \lambda_i \right|$$

or, equivalently, the maximum value of the frequency response of $G_c(\imath\omega)$ occurs at $\omega = 0$. Note also that this last condition can always be satisfied by placing λ_1 'far enough' to the left of the origin on the real line.

The second example we consider is when $G_0(s)$ has the structure of a multivariable first order lag which can be regarded as the natural generalization of the classical first order transfer-function in 1D differential linear systems theory/analysis. The formal definition is as follows.

Definition 7.2.1. *[122] An m-output/m-input strictly proper 1D differential linear system described by the $m \times m$ transfer-function matrix $G(s)$ is said to be a multivariable first order lag if, and only if,* $\det G(s) \neq 0$ *and*

$$G^{-1}(s) = A_0 s + A_1, \ \det A_0 \neq 0$$

where A_0 and A_1 are real $m \times m$ constant matrices.

Suppose now that the particular example under consideration is (for ease of presentation) unit memory and $m = r = n$. Then the natural generalization to the differential repetitive process case is a state-space model of the form

$$\dot{y}_{k+1}(t) = -A_0^{-1}A_1 y_{k+1}(t) + A_0^{-1}u_{k+1}(t) + \hat{B}_0 y_k(t),\ 0 \le t \le \alpha,\ k \ge 0$$

Now select $K(s)$ as

$$K(s) = \rho A_0 - A_1$$

where $\rho > 0$ is a real scalar. Then using the theory of multivariable first order lags it follows immediately that $K(s)$ stabilizes $G_0(s)$ under unity negative feedback control for any choice of $\rho > 0$. Hence stability along the pass of the controlled process requires that the eigenvalues, or characteristic loci, generated by

$$G_c(s) = (I_m + G_0(s)K(s))^{-1}G_1(s) = \frac{\hat{B}_0}{s+\rho}$$

must lie within the unit circle in the complex plane $s = \imath\omega$, $\forall\ \omega$. Hence this holds if, and only if,

$$\rho > r(\hat{B}_0)$$

which can always be satisfied by choosing a 'high enough' value of ρ.

This example also leads to a solution of the problem of designing for a prescribed limit profile (in terms of along the pass dynamic response). In particular, suppose that the controlled process is stable along the pass. Then the resulting limit profile state-space model can be written as

$$\left(sI_m + \rho\left(I_m - \frac{\hat{B}_0}{\rho}\right)\right)Y_\infty(s) = \rho\left(I_m - \frac{A_0^{-1}A_1}{\rho}\right)R_\infty(s)$$

which is stable in the 1D differential linear systems sense if, and only if,

$$\rho > \max_{1\le i\le m}\mathrm{Re}(\lambda_i)$$

where $\lambda_1, \ldots, \lambda_n$ are the eigenvalues of $\hat{B}_0$. Consider also the case when $\rho \to \infty$ ('high gain'). In this case, $y_\infty(t)$ is arbitrarily close to the inverse Laplace transform of

$$Y_\infty(s) = \frac{\rho}{s+\rho}I_m R_\infty(s)$$

This is the response of a totally non-interacting 1D differential linear system with zero steady-state error to a unit step applied at $t = 0$ in any channel.

It is also easy to see that the matrix $[L_0]_p$ of Theorem 2.3.3 or Theorem 2.3.4 for the controlled process, denoted here by $[L_c]_p$, is given by

$$[L_c]_p = \frac{[\hat{B}_0]_p}{\rho}$$

Hence the condition of Theorem 2.3.3 holds for all $\rho > r([\hat{B}_0]_p)$ and the condition of Theorem 2.3.4 holds for all $\rho > ||([\hat{B}_0]_p)||$.

Interpreting Theorem 2.3.5 now yields that the pass profile sequence produced by the controlled process approaches the limit profile at a geometric rate, denoted here by cr, governed by

$$cr \in \left(\frac{r([\hat{B}_0]_p)}{\rho}, 1 \right)$$

where this set is non-empty for any choice of $\rho > r([\hat{B}_0]_p)$ and we also have the error band for each entry in $y_k - y_\infty, \ k \geq 0$. Similarly, interpreting Theorem 2.3.6 now yields that the pass profile sequence produced by the controlled process approaches y_∞ at a geometric rate governed by

$$\frac{||([\hat{B}_0]_p)||}{\rho} < 1$$

for suitable choice of ρ. Also we have the single error band for $y_k - y_\infty, \ k \geq 0$.

Finally, consider the case of $\rho \to \infty$. Then

$$||L_c|| = \frac{||([\hat{B}_0]_p)||}{\rho} \to 0$$

i.e. the limit profile dynamics are reached to within arbitrary accuracy on the first pass.

Next we extend this last analysis to include (a form of) fast sampling control.

7.2.1 Fast Sampling Control of a Class of Differential Linear Repetitive Processes

Consider again a discrete unit memory linear repetitive process of the form defined by Example 1.2.10 whose state-space model we write here (with the same vector dimensions) as

$$\begin{aligned} x_{k+1}(p+1) &= A_d x_{k+1}(p) + B_d u_{k+1}(p) + B_{d0} y_k(p) \\ y_{k+1}(p) &= C_d x_{k+1}(p) + D_{d0} y_k(p) \end{aligned} \tag{7.14}$$

(where the direct feedthrough term on the pass profile updating equation has been deleted without any loss of generality). Also the state initial vector sequence is assumed to be zero, i.e. $d_{k+1} = 0, \ k \geq 0$. Next we show how a model of the form (7.14) results from sampling a model of the form defined by Example 1.2.3 and establish the achievable accuracy.

To start with, first write the model of Example 1.2.3 in the form

$$y_{k+1}(t) = C\int_0^t e^{A(t-\tau)}[Bu_{k+1}(\tau) + B_0 y_k(\tau)]\, d\tau + D_0 y_k(t),\ 0 \le t \le \alpha,\ k \ge 0 \tag{7.15}$$

Suppose also that this model is subject to synchronous sampling with sampling interval h and a piecewise constant input vector, i.e.

$$u_{k+1}(t) = u_{k+1}(ph) =: u_{k+1}^p$$

where $ph \le t < (p+1)h$ and $0 \le p \le \frac{\alpha}{h}$. Also, write $x_{k+1}^p := x_{k+1}(ph),\ y_{k+1}^p := y_{k+1}(p),\ 0 \le p \le \frac{\alpha}{h}$. Then, at $t = ph$, (7.15) can be written for $k \ge 0$ as

$$y_{k+1}(ph) = C\int_0^{ph} e^{A(ph-\tau)}[Bu_{k+1}(\tau) + B_0 y_k(\tau)]\, d\tau + D_0 y_k(ph)$$

and also the state equation takes the form

$$x_{k+1}(ph) = e^{Aph}\int_0^{ph} e^{-A\tau}[Bu_{k+1}(\tau) + B_0 y_k(\tau)]\, d\tau$$

At instant $t = (p+1)h$, the following holds for $k \ge 0$

$$\begin{aligned} x_{k+1}((p+1)h) &= e^{Ah}x_{k+1}(ph) \\ &+ e^{Ah}\int_{ph}^{(p+1)h} e^{Ah}e^{-A\tau}[Bu_{k+1}(\tau) + B_0 y_k(\tau)]\, d\tau \end{aligned} \tag{7.16}$$

and we now consider each term in turn in this last equation.

Since the current pass input vector is assumed to be piecewise constant, i.e. $u_{k+1}(t) = u_{k+1}(ph)$ on $[ph, (p+1)h)$, we can write, for $0 \le p \le \frac{\alpha}{h}$, $k \ge 0$,

$$e^{Ah}e^{Aph}\int_{ph}^{(p+1)h} e^{-A\tau}Bu_{k+1}(\tau)\, d\tau = e^{Ah}\int_0^h e^{-A\tau}B\, d\tau\ u_{k+1}^p$$

Also under the assumption that $h \to 0$, which we term fast sampling, $y_k(t)$ over the interval $[ph, (p+1)h)$ can be approximated by $y_k(ph)$, $0 \le p \le \frac{\alpha}{h}$, $k \ge 0$. The accuracy of this approximation improves with decreasing h and we have that for $\tau \in [ph, (p+1)h)$

$$\lim_{h \to 0} y_k(\tau) = y_k(ph)$$

which is equivalent to the assumption that the previous pass profile is piecewise continuous. Hence, with this assumption, the term involving $y_k(t)$ in (7.16) can be written as

$$e^{Ah}e^{Aph}\int_{ph}^{(p+1)h} e^{-A\tau}B_0 y_k(\tau)\,d\tau = e^{Ah}\int_0^h e^{-A\tau}B_0\,d\tau\; y_k^p$$

We can now also write

$$x_{k+1}^{p+1} = e^{Ah}x_{k+1}^p + e^{Ah}\int_0^h e^{-A\tau}[B\,d\tau\; u_{k+1}^p + B_0\,d\tau\; y_k^p] \tag{7.17}$$

Then it follows from comparison with (7.14) that

$$A_d = e^{Ah},\quad B_d = A_d\int_0^h e^{-A\tau}B\,d\tau,\quad B_{d0} = A_d\int_0^h e^{-A\tau}B_0\,d\tau$$

and also that the matrices defining the output (pass profile vector) updating are unchanged by the sampling operation, i.e. $C_d = C$, $D_{d0} = D_0$.

Now we show that a discrete unit memory linear repetitive process of the form defined here by (7.14) obtained via the above synchronous sampling scheme becomes a differential unit memory linear repetitive process of the form defined by Example 1.2.3 in the limit as $h \to 0$. To start with, first note that from (7.17) we can write

$$\begin{aligned}\lim_{h\to 0}(x_{k+1}^{p+1} - x_{k+1}^p) &= \lim_{h\to 0}\Big\{(e^{Ah} - I_n)x_{k+1}^p + e^{Ah}\int_0^h e^{-A\tau}B\,d\tau\; u_{k+1}^p \\ &+ e^{Ah}\int_0^t e^{-A\tau}B_0 d\tau\; y_k^p\Big\}\end{aligned}$$

and we now individually consider each term in this last equation.

Consider first the term involving x_{k+1}^p. Then, using the power series expansion for e^{Ah}, we can write

$$\begin{aligned}\lim_{h\to 0}\frac{1}{h}(e^{Ah} - I_n)x_{k+1}^p &= \lim_{h\to 0}\frac{1}{h}(I_n + Ah + o(h^2) - I_n)x_{k+1}^p \\ &= Ax_{k+1}^p\end{aligned}$$

where $o(h^2)$ represents terms in h^2 and higher powers. In the case of the term involving u_{k+1}^p, we can write

$$\begin{aligned}(\lim_{h\to 0}\frac{1}{h}e^{Ah}\;\textstyle\int_0^t\; e^{-A\tau}B\,d\tau\; u_{k+1}^p) &= \lim_{h\to 0}\frac{1}{h}((I_n + Ah \\ &+ o(h^2))\int_0^h (I_n + Ah + o(h^2))B\,d\tau\; u_{k+1}^p) \\ &= \lim_{h\to 0}\frac{1}{h}((I_n + Ah + o(h^2))(h + o(h^2))\;B\;u_k^p) \\ &= Bu_k^p\end{aligned}$$

Moreover, an identical analysis for the term involving B_0 shows that it can be replaced by $B_0 y_k^p$. Note also that, as $h \to 0$, u_{k+1}^p and y_k^p become continuous variables.

Combining the above results, and noting that, in the limit,

$$\lim_{h\to 0} \frac{1}{h}(x_{k+1}^{p+1} - x_{k+1}^p) = \dot{x}_{k+1}^p$$

we can write

$$\dot{x}_{k+1}^p = Ax_{k+1}^p + Bu_{k+1}^p + B_0 y_k^p$$

which, as $h \to 0$, approaches

$$\dot{x}_{k+1}(t) = Ax_{k+1}(t) + Bu_{k+1}(t) + B_0 y_k(t)$$

This last equation, when combined with the fact that both the output equation and the boundary conditions in this case are static, is just a differential linear repetitive process state-space model of the form defined by Example 1.2.3 (with $D = 0$).

The construction of discrete approximations to the dynamics of differential linear repetitive processes has been the subject of other research, see, for example, [78]. To-date, however, the use of these approximations in control law design has not been considered. This is left here as a topic for possible future research.

7.2.2 Discrete Multivariable First Order Lag based Control Law Design

Consider first the case of a controllable and observable 1D discrete linear system with m inputs and m outputs defined by the state-space triple (state, input and output matrices respectively) $\{\Xi, \Delta, \Gamma\}$ with inverse transfer-function matrix

$$G^{-1}(z) = (z-1)\hat{B}_0 + \hat{B}_1$$

where here $\hat{B}_0$ and $\hat{B}_1$ are real $m \times m$ matrices and $\det \hat{B}_0 \neq 0$. An equivalent definition is an m-input m-output discrete-time model with state dimension equal to the output vector dimension and $\det \Xi\Delta \neq 0$, i.e. the transfer-function matrix in this case has minimal state-space realization specified by

$$\Xi := I_m - \hat{B}_0^{-1}\hat{B}_1, \quad \Delta := \hat{B}_0^{-1}, \ \Gamma := I_m$$

A 1D discrete linear system with this structure is termed a discrete multivariable first order lag (by analogy with the classical SISO case) and a 'rich' control theory for such processes exists – see [123]. The analysis here is from [22].

In the case of a discrete unit memory linear repetitive process of the form defined by Example 1.2.10, the natural generalization of the multivariable first order lag structure over $0 \le p \le \alpha$, $k \ge 0$, is

$$\begin{aligned}\tilde{x}_{k+1}(p+1) &= (I_m - \hat{B}_0^{-1}\hat{B}_1)\tilde{x}_{k+1}(p) + \hat{B}_0^{-1}u_{k+1}(p) + \hat{\Delta}_0 y_k(p)\\ y_{k+1}(p) &= \tilde{x}_{k+1}(p)\end{aligned}$$

Consider also the use of a memoryless dynamic unity negative feedback control law with proportional repetitive minor loop compensation which constructs the input $u_{k+1}(p),\ k \geq 0$, as

$$u_{k+1}(p) = y_{k+1}^c(p) - K_1 y_k(p) \tag{7.18}$$

where

$$\begin{aligned}x_{k+1}^c(p+1) &= A^c x_{k+1}^c(p) + B^c e_{k+1}(p)\\ y_{k+1}^c(p) &= C^c x_{k+1}(p) + B^c e_{k+1}(p)\end{aligned}$$

$x_{k+1}^c(p)$ denotes the internal state vector, and K_1 is an $r \times m$ matrix to be designed.

Now set

$$K_1 = \hat{B}_0 \hat{\Delta}_0 \tag{7.19}$$

Then the previous pass dynamics are completely decoupled from those of the current one for all passes. In which case, the 2D transfer-function matrix of the controlled process is a function of the along the pass shift variable only and can be written as

$$(I_m + G(z_1^{-1})K(z_1^{-1}))^{-1}G(z_1^{-1})K(z_1^{-1})$$

where

$$G(z_1^{-1}) := (z_1^{-1}I_m - I_m + \hat{B}_0^{-1}\hat{B}_1)^{-1}\hat{B}_0^{-1}$$

which is just the result of applying dynamic unity negative feedback control to a 1D discrete linear system with transfer-function matrix $G(z_1^{-1})$ and forward path controller transfer-function matrix $K(z_1^{-1})$. Moreover, stability along the pass for the controlled process is clearly equivalent to designing $K(z_1^{-1})$ to stabilize $G(z_1^{-1})$.

Suppose now that with (7.18) and (7.19) in place, $K(z_1^{-1})$ is chosen to have the form

$$K(z_1^{-1}) := \hat{B}_0 \operatorname{diag}\{1 - k_j\}_{1 \leq j \leq m} - \hat{B}_1$$

Then it follows from routine analysis that the controlled process is stable along the pass if, and only if,

$$\rho_c(z_1^{-1}) \neq 0$$

$\forall\, |z_1^{-1}| > 1$ where

$$\begin{aligned}\rho_c(z_1^{-1}) &:= \rho(z_1^{-1})\det(I_m + G(z_1^{-1})K(z_1^{-1}))\\ &= \det(\hat{B}_0^{-1})\ \det(G^{-1}(z_1^{-1}) + K(z_1^{-1}))\\ &= \prod_{j=1}^{m}(z_1^{-1} - k_j)\end{aligned}$$

Hence the controlled process is stable along the pass if, and only if, $k_j \in (-1, 1)$, $1 \le j \le m$.

Under this form of control scheme, the limit profile is reached on the first pass, and as in the 1D linear systems case, transient performance for the controlled process can be assessed by evaluating (the transfer-function matrix of the controlled process limit profile)

$$\begin{aligned} H(z_1^{-1}) &:= (I_m + G(z_1^{-1})K(z_1^{-1}))^{-1}G(z_1^{-1})K(z_1^{-1}) \\ &= (G^{-1}(z_1) + K(z_1^{-1}))^{-1}K(z_1^{-1}) \\ &= \text{diag}\left\{\frac{1}{z_1^{-1} - k_j}\right\}(\text{diag}\,\{1 - k_j\}_{1\le j\le m} - \hat{B}_0^{-1}\hat{B}_1) \quad (7.20) \end{aligned}$$

Hence the limit profile here will have small steady-state errors and small interaction effects in response to unit step commands only provided the elements of the matrix $\hat{B}_0^{-1}\hat{B}_1$ are 'sufficiently small'. In particular, it will have these properties if the sampling rate is 'fast enough'.

Suppose (for simplicity) that the matrix A of the underlying differential process has a diagonal canonical form with eigenvalues λ_j, $1 \le j \le m$, and eigenvector matrix $\widetilde{E}$. Then consideration of $\widetilde{E}\text{diag}\{e^{\lambda_j h}\}_{1\le j\le m}\widetilde{E}^{-1}$ suggests that a necessary condition for $\hat{B}_0^{-1}\hat{B}_1$ to be 'small' is that

$$|\lambda_j h| \ll 1,\ 1 \le j \le m$$

Equivalently, the sampling rate must be 'fast' in comparison to the eigenvalues of the matrix A in the underlying differential process.

The above analysis is easily extended to the choice of $K(z_1^{-1})$ as a controllable and observable proportional-plus-summation controller of the general parametric form

$$K(z_1^{-1}) := \hat{B}_0\text{diag}\left\{1 - k_j c_j + \frac{(1-k_j)(1-c_j)z_1^{-1}}{(z_1^{-1}-1)}\right\}_{1\le j\le m} - \hat{B}_1 \quad (7.21)$$

Suppose that only q of the parameters c_j, $1 \le j \le m$, are not equal to unity. Then it follows immediately that the controller $(K(z_1^{-1}))$ has minimal realization of state dimension q and (by a routine extension of the case above) the controlled process is easily shown to be stable along the pass if, and only if,

$$\left(\prod_{j=1}^{m}(z_1^{-1} - k_j)\right)\left(\prod_{c_j\neq 1}(z_1^{-1} - c_j)\right) \neq 0,\ \forall\,|z_1^{-1}| \ge 1$$

i.e. if, and only if, $k_j \in (-1, 1)$ and $c_j \in (-1, 1)$, $1 \le j \le m$.

In practice, the inclusion of summation action would normally be considered in cases where proportional action displayed large steady-state errors. Also the reset times will normally be considerably longer than the rise times, i.e. it is likely that

$$|1 - c_j| \ll |1 - k_j|,\ 1 \le j \le m$$

and hence $H(z_1^{-1})$ of (7.20) is still a good representation of the limit profile dynamics of the controlled process with the exception that steady-state errors in response to step commands will be zero in all channels of the limit profile output vector $y_\infty(p)$ corresponding to parameters $c_j \ne 1$. This fact is easily verified by noting that

$$\begin{aligned}\lim_{z_1 \to 1} H(z_1^{-1}) &= \lim_{z_1^{-1} \to 1} (G^{-1}(z_1^{-1}) + K(z_1^{-1}))^{-1} K(z_1^{-1}) \\ &= I_m - \widetilde{M} \hat{B}_0^{-1} \hat{B}_1\end{aligned}$$

where $\widetilde{M} := \mathrm{diag}\{\widetilde{m}_j\}_{1 \le j \le m}$ and

$$\widetilde{m}_j := \begin{cases} 0, & c_j \ne 1 \\ \frac{1}{1-k_j}, & c_j = 1 \end{cases}$$

7.2.3 An Extension

The analysis given first below makes use of the following result for 1D discrete time linear systems. Here $Q(z)$ is the forward path transfer-function matrix of the unity negative feedback control scheme with plant and controller transfer-function matrices $G(z)$ and $K(z)$ respectively, i.e. $Q(z) = G(z)K(z)$. (Here we use z to denote the shift operator and again follow [22].)

Theorem 7.2.1. *Suppose that the controllable and observable invertible 1D minimum phase m-input m-output discrete linear system $Q(z)$ is to be approximated by the invertible minimum phase 1D discrete linear system $Q_A(z)$. Suppose also that $Q_A(z)$ is stable under unity negative feedback control and that the poles of the resulting closed-loop system generated by $Q(z)$ under such control action lie in the open ball $|z| < R$, where $R > 1$. Then the system $Q(z)$ is stable under unity negative feedback control if the following so-called contraction condition holds*

$$||(I_m + Q_A^{-1}(z))^{-1}(Q_A^{-1}(z) - Q^{-1}(z))|| < 1 \qquad (7.22)$$

where, if $E(z)$ is any $m \times m$ matrix function of z,

$$||E(z)|| := \max_{1 \le j \le m} \max_{|z|=1,\ |z|=R} \sum_{i=1}^{m} |E_{ji}(z)|$$

Proof. This result is proved by noting that the controllability and observability assumptions ensure that the closed loop system is stable if, and only if, the matrix $(I_m + Q(z))^{-1}Q(z) = (I_m + Q^{-1}(z))^{-1}$ is analytic and bounded in the region $1 < |z| < R$, i.e. if, and only if, the solution of

$$y(z) = -Q^{-1}(z)y(z) + \beta$$

is analytic and bounded in $1 < |z| < R$ independent of the choice of $\beta \in \mathbb{R}^m$.

Now write this last equation in the form

$$y(z) = (I_m + Q_A^{-1}(z))^{-1}((Q_A^{-1}(z) - Q^{-1}(z))y(z) + \beta)$$

Then the remainder of the proof is identical to that in [120] for the equivalent 1D differential linear systems case with the Nyquist contour replaced by the closed region $1 \le |z| \le R$. Hence the details are omitted here. □

In practice, R will, of course, be unknown but this does not limit the applicability of the result since we can always take $R \gg 1$. Given $Q(z)$ and $Q_A(z)$, the contraction condition (7.22) can be tested numerically but its major use in this setting is for theoretical closed-loop stability analysis based on some approximation $Q_A(z)$ to $Q(z)$. Next we apply this result to extend the analysis given previously in this chapter on the control of a discrete unit memory linear repetitive process of the form defined by Example 1.2.10.

Return, therefore, to the theme of the previous control law design analysis in this chapter and suppose that the inverse transfer-function matrix $G(z_1^{-1})$ is of the form

$$G^{-1}(z_1^{-1}) = (z_1^{-1} - 1)\hat{B}_0 + \hat{B}_1 + \hat{B}_0\hat{H}(z_1^{-1}) \tag{7.23}$$

where $\hat{H}(z_1^{-1})$ is proper with $\hat{H}(1) = 0$ and, as before, $\det \hat{B}_0 \neq 0$. Also the $m \times m$ multivariable first order lag transfer-function matrix

$$G_a^{-1}(z_1^{-1}) := (z_1^{-1} - 1)\hat{B}_0 + \hat{B}_1$$

is to be used as a reduced order model for control law design purposes. Note that $G_a(z_1^{-1})$ is a good approximation to the high frequency and steady state behavior of the 1D discrete linear system defined by the state-space triple $\{\Xi, \Delta, \Gamma\}$ but contains no information on its zero structure – in fact, $G_a(z_1^{-1})$ has no zeros.

Suppose now that the previous analysis given in this chapter is used to design a proportional-plus-summation controller for unity negative feedback control of the 1D discrete linear system described by $G_a(z_1^{-1})$. Suppose also that this controller is used in the repetitive process control law with K_1 again given by (7.19). Then the contraction condition (7.22) can be used to assess stability along the pass of the resulting controlled discrete unit memory linear repetitive process. The relevant matrix is

$$\begin{aligned} E(z_1^{-1}) &:= (I_m + Q_A^{-1}(z_1^{-1}))^{-1}(Q_A^{-1}(z_1^{-1}) - Q^{-1}(z_1^{-1})) \\ &= (K(z_1^{-1}) + G_a^{-1}(z_1^{-1}))^{-1}(G_a^{-1}(z_1^{-1}) - G^{-1}(z_1^{-1})) \\ &= (-1)\mathrm{diag}\left\{\frac{(z_1^{-1} - 1)}{(z_1^{-1} - k_j)(z_1^{-1} - c_j)}\right\}_{1 \le j \le m} H(z_1^{-1}) \end{aligned}$$

and a sufficient condition for stability along the pass here is that $||E(z_1^{-1})|| < 1$.

It would be possible to directly investigate this last condition if the inverse system was computed from the state-space triple $\{\Xi, \Delta, \Gamma\}$ using known techniques (see, for example, [99]) and noting the fact that the condition $\det \hat{B}_0$ is equivalent to $\det \Gamma\Delta \neq 0$. Note also that the contraction condition here can be replaced by the simpler condition

$$\max_{1\leq i\leq m} \max_{|z_1^{-1}|=1} \sum_{i=1}^{m} |E_{ji}(z_1^{-1})| < 1$$

by letting $R \to \infty$ and noting that $E(z_1^{-1}) = o(z_1^{-1})$.

The question which must now be addressed is whether or not the contraction condition can be satisfied for a 'reasonable' class of examples. In which context, suppose that the discrete unit memory linear repetitive process considered here can be regarded as a discrete representation of a differential unit memory linear repetitive process of the form defined by Example 1.2.3 (with $D_0 = 0$) with sampling period $h > 0$. Then we have the following result.

Theorem 7.2.2. *Suppose that the 1D differential linear system $\{A, B, C\}$ constructed from a differential unit memory linear repetitive process of the form defined by Example 1.2.3 (with $D_0 = 0$) is minimum phase and $\det CB \neq 0$. Suppose also that this repetitive process is to be controlled by a control law of the form (7.18) and (7.19) where $\hat{y}_{k+1}(p)$ is generated such that its transfer-function matrix $K(z_1^{-1})$ is of the proportional plus summation form given by (7.21). Then for each choice of parameters k_j and c_j, $1 \leq j \leq m$, there exists a positive real number h^* such that for $h \in (0, h^*)$, the 1D discrete linear system $\{\Xi, \Delta, \Gamma\}$ constructed from the resulting discrete linear repetitive process is minimum phase with inverse of the form given by (7.23) and the contraction condition for stability along the pass of the controlled process holds.*

Proof. First note that the 1D discrete system $\{\Xi, \Delta, \Gamma)$ has an inverse of the form given in (7.23) if, and only if, $\Gamma\Delta$ $(= \hat{B}_0^{-1})$ is nonsingular. Hence

$$\lim_{h\to 0} \frac{1}{h} \Gamma\Delta = CB \tag{7.24}$$

which, by assumption, is nonsingular. It therefore follows that (7.23) is the correct parametric representation of the required inverse system at all high enough sampling rates.

Using results in [123], note that

$$\begin{aligned} G^{-1}(z_1^{-1}) &= (z_1^{-1} - 1)(\Gamma\Delta)^{-1} - (\Gamma\Delta)^{-1}\Gamma\{\Xi - I_n\}\Delta(\Gamma\Delta)^{-1} \\ &- (\Gamma\Delta)^{-1}\Gamma\{\Xi - I_n\}M(z_1^{-1}I_n - N\Xi M)^{-1} \\ &\times N\{\Xi - I_n\}\Delta(\Gamma\Delta)^{-1} \end{aligned}$$

where here N and M are $(n-m) \times n$ and $n \times (n-m)$ full-rank matrices respectively, which satisfy

$$\Gamma M = 0, \qquad N\Delta = 0, \qquad NM = I_{n-m}$$

It now follows immediately that

$$\begin{aligned} H(z_1^{-1}) &= \Gamma\{\Xi - I_n\}M[\{I_n - N\Xi M\}^{-1} \\ &- (z_1^{-1}I_n - N\Xi M\}^{-1}]N\{\Xi - I_n\}\Delta(\Gamma\Delta)^{-1} \\ &= \Gamma\{\Xi - I_n\}M(z_1^{-1} - 1)(I_n - N\Xi M)^{-1}(z_1^{-1}I_n - N\Xi M)^{-1} \\ &\times N\{\Xi - I_n\}\Delta(\Gamma\Delta)^{-1} \end{aligned}$$

Now consider the behavior of $H(z_1^{-1})$ as $h \to 0$ on the unit circle $|z_1^{-1}| = 1$. Then the eigenvalues $\lambda_1, \cdots, \lambda_{n-m}$ of $N\Xi M$ are simply the $n-m$ zeros of $\{\Xi, \Delta, \Gamma\}$ i.e. the $n-m$ solutions of

$$\det \begin{pmatrix} \lambda I_n - \Xi & -\Delta \\ \Gamma & 0 \end{pmatrix} = 0$$

Then since

$$\begin{aligned} \Xi &= I_n + Ah + o(h^2) \\ \Delta &= hB + o(h^2) \end{aligned} \tag{7.25}$$

it follows that

$$\lambda_j = 1 + \tilde{z}_j h + o(h^2),\ 1 \le j \le n-m$$

where $\tilde{z}_1, \ldots, \tilde{z}_{n-m}$ are the zeros of the differential linear system $\{A, B, C\}$. By the minimum phase assumption, all of these numbers have strictly negative real parts and hence $\{\lambda_j\}_{1 \le j \le n-m}$ lie in the open unit circle in the complex plane for all 'fast enough' sampling rates. Equivalently, $\{\Xi, \Delta, \Gamma\}$ is minimum phase for all 'fast enough' sampling rates.

To prove the last part, suppose, without loss of generality, that M is constant. Then the assumption that $\det CB \neq 0$ ensures that the limit $N_0 := \lim_{h \to 0} N$ exists, and also that $N_0 M = I_{n-m}$ and $N_0 B = 0$. Also the minimum phase nature of $\{A, B, C\}$ ensures that $\det N_0 AM \neq 0$, and this fact, together with (7.24) and (7.25) and the identity

$$(z_1^{-1}I_n - N\Xi M)^{-1} = [(z_1^{-1} - 1)I_n - N_0 AMh + o(h^2)]^{-1}$$

guarantees that there exists $h^* > 0$ and $e > 0$ such that

$$\sup_{0<h<h^*} ||(z_1^{-1} - 1)^{-1} H(z_1^{-1})|| < e$$

In particular, $(z_1^{-1} - 1)^{-1} H(z_1)$ converges uniformly to zero as $h \to 0$ on any relatively open subset $\mathcal{T}$ of the unit circle not containing the point $z_1^{-1} = 1$. Consequently $H(z_1)$ (and hence $E(z_1^{-1})$) converges uniformly to zero on $|z_1^{-1}| = 1$ as $h \to 0$. This completes the proof of the result since suitable choice of h^* ensures that $||E(z_1^{-1})|| < 1,\ 0 < h < h^*$. □

In effect, this last theorem states that the contraction condition for stability along the pass of the controlled process is always satisfied if the 1D differential system $\{A, B, C\}$ is minimum phase, uniform rank one, and the sampling rate is fast enough. Note also that these conditions are sufficient only and do not necessarily prevent application of the technique if they do not hold.

The following is a systematic procedure for the application of Theorem 7.2.2.

Step 1. Evaluate the matrices $\hat{B}_0$ and $\hat{B}_1$. One way of doing this is from the matrices Ξ, Δ, Γ, using known formulas for the inverse of 1D discrete linear systems (see, for example, [123]). Alternatively, if $r(\Xi) < 1$, these matrices may be deduced by obtaining the response of the 1D discrete linear system $\{\Xi, \Delta, \Gamma\}$ to step inputs from zero initial conditions. In particular, the output responses of this system from zero initial conditions to the vector step demand s_d at $p = 1$ and $p = \infty$ are

$$y_1 = \hat{B}_0^{-1} s_d, \; y_\infty = \hat{B}_1^{-1} s_d$$

and $\hat{B}_0$ and $\hat{B}_1$ are easily evaluated from this data by taking m linearly independent step demands.

Step 2. Compute K_1 of (7.19) and choose a proportional-plus-summation controller of the required form to meet the performance specifications on the resulting controlled process limit profile.

Step 3. Check stability along the pass of the actual controlled process by a numerical check of the contraction condition (7.22). If the process is unstable along the pass, consider the possibility of increasing the sampling rate.

The above theory will remain essentially the same in the presence of sensor or actuator delays, provided that the sampling rate is fast enough (relative to the designed limit profile responses of the controlled process). Also the requirement that $\hat{H}(1) = 0$ in (7.23) can be removed without altering the results. To see that this is indeed possible, first note that $\hat{B}_1$ represents steady state response characteristics and hence it is independent of the sampling rate. Suppose also that $\tilde{B}_1$ is some other constant $m \times m$ matrix which is used instead of $\hat{B}_1$. Then since the theoretical development is, with the exception of Theorem 7.2.2, independent of the assumption $\hat{H}(1) = 0$, the analysis follows through with $\hat{H}(z_1^{-1})$ replaced by $H(z_1^{-1}) + \hat{B}_0^{-1}(\hat{B}_1 - \tilde{B}_1)$. Also $\hat{B}_0^{-1} = \Gamma\Delta = o(h)$ yields

$$\lim_{h \to 0} \hat{B}_0^{-1}(\hat{B}_1 - \tilde{B}_1) = 0$$

and it follows immediately that Theorem 7.2.2 is valid if $\hat{B}_1$ is replaced by any constant matrix.

7.3 LMI based Control Law Design for Stability Along the Pass

From this point onwards, $(\star)$ will be used to denote the transpose of matrix blocks in some of the LMIs employed (which are required to be symmetric). (Recall also the notation introduced at the start of Sect. 3.1 for symmetric positive-definite matrices etc.)

7.3.1 Discrete Processes

Consider first a control law of the form (7.9) and apply it to a discrete unit memory linear repetitive process of the form defined by Example 1.2.10. Also let $\mathcal{C}_c(z_1, z_2)$ denote the characteristic polynomial (see (1.52)) for the controlled process. Then by an immediate application of Theorem 2.2.8 this process is stable along the pass if, and only if,

$$\mathcal{C}_c(z_1, z_2) = \det \begin{bmatrix} I - z_1(A + BK_1) & -z_1(B_0 + BK_2) \\ -z_2(C + DK_1) & I - z_2(D_0 + DK_2) \end{bmatrix} \neq 0 \text{ in } \mathbb{U}_c^2$$

For the remainder of this section we will make extensive use of the augmented plant matrix of (3.4), i.e.

$$\Phi = \begin{bmatrix} A & B_0 \\ C & D_0 \end{bmatrix}$$

and we can also write

$$\Phi = \widehat{A}_1 + \widehat{A}_2$$

(where the matrices on the right-hand side are defined in (2.66)).

At this stage, we have the following result which provides easily verified necessary conditions for the existence of a control law of the form (7.9) to guarantee that the resulting controlled process is stable along the pass. The proof of this result is obvious from this last condition and is hence omitted here.

Lemma 7.3.1. *The following three conditions are necessary for stability along the pass of a discrete unit memory linear repetitive process of the form defined by Example 1.2.10 with a control law of the form (7.9) applied.*

(a) (A, B) as a 1D discrete linear system (i.e. state matrix A and input matrix B) can be stabilized by application of a state feedback control law.
(b) (D_0, D) as a 1D linear system (i.e. state matrix D_0 and input matrix D) can be stabilized by application of a state feedback control law.
(c) $(\Phi, \widehat{B})$ as a 1D linear system (i.e. state matrix Φ and input matrix $\widehat{B}$) can be stabilized by application of a state feedback control law where $\widehat{B} := \begin{bmatrix} B^T & D^T \end{bmatrix}^T$.

Clearly these conditions should be verified before proceeding further with a given design example.

Other existence-type results can be obtained by, for example, directly transferring results from Sect. IV of [105] for 2D discrete linear systems described by the Roesser model. The major difficulty with such results is that they are not really feasible as general purpose algorithms for control systems design. In the remainder of this section, we show that the design of control laws of the form (7.9) is highly tractable in an LMI setting (at a cost of sufficient but not necessary conditions for stability along the pass of the controlled process).

Introduce the matrices

$$\widehat{B}_1 := \begin{bmatrix} B \\ 0 \end{bmatrix}, \quad \widehat{B}_2 := \begin{bmatrix} 0 \\ D \end{bmatrix}$$

Then we have the following result which is simply Theorem 5.4.1 applied to the controlled process state-space model.

Theorem 7.3.1. *Suppose that a control law described by (7.9) is applied to a discrete unit memory linear repetitive process of the form defined by Example 1.2.10. Then the resulting process is stable along the pass if ∃ matrices $P > 0$ and $Q > 0$ such that*

$$\begin{bmatrix} \hat{L}_1 & \hat{L}_2 \\ \hat{L}_2^T & \hat{L}_3 \end{bmatrix} < 0$$

where

$$\begin{aligned} \hat{L}_1 &:= (\widehat{A}_1^T + K^T \widehat{B}_1^T) P (\widehat{A}_1 + \widehat{B}_1 K) + Q - P \\ \hat{L}_2 &:= (\widehat{A}_1^T + K^T \widehat{B}_1^T) P (\widehat{A}_2 + \widehat{B}_2 K) \\ \hat{L}_3 &:= (\widehat{A}_2^T + K^T \widehat{B}_2^T) P (\widehat{A}_2 + \widehat{B}_2 K) - Q \end{aligned}$$

The difficulty with the matrix inequality of Theorem 7.3.1 is that it is nonlinear in its parameters. It can, however, be converted into the following result where each of the inequalities is a strict LMI with a linear constraint and a formula for computing K in (7.9) is available.

Theorem 7.3.2. *The condition of Theorem 7.3.1 is equivalent to the requirement that ∃ matrices $Y > 0$, $Z > 0$, and N such that the following LMI holds*

$$\begin{bmatrix} Z - Y & -0 & Y\widehat{A}_1^T + N^T \widehat{B}_1^T \\ 0 & -Z & Y\widehat{A}_2^T + N^T \widehat{B}_2^T \\ \widehat{A}_1 Y + \widehat{B}_1 N & \widehat{A}_2 Y + \widehat{B}_2 N & -Y \end{bmatrix} < 0 \tag{7.26}$$

Also if this last condition holds, a stabilizing K in the control law (7.9) is given by

$$K = NY^{-1} \tag{7.27}$$

Proof. Apply the Schur's complement formula (Lemma 5.4.1) to the condition of the previous theorem, and then pre- and post-multiply the result by diag(P^{-1}, P^{-1}, I). Now introduce the substitutions $Z := P^{-1}QP^{-1} > 0$, $Y := P^{-1} > 0$ to obtain

$$\begin{bmatrix} Z-Y & 0 & Y(\widehat{A}_1^T + K^T\widehat{B}_1^T) \\ 0 & -Z & Y(\widehat{A}_2^T + K^T\widehat{B}_2^T) \\ (\widehat{A}_1 + \widehat{B}_1K)Y & (\widehat{A}_2 + \widehat{B}_2K)Y & -Y \end{bmatrix} < 0$$

Use of (7.27) now completes the proof. □

In many cases the state vector $x_{k+1}(p)$ may not be available or, at best, only some of its entries are. Hence we now consider the use of pass profile (or output) only based control laws to achieve stability along the pass of the controlled process. The first law considered has the form of (7.11). In what follows on this type of control law we mainly follow [167].

This control law is, in general, weaker than that of (7.9) and examples are easily given where stability along the pass can be achieved using (7.9) but not (7.11). It is important to note again here that, by definition, the pass profile produced on each pass is available for control purposes before the start of each new pass. As such, this control law (and extensions) assumes storage of the required previous pass profiles and that they are not corrupted by noise etc.

To consider the effect of a control law of the form (7.11) on the process dynamics, first substitute the pass profile (second) equation of the process state-space model into (7.11) to obtain (assuming the required matrix inverse exists)

$$\begin{aligned} u_{k+1}(p) &= (I_r - \widetilde{K}_1D)^{-1}\widetilde{K}_1Cx_{k+1}(p) \\ &+ (I_r - \widetilde{K}_1D)^{-1}[\widetilde{K}_2 + \widetilde{K}_1D_0]y_k(p) \end{aligned}$$

and hence this control law can be treated as a particular case of (7.9) with

$$\begin{aligned} K_1 &:= (I_r - \widetilde{K}_1D)^{-1}\widetilde{K}_1C \\ K_2 &:= (I_r - \widetilde{K}_1D)^{-1}(\widetilde{K}_2 + \widetilde{K}_1D_0) \end{aligned} \tag{7.28}$$

This route may, however, encounter serious numerical difficulties (arising from the fact that (7.28) is a set of matrix nonlinear algebraic equations) and hence we proceed by rewriting these last equations to obtain

$$\begin{aligned} (I_r - \widetilde{K}_1D)K_1 &= \widetilde{K}_1C \\ (I_r - \widetilde{K}_1D)K_2 &= \widetilde{K}_2 + \widetilde{K}_1D_0 \end{aligned}$$

and assume that

$$K_1 = L_1C \tag{7.29}$$

Note that this assumption imposes no restrictions on the results developed here but could be a source of difficulty in other cases, e.g. in uncertainty analysis where the resulting robust control problem may not be convex.

It now follows immediately that

$$\begin{aligned}\widetilde{K}_1 &= L_1(I_m + DL_1)^{-1}\\ \widetilde{K}_2 &= [I_r - L_1(I_m + DL_1)^{-1}D]K_2 - L_1(I_m + DL_1)^{-1}D_0 \end{aligned}\quad (7.30)$$

for any L_1 such that $I_m + DL_1$ is nonsingular, and we have the following result.

Theorem 7.3.3. *Suppose that a control law described by (7.11) is applied to discrete unit memory linear repetitive process of the form defined by Example 1.2.10 and (7.29) holds. Then the resulting process is stable along the pass if ∃ matrices $Y > 0$, $Z > 0, X > 0$ and N such that the following LMI holds*

$$\begin{aligned}&\begin{bmatrix} Z - Y & (\star) & (\star) \\ 0 & -Z & (\star) \\ \widehat{A}_1 Y + \widehat{B}_1 N\widetilde{C} & \widehat{A}_2 Y + \widehat{B}_2 N\widetilde{C} & -Y \end{bmatrix} < 0\\ &X\widetilde{C} = \widetilde{C}Y \end{aligned}\quad (7.31)$$

where $\widetilde{C} = diag(C, I)$. Also if this condition holds, the control law matrices $\widetilde{K}_1$ and $\widetilde{K}_2$ can be obtained from (7.30), where

$$[L_1 \;\; K_2] = NX^{-1} \quad (7.32)$$

and it is required that $I_m + DL_1$ is nonsingular.

Proof. From (7.32) we have that $N = LX$, $L := \begin{bmatrix} L_1 & K_2 \end{bmatrix}$, and substitution into the LMI of (7.31) now gives with $X\widetilde{C} := \widetilde{C}Y$ applied

$$\begin{bmatrix} Z - Y & (\star) & (\star) \\ 0 & -Z & (\star) \\ \widehat{A}_1 Y + \widehat{B}_1 L\widetilde{C}Y & \widehat{A}_2 Y + \widehat{B}_2 L\widetilde{C}Y & -Y \end{bmatrix} < 0$$

Finally, set $L\widetilde{C} = K_{dd}$ to obtain the LMI stabilization condition (i.e. Theorem 5.4.1 applied to the process resulting from the application of the control law) which completes the proof. □

The design developed above is easily implemented using LMI toolboxes, such as Scilab or Matlab, but has the possible disadvantage that it is based on a sufficient but not necessary stability condition. (Also (7.32) can be a source of numerical difficulties when using the Matlab LMI toolbox but Scilab (see, for example, [30]) avoids such problems and hence is used in the numerical computations reported here.) This means that there could well be a significant degree of conservativeness associated with this design method in the sense that in some cases it will fail to produce a stabilizing control law when one actually exists. To avoid, or lower, the level of conservativeness present, we next develop an extension of the control law considered above based on the additional use of pass profile information from the previous two passes (which have already been generated by the process evolution and are hence available for control purposes).

The control law considered has the following form and is, in effect, (7.11) augmented at point p by additive contributions from points $p-1$ on the previous pass and p on the last but one pass respectively

$$u_{k+1}(p)=\widetilde{K}_1 y_{k+1}(p)+\widetilde{K}_2 y_k(p)+\widetilde{K}_3 y_k(p-1)+\widetilde{K}_4 y_{k-1}(p) \tag{7.33}$$

Substituting the equation defining the process pass profile into (7.33) now yields that this last control law is, in fact, a particular case of the so-called extended, mixed state, pass profile control law

$$u_{k+1}(p)=K_1 x_{k+1}(p)+K_2 y_k(p)+K_3 y_k(p-1)+K_4 y_{k-1}(p) \tag{7.34}$$

Hence this control law is, in effect, another extension of (7.9) but here it is used as an intermediate step in the computation of the matrices $\widetilde{K}_i$, $i = 1,\ldots,4$, through use of the following result.

Theorem 7.3.4. *Suppose that a control law described by (7.34) is a applied to a discrete unit memory linear repetitive process of the form defined by Example 1.2.10 and (7.29) holds. Then the resulting process is stable along the pass if* $\exists$ *matrices* $Y > 0$, $X := diag(X_1, X_2, X_3, X_4) > 0$, $Z > 0$ *and* N *such that*

$$\begin{bmatrix} Z-Y & (\star) & (\star) \\ 0 & -Z & (\star) \\ \widehat{A}_1 Y + \widehat{B}_1 N \widehat{C} & \widehat{A}_2 Y + \widehat{B}_2 N \widehat{C} & -Y \end{bmatrix} < 0 \tag{7.35}$$

$$X\widehat{C} = \widehat{C}Y$$

where

$$\widehat{A}_1 := \begin{bmatrix} A & -I & 0 & B_0 \\ 0 & 0 & 0 & 0 \\ 0 & 0 & 0 & 0 \\ 0 & 0 & 0 & 0 \end{bmatrix}, \widehat{A}_2 := \begin{bmatrix} 0 & 0 & 0 & 0 \\ 0 & 0 & 0 & 0 \\ 0 & 0 & 0 & 0 \\ C & 0 & -I & D_0 \end{bmatrix}, \widehat{B}_1 := \begin{bmatrix} B & 0 & 0 & B \\ 0 & B & 0 & 0 \\ 0 & D & 0 & 0 \\ 0 & 0 & 0 & 0 \end{bmatrix}$$

$$\widehat{B}_2 := \begin{bmatrix} 0 & 0 & 0 & 0 \\ 0 & 0 & B & 0 \\ 0 & 0 & D & 0 \\ D & 0 & 0 & D \end{bmatrix}, N := \begin{bmatrix} N_1 & 0 & 0 & 0 \\ 0 & 0 & 0 & -N_3 \\ 0 & 0 & 0 & -N_4 \\ 0 & 0 & 0 & N_2 \end{bmatrix}, \widehat{C} := \begin{bmatrix} C & 0 & 0 & 0 \\ 0 & I & 0 & 0 \\ 0 & 0 & I & 0 \\ 0 & 0 & 0 & I \end{bmatrix}$$

$$\begin{bmatrix} L_1 & 0 & 0 & 0 \\ 0 & 0 & 0 & K_3 \\ 0 & 0 & 0 & K_4 \\ 0 & 0 & 0 & K_2 \end{bmatrix} =: NX^{-1}$$

Also if (7.35) holds, the control law matrices $\widetilde{K}_1$ and $\widetilde{K}_2$ can be computed using (7.30) and then

$$\begin{aligned} \widetilde{K}_3 &= [I_r - L_1(I_m + DL_1)^{-1}D]K_3 \\ \widetilde{K}_4 &= [I_r - L_1(I_m + DL_1)^{-1}D]K_4 \end{aligned} \tag{7.36}$$

where it is assumed that $I_m + DL_1$ is nonsingular.

Proof. Substitute (7.34) into the model of Example 1.2.10 and use (7.29) to obtain

$$\begin{aligned} x_{k+1}(p+1) &= (A + BL_1C)x_{k+1}(p) + (B_0 + BK_2)y_k(p) \\ &+ BK_3y_k(p-1) + BK_4y_{k-1}(p) \\ y_{k+1}(p) &= (C + DL_1C)x_{k+1}(p) + (D_0 + DK_2)y_k(p) \\ &+ DK_3y_k(p-1) + DK_4y_{k-1}(p) \end{aligned} \tag{7.37}$$

This last description is not in the form to which Theorem 5.4.1 can be applied but it is possible to obtain an equivalent state-space model for which this is the case. Here the route is by using the delay operators of Chap. 1, i.e.

$$x_k(p) := z_1 x_k(p+1), \quad y_k(p) := z_2 y_{k+1}(p) \tag{7.38}$$

and the 2D characteristic polynomial. To begin, apply (7.38) to (7.37) to obtain (after some routine algebraic manipulations which are omitted here)

$$\mathcal{C}_{cm}(z_1, z_2) := det \begin{bmatrix} I_n - z_1\widetilde{A} & -z_1\widetilde{B}_0 - z_1^2F_1 - z_1z_2F_3 \\ -z_2\widetilde{C} & I_m - z_2\widetilde{D}_0 - z_1z_2F_2 - z_2^2F_4 \end{bmatrix}$$

where

$$\begin{aligned} &\widetilde{A} := A + BL_1C, \ \widetilde{B}_0 := B_0 + BK_2, \ F_1 := BK_3, \ F_2 := DK_3 \\ &\widetilde{C} := C + DL_1C, \ \widetilde{D}_0 := D_0 + DK_2, \ F_3 := BK_4, \ F_4 := DK_4 \end{aligned}$$

Application of appropriate elementary operations (which leave the determinant invariant) to the right-hand side of this last expression now yields that it can be replaced by

$$det \begin{bmatrix} I_n - z_1\widetilde{A} & z_1 I_n & 0 & -z_1\widetilde{B}_0 \\ 0 & I_n & 0 & z_1F_1 + z_2F_3 \\ 0 & 0 & I_m & z_1F_2 + z_2F_4 \\ -z_2\widetilde{C} & 0 & z_2 I_m & I_m - z_2\widetilde{D}_0 \end{bmatrix}$$

At this stage, the state-space model of the controlled process has a 2D characteristic polynomial to which Theorem 5.4.1 is applicable. This action together with some obvious algebraic operations now yield directly the LMI of (7.35) as a sufficient condition for stability along the pass. Finally, by an identical argument to that in the previous result we have that $\widetilde{K}_1$ and $\widetilde{K}_2$ can be computed using (7.30) and $\widetilde{K}_3$ and $\widetilde{K}_4$ using (7.36), provided $I_m + DL_1$ nonsingular, and the proof is complete. □

Note that the LMI of (7.35) is of dimension $6(n+m) \times 6(n+m)$ but, since we are dealing with stability along the pass, this does not depend on the number of samples along the pass, i.e. on the pass length α. The existing LMI toolboxes also allow us to solve relatively large problems. For example, we have successfully completed this design for the case when in the basic process state-space model $n = 15$, $m = 3$, which gives an LMI of dimension 108×108.

Example 7.3.1. Consider the following special case of Example 1.2.10 which is asymptotically unstable (and hence unstable along the pass) since $r(D_0) > 1$

$$A = \begin{bmatrix} -1.36 & -1.29 & -0.8 \\ 0.15 & 0.34 & 0 \\ -0.19 & 0 & -1.36 \end{bmatrix}, \quad B_0 = \begin{bmatrix} 0.44 & 0.51 \\ 0.93 & 0.14 \\ 0.65 & 0 \end{bmatrix}$$

$$B = \begin{bmatrix} 0.18 & -2.35 & 0.8 \\ 1.07 & -2.5 & 0.5 \\ -0.43 & 0.8 & 2.82 \end{bmatrix}, \quad C = \begin{bmatrix} -0.38 & 0 & -0.37 \\ 0 & 0 & -0.98 \end{bmatrix}$$

$$D = \begin{bmatrix} -2.85 & -0.65 & -2.5 \\ -0.28 & -2.98 & 1.96 \end{bmatrix}, \quad D_0 = \begin{bmatrix} -1.15 & 0 \\ -0.42 & 1.13 \end{bmatrix}$$

In this case

$$\widetilde{K}_1 = \begin{bmatrix} 49.5 & -40.8 \\ 14.27 & -11.77 \\ -44.49 & 36.46 \end{bmatrix}, \quad \widetilde{K}_2 = \begin{bmatrix} -1.77 & 0.96 \\ -0.18 & 0.31 \\ 1.26 & -0.97 \end{bmatrix}$$

$$\widetilde{K}_3 = 10^{-12} \times \begin{bmatrix} 0.37 & -0.33 \\ 0.11 & -0.1 \\ -0.33 & 0.29 \end{bmatrix}, \quad \widetilde{K}_4 = 10^{-13} \times \begin{bmatrix} 0.68 & -0.6 \\ 0.2 & -0.17 \\ -0.61 & 0.53 \end{bmatrix}$$

Suppose now that the boundary conditions are

$$\begin{aligned} x_{k+1}(0) &= [\ 0.4 \ \ 0.4 \ \ 0.4\ ]^T, \ k \geq 0 \\ y_0(p) &= [\ 1 \ \ 1\ ]^T, \ 1 \leq p \leq 19 \end{aligned}$$

Then Fig. 7.1 shows the response of the first entry in the pass profile vector for this example and demonstrates that it is unstable along the pass. Figure 7.2 shows the corresponding stable along the pass response with the control law of (7.33) applied.

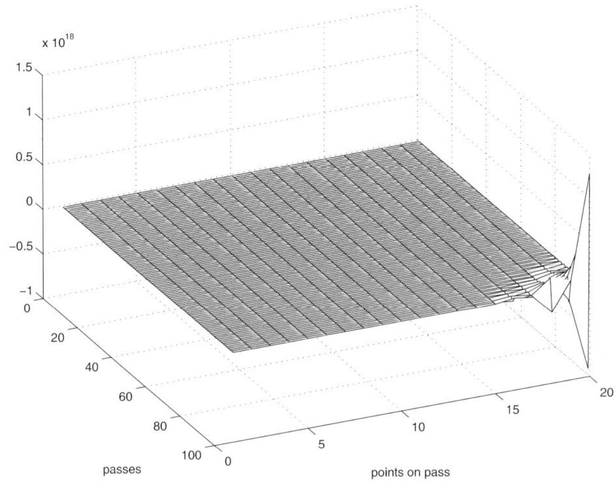

Fig. 7.1. Response of the 1st channel with no control applied

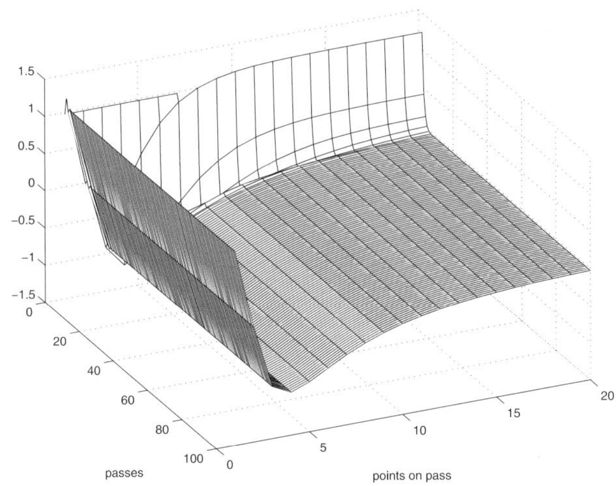

Fig. 7.2. Response of the first channel with the designed control law applied

Note that in the example here the elements of $\widetilde{K}_3$ and $\widetilde{K}_4$ are significantly smaller in magnitude than those in the other control law matrices. Also if these matrices are deleted from the control law then it can be verified that the controlled process is still stable along the pass and there is very little difference in the controlled response. Note also that direct use of the design method referenced in [167] fails to give a stable design. Hence it can be conjectured that this last design method can be exploited to reduce the degree of conservativeness due to the use of a sufficient but not necessary stability along the pass condition.

7.3.2 Discrete Processes with Dynamic Boundary Conditions

We consider the unit memory case and start with processes whose 1D equivalent model takes the form of (1.47)–(1.49), where of the matrices involved, it is only the detailed structures of $\mathfrak{B}_7$ and $\mathfrak{B}_2$ which are required, i.e.

$$\mathfrak{B}_7 = \begin{bmatrix} D_0 + CJ_0 & CJ_1 & \cdots & CJ_{\alpha-1} \\ C(B_0 + AJ_0) & (D_0 + CAJ_1) & \cdots & CAJ_{\alpha-1} \\ CA(B_0 + AJ_0) & C(B_0 + A^2J_1) & \cdots & CA^2J_{\alpha-1} \\ \vdots & \vdots & \ddots & \vdots \\ CA^{\alpha-2}(B_0 + AJ_0) & CA^{\alpha-3}(B_0 + A^2J_1) & \cdots & (D_0 + CA^{\alpha-1}J_{\alpha-1}) \end{bmatrix} \tag{7.39}$$

$$\mathfrak{B}_2 = \begin{bmatrix} D & 0 & 0 & \cdots & 0 \\ CB & D & 0 & \cdots & 0 \\ CAB & CB & D & \cdots & 0 \\ \vdots & \vdots & \vdots & \ddots & \vdots \\ CA^{\alpha-2}B & CA^{\alpha-3}B & CA^{\alpha-4}B & \cdots & D \end{bmatrix}$$

Suppose now that the control law (7.12) is to be used, which (together the natural generalization to the non–unit memory case) is defined only in terms of the output equation (that for updating $Y(l)$) in the 1D equivalent model). The reason for this is that the state equation in this 1D equivalent model has no dynamic updating. Also the condition for asymptotic stability under the action of this control law is just the standard 1D pole placement problem. Using the Lyapunov theory for 1D discrete linear systems (and then making an obvious application of the Schur's complement formula) now gives that the controlled process is asymptotically stable provided there exist matrices $W > 0$ and K_{1D} such that

$$\begin{bmatrix} -W & \mathfrak{B}_7^T W + K_{1D}^T \mathfrak{B}_2^T W \\ W\mathfrak{B}_7 + W\mathfrak{B}_2 K_{1D} & -W \end{bmatrix} < 0$$

This, however, is not in LMI form since it involves the product of two matrices, W and K_{1D} which have to be designed. We have, however, the following result which removes this problem.

Theorem 7.3.5. *A discrete unit memory linear repetitive process described by defined by the state-space model of Example 1.2.13 and pass state initial vector sequence (6.21) is asymptotically stable under a control law of the form (7.12) if, and only if, $\exists$ matrices $W > 0$ and N such that*

$$\begin{bmatrix} -W & W\mathfrak{B}_7^T + N^T\mathfrak{B}_2^T \\ \mathfrak{B}_7 W + \mathfrak{B}_2 N & -W \end{bmatrix} < 0 \tag{7.40}$$

Also if this condition holds, a stabilizing K_{1D} in the control law (7.12) is given by $K_{1D} = NW^{-1}$.

Proof. Immediate on pre- and post-multiplying (7.40) by $\text{diag}(W^{-1}, W^{-1})$ and then replacing W^{-1} by W. □

Write the control law (7.12) in the form

$$u_l(p) = \sum_{j=0}^{\alpha-1} K_{pj} v_l(j)$$

or, on recovering the original process variables $y_k(p)$,

$$u_{k+1}(p) = \sum_{j=0}^{\alpha-1} K_{pj} y_k(j)$$

where $K_{1D} = [K_{pj}]_{j=0\ldots\alpha-1, p=0\ldots\alpha-1}$. Then, in physical terms, we see that the control vector $u_{k+1}(p)$ applied at any point on a given pass computed using this law depends on the complete previous pass profile vector $y_k(p)$, $0 \le p \le \alpha - 1$, and this could lead to implementation difficulties. To simplify the structure, we can require that in Theorem 7.3.5 $W = \text{diag}\,(W_0, W_1, \ldots, W_{\alpha-1}) =: W_p$, $N = \text{diag}\,(N_0, N_1, \ldots, N_{\alpha-1}) =: N_p$, where W_j and N_j are matrices of dimensions $m \times m$ and $r \times m$ respectively. Then the control law takes the form

$$u_{k+1}(p) := K_p y_k(p) = N_p W_p^{-1} y_k(p)$$

which introduces a considerable simplification in implementation terms since now the complete previous pass dependence has been replaced by a point-to-point dependence.

Another possible simplification is to require that the control law is constant along the pass, i.e. assuming that $W_j = \tilde{W}, N_j = \tilde{N}$, which yields

$$u_{k+1}(p) := \widehat{K} y_k(p) = \tilde{N}\tilde{W}^{-1} y_k(p)$$

This requirement is, however, more strict and hence more difficult to satisfy than the previous one.

It is also possible to re–formulate Theorem 7.3.5 into a more computationally attractive version by simplifying the form of the underlying matrices. This first requires two intermediate results for the uncontrolled case (Lemmas 7.3.2 and 7.3.3 below).

First recall from (1.48) that $\mathfrak{B}_7 = \mathfrak{B}_1 + \mathfrak{B}_3 J$. Also the matrix $\mathfrak{B}_3 J$ can be rewritten as $\mathfrak{B}_3 J = \mathbb{CAJ}$ where

$$
\begin{aligned}
\mathbb{C} &:= \left\|\mathrm{diag}\{I, A, A^2, \ldots, A^{\alpha-1}\}\right\| \, \mathrm{diag}\,\{C\}_\alpha \\
\mathbb{A} &:= \left\|\mathrm{diag}\{I, A, A^2, \ldots, A^{\alpha-1}\}\right\|^{-1} \mathrm{diag}\{I, A, A^2, \ldots, A^{\alpha-1}\} \\
\mathbb{J} &:= \begin{bmatrix} J_0 & \cdots & J_{\alpha-1} \\ \vdots & & \vdots \\ J_0 & \cdots & J_{\alpha-1} \end{bmatrix}
\end{aligned} \tag{7.41}
$$

where $||\cdot||$ denotes any convenient matrix norm. Hence, noting that $||\mathbb{A}|| = 1$ allows us to obtain a result similar to the previous theorem.

Lemma 7.3.2. *A discrete unit memory linear repetitive process of the form defined by the state-space model of Example 1.2.13 and pass state initial vector sequence (6.21) is asymptotically stable if, and only if, $\exists$ a real scalar $\varepsilon > 0$ and a matrix $W > 0$ such that*

$$
\begin{bmatrix} -W^{-1} + \varepsilon \mathbb{C}\mathbb{C}^T & \mathfrak{B}_1 \\ \mathfrak{B}_1^T & \varepsilon^{-1}\mathbb{J}^T\mathbb{J} - W \end{bmatrix} < 0
$$

Proof. This is a direct consequence of using (7.41) which enables us to eliminate the matrix $\mathbb{A}$ from the matrix $\mathfrak{B}_7$ in the LMI of the previous theorem (in effect, we use the elimination procedure of [25]). □

Use of the Schur's complement formula (Lemma 5.4.1) and appropriate pre- and post-multiplication by block diagonal matrices enables Lemma 7.3.2 to be restated in the following form which is more attractive in terms of onward analysis.

Lemma 7.3.3. *A discrete unit memory linear repetitive process of the form defined by the state-space model of Example 1.2.13 and pass state initial vector sequence (1.35) is asymptotically stable if, and only if, $\exists$ a real scalar $\mu > 0$ and a matrix $W > 0$ such that*

$$
\begin{bmatrix} -W & W\mathfrak{B}_1 & W\mathbb{C} & 0 \\ \mathfrak{B}_1^T W & -W & 0 & \mu\mathbb{J}^T \\ \mathbb{C}^T W & 0 & -\mu I & 0 \\ 0 & \mu\mathbb{J} & 0 & -\mu I \end{bmatrix} < 0
$$

The following result uses this last LMI to obtain the following alternative to Theorem 7.3.5.

Theorem 7.3.6. *A discrete unit memory linear repetitive process described by defined by the state-space model of Example 1.2.13 and pass state initial vector sequence (1.35) is asymptotically stable under a control law of the form (7.12) if, and only if, $\exists$ a real scalar $\varepsilon > 0$ and matrices $W > 0$ and N such that*

$$
\begin{bmatrix} -W & \mathfrak{B}_1 W + \mathfrak{B}_2 N & \mathbb{C}\varepsilon & 0 \\ (\mathfrak{B}_1 W + \mathfrak{B}_2 N)^T & -W & 0 & W\mathbb{J}^T \\ \mathbb{C}^T\varepsilon & 0 & -\varepsilon I & 0 \\ 0 & \mathbb{J}W & 0 & -\varepsilon I \end{bmatrix} < 0 \tag{7.42}
$$

Also if this condition holds, the control law matrix is given by $K_{1D} = NW^{-1}$.

Proof. First pre- and post-multiply (7.42) by $\mathrm{diag}(W^{-1}, W^{-1}, \varepsilon^{-1}I, \varepsilon^{-1}I)$ and then replace W by W^{-1} and ε by ε^{-1}. Hence, since $K_{1D} = NW^{-1}$, we obtain the result of Lemma 7.3.3 interpreted in terms of the controlled process state-space model. □

Finally, note that if the stronger property of stability along the pass is required, the control law emerging from the asymptotic stability design must also satisfy the characteristic polynomial condition (or any equivalent condition) of (2.44) interpreted in terms of the controlled process. This can obviously be treated using the LMI based result of Theorem 7.3.2 and hence the details are omitted here.

In order to apply the above analysis to the case of a discrete non–unit memory linear repetitive process with dynamic boundary conditions, the basic route is to first construct the 1D equivalent model. Now, however, the dimension of the vector replacing $Y(l)$ is considerably larger, but all of the results given above for the unit memory case above generalize in a natural manner. Hence they are not explicitly given here except to note that the control law matrix K_{1D} now expands to $K_{1D} = \begin{bmatrix} K_0 & K_1 & \cdots & K_{\alpha-1} \end{bmatrix}$.

7.3.3 Differential Processes

Here we show that the LMI setting leads directly to the design of a stabilizing control law for a differential unit linear repetitive process of the form defined by Example 1.2.3. The control law considered is given by (7.8).

Application of this control law to the process state-space model and then interpreting the result of Theorem 2.2.7 (i.e. (2.43)) gives the following necessary and sufficient condition for stability along the pass of the resulting controlled process

$$\mathcal{C}_c(s, z_2) \neq 0 \text{ in } \mathbb{I} \times \mathbb{U}$$

where

$$\mathcal{C}_c(s, z_2) := \det \left[sI_n \oplus I_m - \begin{bmatrix} (A + BK_1) & (B_0 + BK_2) \\ z_2(C + DK_1) & z_2(D_0 + DK_2) \end{bmatrix} \right]$$

Then we have the following theorem which results directly from interpreting (5.70) for the controlled process.

Theorem 7.3.7. *A differential unit memory linear repetitive process described by defined by Example 1.2.3 is stable along the pass under a control law of the form (7.8) if* $\exists$ *matrices* $Y > 0$, $Z > 0$, N *and* M *such that the following LMI holds*

$$\begin{bmatrix} YA^T + N^TB^T + AY + BN & B_0Z + BM & YC^T + N^TD^T \\ ZB_0^T + M^TB^T & -Z & ZD_0^T + M^TD^T \\ CY + DN & D_0Z + DM & -Z \end{bmatrix} < 0 \tag{7.43}$$

Also if this condition holds, K_1 and K_2 in the control law (7.8) are given by

$$K_1 = NY^{-1}, \ K_2 = MZ^{-1} \tag{7.44}$$

Proof. First recall (5.70) and define the matrices $\widehat{B}_1$ and $\widehat{B}_2$ as in the discrete case, i.e.

$$\widehat{B}_1 = \begin{bmatrix} B \\ 0 \end{bmatrix}, \ \widehat{B}_2 = \begin{bmatrix} 0 \\ D \end{bmatrix}$$

Then the controlled process is stable along the pass if $\exists$ matrices $W_1 > 0$ and $W_2 > 0$ such that

$$\begin{bmatrix} -W^{0,1} + (\widehat{A}_1 + \widehat{B}_1 K)^T W^{1,0} + W^{1,0}(\widehat{A}_1 + \widehat{B}_1 K) & (\widehat{A}_2 + \widehat{B}_2 K)^T \widehat{W}_2 \\ \widehat{W}_2(\widehat{A}_2 + \widehat{B}_2 K) & -\widehat{W}_2 \end{bmatrix} < 0$$

The difficulty with the this last matrix inequality is that it is nonlinear in its parameters. To obtain a strict LMI form first note that the above condition can be rewritten, by direct substitution for the relevant block entries, as

$$\begin{bmatrix} (A + BK_1)^T W_1 + W_1(A + BK_1) & W_1(B_0 + BK_2) & 0 & (C + DK_1)^T W_2 \\ (B_0 + BK_2)^T W_1 & -W_2 & 0 & (D_0 + DK_2)^T W_2 \\ 0 & 0 & -W_2 & 0 \\ W_2(C + DK_1) & W_2(D_0 + DK_2) & 0 & -W_2 \end{bmatrix} < 0$$

Next, pre- and post-multiply this last expression by

$$\operatorname{diag}\left\{W_1^{-1}, \ W_2^{-1}, \ W_2^{-1}, \ W_2^{-1}\right\}$$

and then note that the third block row and the third block column of the resulting matrix can be removed without changing the underlying inequality solution. Finally, introduce the substitutions $Y = W_1^{-1}$, $Z = W_2^{-1}$, and use the formulas for K_1 and K_2 in (7.44) to obtain (7.43). □

7.4 Design for Performance and Disturbance Rejection

In this section we consider the problem of control law design to achieve stabilization and disturbance rejection for discrete unit memory linear repetitive processes. The class of processes considered are those described by the following state-space model over $0 \leq p \leq \alpha - 1$, $k \geq 0$

$$
\begin{aligned}
x_{k+1}(p+1) &= Ax_{k+1}(p) + Bu_{k+1}(p) + B_0 y_k(p) + Ew_{k+1}(p) \\
y_{k+1}(p) &= Cx_{k+1}(p) + Du_{k+1}(p) + D_0 y_k(p) + Fw_{k+1}(p)
\end{aligned} \tag{7.45}
$$

Here $x_k(p)$, $y_k(p)$, and $u_k(p)$ are as in Example 1.2.10, and $w_{k+1}(p)$ is a $q \times 1$ disturbance vector which enters on the current pass. Here we assume that the disturbance term is the same on each pass and it is denoted below by $w(p)$. (The case when this assumption does hold is considered in the next chapter.) The boundary conditions of Example 1.2.10 are also assumed and, in particular, $x_{k+1}(0) = d_{k+1}$, $k \geq 0$. The results in this section are mainly from [168].

In what follows we consider the use of control laws to achieve asymptotic and/or stability along the pass coupled with desired pass-to-pass and along the pass performance as required. In the case of the asymptotic stability based analysis, we make use of the 1D equivalent model of the underlying dynamics which is summarized next.

In actual fact, the 1D equivalent model for processes described by (7.45) is that of (1.45) and (1.46) augmented by extra terms to account for the influence of the disturbance. In particular, introduce the so-called disturbance super-vector of dimension $q\alpha \times 1$ as

$$
W_{1D} := \begin{bmatrix} w^T(0) & w^T(1) & w^T(2) & \cdots & w^T(\alpha-1) \end{bmatrix}^T
$$

Then the 1D equivalent state-space model of the dynamics of processes of the form (7.45) is given by

$$
\begin{aligned}
Y(l+1) &= \mathfrak{B}_1 Y(l) + \mathfrak{B}_2 U(l) + \mathfrak{B}_3 d_l + \Omega_y W_{1D} \\
X(l) &= \mathfrak{B}_4 Y(l) + \mathfrak{B}_5 U(l) + \mathfrak{B}_6 d_l + \Omega_x W_{1D}
\end{aligned} \tag{7.46}
$$

where Ω_x and Ω_y are given below (and the rest of the matrices here are as in (1.45) and (1.46))

$$
\Omega_x := \begin{bmatrix}
E & 0 & 0 & \ldots & 0 \\
AE & E & 0 & \ldots & 0 \\
A^2E & AE & E & \ldots & 0 \\
\vdots & \vdots & \vdots & \ddots & \vdots \\
A^{\alpha-1}E & A^{\alpha-2}E & A^{\alpha-3}E & \ldots & E
\end{bmatrix}
$$

$$
\Omega_y := \begin{bmatrix}
F & 0 & 0 & \ldots & 0 \\
CE & F & 0 & \ldots & 0 \\
CAE & CE & F & \ldots & 0 \\
\vdots & \vdots & \vdots & \ddots & \vdots \\
CA^{\alpha-2}E & CA^{\alpha-3}E & CA^{\alpha-4}E & \ldots & F
\end{bmatrix}
$$

7.4.1 Asymptotic Stability with Performance Design

We consider first the use of the control law (7.12) in the disturbance free case. Using well known LMI results for the 1D discrete linear systems case, we immediately obtain the following result for asymptotic stability of the resulting controlled process. (In fact, this result is a generalization of that of Theorem 7.3.5 (applied with boundary conditions of the form defined by Example 1.2.10) and is adapted from [15].)

Theorem 7.4.1. *Suppose that the 1D equivalent model (7.46) is used to design a control law of the form (7.12) for a discrete unit memory linear repetitive process described by (7.45) with $w(p) = 0$. Then the resulting controlled process is asymptotically stable if, and only if, $\exists$ matrices $P > 0$, G and L such that the following LMI holds*

$$\begin{bmatrix} -P & \mathfrak{B}_1 G + \mathfrak{B}_2 L \\ G^T \mathfrak{B}_1^T + L^T \mathfrak{B}_2^T & P - G - G^T \end{bmatrix} < 0$$

If this condition holds, the control law matrix K_{1D} is given by

$$K_{1D} = LG^{-1} \tag{7.47}$$

Next we give a new result which addresses the currently open question of how to design a control law for discrete linear repetitive processes to ensure both stability and performance. We assume that the control law of (7.12) is applied and hence that no reference or tracking vector is specified for each current pass (analogous to 1D linear systems regulator theory). The problem then is control law design for asymptotic stability with the additional requirement that the resulting limit profile vector is identically zero over the pass length.

Model matching control is a long standing technique in 1D systems theory and there has also been some work on this problem for 2D discrete linear systems described by the Roesser and Fornasini-Marchesini state-space models [178]. In the case of the discrete unit memory linear repetitive processes treated here, first note again that the 1D equivalent model matrix $\mathfrak{B}_1$ describes the contribution of the previous pass profile to the current one and asymptotic stability requires that $r(\mathfrak{B}_1) < 1$. Also under the action of the control law (7.12) this matrix is 'mapped' as follows

$$\mathfrak{B}_1 \mapsto \mathfrak{B}_1 + \mathfrak{B}_2 K_{1D}$$

Suppose now that we want to assign the matrix resulting from the application of the control law here to $\widehat{\mathfrak{B}}_1$ where this matrix is selected to give a state-space model whose behavior the controlled process is required to follow. Then we have the following result.

Theorem 7.4.2. *Suppose that the 1D equivalent model (7.46) is used to design a control law of the form (7.12) for a discrete linear repetitive process described by (7.45) with $w(p) = 0$. Then the resulting process is asymptotically stable and reaches the required form (as expressed above in terms of the matrix $\widehat{\mathfrak{B}}_1$) if, and only if, $\exists$ matrices $P > 0$, G and L such that the following LMI holds*

$$\begin{bmatrix} -P & (\mathfrak{B}_1 - \widehat{\mathfrak{B}}_1)\, G + \mathfrak{B}_2 L \\ G^T(\mathfrak{B}_1 - \widehat{\mathfrak{B}}_1)^T + L^T \mathfrak{B}_2^T & P - G - G^T \end{bmatrix} < 0 \tag{7.48}$$

If this condition holds, the control law matrix K_{1D} is computed using (7.47) to ensure that

$$\mathfrak{B}_1 - \widehat{\mathfrak{B}}_1 + \mathfrak{B}_2 K_{1D} = 0 \tag{7.49}$$

Proof. First note again that if the LMI (7.48) holds then the control law matrix K_{1D} of (7.12) is given by (7.47). Also it is a standard fact that it is possible to obtain from the LMI solver a K_{1D} such that (7.48) holds. In such a case

$$\widehat{\mathfrak{B}}_1 = \mathfrak{B}_1 + \mathfrak{B}_2 K_{1D}$$

which completes the proof. □

Suppose now that the matrix K_{1D} can be designed such that (7.49) of Theorem 7.4.2 holds. Then, since the control law does not contain a tracking or reference vector, the resulting controlled process produces the zero limit profile, i.e. the process has been regulated to produce zero output over the pass length on the limit profile.

It is essential to note here that it is impossible to obtain an arbitrarily specified 1D equivalent model matrix $\widehat{\mathfrak{B}}_1$ starting from an arbitrarily specified $\mathfrak{B}_1$. However, conditions under which (7.49) has a solution can be characterized easily using, for example, Cramer's rule for simultaneous linear algebraic equations and the matrix Kronecker product. This is clearly an area for further research and in what follows we focus on the design of control laws for stability along the pass and desired performance, as opposed to just stability along the pass in the (relevant) analysis already given in this chapter.

7.4.2 Stability Along the Pass with Performance Design

Here we consider a control law of the form (7.9). Note again that the state-space quadruple $\{A, B_0, C, D_0\}$ describes the contribution of the previous pass profile to the current one and also determines stability along the pass (via the 2D characteristic polynomial – see Theorem 2.2.8). Also under the action of the control law (7.9) this quadruple is 'mapped' as follows

$$\begin{bmatrix} A & B_0 \\ C & D_0 \end{bmatrix} \rightarrow \begin{bmatrix} A + BK_1 & B_0 + BK_2 \\ C + DK_1 & D_0 + DK_2 \end{bmatrix}$$

and since there is no tracking (or target) vector for the resulting process, its dynamics are completely described by the matrices on the right-hand side of this last expression. Suppose also that we want to use control action to assign $\{A, B_0, C, D_0\}$ to $\{\mathcal{A}, \mathcal{B}_0, \mathcal{C}$, and $\mathcal{D}_0\}$ respectively, where these latter matrices are selected to give a state-space model whose behavior the controlled process is required to follow. Then the following result is relevant.

Theorem 7.4.3. *Suppose that a control law of the form (7.9) is applied to discrete unit memory linear repetitive process described by (7.45) with $w(p) = 0$. Suppose also that the design of Theorem 7.3.2 gives control law matrices K_1, K_2 such that*

$$\begin{bmatrix} A-\mathcal{A} & B_0-\mathcal{B}_0 \\ C-\mathcal{C} & D_0-\mathcal{D}_0 \end{bmatrix} + \begin{bmatrix} B \\ D \end{bmatrix} [K_1\ K_2] = 0 \tag{7.50}$$

Then the resulting process is stable along the pass and its dynamics are defined by the (previously specified) state-space quadruple $\{\mathcal{A}, \mathcal{B}_0, \mathcal{C}, \mathcal{D}_0\}$ if $\exists$ matrices $Y > 0$, $Z > 0$ and N such that

$$\begin{bmatrix} Z-Y & 0 & Y\widetilde{A}_1^T + N^T\widehat{B}_1^T \\ 0 & -Z & Y\widetilde{A}_2^T + N^T\widehat{B}_2^T \\ \widetilde{A}_1 Y + \widehat{B}_1 N & \widetilde{A}_2 Y + \widehat{B}_2 N & -Y \end{bmatrix} < 0$$

where

$$\widetilde{A}_1 := \begin{bmatrix} A-\mathcal{A} & B_0-\mathcal{B}_0 \\ 0 & 0 \end{bmatrix}, \widetilde{A}_2 := \begin{bmatrix} 0 & 0 \\ C-\mathcal{C} & D_0-\mathcal{D}_0 \end{bmatrix}$$

(and the matrices $\widehat{B}_1$ and $\widehat{B}_2$ are as before). If this condition holds, then the control law matrices are again computed using (7.27).

Proof. First note again that if the LMI of Theorem 7.3.2 holds then the control law matrix K ($= \begin{bmatrix} K_1 & K_2 \end{bmatrix}$) is given by (7.27). Also it is a standard fact that it is possible to obtain from the LMI solver a K such that (7.50) holds. In which case, the matrices of the process with the control law applied are such that

$$\begin{bmatrix} \tilde{A} & \tilde{B}_0 \\ \tilde{C} & \tilde{D}_0 \end{bmatrix} := \begin{bmatrix} A+BK_1 & B_0+BK_2 \\ C+DK_1 & D_0+DK_2 \end{bmatrix} = \begin{bmatrix} \mathcal{A} & \mathcal{B}_0 \\ \mathcal{C} & \mathcal{D}_0 \end{bmatrix}$$

which completes the proof. □

The problem encountered in the asymptotic stability analysis earlier in this section also arises here. In particular, it is impossible to obtain arbitrarily specified model matrices $\{\mathcal{A}, \mathcal{B}_0, \mathcal{C}, \mathcal{D}_0\}$ starting from an arbitrary specified $\{A, B_0, C, D_0\}$. However, conditions under which (7.50) has a solution can be characterized easily using, for example, Cramer's rule for simultaneous linear algebraic equations and the matrix Kronecker product.

As noted in Sect. 7.1, a natural approach to the control of repetitive processes is to specify a reference, or tracking vector, representing desired performance on each pass and then attempt to use control action to achieve this goal. Here we consider the case when the term $K_3 r_{k+1}(p), 0 \leq p \leq \alpha - 1$, is added to the control law where $r_{k+1}(p)$ is an $m \times 1$ column vector representing desired behavior on pass $k+1$, $k \geq 0$, and K_3 is an $r \times m$ control law matrix to be selected. The resulting controlled process state-space model is

$$\begin{aligned} x_{k+1}(p+1) &= \mathcal{A} x_{k+1}(p) + B K_3 r_{k+1}(p) + \mathcal{B}_0 y_k(p) \\ y_{k+1}(p) &= \mathcal{C} x_{k+1}(p) + D K_3 r_{k+1}(p) + \mathcal{D}_0 y_k(p) \end{aligned}$$

The questions which now arise are: (i) what is a suitable choice for $r_{k+1}(p)$?; and (ii) how can we design the control law to give stability along the pass plus acceptable transient behavior?

Example 7.4.1. To illustrate what can be achieved here, we focus on the SISO case and use the numerical data given next (for the metal rolling process associated with Example 1.2.10) with $\alpha = 20$ and select the current pass reference signal as $r_{k+1}(p) = -1$, $0 \leq p \leq 19$, $k \geq 0$, i. e. a downward unit step applied at $p = 0$ on each pass

$$A = \begin{bmatrix} 1.9118 & -0.9559 & -1.4706 & 0.7353 \\ 1 & 0 & 0 & 0 \\ 0 & 0 & 0 & 0 \\ 0 & 0 & 1 & 0 \end{bmatrix}$$

$$B = \begin{bmatrix} -2.2059 \times 10^{-5} \\ 0 \\ 0 \\ 0 \end{bmatrix}$$

$$\begin{aligned} B_0 &= \begin{bmatrix} 0.7794 & 0 & 1 & 0 \end{bmatrix}^T, \quad E = \begin{bmatrix} 1 & 0 & 0 & 0 \end{bmatrix} \\ C &= \begin{bmatrix} 1.9118 & -0.9559 & -1.4706 & 0.7353 \end{bmatrix}, \quad D = 2.2059 \times 10^{-5} \\ D_0 &= 0.7794, \quad F = 1 \end{aligned}$$

One possible way of designing the control law is to note that K_3 does not influence stability along the pass. Hence we can execute the LMI design of Theorem 7.4.3 to obtain control law matrices K_1 and K_2 which ensure stability along the pass and then attempt to select a suitable K_3 to meet the performance requirements by 'tuning' the response of the resulting controlled process state-space model.

In the case of this numerical example, it is easily checked that stability along the pass does not hold. Also the stabilization procedure of Theorem 7.4.3 provides the control law matrices K_1 and K_2 as

$$\begin{aligned} K_1 &= \begin{bmatrix} 83204.1343 & -38895.6888 & -67555.1928 & 29928.8272 \end{bmatrix} \\ K_2 &= -2883.1769 \end{aligned}$$

and in the resulting stable along the pass process

$$\left[\begin{array}{c|c} \mathcal{A} & \mathcal{B}_0 \\ \hline \mathcal{C} & \mathcal{D}_0 \end{array}\right] = \left[\begin{array}{cccc|c} 0.0764 & -0.0979 & 0.0196 & 0.0751 & 0.8430 \\ 1.0 & 0.0 & 0.0 & 0.0 & 0.0 \\ 0.0 & 0.0 & 0.0 & 0.0 & 1.0 \\ 0.0 & 0.0 & 1.0 & 0.0 & 0.0 \\ \hline 0.0764 & -0.0979 & 0.0196 & 0.0751 & 0.8430 \end{array}\right]$$

(It was also found that $K_3 = -3765.8$ gives acceptable performance.)

The empirical nature of the above approach means that it is clearly not feasible in the general case. Instead, the method developed next can be used. This new approach is based on a (simple structure) re-formulation of the problem and is the first step which also allows us to extend the design analysis to the case when external disturbances are present. The starting point is the fact that the control task here is to drive the process pass profiles to some prescribed reference signal $y_{\text{ref}}(p)$, $0 \leq p \leq \alpha - 1$ where, if asymptotic stability holds, the resulting limit profile is described by a 1D linear systems state-space model. Here what we are aiming to do is to specify this limit profile as $y_{\text{ref}}(p)$.

7.4.3 Design for Disturbance Rejection

Consider first asymptotic stability. Then the aim here is full decoupling of the influence of the disturbance $w(p)$ from the state and pass profile dynamics of (7.45) together with simultaneously driving the process to the required reference signal denoted in this section by Y_{ref}. In this case, it means that the resulting limit profile defined in terms of the 1D equivalent model as $Y_\infty := \lim_{l \to \infty} Y(l)$ must be the a-priori specified vector Y_{ref}, written in component form as

$$Y_{\text{ref}} := \left[\begin{array}{ccccc} y_{\text{ref}}^T(0) & y_{\text{ref}}^T(1) & y_{\text{ref}}^T(2) & \ldots & y_{\text{ref}}^T(\alpha - 1) \end{array}\right]^T$$

and note again that asymptotic stability of (7.45) is equivalent to $r(\mathfrak{B}_1) < 1$. The overall design task at this stage is to achieve the following.

(a) Asymptotic stability, and hence the limit profile exists.
(b) The resulting limit profile equals the a-priori specified vector, i.e.

$$Y(l) \to Y_{\text{ref}}, \ l \to \infty$$

(c) Complete decoupling of the effects of the disturbance vector $w(p)$ from the dynamics of the controlled process.

Suppose now that asymptotic stability holds, i.e. $r(\mathfrak{B}_1) < 1$, then

$$Y(l+1) - Y(l) \to 0, \ l \to \infty$$

which is equivalent to

$$Y(l+1) = Y(l) = Y_{\text{ref}}, \ l \to \infty$$

and hence the resulting limit profile is described in this representation by

$$Y_{\text{ref}} = \mathfrak{B}_1 Y_{\text{ref}} + \mathfrak{B}_2 U_\infty + \mathfrak{B}_3 d_\infty + \Omega_y W_{1D} \tag{7.51}$$

where U_∞ denotes the 1D equivalent model representation of the strong limit of the control sequence $\{u_k\}_{k\geq 1}$ applied to the process (and d_∞ again denotes the strong limit of the pass state initial vector sequence).

Subtracting (7.51) from the first equation in (7.46) yields

$$Y(l+1) - Y_{ref} = \mathfrak{B}_1\big(Y(l) - Y_{\text{ref}}\big) + \mathfrak{B}_2\big(U(l) - U_\infty\big) + \mathfrak{B}_3(d_l - d_\infty) \tag{7.52}$$

Now introduce the so-called residual variables

$$\begin{aligned} \hat{Y}(l+1) &:= Y(l+1) - Y_{\text{ref}} \\ \hat{U}(l+1) &:= U(l+1) - U_\infty \\ \hat{d}_l &:= d_l - d_\infty \end{aligned}$$

Then (7.52) can be written as

$$\hat{Y}(l+1) = \mathfrak{B}_1 \hat{Y}(l) + \mathfrak{B}_2 \hat{U}(l) + \mathfrak{B}_3 \hat{d}_l \tag{7.53}$$

and the external disturbance vector $w(p)$ (or W_{1D} in the 1D equivalent model) has been decoupled from the process dynamics.

If the uncontrolled process is asymptotically unstable, then control action can be employed which, in effect, is a modified form of (7.12). In particular, apply the following control law to (7.53)

$$\hat{U}(l) = K_{1D}\hat{Y}(l) \Rightarrow U(l) - U_\infty = K_{1D}(Y(l) - Y_{\text{ref}}) \tag{7.54}$$

and choose K_{1D} such that asymptotic stability holds under the control law, i.e. $r(\mathfrak{B}_1 + \mathfrak{B}_2 K_{1D}) < 1$. Then it follows immediately that this control law gives the required limit profile and also complete decoupling of the disturbance vector $w(p)$ (or W_{1D} in the 1D equivalent model).

The only difficulty here is that to apply $U(l)$, i.e.

$$U(l) = K_{1D}Y(l) - K_{1D}Y_{ref} + U_\infty$$

it is necessary to compute U_∞ explicitly. To do this, rewrite (7.51) in the form

$$\mathfrak{B}_2 U_\infty = (I - \mathfrak{B}_1)Y_{\text{ref}} - \Omega_y W_{1D} - \mathfrak{B}_3 d_\infty \tag{7.55}$$

to obtain the condition which must hold for the control task to be achieved. This, in turn, means that not all choices for Y_{ref} can be achieved and, in particular, only those which satisfy

$$\text{rank}[\ \mathfrak{B}_2 \quad (I-\mathfrak{B}_1)Y_{\text{ref}} \quad -\Omega_y W_{1D} \quad -\mathfrak{B}_3 d_\infty\ ] = \text{rank}\mathfrak{B}_2 \tag{7.56}$$

are feasible.

It it now follows immediately that if D is a full row rank matrix then (7.55) is solvable and hence

$$U_\infty = \mathfrak{B}_2^\sharp\Big((I-\mathfrak{B}_1)Y_{ref} - \Omega_y W_{1D} - \mathfrak{B}_3 d_\infty\Big) \tag{7.57}$$

where the superscript $\sharp$ denotes the matrix pseudo inverse.

Note that if D is square and nonsingular, the pseudo inverse can be replaced by the matrix inverse. In which case (7.57) becomes

$$U_\infty = \mathfrak{B}_2^{-1}\Big((I-\mathfrak{B}_1)Y_{ref} - \Omega_y W_{1D} - \mathfrak{B}_3 d_\infty\Big)$$

In general, however, it will often be the case that the dimension of the pass profile vector exceeds that of the input vector and hence neither D or $\mathfrak{B}_2$ are full row rank matrices. There is, however, a means of reducing the limitations this may impose. This is due to the fact that it may be possible to choose d_∞ to ensure that (7.56) holds.

Example 7.4.2. Return to the numerical data of the last example and suppose that to meet other performance requirements we must have

$$\left[\begin{array}{c|c} \mathcal{A} & \mathcal{B}_0 \\ \hline \mathcal{C} & \mathcal{D}_0 \end{array}\right] = \left[\begin{array}{cccc|c} 0.0764 & -0.0979 & 0.0196 & 0.0751 & 0.8430 \\ 1.0 & 0.0 & 0.0 & 0.0 & 0.0 \\ 0.0 & 0.0 & 0.0 & 0.0 & 1.0 \\ 0.0 & 0.0 & 1.0 & 0.0 & 0.0 \\ \hline 0.0764 & -0.0979 & 0.0196 & 0.0751 & 0.8430 \end{array}\right]$$

and $y_{ref}(p) = -1$, $0 \leq p \leq 19$. Also we consider the case when $w(p)$ is the sampled sine wave $w(p) = -0.5\sin\left(\frac{2\pi p}{\alpha}\right)$.

The control law matrix K_{1D} in (7.54) for this case is

$$K_{1D} = \left[\ K^1 \quad K^2 \quad K^3 \quad \ldots \quad K^{20}\ \right] = \begin{bmatrix} k_1 & 0 & 0 & \ldots & 0 \\ k_2 & k_1 & 0 & \ldots & 0 \\ k_3 & k_2 & k_1 & \ldots & 0 \\ \vdots & \vdots & \vdots & \ddots & \vdots \\ k_{20} & k_{19} & k_{18} & \ldots & k_1 \end{bmatrix}$$

where

$$\begin{aligned} K^1 = \Big[& -2883.1769 \quad 2585.8924 \quad 4129.3415 \quad -3351.6179 \quad -651.3602 \\ & 278.3595 \quad 85.0348 \quad -20.7547 \quad -9.9106 \quad 1.2747 \\ & 1.0676 \quad -0.0432 \quad -0.1078 \quad -0.0040 \quad 0.0102 \\ & 0.0012 \quad -0.0009 \quad -0.0002 \quad 0.0001 \quad 0.0\ \Big]^T \end{aligned}$$

Next we show how to solve the design problem considered here under stability along the pass, i.e. when asymptotic stability in the design objectives listed under (a)–(c) earlier in this sub-section is replaced by the stronger requirement of stability along the pass.

Suppose that as $k \to \infty$, $x_k(p) \to x_\infty(p)$, $u_k(p) \to u_\infty(p)$ and $y_k(p) \to y_{ref}(p)$. Then

$$\begin{aligned} x_\infty(p+1) &= Ax_\infty(p) + Bu_\infty(p) + B_0 y_{\text{ref}}(p) + Ew(p) \\ y_{ref}(p) &= Cx_\infty(p) + Du_\infty(p) + D_0 y_{\text{ref}}(p) + Fw(p) \end{aligned} \tag{7.58}$$

Now introduce (the residual variables)

$$\begin{aligned} \hat{y}_k(p) &:= y_k(p) - y_{\text{ref}}(p) \\ \hat{x}_k(p) &:= x_k(p) - x_\infty(p) \\ \hat{u}_k(p) &:= u_k(p) - u_\infty(p) \end{aligned}$$

Then using (7.58) and (7.45) we obtain

$$\begin{aligned} \hat{x}_{k+1}(p+1) &= A\hat{x}_{k+1}(p) + B\hat{u}_{k+1}(p) + B_0\hat{y}_k(p) \\ \hat{y}_{k+1}(p) &= C\hat{x}_{k+1}(p) + D\hat{u}_{k+1}(p) + D_0\hat{y}_k(p) \end{aligned}$$

and in this model the influence of the disturbance has been decoupled.

Suppose now that a stabilizing control law is required. Then this can be achieved by writing

$$\hat{u}_{k+1}(p) = K_1\hat{x}_{k+1}(p) + K_2\hat{y}_k(p) \tag{7.59}$$

or, equivalently,

$$u_{k+1}(p) = u_\infty(p) + K_1 x_{k+1}(p) - K_1 x_\infty(p) + K_2 y_k(p) - K_2 y_{\text{ref}}(p)$$

and the task now is to construct $x_\infty(p)$ and $u_\infty(p)$ when $w(p)$ and $y_{\text{ref}}(p)$ are known.

From the algebraic equation of (7.58) for $y_{ref}(p)$ we have that

$$Du_\infty(p) = (I - D_0)y_{\text{ref}}(p) - Cx_\infty(p) - Fw(p)$$

Here we see again that not all $y_{\text{ref}}(p)$ are available under this design. In particular, only those which satisfy

$$\text{rank}[\begin{array}{cccc} D & (I - D_0)y_{\text{ref}} & -Cx_\infty(p) & -Fw(p) \end{array}] = \text{rank}D$$

are allowed. Suppose now that this condition holds (as is the case when D is full row rank). Then we have that

$$u_\infty(p) = D^\sharp((I - D_0)y_{\text{ref}}(p) - Cx_\infty(p) - Fw(p))$$

Also if D is square and nonsingular, $\sharp$ can be replaced by the matrix inverse and then

$$u_\infty(p) = D^{-1}((I - D_0)y_{\text{ref}}(p) - Cx_\infty(p) - Fw(p))$$

Once $u_\infty(p)$ has been computed over $0 \le p \le \alpha - 1$ then (when D^{-1} exists)

$$\begin{aligned} x_\infty(p+1) &= (A - BD^{-1}C)x_\infty(p) + (B_0 + BD^{-1}(I - D_0))y_{\text{ref}}(p) \\ &+ (E - BD^{-1}F)w(p) \end{aligned}$$

which is simply a 1D discrete linear systems state equation which can be computed given d_∞ – the strong limit of the known pass state initial vector sequence. Also the overall control law is of the form (7.59) where K_1 and K_2 are the control law matrices computed by the LMI design method to give stability along the pass of the controlled process.

7.5 PI Control Laws

7.5.1 Asymptotic Stability Using the 1D Equivalent Model for Discrete Processes

We begin in this section by generalizing proportional plus integral (PI) control to discrete unit memory linear repetitive processes. The objective is to force the controlled process to produce a pre-specified pass profile vector, say Y_{ref} as in the previous section, on a pre-specified pass. Also this pass could be chosen as any finite value of k but initially, since we will only require asymptotic stability of the controlled process, the target is to make Y_{ref} the resulting limit profile. The notation is as in the last section and we follow [169].

If it is assumed that Y_{ref} has only zero entries and there are no disturbances present, then asymptotic stability alone is enough to ensure this property. Here we consider the more general case when disturbances are present and also the target limit profile contains constant but non-zero entries. The method is to introduce a form of proportional plus integral (PI) control for these processes in the pass-to-pass direction.

On any pass, say j, the tracking error in terms of the 1D equivalent model is given by $Y(j) - Y_{\text{ref}}$ and summing this over passes $j = 0$ to $j = l$ defines the so-called total tracking error $\beta(l)$ as

$$\beta(l) := \sum_{j=0}^{l}(Y(j) - Y_{\text{ref}})$$

and it follows immediately that

$$\beta(l+1) = \beta(l) + Y(l+1) - Y_{\text{ref}} \tag{7.60}$$

Now, introduce the so-called extended state vector, which incorporates the total tracking error, as $Z(l) := \left[\begin{array}{cc} Y^T(l) & \beta^T(l) \end{array}\right]^T$. Then, on using the first equation in the 1D equivalent model (7.46) and (7.60), the evolution of $Z(l)$ is governed by

$$\begin{aligned} Z(l+1) &= \begin{bmatrix} \mathfrak{B}_1 & 0 \\ \mathfrak{B}_1 & I \end{bmatrix} Z(l) + \begin{bmatrix} \mathfrak{B}_2 \\ \mathfrak{B}_2 \end{bmatrix} U(l) + \begin{bmatrix} 0 \\ -I \end{bmatrix} Y_{\text{ref}} + \begin{bmatrix} \mathfrak{B}_3 \\ \mathfrak{B}_3 \end{bmatrix} d_l \\ &+ \begin{bmatrix} \Omega_y \\ \Omega_y \end{bmatrix} W_{1D} \end{aligned} \tag{7.61}$$

Suppose now that asymptotic stability holds. Then as $l \to \infty$, $Z(l) \to Z_\infty$ and also from (7.61)

$$Z_\infty = \begin{bmatrix} \mathfrak{B}_1 & 0 \\ \mathfrak{B}_1 & I \end{bmatrix} Z_\infty + \begin{bmatrix} \mathfrak{B}_2 \\ \mathfrak{B}_2 \end{bmatrix} U_\infty + \begin{bmatrix} 0 \\ -I \end{bmatrix} Y_{ref} + \begin{bmatrix} \mathfrak{B}_3 \\ \mathfrak{B}_3 \end{bmatrix} d_\infty + \begin{bmatrix} \Omega_y \\ \Omega_y \end{bmatrix} W_{1D} \tag{7.62}$$

Now introduce $\hat{Z}(l) := Z(l) - Z_\infty$, $\hat{U}(l) := U(l) - U_\infty$, $\hat{Y}(l) := Y(l) - Y_{\text{ref}}$, $\hat{\beta}(l) := \beta(l) - \beta_\infty$, $\hat{d}_l := d_l - d_\infty$, where β_∞ denotes limit as $l \to \infty$ of $\beta(l)$, and subtract (7.62) from (7.61) to obtain the so-called incremental process dynamics as

$$\begin{aligned} \hat{Z}(l+1) &= \begin{bmatrix} \mathfrak{B}_1 & 0 \\ \mathfrak{B}_1 & I \end{bmatrix} \hat{Z}(l) + \begin{bmatrix} \mathfrak{B}_2 \\ \mathfrak{B}_2 \end{bmatrix} \hat{U}(l) + \begin{bmatrix} \mathfrak{B}_3 \\ \mathfrak{B}_3 \end{bmatrix} \hat{d}_l \\ &=: \hat{\mathfrak{B}}_1 \hat{Z}(l) + \hat{\mathfrak{B}}_2 \hat{U}(l) + \hat{\mathfrak{B}}_3 \hat{d}_l \end{aligned} \tag{7.63}$$

This last linear system is unstable as a 1D equivalent model due to the structure of the matrix $\hat{\mathfrak{B}}_1$ but the external disturbance vector W_{1D} has been completely decoupled. Hence if we can make this model asymptotically stable then we will have both disturbance rejection and the required limit profile. Consider, therefore, the following control law for (7.63)

$$\hat{U}(l) := \hat{K}\hat{Z}(l) = \begin{bmatrix} \hat{K}_1 & \hat{K}_2 \end{bmatrix} \begin{bmatrix} \hat{Y}(l) \\ \hat{\beta}(l) \end{bmatrix} \tag{7.64}$$

or, equivalently,

$$\begin{aligned} U(l) &= \hat{K}\big(Z(l) - Z_\infty\big) + U_\infty \\ &= \begin{bmatrix} \hat{K}_1 & \hat{K}_2 \end{bmatrix} \begin{bmatrix} Y(l) - Y_{\text{ref}} \\ \beta(l) - \beta_\infty \end{bmatrix} + U_\infty \end{aligned} \tag{7.65}$$

To implement the control law (7.65), we require β_∞ and U_∞. To provide these, first note that the second block row in (7.62) can be rewritten as

$$(I - \mathfrak{B}_1)Y_{\text{ref}} - \mathfrak{B}_3 d_\infty - \Omega_y W_{1D} = \mathfrak{B}_2 U_\infty \tag{7.66}$$

and also (on letting $l \to \infty$ in (7.64)) we must have that

$$U_\infty - \hat{K}_1 Y_{\text{ref}} - \hat{K}_2 \beta_\infty = 0 \tag{7.67}$$

Now substitute (7.67) into (7.66) to obtain

$$[I - \mathfrak{B}_1 - \mathfrak{B}_2 \hat{K}_1] Y_{\text{ref}} - \mathfrak{B}_3 d_\infty - \Omega_y W_{1D} - \mathfrak{B}_2 \hat{K}_2 \beta_\infty = 0$$

and this last expression gives the condition which must hold to achieve Y_{ref} under the control law (7.65), i.e.

$$\begin{aligned} &\text{rank}\big[I - \mathfrak{B}_1 - \mathfrak{B}_2 \hat{K}_1\big] \\ &= \text{rank}\big[\; I - \mathfrak{B}_1 - \mathfrak{B}_2 \hat{K}_1 \quad \mathfrak{B}_3 d_\infty + \Omega_y W_{1D} + \mathfrak{B}_2 \hat{K}_2 \beta_\infty \;\big] \end{aligned} \tag{7.68}$$

Note also that if $\hat{K}_1$ stabilizes $\mathfrak{B}_1$, i.e. ensures that $r(\mathfrak{B}_1 + \mathfrak{B}_2 \hat{K}_1) < 1$, then the matrix $I - \mathfrak{B}_1 - \mathfrak{B}_2 \hat{K}_1$ is invertible and this last condition always holds.

If (7.68) holds, it can be used to compute β_∞ and hence, using (7.67), U_∞. Moreover, using (7.67) it is possible to rewrite the control law (7.65) as

$$U(l) = \hat{K}_1 Y(l) + \hat{K}_2 \beta(l) = \hat{K} Z(l) \tag{7.69}$$

Also the stabilizing control law matrix $\hat{K}$ can be computed using Theorem 7.4.1 for the state-space model (7.63).

Theorem 7.5.1. *Suppose that the 1D equivalent model is used to design a control law which can be written in the form (7.69) for a discrete unit memory linear repetitive process described by (7.45). Then the resulting process is asymptotically stable with prescribed limit profile Y_{ref} if, and only if, $\exists$ matrices $P > 0$, G and L such that the following LMI holds*

$$\begin{bmatrix} -P & \mathfrak{B}_1 G + \mathfrak{B}_2 L \\ G^T {\mathfrak{B}_1}^T + L^T {\mathfrak{B}_2}^T & P - G - G^T \end{bmatrix} < 0$$

Also if this condition holds, the control law matrix is given by $\hat{K} = LG^{-1}$.

7.5.2 Stability Along the Pass with Performance for Discrete Processes

The control law (7.69) has the extensively used 1D proportional plus integral (PI) (or summation in the discrete case) two term structure. In particular, the term $\hat{K}_1 Y(l)$ provides proportional action (from the previous pass profile) for stability and the term $\hat{K}_2 \beta(l)$ summation action (in the pass-to pass direction) to ensure that the desired limit profile is reached. Here we consider the use of a PI control law in the case when stability along the pass replaces asymptotic stability in the design criterion.

Acceptable, or desired, performance from a given example requires, in general, stability along the pass (this will guarantee the existence of a limit profile which is stable as a 1D discrete linear system). We have already shown in this chapter (Sect. 7.3.1) that a control law of the form

$$u_{k+1}(p) = K_x x_{k+1}(p) + K_{z1} y_k(p)$$

can be used to give this property (for discrete linear repetitive processes of the form defined by Example 1.2.10). Moreover, the design of the control law matrices can be implemented using LMIs. Using such a control law, a stable along the pass process results with a limit profile described by a stable 1D discrete linear systems state-space model. The new question solved below is how to obtain a specified limit profile $y_{\mathrm{ref}}(p)$ even in the presence of disturbance terms. Since the limit profile is described by a 1D discrete linear system state-space model, its transient characteristics are most naturally defined in terms of standard 1D measures.

Consider the state-space model (7.45) at point p on pass k. Then the total tracking error at this point is defined as

$$\chi_k(p) := \sum_{j=0}^{k} \Big(y_j(p) - y_{\mathrm{ref}}(p) \Big)$$

i.e. the error at point p summed from pass 0 to k. Substitution from the process state-space model now gives

$$\begin{aligned} \chi_{k+1}(p) &= \chi_k(p) + y_{k+1}(p) - y_{\mathrm{ref}}(p) \\ &= \chi_k(p) + Cx_{k+1}(p) + Du_{k+1}(p) + D_0 y_k(p) \\ &+ Fw(p) - y_{\mathrm{ref}}(p) \end{aligned} \tag{7.70}$$

Now, introduce the so-called extended output vector

$$\widetilde{z}_{k+1}(p) := \begin{bmatrix} y_{k+1}(p) \\ \chi_{k+1}(p) \end{bmatrix}$$

Then (7.70) yields

$$\begin{aligned} \widetilde{z}_{k+1}(p) &= \begin{bmatrix} C \\ C \end{bmatrix} x_{k+1}(p) + \begin{bmatrix} D_0 & 0 \\ D_0 & I \end{bmatrix} \widetilde{z}_k(p) \\ &+ \begin{bmatrix} D \\ D \end{bmatrix} u_{k+1}(p) + \begin{bmatrix} 0 \\ -I \end{bmatrix} y_{\mathrm{ref}}(p) + \begin{bmatrix} F \\ F \end{bmatrix} w(p) \end{aligned} \tag{7.71}$$

Suppose now that as $k \to \infty$, $x_k(p) \to x_\infty(p)$, $u_k(p) \to u_\infty(p) \to x_\infty(p)$, $y_k(p) \to y_{\mathrm{ref}}(p)$, and let $\chi_\infty(p)$ denote $\lim_{k\to\infty} \chi_k(p)$. Also introduce the following so-called incremental vectors $\hat{z}_k(p) := \widetilde{z}_k(p) - \widetilde{z}_\infty(p)$, $\hat{u}_k(p) := u_k(p) - u_\infty(p)$, and $\hat{x}_k(p) := x_k(p) - x_\infty(p)$. Then using (7.45) and (7.71) we obtain

$$\begin{aligned} \hat{x}_{k+1}(p+1) &= A\hat{x}_{k+1}(p) + \hat{B}_0 \hat{z}_k(p) + B\hat{u}_{k+1}(p) \\ \hat{z}_{k+1}(p) &= \hat{C}\hat{x}_{k+1}(p) + \hat{D}_0 \hat{z}_k(p) + \hat{D}\hat{u}_{k+1}(p) \end{aligned} \tag{7.72}$$

where

$$\hat{B}_0 := \begin{bmatrix} B_0 & 0 \end{bmatrix}, \quad \hat{C} := \begin{bmatrix} C \\ C \end{bmatrix}, \quad \hat{D}_0 := \begin{bmatrix} D_0 & 0 \\ D_0 & I \end{bmatrix}, \quad \hat{D} := \begin{bmatrix} D \\ D \end{bmatrix}$$

and the key point now is that the influence of the disturbance has been completely rejected. The remaining task is to meet the specification that the limit profile (for the original process) be equal to the prescribed vector $y_{\text{ref}}(p)$.

The matrix $\hat{D}_0$ in (7.72) always has eigenvalues with modulus at least equal to unity and hence this discrete unit memory linear repetitive process state-space model is asymptotically unstable and hence unstable along the pass. Consequently to obtain any limit profile from it, control action must be applied. Moreover, in order to make this limit profile equal to $y_{\text{ref}}(p)$ with the specified 1D transient performance specifications we will also require that the controlled process is stable along the pass.

Now consider applying the control law

$$\begin{aligned} \hat{u}_{k+1}(p) &= K_x \hat{x}_{k+1}(p) + K_{z1}\hat{y}_k(p) + K_{z2}\hat{\chi}_k(p) \\ &= \begin{bmatrix} K_x & K_{z1} & K_{z2} \end{bmatrix} \begin{bmatrix} \hat{x}_{k+1}(p) \\ \hat{y}_k(p) \\ \hat{\chi}_k(p) \end{bmatrix} \end{aligned} \tag{7.73}$$

to (7.72). Then the following result gives an LMI based sufficient condition for stability along the pass of the controlled process together with a formula for computing the control law matrices. The proof of this result follows immediately on interpreting the basic LMI design result (Theorem 7.3.2) for the controlled process state-space model and hence the details are omitted here.

Theorem 7.5.2. *Suppose that a control law of the form (7.73) is applied to a discrete unit memory linear repetitive process described by the state-space model (7.72). Then the resulting process is stable along the pass if ∃ matrices $Y > 0$, $Z > 0$, and N such that the following LMI holds*

$$\begin{bmatrix} Z - Y & 0 & Y\widetilde{A}_1^T + N^T\widetilde{B}_1^T \\ 0 & -Z & Y\widetilde{A}_2^T + N^T\widetilde{B}_2^T \\ \widetilde{A}_1 Y + \widetilde{B}_1 N & \widetilde{A}_2 Y + \widetilde{B}_2 N & -Y \end{bmatrix} < 0$$

where

$$\widetilde{A}_1 := \begin{bmatrix} A & \hat{B}_0 \\ 0 & 0 \end{bmatrix}, \quad \widetilde{A}_2 := \begin{bmatrix} 0 & 0 \\ \hat{C} & \hat{D}_0 \end{bmatrix}, \quad \widetilde{B}_1 := \begin{bmatrix} B \\ 0 \end{bmatrix}, \quad \widetilde{B}_2 := \begin{bmatrix} 0 \\ \hat{D} \end{bmatrix}$$

Also if this condition holds the control law matrices are given by

$$[K_x \; K_{z1} \; K_{z2}] = NY^{-1}$$

Suppose now that this last result holds. Then it follows immediately that $y_\infty(p) = y_{\text{ref}}(p)$ as required. Moreover

$$\begin{aligned} u_{k+1}(p) &= K_x\Big(x_{k+1}(p) - x_\infty(p)\Big) + K_{z1}\Big(y_k(p) - y_{\text{ref}}(p)\Big) \\ &+ K_{z2}\Big(\chi_k(p) - \chi_\infty(p)\Big) + u_\infty(p) \end{aligned}$$

and also

$$-K_x x_\infty(p) - K_{z1} y_{\text{ref}}(p) - K_{z2}\chi_\infty(p) + u_\infty(p) = 0$$

Hence the final form of the control law to be applied to the original process is

$$u_{k+1}(p) = K_x x_{k+1}(p) + K_{z1} y_k(p) + K_{z2}\chi_k(p) \tag{7.74}$$

Suppose now that we rewrite the first two terms on the right-hand side of (7.74) in the form

$$\begin{aligned} K_x x_{k+1}(p) + K_{z1} y_k(p) &= \begin{bmatrix} K_x & K_{z1} \end{bmatrix} \begin{bmatrix} x_{k+1}(p) \\ y_k(p) \end{bmatrix} \\ &=: K_{PI} X^a_{k+1}(p) \end{aligned}$$

where $X^a_{k+1}(p)$ is termed the augmented state vector. Then we immediately see that the final control law (7.74) has (in common with its counterpart from the asymptotic stability analysis above) a two term structure, where the first of these, $KX^a_{k+1}(p)$, is static, or proportional, control action for stability and the second $K_{z2}\chi_k(p)$ is integral action to enforce tracking of the desired limit profile $y_{\text{ref}}(p)$.

Example 7.5.1. Consider the following special case of Example 1.2.10

$$A = \begin{bmatrix} 0.92 & 0.14 & -0.98 & 0.41 \\ -0.76 & -0.93 & -0.62 & 0.13 \\ 0.68 & -0.65 & 1.02 & -0.81 \\ 0.94 & 0.04 & 0.83 & 0.2 \end{bmatrix}$$

$$B = \begin{bmatrix} 0.99 & -0.99 & 0.07 \\ 0.07 & -0.94 & -0.63 \\ 0.98 & -0.73 & 0.02 \\ -0.37 & 0.19 & -0.65 \end{bmatrix}$$

$$B_0 = \begin{bmatrix} -0.01 & -0.43 \\ 0.29 & -0.13 \\ 0.98 & 1.09 \\ 1.09 & 0.17 \end{bmatrix}, \quad E = \begin{bmatrix} 0.908 \\ 0.5931 \\ 0.183 \\ 0.3094 \end{bmatrix}$$

$$C = \begin{bmatrix} -0.75 & 0.75 & 0.31 & 0.84 \\ -0.86 & 0.99 & 0.33 & -0.84 \end{bmatrix}$$

$$D = \begin{bmatrix} -0.33 & -0.14 & 0.59 \\ -0.18 & 0.94 & -0.17 \end{bmatrix}$$

$$D_0 = \begin{bmatrix} 1.11 & -0.66 \\ 0.46 & 1.23 \end{bmatrix}, \quad F = \begin{bmatrix} -0.0611 \\ 0.3649 \end{bmatrix}$$

over the pass length $\alpha = 100$. The disturbance $w(p)$ is two full periods of sine wave.

This example is asymptotically unstable, and hence unstable along the pass, since $r(D_0) = 1.2919$. Application of Theorem 7.5.2 now gives the following control law matrices

$$K_x = \begin{bmatrix} -1.7945 & 1.0158 & 0.134 & 1.1219 \\ 0.5075 & -0.5836 & -0.5438 & 1.1854 \\ 0.3834 & -0.7341 & -0.6099 & -0.4623 \end{bmatrix}$$

$$K_{z1} = \begin{bmatrix} 1.8756 & -0.6793 \\ 0.0952 & -1.3957 \\ -0.7203 & 0.3862 \end{bmatrix}, \quad K_{z2} = \begin{bmatrix} 0.0673 & 0.0407 \\ 0.0441 & -0.0559 \\ -0.0606 & 0.0056 \end{bmatrix}$$

In order to assess the quality of the performance available from the controlled process we focus on the fact that the limit profile is a 1D discrete linear system and follow the standard route of using a step signal applied in each of the two channels in turn. Figures 7.3 and 7.4 respectively show the controlled process responses in the case when $y_{\text{ref}}(p) = \begin{bmatrix} -3 & 0 \end{bmatrix}^T$, $0 \leq p \leq 99$. Here the interaction in the second channel is initially quite large but the key feature is that the process converges relatively quickly to the limit profile which has exactly the along the pass dynamics predicted and, in particular, the integral term completely kills off the interaction. Figures 7.5 and 7.6 show the controlled process responses in the case when $y_{\text{ref}}(p) = \begin{bmatrix} 0 & 3 \end{bmatrix}^T$, $0 \leq p \leq 99$, and the same general comments hold. (These simulations are for the boundary conditions $d_{k+1} = \begin{bmatrix} -1.3289 & -0.3243 & 1.1274 & -1.5647 \end{bmatrix}^T$ and zero initial pass profile $y_0(p)$.)

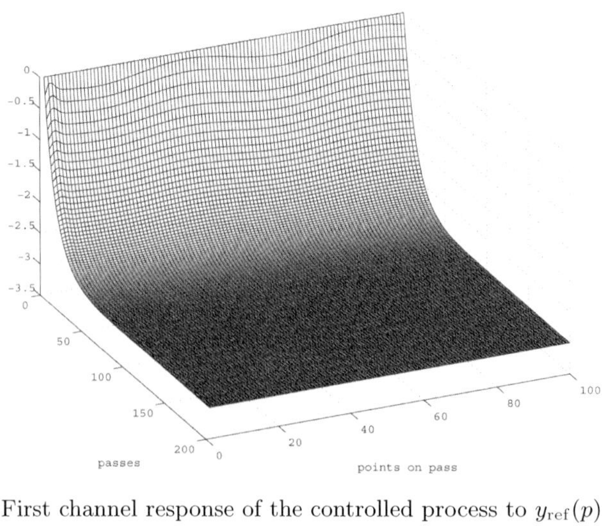

Fig. 7.3. First channel response of the controlled process to $y_{\text{ref}}(p) = [-3\ 0]^T$

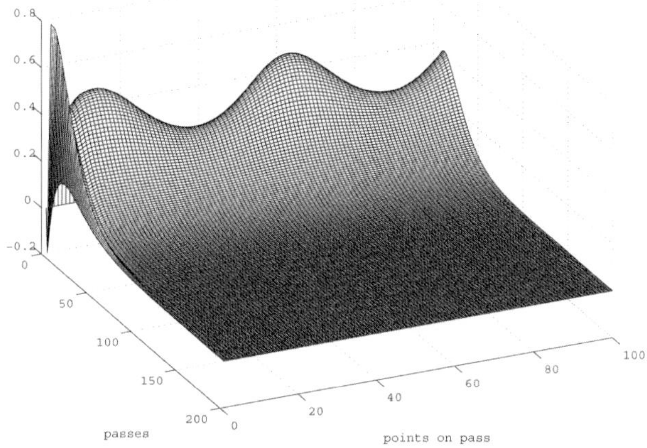

Fig. 7.4. Second channel response of the controlled process to $y_{\text{ref}}(p) = [-3 \ \ 0]^T$

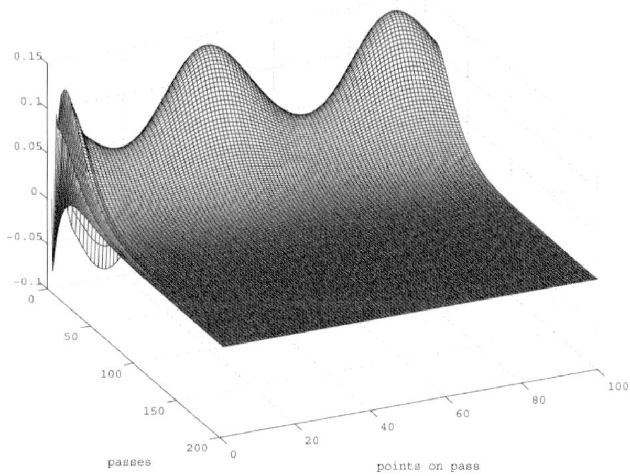

Fig. 7.5. First channel response of the controlled process to $y_{\text{ref}}(p) = [0 \ \ 3]^T$

7.5.3 Differential Processes

We now consider differential linear repetitive processes described by the following state-space model over $\leq t \leq \alpha$, $k \geq 0$

$$\begin{aligned} \dot{x}_{k+1}(t) &= Ax_{k+1}(t) + Bu_{k+1}(t) + B_0 y_k(t) + Ew_{k+1}(t) \\ y_{k+1}(t) &= Cx_{k+1}(t) + Du_{k+1}(t) + D_0 y_k(t) + Fw_{k+1}(t) \end{aligned} \tag{7.75}$$

Here $x_k(t)$, $y_k(t)$, and $u_k(t)$ are as in Example 1.2.3, and $w_{k+1}(t)$ is a $q \times 1$ disturbance vector which enters on the current pass and, as in the discrete

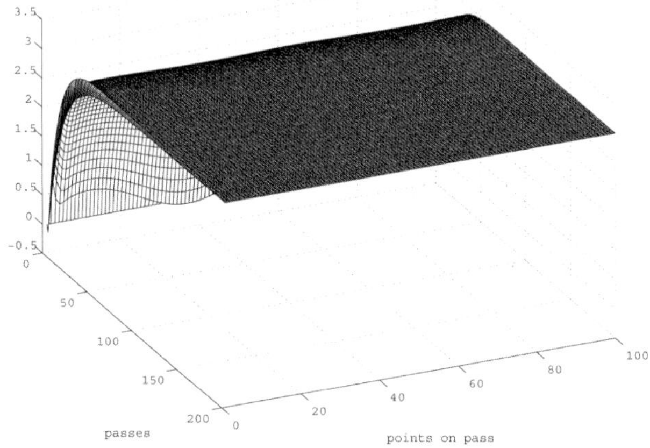

Fig. 7.6. Second channel response of the controlled process to $y_{\text{ref}}(p) = [0 \;\; 3]^T$

process analysis given above, is assumed to be the same on each pass and denoted from this point onwards by $w(t)$. (The case when this assumption does not hold is considered in the next chapter.) No loss of generality results from assuming zero boundary conditions. The design task is again stability along the pass coupled with the resulting limit profile equal to the prescribed vector $y_{\text{ref}}(t)$.

There is no 1D equivalent model for the differential case so we move immediately to stability along the pass as the basic stability condition. In particular, following the discrete analysis of the previous sub-section, define the so-called total tacking error $\chi_k(t)$ at point t on pass k as

$$\chi_k(t) := \sum_{j=0}^{k} (y_j(t) - y_{ref}(t))$$

and hence

$$\chi_{k+1}(t) = \chi_k(t) + y_{k+1}(t) - y_{ref}(t)$$

or, on substituting from (7.75),

$$\begin{aligned} \chi_{k+1}(t) &= \chi_k(t) + Cx_{k+1}(t) + Du_{k+1}(t) \\ &+ D_0 y_k(t) + Fw(t) - y_{ref}(t) \end{aligned} \tag{7.76}$$

Now introduce so-called extended state vector as

$$\tilde{z}_k(t) := \left[\begin{array}{cc} y_k^T(t) & \chi_k^T(t) \end{array} \right]^T$$

Then on using the pass profile updating equation from (7.75) and (7.76) we have that

$$\begin{aligned}
\dot{x}_{k+1}(t) &= Ax_{k+1}(t) + [B_0 \;\; 0\,]\,\widetilde{z}_k(t) \\
&+ Bu_{k+1}(t) + Ew(t) \\
\widetilde{z}_{k+1}(t) &= \begin{bmatrix} C \\ C \end{bmatrix} x_{k+1}(t) + \begin{bmatrix} D_0 & 0 \\ D_0 & I \end{bmatrix} \widetilde{z}_k(t) \\
&+ \begin{bmatrix} D \\ D \end{bmatrix} u_{k+1}(t) + \begin{bmatrix} 0 \\ -I \end{bmatrix} y_{ref}(t) \\
&+ \begin{bmatrix} F \\ F \end{bmatrix} w(t)
\end{aligned}$$

Suppose also that as $k \to \infty$, $x_k(t) \to x_\infty(t)$, $u_k(t) \to u_\infty(t)$, $y_k(t) \to y_{\mathrm{ref}}(t)$, and let $\chi_\infty(t)$ denote $\lim_{k\to\infty} \chi_k(t)$. Also introduce the following incremental vectors $\hat{z}_k(t) := \widetilde{z}_k(t) - \widetilde{z}_\infty(t)$, $\hat{u}_k(t) := u_k(t) - u_\infty(t)$, and $\hat{x}_k(t) := x_k(t) - x_\infty(t)$. Then

$$\begin{aligned}
\dot{\hat{x}}_{k+1}(t) &= A\hat{x}_{k+1}(t) + \hat{B}_0\hat{z}_k(t) + B\hat{u}_{k+1}(t) \\
\hat{z}_{k+1}(t) &= \hat{C}\hat{x}_{k+1}(t) + \hat{D}_0\hat{z}_k(t) + \hat{D}\hat{u}_{k+1}(t)
\end{aligned} \tag{7.77}$$

and

$$\hat{B}_0 := \begin{bmatrix} B_0 & 0 \end{bmatrix}, \quad \hat{C} := \begin{bmatrix} C \\ C \end{bmatrix} \quad \hat{D}_0 := \begin{bmatrix} D_0 & 0 \\ D_0 & I \end{bmatrix}, \quad \hat{D} := \begin{bmatrix} D \\ D \end{bmatrix}$$

where as in the discrete case, the influence of the disturbance has been completely rejected.

This last differential linear repetitive process is unstable along the pass since the matrix $\hat{D}_0$ has eigenvalues of modulus unity. To provide this property, consider the application of a control law of the form

$$\begin{aligned}
\hat{u}_{k+1}(t) &= K_x\hat{x}_{k+1}(t) + K_z\hat{z}_k(t) \\
&= K_x\hat{x}_{k+1}(t) + K_{z1}\hat{y}_k(t) + K_{z2}\hat{\chi}_k(t) \\
&=: \begin{bmatrix} K_x & K_{z1} & K_{z2} \end{bmatrix} \begin{bmatrix} \hat{x}_{k+1}(t) \\ \hat{y}_k(t) \\ \hat{\chi}_k(t) \end{bmatrix}
\end{aligned} \tag{7.78}$$

Then we have the following result whose proof follows identical steps to that for the discrete case of the previous sub-section and is hence omitted here.

Theorem 7.5.3. *Suppose that a control law of the form (7.78) is applied to a differential linear repetitive process described by the state-space model (7.77). Then the resulting process is stable along the pass if* $\exists$ *matrices* $\hat{Y} > 0$, $\hat{Z} > 0$, $\hat{M}$ *and* $\hat{N}$ *such that the following LMI holds*

$$\begin{bmatrix}
\hat{Y}A^T + A\hat{Y} + \hat{N}^TB^T + B\hat{N} & \hat{B}_0\hat{Z} + B\hat{M} & \hat{Y}\hat{C}^T + \hat{N}^T\hat{D}^T \\
\hat{Z}\hat{B}_0^T + \hat{M}^TB^T & -\hat{Z} & \hat{Z}\hat{D}_0^T + \hat{M}^T\hat{D}^T \\
\hat{C}\hat{Y} + \hat{D}\hat{N} & \hat{D}_0\hat{Z} + \hat{D}\hat{M} & -\hat{Z}
\end{bmatrix} < 0$$

If this condition holds then the stabilizing control law matrices are given by

$$K_x = \hat{N}\hat{Y}^{-1}, \quad K_z = \hat{M}\hat{Z}^{-1}$$

Suppose now that this last result holds. Then it follows immediately that $y_\infty(t) = y_{\mathrm{ref}}(t)$ as required. Moreover

$$\begin{aligned} u_{k+1}(t) &= K_x\Big(x_{k+1}(t) - x_\infty(t)\Big) + K_{z1}\Big(y_k(t) - y_{\mathrm{ref}}(t)\Big) \\ &+ K_{z2}\Big(\chi_k(t) - \chi_\infty(t)\Big) + u_\infty(t) \end{aligned}$$

and also

$$-K_x x_\infty(t) - K_{z1} y_{ref}(t) - K_{z2}\chi_\infty(t) + u_\infty(t) = 0$$

Hence the final form of the control law to be applied to the original process is

$$u_{k+1}(t) = K_x x_{k+1}(t) + K_{z1} y_k(t) + K_{z2}\chi_k(t)$$

i.e. again we have the two term structure with an identical interpretation to its discrete counterpart considered previously in this chapter.

7.6 Direct Application of Delay Differential Stability Theory to Differential Processes

Consider again the metal rolling model of Sect. 1.1.2. Then in this section we first apply existing delay differential systems stability theory and follow this by establishing links with that for a differential unit memory linear repetitive process of the form defined by Example 1.2.3. The proofs of the results for delay differential systems used in this section can be found (except where references are given) in standard references in the delay differential systems area, such as [82]. Moreover, we follow the notation in this area and hence the interpretation of the shift operators z_1 and z_2 is different in this section from other places in the monograph.

First note that the repetitive process model of (1.10) through to (1.11) can also be treated as a delay differential system of the neutral type with two noncommensurate delays, h_1 and h_2. (Delays $h_1, \cdots, h_q$ are termed noncommensurate if $\exists$ no integers $l_1, \cdots, l_q$ (not all of them zero) such that $\sum_{i=1}^{q} l_i h_i = 0$. The underlying delay differential system is termed commensurate if $q = 1$.) Also introduce $z_i := e^{-h_i s}$, $i = 1, 2$, i.e. z_i is a left shift operator of duration h_i, and s denotes the Laplace transform variable. Then the characteristic polynomial associated with this model is a two-variable polynomial of the form

$$\rho(s, z_1, z_2) := s^2 + \sum_{i=0}^{2}\sum_{j_1=0}^{1}\sum_{j_2=0}^{1} c_{ij_1j_2} s^i z_1^{j_1} z_2^{j_2}$$

It is also possible to treat (1.10) through to (1.11) as a special case of the generalized linear system

$$G(z_1, z_2)\dot{x}(t) = H_1(z_1, z_2)x(t) + H_2(z_1, z_2)u(t)$$

where $G, H_1 \in \mathbb{R}^{p\times p}[z_1, z_2]$, $H_2 \in \mathbb{R}^{p\times b}[z_1, z_2]$, z_i is (in this representation) a delay operator of duration h_i, $i = 1, 2$, and $\mathbb{R}[z_1, z_2]$ denotes the ring of polynomials in (z_1, z_2) with coefficients in $\mathbb{R}$. Also $\mathbb{R}[z_1, z_2]$ is, in general, a commutative ring and in the commensurate case it is also a principal ideal domain.

To detail this representation for the particular case considered here, take z_1, z_2 to be the shift operators defined by $z_i y_k(t) := y_k(t - h_i)$, $i = 1, 2$, and also introduce $x_1(t) := y_k(t)$, $x_2(t) := \dot{y}_k(t)$. Then, with $x(t) := [x_1(t), x_2(t)]^T$, (1.10) through to (1.11) can be modelled by

$$G(z_1, z_2)\dot{x}(t) = H_1(z_1, z_2)x(t) + H_2(z_1, z_2)y_r(t) \tag{7.79}$$

where G, H_1 and H_2 are over the ring $\mathbb{R}[z_1, z_2]$ and, in detail,

$$\begin{aligned} G(z_1, z_2) &:= \begin{bmatrix} 1 & 0 \\ 0 & (1 - c_3 z_1) \end{bmatrix} \\ H_1(z_1, z_2) &:= \begin{bmatrix} 0 & 1 \\ g_1 & -2\zeta\omega_n(1 - c_3 z_1) \end{bmatrix} \\ H_2(z_1, z_2) &:= \begin{bmatrix} 0 \\ \frac{c_1 k_a k_c}{\widetilde{M}} \end{bmatrix} \end{aligned} \tag{7.80}$$

$$g_1 := \omega_n^2 - \left(\frac{c_2}{\widetilde{M}} + \omega_n^2 c_3\right) z_1 + \frac{c_1 k_a k_c}{\widetilde{M}} z_2$$

Now we require the following definition and result (for the general case)

Definition 7.6.1. *A matrix* $G(z_1, z_2) \in \mathbb{R}^{p\times p}[z_1, z_2]$ *is said to be atomic at zero if* $\det G(0) \neq 0$ *over the field of real numbers.*

Here we will always assume that this property holds. Also introduce the closed bidisc $\mathbb{U}_\delta^2 := \{(z_1, z_2) \in \mathbb{C}^2 : |z_i| \leq 1 + \delta,\ 1 \leq i \leq 2\}$. Then formal stability of $G(z_1, z_2)$ is defined as follows.

Definition 7.6.2. $G(z_1, z_2)$ *is said to be formally stable provided*

$$\det G(z_1, z_2) \neq 0 \text{ in } \mathbb{U}_\delta^2 \text{ forsome } \delta > 0$$

Clearly the particular case of $G(z_1, z_2)$ defined by (7.80) is atomic at zero and is formally stable since $c_3 = \frac{\lambda_2}{\lambda_1+\lambda_2} < 1$. Hence we can invert $G(z_1, z_2)$ over the closed unit bidisc and (7.79) can be rewritten as

$$\dot{x}(t) = H_3(z_1, z_2)x(t) + H_4(z_1, z_2)r_d(t)$$

where

$$H_3(z_1, z_2) := \begin{bmatrix} 0 & 1 \\ \hat{g}_1 & \hat{g}_2 \end{bmatrix}, \quad H_4(z_1, z_2) := \begin{bmatrix} 0 \\ \hat{g}_3 \end{bmatrix}$$

with

$$\begin{aligned} \hat{g}_1 &:= \left[\omega_n^2 - \left(\frac{c_2}{\widetilde{M}} + \omega_n^2 c_3 \right) z_1 + \frac{c_1 k_a k_c}{\widetilde{M}} z_2 \right] / (1 - c_3 z_1) \\ \hat{g}_2 &:= -2\zeta\omega_n, \quad \hat{g}_3 := \frac{c_1 k_a k_c}{\widetilde{M}(1 - c_3 z_1)} \end{aligned}$$

Now set $h = \max h_i$, $i = 1, 2$, and let B denote the Banach space of continuous functions $[-h, 0] \to \mathbb{R}$ with norm defined by $||f|| = \sup_{\sigma \in [-h,0]} |f(\sigma)|$, for any $f \in B$. Suppose also that $y(t)$ denotes the output of an autonomous delay differential system of the form considered here. Also let $y_t \in B$ denote the function segment defined by $y_t(\sigma) = y(t + \sigma)$, $\sigma \in [-h, 0]$, and take the initial condition as $f(t)$, $t \in [-h, 0]$, $f \in B$. Then asymptotic stability of this delay differential system (not to be confused with the repetitive process property) is defined and characterized as follows.

Definition 7.6.3. *A delay differential system of the form considered here is said to asymptotically stable if $\exists$ finite real scalars $M_{as} > 0$, $\gamma > 0$, such that, for each $f \in B$, the solution (or output) $y(t)$ with initial condition f satisfies*

$$||y(t)|| \leq M_{as}|f|e^{-\gamma t}, \; \forall \, t \geq 0$$

Also if this condition holds $\forall \; h_i \geq 0$, $1 \leq i \leq 2$, then the system under consideration is said to be stable independent of delay (denoted i.o.d.).

Theorem 7.6.1. *A delay differential system of the form considered here is asymptotically stable if, and only if, its characteristic polynomial satisfies*

$$\rho(s, e^{-h_1 s}, e^{-h_2 s}) \neq 0, \; \text{Re } s \geq 0$$

and is asymptotically stable independent of delay (i.o.d.) if, and only if, this condition holds $\forall \, h_i \geq 0$, $i = 1, 2$.

Corollary 7.6.1. *A delay differential system of the form considered here is asymptotically stable if, and only if,*

$$\det(sI_2 - H_3(e^{-h_1 s}, e^{-h_2 s})) \neq 0, \; \forall \, s \in \mathbb{I}$$

and asymptotically stable i.o.d. if and only if, this condition holds $\forall \; h_i \geq 0$, $i = 1, 2$, where as before (see the notation of Theorem 2.2.7) $\mathbb{I}$ denotes the closed right-half of the s plane.

The following result [86] gives an analytic test for asymptotic stability i.o.d.

Theorem 7.6.2. *A delay differential system of the form considered here is asymptotically stable i.o.d. if, and only if,*

(a) $\rho(s,1) \neq 0, \ s \in \mathbb{I}$

(b) $\rho(s,-1) \neq 0, s \in \mathbb{I}_{ia} \ (s \neq 0)$; *and*
(c)

$$(1+sT)^2 \rho\left(s, \frac{1-sT}{1+sT}\right) \neq 0, \ s \in \mathbb{I}_{ia}, \ \forall\, T > 0$$

where here $\mathbb{I}_{ia}$ *denotes the imaginary axis of the s-plane.*

Pointwise asymptotic stability is a stronger concept than that above and is defined as follows.

Definition 7.6.4. *A delay differential system of the form considered here with characteristic polynomial* $\rho(s, z_1, z_2)$ *is said to be pointwise asymptotically stable if, and only if,*

$$\rho(s, z_1, z_2) \neq 0, \ \forall\, (s, z_1, z_2) \in \mathbb{I} \times \mathbb{U}_c^2$$

where as before (see the notation of Theorem 2.2.8) $\mathbb{U}_c^2$ *denotes the closed unit biidsc.*

It now follows immediately that a delay differential system of the form considered here is pointwise asymptotically stable if, and only if,

$$\det(sI_2 - H_3(z_1, z_2)) \neq 0, \ \forall\, (s, z_1, z_2) \in \mathbb{I} \times \mathbb{U}_c^2 \tag{7.81}$$

The analysis which follows is due to [60, 61] and the starting point is the calculation of bounds for the PD controller gains k_a, k_b and k_c to guarantee stability of the processes described by (1.10) through to (1.11). Here h_1 and h_2 are, in general, not constant and hence asymptotic stability i.o.d. should be considered in this case. Note also that, in general, it is very difficult to check any of the resulting stability conditions for the noncommensurate case. Consequently we have to consider pointwise asymptotic stability using the results given earlier in this section.

To proceed, note that the condition of (7.81) in this particular case can be further simplified due to the fact that the regularity of $H_3(z_1, z_2)$ over $\mathbb{U}_c^2$ implies that its eigenvalues, denoted by λ_i, are regular functions of (z_1, z_2). Hence if Re $\lambda_i > 0$ for some $(z_1^o, z_2^o) \in \mathbb{U}_c^2$, then Re $\lambda_i > 0, \forall (z_1, z_2)$ in an open neighborhood of (z_1^o, z_2^o). Hence instability in the pointwise asymptotic sense here can be detected by checking the eigenvalues of $H_3(z_1, z_2)$ on the distinguished boundary of the unit bidisc, i.e. $T^2 := \{(z_1, z_2) : |z_i| = 1, \ i = 1, 2\}$. Also this can be implemented for a given example by a suitably fine partition of T^2 which is a finite computation. Alternatively, eigenvalue location tests can be implemented by applying known stability results (see, for example, [17]) for polynomials with complex coefficients to the characteristic polynomial of this matrix.

Suppose now that the pass delay is an integer multiple of the output sensor measurement delay or the sensor delay can be neglected. Then in such a case the delay differential model will be commensurate and here we assume that the second of these holds. Then, with additional assumption of zero reference input (or signal), the delay differential model of (1.10) through to (1.11) becomes

$$\begin{aligned}\ddot{y}_k(t) &+ 2\zeta\omega_n\dot{y}_k(t) + \left(\omega_n^2 + \frac{c_1k_ak_c}{\widetilde{M}}\right)y_k(t) \\ &- c_3\ddot{y}_k(t-h_1) - 2\zeta\omega_nc_3\dot{y}_k(t-h_1) \\ &- \left(\omega_n^2c_3 + \frac{c_2}{\widetilde{M}}\right)y_k(t-h_1) = 0 \qquad (7.82)\end{aligned}$$

with corresponding characteristic polynomial

$$\begin{aligned}\rho(s,z) &= s^2 + 2\zeta\omega_ns + \left(\omega_n^2 + \frac{c_1k_ak_c}{\widetilde{M}}\right) \\ &- c_3s^2z_1 - 2\zeta\omega_nc_3sz_1 \\ &- \left(\omega_n^2c_3 + \frac{c_2}{\widetilde{M}}\right)z_1 = 0 \qquad (7.83)\end{aligned}$$

where we also assume that the gains k_a, k_b and k_c are positive.

The following result is now obtained by application of Theorem 7.6.2.

Theorem 7.6.3. *Consider the repetitive process (7.82) with characteristic polynomial (7.83) where the gains k_a, k_b and k_c are assumed to be positive. Then this commensurate metal rolling process model is asymptotically stable i.o.d. if, and only if,*

(a)

$$k_a > \frac{c_2 - \lambda(1-c_3)}{(\frac{k_c\lambda}{\lambda_2}) + 1 - c_3}$$

and

(b)

$$k_ak_c < \lambda_1 + \frac{\lambda_1(1-c_3^2)}{2c_2\widetilde{M}}k_b^2$$

Proof. To prove (a), the first condition of Theorem 7.6.2 is easily seen to be equivalent to

$$\omega_n^2 + \frac{c_1k_ak_c - c_2}{(1-c_3)\widetilde{M}} > 0$$

and the condition here follows immediately.

To prove (b), the second condition of Theorem 7.6.2 in this case requires that

$$\omega_n^2 + \frac{c_1 k_a k_c + c_2}{(1+c_3)\widetilde{M}} > 0$$

which holds for all $k_a > 0$ and $k_c > 0$. Finally, the third condition of Theorem 7.6.2 requires (after some routine analysis) in this case that

$$s^3 + \hat{a}_1 s^2 + \hat{a}_2 s + \hat{a}_3 \neq 0, \ \forall\, s \in \mathbb{I}$$

where

$$\begin{aligned}
\hat{a}_1 &:= 2\zeta\omega_n + \frac{(1-c_3)}{(1+c_3)T} \\
\hat{a}_2 &:= \omega_n^2 + 2\zeta\omega_n \frac{(1-c_3)}{(1+c_3)T} + \frac{c_1 k_a k_c + c_2}{(1+c_3)\widetilde{M}} \\
\hat{a}_3 &:= \omega_n^2 \frac{(1-c_3)}{(1+c_3)T} + \frac{c_1 k_a k_c - c_2}{(1+c_3)\widetilde{M}T}
\end{aligned}$$

Some routine analysis now yields that this last requirement is equivalent to (b) here and the proof is complete. □

An autonomous delay differential system of the commensurate type can also be modelled by a 2D state-space model of the form

$$\begin{bmatrix} \dot{x}_1(t) \\ x_2(t+\gamma) \end{bmatrix} = \begin{bmatrix} A_1 & A_2 \\ A_3 & A_4 \end{bmatrix} \begin{bmatrix} x_1(t) \\ x_2(t) \end{bmatrix} \tag{7.84}$$

where the $n_1 \times 1$ vector $x_1(t)$ denotes the differential state vector, the $n_2 \times 1$ vector $x_2(t)$ the delay state vector, and for differential unit memory linear repetitive processes of the form considered here γ is equal to the pass length α. In particular, the processes under consideration here is a special case of (7.84) one above with

$$\begin{aligned}
A_1 &:= \begin{bmatrix} -\zeta\omega_n & (\zeta^2\omega_n^2 - \omega_n^2 - \frac{c_1 k_a k_c}{\widetilde{M}}) \\ 1 & -\zeta\omega_n \end{bmatrix} \\
A_2 &:= \begin{bmatrix} 0 \\ \hat{a} \end{bmatrix}, \ A_3 = \begin{bmatrix} 1 & 0 \end{bmatrix}, \ A_4 = c_3 \\
\hat{a} &:= c_3(c_2 - c_1 k_a k_c)/(\omega_n^2 \widetilde{M} + c_1 k_a k_c - \zeta^2 \omega_n^2 \widetilde{M})
\end{aligned}$$

Suppose now that all eigenvalues of the matrix A_1 have strictly negative real parts and note that $A_4 < 1$. Then, on assuming zero initial conditions, the required equations (with $\gamma = \alpha$) can be rewritten in the form

$$\begin{aligned}
x_2(t+\alpha) &= (A_3(sI - A_1)^{-1}A_2 + A_4)x_2(t) \\
&= G_1(s)x_1(t)
\end{aligned} \tag{7.85}$$

In this case, it is easily shown (by inspection of the roots of the characteristic polynomial (i.e. $\det(sI_2 - A_1)$) that $G_1(s)$ is analytic in $\mathbb{I}$. Moreover, the characteristic polynomial $\rho(s, z)$ is first order in the delay operator. Recall also the definition of the SBR property of Definition 4.2.1 and the (SISO processes only) necessary and sufficient condition for stability along the pass of (4.19), i.e.

$$||G_1^k(\imath\omega)|| \to 0, \text{ as } k \to \infty, \; 0 \leq \omega < \infty$$

where $||G_1(\imath\omega)|| := \sup_{0\leq\omega<\infty} |G_1(\imath\omega)|$, i.e. each frequency component is attenuated from pass-to-pass. Hence it follows that pointwise asymptotic stability is equivalent to $G_1(s)$ being SBR. Hence stability along the pass for SISO differential linear repetitive processes is equivalent to pointwise asymptotic stability for delay differential systems – which is a strong concept of stability i.o.d.

7.7 Linear Quadratic Control of Differential Processes

The state-space model considered in this section is a special case of that in Example 1.2.3 where the pass profile vector is equal the state vector, i.e. the dynamics are described by

$$\dot{x}_k(t) = Ax_k(t) + Bu_k(t) + B_0 x_{k-1}(t) \tag{7.86}$$

over $0 \leq t \leq \alpha$, $k \geq 1$. The boundary conditions are those of Example 1.2.3 specialized to this case, i.e.

$$\begin{aligned} x_k(0) &= d_k, \; k \geq 1 \\ x_0(0) &= f(t), \; 0 \leq t \leq \alpha \end{aligned}$$

In the 1D linear systems case, there are two main types of controller required in quadratic optimization problems, namely terminal controllers and regulators. A terminal controller is designed to bring a system close to a desired terminal condition (usually taken as zero) whilst exhibiting 'acceptable' behavior on the way. A regulator is designed to keep a stationary system within 'acceptable' deviation from a reference vector using 'acceptable' amounts of control action. Hence given a process of the form (7.86), we aim to steer it from an initial state $x_k(0) = d_k$ to a terminal state $x_k(\alpha) = 0$ on each pass k, $1 \leq k \leq N$, using 'acceptable' levels of control action given also the initial pass profile $x_0(t)$. The analysis and notation in this section is from [175].

The obvious first attempt at this problem is (by analogy with the 1D linear systems case) to minimize a quadratic cost function constructed from terms involving a quadratic form in the terminal state added to an integral of quadratic forms in the state and control vectors over $0 \leq t \leq \alpha$ for each pass k and then summed over all passes $k = 1, 2, \cdots, N$. In particular, we consider first the problem of minimizing the cost function

$$J := \frac{1}{2}\sum_{k=1}^{N}\left\{x_k^T(\alpha)Gx_k(\alpha) + \int_0^{\alpha}(x_k^T(t)Qx_k(t) + u_k^T(t)Ru_k(t))\,dt\right\} \quad (7.87)$$

Here $G \geq 0$ and $Q \geq 0$ are $n \times n$ matrices and $R > 0$ is an $r \times r$ matrix.

At this stage, we again note links to classes of delay differential systems. In particular, consider the problem of minimizing the cost function

$$J_d := \frac{1}{2}x^T(T_f)Gx(T_f) + \frac{1}{2}\int_0^{T_f}(x^T(t)Qx(t) + u^T(t)Ru(t))\,dt$$

subject to

$$\begin{aligned} \dot{x}(t) &= A_d x(t) + A_{d1}x(t-\tau) + B_d u(t),\ 0 \leq t \leq T_f \\ x(\theta) &= \psi(\theta),\ -\tau \leq \theta \leq 0 \end{aligned}$$

where G, Q, and R are constant matrices with the same dimensions and properties as in (7.87), T_f is the fixed terminal time of the process and τ the fixed length of the delay. The repetitive process problem we are considering here reduces to the above form if we set $d_k = x_{k-1}(\alpha),\ 1 \leq k \leq N$ (see also Example 1.2.6). In this case, the delay length τ becomes α and the final time becomes $T_f = N \times \alpha$. A solution to this delay differential problem is known in the literature.

Next we need to introduce some notation starting with $C_n := C([0,\alpha];\mathbb{R}^n)$; which is used here to denote the Banach space of continuous functions with domain $[0,\alpha]$ and range $\mathbb{R}^n$. The norm on this space is defined as $||f|| := \sup_{0\leq t\leq\alpha}|f|$, where here $|\cdot|$ denotes the usual Euclidean norm on $\mathbb{R}^n$. This norm on $\mathbb{R}^n$ gives arise to the spectral matrix norm $||\mathcal{M}||_n := (r(\mathcal{M}^T\mathcal{M}))^{\frac{1}{2}}$, where $\mathcal{M}$ is an $n \times n$ matrix. The notation $\hat{C}_F := \hat{C}_F([0,\alpha] \times [0,\alpha];\mathbb{R}^n)$ is used here to denote the Banach space of real valued $n \times n$ matrix functions of two variables with domain $[0,\alpha] \times [0,\alpha]$. These functions are piecewise continuous with finite discontinuity along the diagonal of the domain such that

$$H(t,t+) - H(t,t-) = F(t),\ 0 \leq t \leq \alpha,\ \forall\, H \in \hat{C}_F$$

where $F(t)$ is a real valued continuous $n \times n$ matrix function defined on $[0,\alpha]$. The norm on this space is defined as

$$||H||_F := \sup_{t,\theta\in[0,\alpha],\ t\neq\theta}(||H(t,\theta)||_n),\ H \in \hat{C}_F$$

Finally, we need the Hilbert spaces $H_1 := H([0,\alpha];\mathbb{R}^n)$ and $H_2 := H([0,\alpha];\mathbb{R}^r)$ of Lebesgue integratable vector functions defined on $[0,\alpha]$ with ranges in $\mathbb{R}^n$ and $\mathbb{R}^r$ respectively. The inner products for these spaces are

$$\langle v, w\rangle_1 := \int_0^{\alpha} v^T(t)\tilde{Q}w(t)\,dt,\ v, w \in H_1$$

and

$$\langle v, w \rangle_2 := \int_0^\alpha v^T(t)\widetilde{R}w(t)\,dt,\ v, w \in H_2$$

respectively, where $\widetilde{Q} > 0$ and $\widetilde{R} > 0$ are of dimensions $n \times n$ and $r \times r$ respectively. These inner products induce the following norms

$$||v||_1 = \left(\int_0^\alpha v^T(t)Qv(t)\,dt\right)^{\frac{1}{2}}$$

and

$$||w||_2 = \left(\int_0^\alpha w^T(t)Rw(t)\,dt\right)^{\frac{1}{2}}$$

respectively.

Now we define (one type of) an admissible control vector for (7.86) (in the context of the current analysis) as follows.

Definition 7.7.1. *An admissible control vector for a differential unit memory linear repetitive process of the form described by (7.86) must only satisfy the requirement that it is a continuous time-dependent functional of the process states, i.e.* $u(t) = u(t, x)$ *where*

$$x := \begin{bmatrix} x_0^T(t) & x_1^T(t) & \cdots & x_n^T(t) \end{bmatrix}^T,\ 0 \le t \le \alpha$$

The control vector is otherwise unconstrained.

Also we will clearly require a solution to be causal, i.e. it must not require access to information which has not yet been generated by the process. The following analysis shows that the existence, and hence computation, of such a solution cannot be found by directly mimicking a 1D linear quadratic optimal control problem solution (as may at first be expected). In particular, we next show how a non-causal solution to this repetitive process problem can arise.

Introduce the following super-vectors and matrices

$$x(t) := \begin{bmatrix} x_1(t) \\ x_2(t) \\ \vdots \\ x_N(t) \end{bmatrix},\quad u(t) := \begin{bmatrix} u_1(t) \\ u_2(t) \\ \vdots \\ u_N(t) \end{bmatrix},\quad d := \begin{bmatrix} d_1 \\ d_2 \\ \vdots \\ d_N \end{bmatrix}$$

$$\tilde{A} := \begin{bmatrix} A & 0 & 0 & \cdots & 0 \\ B_0 & A & 0 & \cdots & 0 \\ 0 & B_0 & A & \cdots & 0 \\ \vdots & \ddots & \ddots & \cdots & 0 \\ 0 & \cdots & 0 & B_0 & A \end{bmatrix},\quad \tilde{B} := \begin{bmatrix} B & 0 & 0 & \cdots & 0 \\ 0 & B & 0 & \cdots & 0 \\ 0 & 0 & B & \cdots & 0 \\ \vdots & \vdots & \vdots & \cdots & \vdots \\ 0 & \cdots & 0 & 0 & B \end{bmatrix}$$

$$\tilde{Q} := \begin{bmatrix} Q & 0 & 0 & \cdots & 0 \\ 0 & Q & 0 & \cdots & 0 \\ 0 & 0 & Q & \cdots & 0 \\ \vdots & \vdots & \vdots & \ddots & \vdots \\ 0 & \cdots & 0 & 0 & Q \end{bmatrix}, \quad \tilde{R} := \begin{bmatrix} R & 0 & 0 & \cdots & 0 \\ 0 & R & 0 & \cdots & 0 \\ 0 & 0 & R & \cdots & 0 \\ \vdots & \vdots & \vdots & \ddots & \vdots \\ 0 & \cdots & 0 & 0 & R \end{bmatrix}$$

$$\tilde{D} := \begin{bmatrix} B_0 \\ 0 \\ 0 \\ \vdots \\ 0 \end{bmatrix}, \quad \tilde{G} := \begin{bmatrix} G & 0 & 0 & \cdots & 0 \\ 0 & G & 0 & \cdots & 0 \\ 0 & 0 & G & \cdots & 0 \\ \vdots & \vdots & \vdots & \ddots & \vdots \\ 0 & \cdots & 0 & 0 & G \end{bmatrix}, \quad d := \begin{bmatrix} d_1 \\ d_2 \\ d_3 \\ \vdots \\ d_N \end{bmatrix}$$

Then the optimal control problem considered here can be stated as minimizing the cost function

$$\min\ J = \frac{1}{2}\left(x^T(\alpha)\tilde{G}x(\alpha) + \int_0^\alpha x^T(t)\tilde{Q}x(t) + u^T(t)\tilde{R}u(t)\,dt\right)$$

subject to

$$\dot{x}(t) = \tilde{A}x(t) + \tilde{B}u(t) + \tilde{D}f(t),\ x(0) = d$$

Use of well known 1D linear systems optimal control theory (see, for example, [9]) now shows that this problem has solution

$$u(t) = -\tilde{R}^{-1}\tilde{B}^T p(t) \tag{7.88}$$

where $p(t)$ is a differentiable $nN \times 1$ vector which is a solution of the two point boundary value problem

$$\begin{aligned} \dot{p}(t) &= -\tilde{A}^T p(t) - \tilde{Q}x(t),\ 0 \le t \le \alpha,\ p(\alpha) = \tilde{G}x(\alpha) \\ \dot{x}(t) &= \tilde{A}x(t) + \tilde{B}u(t) + \tilde{D}f(t),\ 0 \le t \le \alpha,\ x(0) = d \end{aligned} \tag{7.89}$$

In the absence of the term $\tilde{D}y(t)$ in (7.89), the two point boundary value problem considered here has the well known solution

$$p(t) = K(t)x(t) \tag{7.90}$$

where the $nN \times nN$ differentiable time-dependent matrix $K(t)$ satisfies the Riccati differential equation

$$\dot{K}(t) = -K(t)\tilde{A} - \tilde{A}^T K(t) - \tilde{Q} + K(t)\tilde{B}\tilde{R}^{-1}\tilde{B}^T K(t),\ K(\alpha) = \tilde{G}$$

This yields a linear feedback control law given by

$$u(t) = -\tilde{R}^{-1}\tilde{B}^T K(t)x(t)$$

Also, since we seek a linear feedback control law for the actual case under consideration here, we first try a solution for (7.89) of the form (7.90) but with an added 'tracking' vector to take account of the $\tilde{D}f(t)$ term. Hence we assume a trial solution of (7.89) of the form

$$p(t) = K(t)x(t) + g(t) \tag{7.91}$$

where $K(t)$ is an $nN \times nN$ time-dependent matrix and $g(t)$ a differentiable $nN \times 1$ time-dependent vector.

We now have (on differentiating (7.91) with respect to t and substituting from the differential equation above defining $x(t)$)

$$\dot{p}(t) = \dot{K}(t)x(t) + K(t)(\tilde{A}x(t) + \tilde{B}u(t) + \tilde{D}f(t)) + \dot{g}(t)$$

or, on using (7.88) and (7.91),

$$\begin{aligned} \dot{p}(t) &= (\dot{K}(t) + K(t)\tilde{A} - K(t)\tilde{B}\tilde{R}^{-1}\tilde{B}^T K(t))x(t) + \dot{g}(t) \\ &- K(t)\tilde{B}\tilde{R}^{-1}\tilde{B}^T g(t) + K(t)\tilde{D}f(t) \end{aligned}$$

Also from (7.89)

$$\dot{p}(t) = -\tilde{A}^T p(t) - \tilde{Q}x(t) = -\tilde{A}^T K(t)x(t) - \tilde{A}^T g(t) - \tilde{Q}x(t)$$

Equating the right-hand sides of these last two equations and rearranging gives for $0 \le t \le \alpha$

$$\begin{aligned} (\dot{K}(t) &+ K(t)\tilde{A} + \tilde{A}^T K(t) - K(t)\tilde{B}\tilde{R}^{-1}\tilde{B}^T K(t) + \tilde{Q})x(t) \\ &+ (\dot{g}(t) + (\tilde{A}^T - K(t)\tilde{B}\tilde{R}^{-1}\tilde{B}^T)g(t) + K(t)\tilde{D}f(t)) = 0 \end{aligned}$$

and we can satisfy this last equation identically if we set both brackets to zero and also require that $p(\alpha) = \tilde{G}x(\alpha)$, i.e.

$$\tilde{G}x(\alpha) = K(\alpha)x(\alpha) + g(\alpha)$$

Hence we choose $K(\alpha) = \tilde{G}$, $g(\alpha) = 0$.

In summary, therefore, for a solution of the form (7.91) to the optimal control problem considered, the matrix K and vector g must satisfy

$$\begin{aligned} \dot{K}(t) &+ K(t)\tilde{A} + \tilde{A}^T K(t) - K(t)\tilde{B}\tilde{R}^{-1}\tilde{B}^T K(t) \\ &+ \tilde{Q} = 0,\ 0 \le t \le \alpha,\ K(\alpha) = \tilde{G} \\ \dot{g}(t) &+ (\tilde{A}^T - K(t)\tilde{B}\tilde{R}^{-1}\tilde{B}^T)g(t) \\ &+ K(t)\tilde{D}f(t) = 0,\ 0 \le t \le \alpha,\ g(\alpha) = 0 \end{aligned}$$

Now partition $K(t)$, $g(t)$, and $p(t)$ respectively as

$$K(t) = \begin{bmatrix} K_{11}(t) & \cdots & \cdots & K_{1N}(t) \\ K_{21}(t) & \ddots & \ddots & K_{2N}(t) \\ \vdots & & \ddots & \vdots \\ K_{N1}(t) & \cdots & \cdots & K_{NN}(t) \end{bmatrix}, \quad g(t) = \begin{bmatrix} g_1(t) \\ g_2(t) \\ \vdots \\ g_N(t) \end{bmatrix}$$

$$p(t) = \begin{bmatrix} p_1(t) \\ p_2(t) \\ \vdots \\ p_N(t) \end{bmatrix}$$

Then

$$p_i(t) = \sum_{j=1}^{N} K_{ij}(t)x_j(t) + g_i(t), \; 1 \leq i \leq N, \; 0 \leq t \leq \alpha$$

and hence from (7.88)

$$u_i(t) = -R^{-1}B^T \left[\sum_{j=1}^{N} K_{ij}(t)x_j(t) + g_i(t) \right], \; 1 \leq i \leq N, \; 0 \leq t \leq \alpha \tag{7.92}$$

Since, in general, $K_{ij}(t) \neq 0$, $1 \leq i, j \leq N$, $0 \leq t \leq \alpha$, it follows that the control law at position t on pass i feeds back the state vector at the corresponding position t from all passes $1, 2, \cdots, N$. Hence it is non-causal. It is, however, linear and this immediately poses the following question:

Is there a causal linear feedback solution of the optimal control problem posed here?

To show that the answer to this question is yes, consider again the Hilbert spaces H_1 and H_2 introduced earlier in this section. Then in this setting, the problem considered becomes that of minimizing the cost function

$$J := \frac{1}{2}\sum_{k=1}^{N} \{||x_k||_1^2 + ||u_k||_2^2\} \tag{7.93}$$

subject to

$$x_{k+1} = \mathcal{A}x_k + \mathcal{B}u_{k+1}, \; x_k \in H_1, \; u_{k+1} \in H_2, \; 1 \leq k \leq N \tag{7.94}$$

where here $\mathcal{A}$ and $\mathcal{B}$ are specified bounded linear operators. Now we can establish the following result.

Theorem 7.7.1. *There exist bounded operators $\mathcal{L}_i$ and $\tilde{\mathcal{L}}_i$ for which the choices*

$$u_k = \mathcal{L}_k x_{k-1} \text{ or } u_k = \tilde{\mathcal{L}}_k x_k, \; 1 \leq k \leq N$$

both minimize J of (7.93).

Proof. This is based on dynamic programming principles and proceeds as follows.

First define J_k as

$$J_k = \frac{1}{2} \sum_{j=N-k+1}^{N} \{||x_j||_1^2 + ||u_j||_2^2\}$$

i.e. J_k is the value of the performance index for the process restricted to the last $k-1$ passes. We now make the following induction hypothesis

$$\min J_k = \langle x_{N-k}, K_k x_{N-k} \rangle_1 \tag{7.95}$$

where here K_k denotes a self-adjoint operator for $k = 1, 2, \cdots, N$. Note also that since the process only completes N passes then $K_0 = 0$.

Suppose now that (7.95) holds for some value of $k > 0$. Then, using the principle of optimality,

$$\min J_{k+1} = \min \left\{ \frac{1}{2}(||x_{N-k}||_1^2 + ||u_{N-k}||_2^2 + \langle x_{N-k}, K_k x_{N-k} \rangle_1) \right\}$$

Substitution from (7.94) and some rearrangement now gives

$$\begin{aligned} \min J_{k+1} = \Big\{ (\frac{1}{2}((\langle \mathcal{A}x_{N-k-1} + \mathcal{B}u_{N-k}, \mathcal{A}x_{N-k-1} + \mathcal{B}u_{N-k} \rangle_1 \\ + \langle u_{N-k}, u_{N-k} \rangle_2 \\ + \langle \mathcal{A}x_{N-k-1} + \mathcal{B}u_{N-k}, K_k(\mathcal{A}x_{N-k-1} + \mathcal{B}u_{N-k}) \rangle_1) \Big\} \end{aligned}$$

For ease notation, now write $x_{N-k-1} = x_o$, $u_{N-k} = u$, and $K_k = K$. Then

$$\min J_{k+1} = \min \left\{ \frac{1}{2}((\langle \mathcal{A}x_o + \mathcal{B}u, (I+K)(\mathcal{A}x_o + \mathcal{B}u) \rangle_1 + \langle u, u \rangle_2) \right\}$$

and denote this minimum value by $\tilde{J}$.

Now we have that

$$\begin{aligned} \tilde{J} &= \frac{1}{2}\langle x_o, \mathcal{A}^*(I+K)\mathcal{A}x_o \rangle_1 + \frac{1}{2}\langle \mathcal{A}x_o, (I+K)\mathcal{B}u \rangle_1 \\ &+ \frac{1}{2}\langle \mathcal{B}u, (I+K)\mathcal{A}x_o \rangle_1 + \frac{1}{2}\langle u, (I + \mathcal{B}^*(I+K)\mathcal{B})u \rangle_2 \\ &= \frac{1}{2}\langle x_o, \mathcal{A}^*(I+K)\mathcal{A}x_o \rangle_1 + \langle \mathcal{B}^*(I+K)\mathcal{A}x_o, u \rangle_1 \\ &+ \frac{1}{2}\langle u, (I + \mathcal{B}^*(I+K)\mathcal{B})u \rangle_2 \end{aligned} \tag{7.96}$$

Note also that since x_o is not determined by u, $\frac{1}{2}\langle x_o, \mathcal{A}^*(I+K)\mathcal{A}x_o \rangle_1$ is fixed. Also by writing $u = u_o + \delta u$, it follows immediately that a necessary condition for a minimum of

$$\left\{\frac{1}{2}(\langle u, Lu\rangle_2 + \langle b, u\rangle_2\right\}$$

is that $Lu + b = 0$.

Using this last fact, a minimum for (7.96) requires that

$$(I + \mathcal{B}^*(I + K)\mathcal{B})u + \mathcal{B}^*(I + K)\mathcal{A}x_o = 0$$

or, after some routine manipulations,

$$u_{N-k} = L_{N-k}x_{N-k-1}$$

where

$$L_{N-k} := -(I + \mathcal{B}^*(I + K)\mathcal{B})^{-1}\mathcal{B}^*(I + K)\mathcal{A}$$

At this stage, note that

$$\mathcal{B}^*(I + K)(I + \mathcal{B}\mathcal{B}^*(I + K)) = (I + \mathcal{B}^*(I + K)\mathcal{B})\mathcal{B}^*(I + K)$$

and therefore

$$\mathcal{B}^*(I + K)(I + \mathcal{B}\mathcal{B}^*(I + K))^{-1} = (I + \mathcal{B}^*(I + K)\mathcal{B})^{-1}\mathcal{B}^*(I + K)$$

Hence

$$u = -\mathcal{B}^*(I + K)(I + \mathcal{B}\mathcal{B}^*(I + K))^{-1}\mathcal{A}x_o$$

and on substitution into $x = \mathcal{A}x_o + \mathcal{B}u$

$$\begin{aligned} x &= \mathcal{A}x_o + \mathcal{B}u \\ &= \left[I - \mathcal{B}\mathcal{B}^*(I + K)(I + \mathcal{B}\mathcal{B}^*(I + K))^{-1}\right]\mathcal{A}x_o \\ &= (I + \mathcal{B}\mathcal{B}^*(I + K) - \mathcal{B}\mathcal{B}^*(I + K))(I + \mathcal{B}\mathcal{B}^*(I + K))^{-1}\mathcal{A}x_o \\ &= (I + \mathcal{B}\mathcal{B}^*(I + K))^{-1}\mathcal{A}x_o \end{aligned} \tag{7.97}$$

and we now have

$$u = -\mathcal{B}^*(I + K)x \tag{7.98}$$

Now

$$\min J_{k+1} = \min \left\{\frac{1}{2}(\langle x, x\rangle_1 + \langle u, u\rangle_2 + \langle x, Kx\rangle_1)\right\}$$

and substituting from (7.98) for u gives

$$\begin{aligned} \min J_{k+1} &= \frac{1}{2}\langle x, (I + K)x\rangle_1 + \frac{1}{2}\langle \mathcal{B}^*(I + K)x, \mathcal{B}^*(I + K)x\rangle_2 \\ &= \frac{1}{2}\langle x, (I + K)(I + \mathcal{B}\mathcal{B}^*(I + K))(I + K)x\rangle_1 \end{aligned}$$

Substituting (7.97) into this last equation now yields

$$\min J_{k+1} = \frac{1}{2}\langle x_o, \mathcal{A}^*(I + \mathcal{B}\mathcal{B}^*(I + K))^{-1}(I + K)\mathcal{A}x_o\rangle_1$$

Hence

$$K_{k+1} = \mathcal{A}^*(I + \mathcal{B}\mathcal{B}^*(I + K_k))^{-1}(I + K_k)\mathcal{A}$$

and the proof is complete by induction. □

Note now that

$$L_k = -(I + \mathcal{B}^*(I + K_{N-k})\mathcal{B})^{-1}\mathcal{B}^*(I + K_{N-k})\mathcal{A}$$

and

$$\tilde{L}_k = -\mathcal{B}^*(I + K_{N-k})$$

in the statement of Theorem 7.7.1. The sequence formed by K_j satisfies the recurrence relation

$$K_{k+1} = \mathcal{A}^*(I + \mathcal{B}\mathcal{B}^*(I + K_k)^{-1}(I + K_k)\mathcal{A},\ 0 \le k \le N$$

with $K_0 = 0$.

This last result answers the question of the existence of a linear causal feedback control law for the differential unit memory linear repetitive processes under consideration here but does not give the conditions to be satisfied by such a solution. Further development of this approach is possible, see [175].

The clearly remains much to be done in this general area beyond the quadratic cost function solution developed here. In which context, a model following approach for 2D discrete linear systems described by the Fornasini-Marchesini model has been developed in [178] and it has been shown that this approach can also be applied to a discrete non-unit memory linear repetitive process of the form defined by Example 1.2.9.

8 Control Law Design for Robustness and Performance

Summary. In this chapter the theme is again control law (or controller) design but here we allow uncertainty in the model structure. The uncertainty allowed is described by particular model structures, some of which have already been introduced in Chap. 5. The resulting control law design algorithms can be computed using LMIs and are based on H_∞, H_2, and mixed H_∞/H_2 techniques. Illustrative examples are given as appropriate. The results in this chapter represent some of the latest available in the subject area but by no means exhaust the possibilities. Rather they provide an essential basis for the eventual development of a comprehensive control systems theory, coupled with numerically reliable control law design algorithms, for these processes.

8.1 LMI Based Design

8.1.1 Discrete Processes

In this section, we develop an LMI approach to control law design in the presence of uncertainty in the process definition. As a starting point, we consider the case as in Sect. 5.4.1 when the augmented plant matrix Φ is subject to additive perturbations as per (5.63), i.e.

$$\Phi_d = \Phi + \Delta\Phi$$

where

$$\Phi = \begin{bmatrix} A & B_0 \\ C & D_0 \end{bmatrix}, \quad \Delta\Phi = \begin{bmatrix} \Delta A & \Delta B_0 \\ \Delta C & \Delta D_0 \end{bmatrix}$$

and for convenience we now repeat the uncertainty structure assumed. In particular, we assume that the uncertainties considered here have the following typical norm bounded structure

$$\Delta\Phi = \begin{bmatrix} H_1 \\ H_2 \end{bmatrix} \mathcal{F} \begin{bmatrix} E_1 & E_2 \end{bmatrix} =: \begin{bmatrix} H_1 \\ H_2 \end{bmatrix} \mathcal{F} E$$

where

$$\mathcal{F}^T \mathcal{F} \leq I$$

Also using (as defined in Sect. 5.4.1)

$$\widehat{\Delta A}_1 = \begin{bmatrix} \Delta A & \Delta B_0 \\ 0 & 0 \end{bmatrix}, \quad \widehat{\Delta A}_2 = \begin{bmatrix} 0 & 0 \\ \Delta C & \Delta D_0 \end{bmatrix} \tag{8.1}$$

we can write $\Delta\Phi$ in the form

$$\Delta\Phi = \widehat{\Delta A}_1 + \widehat{\Delta A}_2 = \widehat{H}_1 \mathcal{F} E + \widehat{H}_2 \mathcal{F} E$$

where

$$\widehat{H}_1 = \begin{bmatrix} H_1 \\ 0 \end{bmatrix}, \quad \widehat{H}_2 = \begin{bmatrix} 0 \\ H_2 \end{bmatrix}$$

The sufficient condition for stability along the pass of Theorem 5.4.1 in the absence of uncertainty in the process state-space model requires the existence of matrices $P > 0$ and $Q > 0$ such that

$$\widehat{A}^T P \widehat{A} + \widehat{Q} < 0$$

where

$$\widehat{A} := \begin{bmatrix} \widehat{A}_1 & \widehat{A}_2 \end{bmatrix}, \quad \widehat{Q} := \begin{bmatrix} Q - P & 0 \\ 0 & -Q \end{bmatrix}$$

(and $\widehat{A}_1$ and $\widehat{A}_2$ are as before (see also (2.66)).

This condition extends in a straightforward manner to the case when uncertainty is present and results in the following sufficient condition for stability along the pass (a restatement of Theorem 5.4.2).

Theorem 8.1.1. *Consider a discrete unit memory linear repetitive process of the form defined by Example 1.2.10 in the presence of uncertainty which satisfies the structure given above. Then stability along the pass holds if* $\exists$ *matrices* $P > 0$ *and* $Q > 0$ *such that*

$$\left(\widehat{A} + \widehat{H}\widehat{\mathcal{F}}\widehat{E}\right)^T P \left(\widehat{A} + \widehat{H}\widehat{\mathcal{F}}\widehat{E}\right) + \widehat{Q} < 0 \tag{8.2}$$

where

$$\widehat{H} := \begin{bmatrix} \widehat{H}_1 & \widehat{H}_2 \end{bmatrix}, \quad \widehat{\mathcal{F}} := I_2 \otimes \mathcal{F}, \quad \widehat{E} := I_2 \otimes E$$

(and $\otimes$ *again denotes the matrix Kronecker product).*

Also it can be shown that (as used in the proof of Theorem 5.4.2) $\exists$ $P > 0$ and $Q > 0$ such that (8.2) holds if, and only if, $\exists$ a scalar $\epsilon > 0$ such that

$$\begin{bmatrix} -P^{-1} + \epsilon \widehat{H}\widehat{H}^T & \widehat{A} \\ \widehat{A}^T & \epsilon^{-1}\widehat{E}^T\widehat{E} + \widehat{Q} \end{bmatrix} < 0$$

Suppose now that we wish to control a process with uncertainty present and hence we also need to deal with the uncertainty in the input terms of the state-space model. To which end, introduce the so-called augmented plant input matrix as

$$\Psi_{ic} := \begin{bmatrix} B^T & D^T \end{bmatrix}^T$$

Then under the additive perturbation assumption we write

$$\Psi_{icp} := \Psi_{ic} + \Delta\Psi_{ic}$$

where

$$\Delta\Psi_{ic} := \begin{bmatrix} \Delta B \\ \Delta D \end{bmatrix}$$

and assume that $\Delta\Psi_{ic}$ can be written in the form

$$\Delta\Psi_{ic} = \begin{bmatrix} H_1 \\ H_2 \end{bmatrix} \mathcal{F} E_3$$

where again $\mathcal{F}^T\mathcal{F} \leq I$.

Consider now the application of a control law of the form (7.9), i.e.

$$u_{k+1}(p) = K_1 x_{k+1}(p) + K_2 y_k(p) =: K \begin{bmatrix} x_{k+1}(p) \\ y_k(p) \end{bmatrix} \tag{8.3}$$

to the uncertain process. Then we have the following result.

Theorem 8.1.2. *Suppose that a control law of the form (8.3) is applied to a discrete unit memory linear repetitive process of the form defined by Example 1.2.10 in the presence of uncertainty which satisfies the structure detailed above. Then the resulting process is stable along the pass if* $\exists$ *a scalar* $\epsilon > 0$ *and matrices* $Y > 0$, $Z > 0$, *and* N *such that the following LMI holds*

$$\begin{bmatrix} -Y + \varepsilon \widehat{H}_1 \widehat{H}_1^T + \varepsilon \widehat{H}_2 \widehat{H}_2^T & (\star) & (\star) & (\star) & (\star) \\ Y\widehat{A}_1^T + N^T \widehat{B}_1^T & Z - Y & (\star) & (\star) & (\star) \\ Y\widehat{A}_2^T + N^T \widehat{B}_2^T & 0 & -Z & (\star) & (\star) \\ 0 & EY + E_3 N & 0 & -\varepsilon I & (\star) \\ 0 & 0 & EY + E_3 N & 0 & -\varepsilon I \end{bmatrix} < 0$$

If this condition holds then a stabilizing control law matrix is given by $K = NY^{-1}$.

Proof. This involves extensive, but routine, manipulations and hence here we only note the main steps. The first of these is to interpret (8.2) in terms of the state-space model resulting from application of (8.3). The result then follows immediately as in [73]. □

8.1.2 Differential Processes

Here we expand on the results of Sect. 5.4.2 and show that an LMI setting leads directly to stabilizing control law design for differential unit memory linear repetitive processes of the form defined by Example 1.2.3 when the model has a norm bounded perturbation in both the current pass state and pass profile updating equations (case (b) in Sect. 5.4.2). In particular, the state-space model considered has the form

$$\begin{aligned} \dot{x}_{k+1}(t) &= (A+\Delta A)x_{k+1}(t) + (B+\Delta B)u_{k+1}(t) + (B_0+\Delta B_0)y_k(t) \\ y_{k+1}(t) &= (C+\Delta C)x_{k+1}(t) + (D+\Delta D)u_{k+1}(t) + (D+\Delta D_0)y_k(t) \end{aligned}$$

where

$$\begin{bmatrix} \Delta A & \Delta B_0 & \Delta B \\ \Delta C & \Delta D_0 & \Delta D \end{bmatrix} = \begin{bmatrix} H_1 \\ H_2 \end{bmatrix} \mathcal{F} \begin{bmatrix} E_1 & E_2 & E_3 \end{bmatrix}$$

and again $\mathcal{F}^T\mathcal{F} \leq I$.

Suppose that a control law of the form (7.8), i.e.

$$u_{k+1}(t) = K_1 x_{k+1}(t) + K_2 y_k(t) =: \widetilde{K} \begin{bmatrix} x_{k+1}(t) \\ y_k(t) \end{bmatrix} \tag{8.4}$$

is to be applied. Suppose also that the uncertainty structure in the process model satisfies the same assumptions as in Sect. 8.1.1. Then it is straightforward to show that the results established in Sect. 5.4.2 can also be applied to the resulting state-space model to give a (sufficient) condition for stability along the pass of the controlled process and (as for the discrete case) a formula for designing the corresponding control law. Hence the details are omitted here. One point to note, however, is that the LMIs in this analysis (and for the corresponding discrete case) could well be large and hence the possibility of numerical difficulties – this is left here as an area for further research.

8.2 H_∞ Control of Discrete Processes

Consider the sequence $\{w_i(j)\}$ whose elements are $q \times 1$ vectors $w_i(j)$ defined over nonnegative integers i and j, i.e. $0 \leq i \leq \infty$ and $0 \leq j \leq \infty$, which is written as $\{[0,\infty],\ [0,\infty]\}$. Then the ℓ_2 norm of this vector sequence is given by

$$\|w\|_2 := \sqrt{\sum_{i=0}^{\infty}\sum_{j=0}^{\infty} w_i^T(j) w_i(j)}$$

and this sequence is said to be a member of $\ell_2^q\{[0,\infty],[0,\infty]\}$, or ℓ_2^q for short, if $||w||_2 < \infty$.

Some application areas will clearly require the design of control laws which guarantee stability along the pass and also have the maximum possible disturbance attenuation. Here we will show how to address this question in an H_∞ setting, for which we now give the relevant background. The results in this section and the following one (on guaranteed cost control) are, in the main, from [133].

The process state-space model considered has the following form over $0 \leq p \leq \alpha$, $k \geq 0$

$$\begin{aligned} x_{k+1}(p+1) &= Ax_{k+1}(p) + Bu_{k+1}(p) + B_0 y_k(p) + B_{11} w_{k+1}(p) \\ y_{k+1}(p) &= Cx_{k+1}(p) + Du_{k+}(p) + D_0 y_k(p) + B_{12} w_{k+1}(p) \end{aligned} \quad (8.5)$$

where $w_{k+1}(p)$ is an $q \times 1$ disturbance vector which belongs to ℓ_2^q, i.e. the model of Example 1.2.10 with disturbance terms added to the state and pass profile vector updating equations. Also it is easy to see that stability along the pass here holds under the same conditions as those for a process described by Example 1.2.10.

Note 8.2.1. The model considered is the same as that in Sect. 7.4 but here we can also allow the disturbance vector to vary from pass-to-pass.

Disturbance attenuation is defined as follows.

Definition 8.2.1. *A discrete unit memory linear repetitive process described by (8.5) is said to have H_∞ disturbance attenuation $\gamma > 0$ if it is stable along the pass and the induced norm from w to y is bounded by γ, i.e.*

$$\sup_{0 \neq w \in \ell_2^q} \frac{\|y\|_2}{\|w\|_2} < \gamma \quad (8.6)$$

The relevance of control law design to reject the effects of disturbances on measurements (and subsequent computations) of variables is well founded physically by noting the conditions in which physical examples have to operate, e.g. in long-wall coal cutting and iterative learning control applications, such as using a gantry robot to synchronously place objects on a chain conveyor [19].

Next, we require the 2D transfer-function matrix coupling the disturbance and current pass profile vectors, which (following the derivation in Sect. 1.4 for the disturbance free case) is easily shown to be given by

$$G_{yw}(z_1, z_2) := \begin{bmatrix} 0 & I \end{bmatrix} \left(\begin{bmatrix} I - z_1 A & -z_1 B_0 \\ -z_2 C & I - z_2 D_0 \end{bmatrix} \right)^{-1} \begin{bmatrix} B_{11} \\ B_{12} \end{bmatrix}$$

The route to using this setup to develop practically efficient design methods for control laws to achieve disturbance rejection is via the 2D Parseval theorem [109] under which (8.6) is equivalent to the requirement that $\|G_{yw}(z_1, z_2)\|_\infty < \gamma$, where

$$\|G_{yw}(z_1, z_2)\|_\infty := \sup_{\omega_1, \omega_2 \in [0, 2\pi]} \overline{\sigma}[G_{yw}(e^{j\omega_1}, e^{j\omega_2})]$$

and $\overline{\sigma}(\cdot)$ again denotes the maximum singular value of its matrix argument.

Consider now the following Lyapunov function for the processes considered here (see also (5.27) (or (5.28)))

$$V(k,p) = x_{k+1}^T(p)P_1x_{k+1}(p) + y_k^T(p)P_2y_k(p) \tag{8.7}$$

with associated increment

$$\begin{aligned}\Delta V(k,p) &= x_{k+1}^T(p+1)P_1x_{k+1}(p+1) - x_{k+1}^T(p)P_1x_{k+1}(p) \\ &+ y_{k+1}^T(p)P_2y_{k+1}(p) - y_k^T(p)P_2y_k(p)\end{aligned}$$

where $P_1 > 0$ and $P_2 > 0$. Then we have the following results.

Lemma 8.2.1. *A discrete unit memory linear repetitive process described by (8.5) is stable along the pass if*

$$\Delta V(k,p) < 0$$

Proof. This follows immediately from the analysis of Chap. 5 (see, in particular, (5.26)–(5.29)) and essentially uses the 2D Lyapunov equation characterization of stability along the pass of Chap. 3. (See also the proof of Lemma 8.4.1 in this chapter for the differential unit memory case.) □

Theorem 8.2.1. *A discrete unit memory linear repetitive process described by (8.5) is stable along the pass and has H_∞ disturbance attenuation $\gamma > 0$ if ∃ matrices $P_1 > 0$ and $P_2 > 0$ such that the following LMI with $P :=$ diag$\{P_1, P_2\}$ holds*

$$\begin{bmatrix} -P & P\Phi & PB_1 & 0 \\ \Phi^T P & -P & 0 & C_2^T \\ B_1^T P & 0 & -\gamma^2 I & 0 \\ 0 & C_2 & 0 & -I \end{bmatrix} < 0 \tag{8.8}$$

where

$$C_2 := \begin{bmatrix} 0 & I \end{bmatrix}, \quad B_1 := \begin{bmatrix} B_{11} \\ B_{12} \end{bmatrix}$$

Proof. It is easily shown that the required H_∞ disturbance attenuation holds provided the associated Hamiltonian

$$H(k,p) := \Delta V(k,p) + y_{k+1}^T(p)y_{k+1}(p) - \gamma^2 w_{k+1}^T(p)w_{k+1}(p)$$

satisfies

$$H(k,p) < 0 \tag{8.9}$$

This requires that $\Delta V(k,p) < 0$ and hence by Lemma 8.2.1 stability along the pass.

Using the process state-space model (8.5) (with the input terms deleted), it is easily shown that

$$H(k,p) = \begin{bmatrix} \zeta_k^T(p) & w_{k+1}^T(p) \end{bmatrix} \begin{bmatrix} \Phi^T P\Phi - P + C_2^T C_2 & \Phi^T PB_1 \\ B_1^T P\Phi & B_1^T PB_1 - \gamma^2 I \end{bmatrix} \begin{bmatrix} \zeta_k(p) \\ w_{k+1}(p) \end{bmatrix}$$

where $\zeta_k(p) := \begin{bmatrix} x_{k+1}^T(p) & y_k^T(p) \end{bmatrix}^T$, and (8.9) holds provided

$$\begin{bmatrix} \Phi^T P \Phi - P + C_2^T C_2 & \Phi^T P B_1 \\ B_1^T P \Phi & B_1^T P B_1 - \gamma^2 I \end{bmatrix} < 0$$

(for any $\begin{bmatrix} \zeta_k^T(p) & w_{k+1}^T(p) \end{bmatrix}^T \neq 0$)

Finally, an obvious application of the Schur's complement formula shows that this last condition is equivalent to stability along the pass and the proof is complete. □

In many cases of practical interest it will be necessary to compute the minimum disturbance attenuation level γ. This can be obtained by solving a linear objective minimization problem of the following form

$$\min_{P_1>0, P_2>0} \quad \mu$$
$$\text{subject to (8.8) with } \mu = \gamma^2$$

Note 8.2.2. Consider the Roesser model with augmented plant matrix Φ. Then, as noted in Chap. 2, it is known from the work of [24] that BIBO stability of this model is equivalent to stability along the pass of discrete unit memory linear repetitive processes described by (8.5). Hence an alternative proof of this last result is to follow the method in [46].

Next we develop a solution to the disturbance attenuation problem which can be summarized as finding an implementable control law which will give stability along the pass of the controlled process together with a prescribed degree of disturbance attenuation, including the case when there is uncertainty in the model structure.

8.2.1 H_∞ Control with a Static Control Law

Suppose that a control law defined by (8.3) is applied to a process described by (8.5). Then we have the following result.

Theorem 8.2.2. *Suppose that a control law of the form (8.3) is applied to a discrete unit memory linear repetitive process described by (8.5). Then the resulting process is stable along the pass and has H_∞ disturbance attenuation $\gamma > 0$ if $\exists$ matrices $W_1 > 0$, $W_2 > 0$, N_1 and N_2 such that the following LMI holds*

$$\begin{bmatrix} -W_1 & (\star) & (\star) & (\star) & (\star) & (\star) \\ 0 & -W_2 & (\star) & (\star) & (\star) & (\star) \\ W_1 A^T + N_1^T B^T & W_1 C^T + N_1^T D^T & -W_1 & (\star) & (\star) & (\star) \\ W_2 B_0^T + N_2^T B^T & W_2 D_0^T + N_2^T D^T & 0 & -W_2 & (\star) & (\star) \\ B_{11}^T & B_{12}^T & 0 & 0 & -\gamma^2 I & (\star) \\ 0 & 0 & 0 & W_2 & 0 & -I \end{bmatrix} < 0 \tag{8.10}$$

If this condition holds, then stabilizing control law matrices are given by

$$K_1 = N_1 W_1^{-1}, \quad K_2 = N_2 W_2^{-1} \tag{8.11}$$

Proof. Applying the LMI of Theorem 8.2.1 it follows immediately that stability along the pass holds for the controlled process if there exist matrices $P_1 > 0$ and $P_2 > 0$ such that

$$\begin{bmatrix} -P_1 & (\star) & (\star) & (\star) & (\star) & (\star) \\ 0 & -P_2 & (\star) & (\star) & (\star) & (\star) \\ A^T P_1 + K_1^T B^T P_1 & C^T P_2 + K_1^T D^T P_2 & -P_1 & (\star) & (\star) & (\star) \\ B_0^T P_1 + K_2^T B^T P_1 & D_0^T P_2 + K_2^T D^T P_2 & 0 & -P_2 & (\star) & (\star) \\ B_{11}^T P_1 & B_{12}^T P_2 & 0 & 0 & -\gamma^2 I & (\star) \\ 0 & 0 & 0 & I & 0 & -I \end{bmatrix} < 0$$

This last inequality is not in LMI form because it is nonlinear with respect to its parameters. Consequently, set $P_1 = W_1^{-1}$, $P_2 = W_2^{-1}$ in this expression and pre- and post-multiply the result by $\text{diag}\{W_1, W_2, W_1, W_2, I, I\}$, to obtain

$$\begin{bmatrix} -W_1 & (\star) & (\star) & (\star) & (\star) & (\star) \\ 0 & -W_2 & (\star) & (\star) & (\star) & (\star) \\ W_1 A^T + W_1 K_1^T B^T & W_1 C^T + W_1 K_1^T D^T & -W_1 & (\star) & (\star) & (\star) \\ W_2 B_0^T + W_2 K_2^T B^T & W_2 D_0^T + W_2 K_2^T D^T & 0 & -W_2 & (\star) & (\star) \\ B_{11}^T & B_{12}^T & 0 & 0 & -\gamma^2 I & (\star) \\ 0 & 0 & 0 & W_2 & 0 & -I \end{bmatrix} < 0$$

Now set $N_1 = K_1 W_1$ and $N_2 = K_2 W_2$ in this last expression to obtain the LMI of (8.10) and the proof is complete. □

Note that the H_∞ disturbance attenuation here can be minimized using the following linear objective minimization procedure which leads, in effect, to minimization of the effects of the disturbance

$$\begin{aligned} &\min_{W_1 > 0, W_2 > 0, N_1, N_2} \quad \mu \\ &\text{subject to (8.10) with } \mu = \gamma^2 \end{aligned} \tag{8.12}$$

Next we extend the above analysis to the case of robust H_∞ control.

8.2.2 H_∞ Control of Uncertain Discrete Processes

Consider a discrete unit memory linear repetitive process with uncertainty modelled as additive perturbations to the nominal model matrices in (8.5) with resulting state-space model

$$\begin{aligned} x_{k+1}(p+1) &= (A + \Delta A) x_{k+1}(p) + (B + \Delta B) u_{k+1}(p) \\ &+ (B_0 + \Delta B_0) y_k(p) + (B_{11} + \Delta B_{11}) w_{k+1}(p) \\ y_{k+1}(p) &= (C + \Delta C) x_{k+1}(p) + (D + \Delta D) u_{k+1}(p) \\ &+ (D_0 + \Delta D_0) y_k(p) + (B_{12} + \Delta B_{12}) w_{k+1}(p) \end{aligned} \tag{8.13}$$

The matrices describing the admissible uncertainties are assumed to satisfy

$$\begin{bmatrix} \Delta A & \Delta B_0 & \Delta B & \Delta B_{11} \\ \Delta C & \Delta D_0 & \Delta D & \Delta B_{12} \end{bmatrix} = \begin{bmatrix} H_1 \\ H_2 \end{bmatrix} \mathcal{F} \begin{bmatrix} E_1 & E_2 & E_3 & E_4 \end{bmatrix} \tag{8.14}$$

Here H_1, H_2, E_1, E_2, E_3, E_4 are known constant matrices with compatible dimensions and $\mathcal{F}$ is an unknown constant matrix which satisfies

$$\mathcal{F}^T \mathcal{F} \leq I \tag{8.15}$$

Now we have the following result.

Theorem 8.2.3. *Suppose that a control law of the form (8.3) is applied to a discrete unit memory linear repetitive process described by (8.13) with the uncertainty structure satisfying (8.14) and (8.15). Then the resulting process is stable along the pass with H_∞ disturbance attenuation $\gamma > 0$ if $\exists$ a scalar $\epsilon > 0$ and matrices $W_1 > 0$, $W_2 > 0$, N_1 and N_2 such that the following LMI holds*

$$\begin{bmatrix} -W_1 + 3\epsilon H_1 H_1^T & (\star) & (\star) & (\star) & (\star) & (\star) & (\star) & (\star) & (\star) \\ 3\epsilon H_1 H_2^T & -W_2 + 3\epsilon H_2 H_2^T & (\star) & (\star) & (\star) & (\star) & (\star) & (\star) & (\star) \\ W_1 A^T + N_1^T B^T & W_1 C^T + N_1^T D^T & -W_1 & (\star) & (\star) & (\star) & (\star) & (\star) & (\star) \\ W_2 B_0^T + N_2^T B^T & W_2 D_0^T + N_2^T D^T & 0 & -W_2 & (\star) & (\star) & (\star) & (\star) & (\star) \\ B_{11}^T & B_{12}^T & 0 & 0 & -\gamma^2 I & (\star) & (\star) & (\star) & (\star) \\ 0 & 0 & 0 & W_2 & 0 & -I & (\star) & (\star) & (\star) \\ 0 & 0 & 0 & W_1 E_1^T + N_1^T E_3^T & 0 & 0 & -\epsilon I & (\star) & (\star) \\ 0 & 0 & 0 & 0 & W_2 E_2^T + N_2^T E_3^T & 0 & 0 & -\epsilon I & (\star) \\ 0 & 0 & 0 & 0 & 0 & E_4^T & 0 & 0 & -\epsilon I \end{bmatrix} < 0 \tag{8.16}$$

If this condition holds, then stabilizing control law matrices are again given by (8.11).

Proof. Stability along the pass for the controlled process holds if

$$\Gamma + \widetilde{H}\widetilde{\mathcal{F}}\widetilde{E} + \widetilde{E}^T \widetilde{\mathcal{F}}^T \widetilde{H}^T < 0$$

where

$$\Gamma := \begin{bmatrix} -W_1 & (\star) & (\star) & (\star) & (\star) & (\star) \\ 0 & -W_2 & (\star) & (\star) & (\star) & (\star) \\ W_1A^T + N_1^TB^T & W_1C^T + N_1^TD^T & -W_1 & (\star) & (\star) & (\star) \\ W_2B_0^T + N_2^TB^T & W_2D_0^T + N_2^TD^T & 0 & -W_2 & (\star) & (\star) \\ B_{11}^T & B_{12}^T & 0 & 0 & -\gamma^2 I & (\star) \\ 0 & 0 & 0 & W_2 & 0 & -I \end{bmatrix}$$

$$\widetilde{H} := \begin{bmatrix} 0 & 0 & H_1 & H_1 & H_1 & 0 \\ 0 & 0 & H_2 & H_2 & H_2 & 0 \\ 0 & 0 & 0 & 0 & 0 & 0 \\ 0 & 0 & 0 & 0 & 0 & 0 \\ 0 & 0 & 0 & 0 & 0 & 0 \\ 0 & 0 & 0 & 0 & 0 & 0 \end{bmatrix}, \ \widetilde{\mathcal{F}} := \mathrm{diag}\{\mathcal{F}, \mathcal{F}, \mathcal{F}, \mathcal{F}, \mathcal{F}, \mathcal{F}\}$$

$$\widetilde{E} := \mathrm{diag}\{0, 0, E_1W_1 + E_3N_1, E_2W_2 + E_4N_2, E_4, 0\}$$

The LMI of (8.16) is now obtained by first making use of the inequality of Lemma 5.4.2 followed by an obvious application of the Schur's complement formula to the result of this last step. □

To reduce the effects of the disturbance vector, the following linear objective minimization procedure can be used

$$\min_{W_1>0, W_2>0, \epsilon>0, N_1, N_2} \mu$$

$$\text{subject to (8.16) with } \mu = \gamma^2$$

Consider now the case when the uncertainty in the process state-space model is of the additive structure defined above but the disturbance terms are absent (i.e. (8.14) with the corresponding matrices deleted on each side). Then on applying the control law (8.3), the resulting process can be written in the form

$$\begin{bmatrix} x_{k+1}(p+1) \\ y_{k+1}(p) \end{bmatrix} = \overline{A} \begin{bmatrix} x_{k+1}(p) \\ y_k(p) \end{bmatrix} \tag{8.17}$$

where

$$\overline{A} := \begin{bmatrix} A + BK_1 & B_0 + BK_2 \\ C + DK_1 & D_0 + DK_2 \end{bmatrix} + \begin{bmatrix} \Delta A + \Delta BK_1 & \Delta B_0 + \Delta BK_2 \\ \Delta C + \Delta DK_1 & \Delta D_0 + \Delta DK_2 \end{bmatrix}$$

Suppose also that the matrices describing the uncertainty in this last model can be written in the form

$$\begin{aligned} \begin{bmatrix} \Delta A + \Delta BK_1 & \Delta B_0 + \Delta BK_2 \\ \Delta C + \Delta DK_1 & \Delta D_0 + \Delta DK_2 \end{bmatrix} &= \begin{bmatrix} H_1 \\ H_2 \end{bmatrix} \gamma^{-1} \mathcal{F} \begin{bmatrix} E_1 + E_3K_1 & E_2 + E_3K_2 \end{bmatrix} \\ &=: \gamma^{-1} H\mathcal{F}E_{df} \end{aligned} \tag{8.18}$$

where the matrices H_1, H_2, E_1, E_2, E_3 have known constant entries, $\gamma > 0$ is a given scalar, and the matrix $\mathcal{F}$ satisfies (8.15). Then in this case the design parameter γ is available for in attenuating the effects of the uncertainty and we have the following result.

Theorem 8.2.4. *Suppose that application of a control law of the form (8.3) results in a discrete unit memory linear repetitive process described by (8.17) with uncertainty structure satisfying (8.18) and (8.15). Then this process is stable along the pass if* $\exists$ *matrices* $W_1 > 0$, $W_2 > 0$, N_1 *and* N_2 *such that the following LMI holds*

$$\begin{bmatrix} -W_1 & (\star) & (\star) & (\star) & (\star) & (\star) \\ 0 & -W_2 & (\star) & (\star) & (\star) & (\star) \\ W_1A^T + N_1^TB^T & W_1C^T + N_1^TD^T & -W_1 & (\star) & (\star) & (\star) \\ W_2B_0^T + N_2^TB^T & W_2D_0^T + N_2^TD^T & 0 & -W_2 & (\star) & (\star) \\ H_1^T & H_2^T & 0 & 0 & -\gamma^2 I & (\star) \\ 0 & 0 & E_1W_1 + E_3N_1 & E_2W_2 + E_3N_2 & 0 & -I \end{bmatrix} < 0 \tag{8.19}$$

If this condition holds, then stabilizing control law matrices are again given by (8.11).

Proof. Applying the result of Theorem 8.2.1 to the controlled process state-space model leads immediately to the conclusion that it is stable along the pass if there exists a block-diagonal matrix $P := \mathrm{diag}\{P_1, P_2\} > 0$ such that the following LMI holds

$$\overline{A}^T P \overline{A} - P < 0$$

An obvious application of the Schur's complement formula now yields

$$\begin{bmatrix} -P^{-1} & (\Omega + \gamma^{-1} H \mathcal{F} E_{df}) \\ (\Omega + \gamma^{-1} H \mathcal{F} E_{df})^T & -P \end{bmatrix} < 0$$

where E_{df} is defined by (8.18) and

$$\Omega := \begin{bmatrix} A + BK_1 & B_0 + BK_2 \\ C + DK_1 & D_0 + DK_2 \end{bmatrix}$$

Applying the result of Lemma 5.4.2 to this last condition and then pre- and post-multiplying the result by $\mathrm{diag}\{\epsilon^{-\frac{1}{2}} P, \epsilon^{-\frac{1}{2}} I\}$ yields

$$\begin{bmatrix} -\overline{P} + \overline{P}\gamma^{-2} H H^T \overline{P} & \overline{P}\Omega \\ \Omega^T \overline{P} & -\overline{P} + E_{df}^T E_{df} \end{bmatrix} < 0$$

where $\overline{P} = \epsilon^{-1}P$. Finally another obvious application of the Schur's complement formula gives the following LMI

$$\begin{bmatrix} -\overline{P} & \overline{P}\Omega & \overline{P}H & 0 \\ \Omega^T\overline{P} & -\overline{P} & 0 & E_{df}^T \\ H^T\overline{P} & 0 & -\gamma^2 I & 0 \\ 0 & E_{df} & 0 & -I \end{bmatrix} < 0$$

which is equivalent to that of (8.8) interpreted for the case under consideration here. Finally, applying the same procedure as in the proof of Theorem 8.2.2 yields the required LMI. □

Note also that the parameter γ in this last result can be minimized using the following linear objective minimization procedure and hence increased robustness

$$\min_{W_1>0, W_2>0, N_1, N_2} \quad \mu$$
$$\text{subject to (8.19) with } \mu = \gamma^2$$

At this stage, some comments on the relationship with Roesser model analysis can be made. The first point is that for repetitive processes the static control law applied here is well defined physically as the pass profile vector is also a process output and hence can be directly measured. Hence it has structure for discrete linear repetitive processes alone which, as the analysis here shows, can be exploited to powerful effect.

As in 1D systems theory, there will be cases when all elements in the current pass state vector cannot be directly measured. If this is the case then one option is to use the dynamic pass profile controller as follows below, where again the structure of the underlying dynamics (as opposed to assumption(s) made) allows us to obtain, relative to Roesser model analysis, simpler results. Note also that it is possible to replace the current pass state vector in the control law here with the current pass profile vector – see Sect. 8.3.1 (and [167]) where algorithms for solving this problem to ensure stability along the pass of the controlled process have been given.

8.2.3 H_∞ Control with a Dynamic Pass Profile Controller

The following well known results will be extensively used in what follows.

Lemma 8.2.2. *[66] Let Ψ be a $q \times q$ symmetric matrix and let P and Q be real $h \times q$ matrices. Then $\exists$ an $h \times q$ matrix Θ such that*

$$\Psi + P^T\Theta^T Q + Q^T \Theta P < 0$$

if, and only if, the inequalities

$$\mathcal{N}_p^T \Psi \mathcal{N}_p < 0 \text{ and } \mathcal{N}_q^T \Psi \mathcal{N}_q < 0$$

both hold, where $\mathcal{N}_p \in \ker(P)$ and $\mathcal{N}_q \in \ker(Q)$

Lemma 8.2.3. *[45] Suppose that the $n \times n$ matrices $\Sigma > 0$ and $\Gamma > 0$ are given and n_c is a positive integer. Then $\exists$ $n \times n_c$ matrices Σ_2, Γ_2 and $n_c \times n_c$ matrices $\Sigma_3 > 0$ and $\Gamma_3 > 0$, such that*

$$\begin{bmatrix} \Sigma & \Sigma_2 \\ \Sigma_2^T & \Sigma_3 \end{bmatrix} > 0 \text{ and } \begin{bmatrix} \Sigma & \Sigma_2 \\ \Sigma_2^T & \Sigma_3 \end{bmatrix}^{-1} = \begin{bmatrix} \Gamma & \Gamma_2 \\ \Gamma_2^T & \Gamma_3 \end{bmatrix}$$

if, and only if,

$$\begin{bmatrix} \Sigma & I \\ I & \Gamma \end{bmatrix} \geq 0$$

The control law (8.3) requires that the complete current pass state vector is available for measurement. If this is not the case then one option (as noted above here) is to use an observer to reconstruct it. In this section, we consider an alternative where the control of processes described by (8.5) is by the use of a dynamic pass profile controller as defined by (7.13) i.e.

$$\begin{aligned} \begin{bmatrix} x_{k+1}^c(p+1) \\ y_{k+1}^c(p) \end{bmatrix} &= \begin{bmatrix} A_{c11} & A_{c12} \\ A_{c21} & A_{c22} \end{bmatrix} \begin{bmatrix} x_{k+1}^c(p) \\ y_k^c(p) \end{bmatrix} + \begin{bmatrix} B_{c1} \\ B_{c2} \end{bmatrix} y_{k+1}(p) \\ u_{k+1}(p) &= \begin{bmatrix} C_{c1} & C_{c2} \end{bmatrix} \begin{bmatrix} x_{k+1}^c(p) \\ y_k^c(p) \end{bmatrix} + D_c y_{k+1}(p) \end{aligned} \tag{8.20}$$

where $x_k^c(p)$ and $y_k^c(p)$ are state vectors for the controller.

To obtain the state-space model of the controlled process, introduce the extra notation

$$B_2 := \begin{bmatrix} B \\ D \end{bmatrix}, \; A_c := \begin{bmatrix} A_{c11} & A_{c12} \\ A_{c21} & A_{c22} \end{bmatrix}, \; B_c := \begin{bmatrix} B_{c1} \\ B_{c2} \end{bmatrix}, \; C_c := \begin{bmatrix} C_{c1} & C_{c2} \end{bmatrix}$$

(and recall from the notation of Theorem 8.2.1 the definitions of B_1 and C_2.) Also introduce the so-called augmented state and pass profile vectors respectively as

$$\overline{x}_{k+1}(p) := \begin{bmatrix} x_{k+1}(p) \\ x_{k+1}^c(p) \end{bmatrix}, \; \overline{y}_k(p) := \begin{bmatrix} y_k(p) \\ y_k^c(p) \end{bmatrix}$$

Then the resulting state-space model is

$$\begin{aligned} \begin{bmatrix} \overline{x}_{k+1}(p+1) \\ \overline{y}_{k+1}(p) \end{bmatrix} &= \overline{\mathcal{A}} \begin{bmatrix} \overline{x}_{k+1}(p) \\ \overline{y}_k(p) \end{bmatrix} + \overline{\mathcal{B}} w_{k+1}(p) \\ y_{k+1}(p) &= \overline{\mathcal{C}} \begin{bmatrix} \overline{x}_{k+1}(p) \\ \overline{y}_{k+1}(p) \end{bmatrix} \end{aligned} \tag{8.21}$$

where

$$\overline{\mathcal{A}} := \Pi \begin{bmatrix} \Phi + B_2 D_c C_2 & B_2 C_c \\ B_c C_2 & A_c \end{bmatrix} \Pi, \; \overline{\mathcal{B}} := \Pi \begin{bmatrix} B_1 \\ 0 \end{bmatrix}, \; \overline{\mathcal{C}} := \begin{bmatrix} C_2 & 0 \end{bmatrix} \Pi$$

$$\Pi := \begin{bmatrix} I & 0 & 0 & 0 \\ 0 & 0 & I & 0 \\ 0 & I & 0 & 0 \\ 0 & 0 & 0 & I \end{bmatrix}$$

and also $\Pi = \Pi^T = \Pi^{-1}$.

Introduce the so-called matrix of controller data as

$$\Theta := \begin{bmatrix} D_c & C_c \\ B_c & A_c \end{bmatrix}$$

and

$$\mathcal{A} := \begin{bmatrix} \Phi & 0 \\ 0 & 0 \end{bmatrix}, \ \mathcal{B} := \begin{bmatrix} B_1 \\ 0 \end{bmatrix}, \ \mathcal{B}_2 := \begin{bmatrix} B_2 & 0 \\ 0 & I \end{bmatrix}, \ \mathcal{C} := \begin{bmatrix} C_2 & 0 \end{bmatrix}, \ \mathcal{C}_2 := \begin{bmatrix} C_2 & 0 \\ 0 & I \end{bmatrix}$$

Hence the controlled process state-space model matrices can be written in the following form which is affine in the controller data matrix Θ

$$\overline{\mathcal{A}} = \Pi \left[\ \mathcal{A} + \mathcal{B}_2 \Theta \mathcal{C}_2 \ \right] \Pi, \ \overline{\mathcal{B}} = \Pi \mathcal{B}, \ \overline{\mathcal{C}} = \mathcal{C} \Pi \tag{8.22}$$

Now we have the following result which gives an existence condition for the controller matrices A_c, B_c, C_c, D_c to ensure stability along the pass and then enables controller design.

Theorem 8.2.5. *Suppose that a controller of the form (8.20) is applied to a discrete unit memory linear repetitive process described by (8.5), with resulting state-space model (8.21). Suppose also that* $\exists$ *matrices* $P_{11} > 0$, *(*$P_{11} := \mathrm{diag}\{P_{h11}, P_{v11}\}$*) and* $R_{11} > 0$, *(*$R_{11} := \mathrm{diag}\{R_{h11}, R_{v11}\}$*) such that the LMIs defined by (8.23)–(8.25) below hold. Then the controlled process is stable along the pass and has* H_∞ *disturbance attenuation* $\gamma > 0$

$$\begin{bmatrix} \mathcal{N}_1 & 0 & 0 \\ 0 & I & 0 \\ 0 & 0 & I \end{bmatrix}^T \begin{bmatrix} \Phi R_{11} \Phi^T - R_{11} & B_1 & \Phi R_{11} C_2^T \\ B_1^T & -\gamma^2 I & 0 \\ C_2 R_{11} \Phi^T & 0 & -I + C_2 R_{11} C_2^T \end{bmatrix}$$

$$\times \begin{bmatrix} \mathcal{N}_1 & 0 & 0 \\ 0 & I & 0 \\ 0 & 0 & I \end{bmatrix} < 0 \tag{8.23}$$

$$\begin{bmatrix} \mathcal{N}_2 & 0 & 0 \\ 0 & I & 0 \\ 0 & 0 & I \end{bmatrix}^T \begin{bmatrix} \Phi^T P_{11} \Phi - P_{11} & \Phi^T P_{11} B_1 & C_2^T \\ B_1^T P_{11} \Phi & B_1^T P_{11} B_1 - \gamma^2 I & 0 \\ C_2 & 0 & -I \end{bmatrix}$$

$$\times \begin{bmatrix} \mathcal{N}_2 & 0 & 0 \\ 0 & I & 0 \\ 0 & 0 & I \end{bmatrix} < 0 \tag{8.24}$$

$$\begin{bmatrix} P_{h11} & I \\ I & R_{h11} \end{bmatrix} \geq 0, \quad \begin{bmatrix} P_{v11} & I \\ I & R_{v11} \end{bmatrix} \geq 0 \tag{8.25}$$

where $\mathcal{N}_1$ and $\mathcal{N}_2$ are full column rank matrices whose images satisfy

$$\mathrm{Im}\mathcal{N}_1 = \ker(B_2^T), \quad \mathrm{Im}\mathcal{N}_2 = \ker(C_2)$$

Proof. Interpreting the inequality (8.8) in terms of the matrices of (8.22) yields

$$\begin{bmatrix} -P & (\star) & (\star) & (\star) \\ \Pi\mathcal{A}^T\Pi P + \Pi\mathcal{C}_2^T\Theta^T\mathcal{B}_2^T\Pi P & -P & (\star) & (\star) \\ \mathcal{B}^T\Pi P & 0 & -\gamma^2 I & 0 \\ 0 & \mathcal{C}\Pi & 0 & -I \end{bmatrix} < 0 \tag{8.26}$$

where $P := \mathrm{diag}\{P_h,\ P_v\}$. Also pre- and post-multiplying (8.26) by

$$\mathrm{diag}\{\Pi, \Pi, I, I\}$$

and then setting $\mathcal{R} = \Pi P \Pi$ gives

$$\begin{bmatrix} -\mathcal{R} & (\star) & (\star) & (\star) \\ \mathcal{A}^T\mathcal{R} + \mathcal{C}_2^T\Theta^T\mathcal{B}_2^T\mathcal{R} & -\mathcal{R} & (\star) & (\star) \\ \mathcal{B}^T\mathcal{R} & 0 & -\gamma^2 I & (\star) \\ 0 & \mathcal{C} & 0 & -I \end{bmatrix} < 0 \tag{8.27}$$

Now, introduce the matrices

$$\Psi := \begin{bmatrix} -\mathcal{R} & \mathcal{R}\mathcal{A} & \mathcal{R}\mathcal{B} & 0 \\ \mathcal{A}^T\mathcal{R} & -\mathcal{R} & 0 & \mathcal{C}^T \\ \mathcal{B}^T\mathcal{R} & 0 & -\gamma^2 I & 0 \\ 0 & \mathcal{C} & 0 & -I \end{bmatrix}, \quad \mathcal{M}^T := \begin{bmatrix} \mathcal{R}\mathcal{B}_2 \\ 0 \\ 0 \\ 0 \end{bmatrix}, \quad \mathcal{N} := \begin{bmatrix} 0 & \mathcal{C}_2 & 0 & 0 \end{bmatrix}$$

and hence write (8.27) in the form

$$\Psi + \mathcal{M}^T\Theta\mathcal{N} + \mathcal{N}^T\Theta^T\mathcal{M} < 0 \tag{8.28}$$

Use of Lemma 8.2.2 now yields the following two matrix inequalities which are equivalent to (8.28)

$$W_{\mathcal{M}}^T\Psi W_{\mathcal{M}} < 0 \quad \text{and} \quad W_{\mathcal{N}}^T\Psi W_{\mathcal{N}} < 0$$

where

$$W_{\mathcal{M}} \in \ker(\mathcal{M}), \quad W_{\mathcal{N}} \in \ker(\mathcal{N}) \tag{8.29}$$

and

$$\mathcal{M} = \mathcal{M}_n S =: \begin{bmatrix} \mathcal{B}_2^T & 0 & 0 & 0 \end{bmatrix} \begin{bmatrix} \mathcal{R} & 0 & 0 & 0 \\ 0 & I & 0 & 0 \\ 0 & 0 & I & 0 \\ 0 & 0 & 0 & I \end{bmatrix}$$

Note now that

$$\ker(\mathcal{M}_n S) = S^{-1} \ker \mathcal{M}_n$$

and using (8.29) yields

$$W_{\mathcal{M}} = S^{-1} W_{\mathcal{M}_n}$$

Therefore we have that

$$W_{\mathcal{M}}^T \Psi W_{\mathcal{M}} < 0 \Leftrightarrow W_{\mathcal{M}_n}^T (S^{-1})^T \Psi S^{-1} W_{\mathcal{M}_n} < 0$$

and also

$$\widetilde{\Phi} := (S^{-1})^T \Psi S^{-1} = \begin{bmatrix} -\mathcal{R}^{-1} & \mathcal{A} & \mathcal{B} & 0 \\ \mathcal{A}^T & -R & 0 & \mathcal{C}^T \\ \mathcal{B}^T & 0 & -\gamma^2 I & 0 \\ 0 & \mathcal{C} & 0 & -I \end{bmatrix}$$

Hence we have the following two inequalities which are not in LMI form

$$W_{\mathcal{M}_n}^T \widetilde{\Phi} W_{\mathcal{M}_n} < 0 \quad \text{and} \quad W_{\mathcal{N}}^T \Psi W_{\mathcal{N}} < 0$$

To obtain these inequalities in the required LMI form, first write the defining matrices out in full, i.e.

$$\mathcal{M}_n = \left[\ \mathcal{B}_2^T\, 0\, 0\, 0\ \right] = \left[\begin{array}{cc|cc|c|c} B_2^T & 0 & 0 & 0 & 0 & 0 \\ 0 & I & 0 & 0 & 0 & 0 \end{array}\right]$$

$$\mathcal{N} = \left[\ 0\, \mathcal{C}_2\, 0\, 0\ \right] = \left[\begin{array}{cc|cc|c|c} 0 & 0 & C_2 & 0 & 0 & 0 \\ 0 & 0 & 0 & I & 0 & 0 \end{array}\right]$$

Also it can be shown that the kernels of $\mathcal{M}_n$ and $\mathcal{N}$ are images of

$$W_{\mathcal{M}_n} := \begin{bmatrix} \mathcal{N}_1 & 0 & 0 & 0 & 0 \\ 0 & 0 & 0 & 0 & 0 \\ 0 & I & 0 & 0 & 0 \\ 0 & 0 & I & 0 & 0 \\ 0 & 0 & 0 & I & 0 \\ 0 & 0 & 0 & 0 & I \end{bmatrix}, \quad W_{\mathcal{N}} := \begin{bmatrix} I & 0 & 0 & 0 & 0 \\ 0 & I & 0 & 0 & 0 \\ 0 & 0 & \mathcal{N}_2 & 0 & 0 \\ 0 & 0 & 0 & 0 & 0 \\ 0 & 0 & 0 & I & 0 \\ 0 & 0 & 0 & 0 & I \end{bmatrix}$$

where

$$\mathcal{N}_1 \in \ker(B_2^T), \quad \mathcal{N}_2 \in \ker(C_2)$$

Now, rewrite the matrix $\mathcal{R}$ as

$$\mathcal{R} = \Pi P \Pi = \left[\begin{array}{cc|cc} P_{h_{11}} & 0 & P_{h_{12}} & 0 \\ 0 & P_{v_{11}} & 0 & P_{v_{12}} \\ \hline P_{h_{12}}^T & 0 & P_{h_{22}} & 0 \\ 0 & P_{v_{12}}^T & 0 & P_{v_{22}} \end{array}\right] = \left[\begin{array}{c|c} P_{11} & P_{12} \\ \hline P_{12}^T & P_{22} \end{array}\right]$$

where

$$P_h = \begin{bmatrix} P_{h_{11}} & P_{h_{12}} \\ P_{h_{12}}^T & P_{h_{22}} \end{bmatrix}, \quad P_v = \begin{bmatrix} P_{v_{11}} & P_{v_{12}} \\ P_{v_{12}}^T & P_{v_{22}} \end{bmatrix}$$

Note also that

$$P_h^{-1} = \begin{bmatrix} R_{h_{11}} & R_{h_{12}} \\ R_{h_{12}}^T & R_{h_{22}} \end{bmatrix}, \quad P_v^{-1} = \begin{bmatrix} R_{v_{11}} & R_{v_{12}} \\ R_{v_{12}}^T & R_{v_{22}} \end{bmatrix}$$

and

$$\mathcal{R}^{-1} = \Pi P^{-1} \Pi = \left[\begin{array}{cc|cc} \mathcal{R}_{h_{11}} & 0 & \mathcal{R}_{h_{12}} & 0 \\ 0 & \mathcal{R}_{v_{11}} & 0 & \mathcal{R}_{v_{12}} \\ \hline \mathcal{R}_{h_{12}}^T & 0 & \mathcal{R}_{h_{22}} & 0 \\ 0 & \mathcal{R}_{v_{12}}^T & 0 & \mathcal{R}_{v_{22}} \end{array}\right]$$

$$= \left[\begin{array}{c|c} \mathcal{R}_{11} & \mathcal{R}_{12} \\ \hline \mathcal{R}_{12}^T & \mathcal{R}_{22} \end{array}\right]$$

Hence the matrix $\widetilde{\Phi}$ can be rewritten as

$$\widetilde{\Phi} = \begin{bmatrix} -\mathcal{R}_{11} & -\mathcal{R}_{12} & \Phi & 0 & B_1 & 0 \\ -\mathcal{R}_{12}^T & -\mathcal{R}_{12} & 0 & 0 & 0 & 0 \\ \Phi^T & 0 & -P_{11} & -P_{12} & 0 & C_2^T \\ 0 & 0 & -P_{12}^T & -P_{22} & 0 & 0 \\ B_1^T & 0 & 0 & 0 & -\gamma^2 I & 0 \\ 0 & 0 & C_2 & 0 & 0 & -I \end{bmatrix}$$

and on using the inequality

$$W_{\mathcal{M}_n}^T \widetilde{\Phi} W_{\mathcal{M}_n} < 0$$

we have (note that the second block row of $W_{\mathcal{M}_n}$ is zero)

$$\Upsilon^T \left[\begin{array}{c|cc|cc} -\mathcal{R}_{11} & \Phi & 0 & B_1 & 0 \\ \hline \Phi^T & -P_{11} & -P_{12} & 0 & C_2^T \\ 0 & -P_{12}^T & -P_{22} & 0 & 0 \\ \hline B_1^T & 0 & 0 & -\gamma^2 I & 0 \\ 0 & C_2 & 0 & 0 & -I \end{array}\right] \Upsilon < 0$$

where $\Upsilon := \text{diag}\{\mathcal{N}_1, I, I, I, I\}$. Next by an obvious application of the Schur's complement formula the LMI of (8.23) is obtained.

To obtain the inequality (8.24), rewrite the matrix Ψ as

$$\Psi = \begin{bmatrix} -P_{11} & (\star) & (\star) & (\star) & (\star) & (\star) \\ -P_{12}^T & -P_{22} & (\star) & (\star) & (\star) & (\star) \\ \Phi^T P_{11} & \Phi^T P_{12} & -P_{11} & (\star) & (\star) & (\star) \\ 0 & 0 & -P_{12}^T & -P_{22} & (\star) & (\star) \\ B_1^T P_{11} & B_1^T P_{12} & 0 & 0 & -\gamma^2 I & (\star) \\ 0 & 0 & C_2 & 0 & 0 & -I \end{bmatrix}$$

Then by an obvious application of the Schur's complement formula, the inequality

$$W_{\mathcal{N}}^T \Psi W_{\mathcal{N}} < 0$$

becomes

$$\begin{bmatrix} \mathcal{N}_2^T & 0 & 0 & 0 \\ 0 & 0 & I & 0 \\ 0 & 0 & 0 & I \end{bmatrix} \begin{bmatrix} \Phi^T P_{11}\Phi - P_{11} & -P_{12} & \Phi^T P_{11} B_1 & C_2^T \\ -P_{12}^T & -P_{22} & 0 & 0 \\ B_1^T P_{11}\Phi & 0 & B_1^T P_{11} B_1 - \gamma^2 I & 0 \\ C_2 & 0 & 0 & -I \end{bmatrix}$$

$$\times \begin{bmatrix} \mathcal{N}_2 & 0 & 0 \\ 0 & 0 & 0 \\ 0 & I & 0 \\ 0 & 0 & I \end{bmatrix} < 0$$

which is equivalent to (8.24).

The last problem is to provide the conditions which allow us to find the matrix P and its inverse. To begin, first note again that $P = \text{diag}\{P_h, P_v\}$ and that only P_{11} and R_{11} appear in the first two LMIs to be satisfied. Application of Lemma 8.2.3 now gives the required conditions in the form of (8.25) and the proof is complete. □

If this last result holds then the stabilizing controller can be designed using the following algorithm.

1. Compute the matrices P_{h12}, P_{v12} using

$$\begin{aligned} P_{h11} - R_{h11}^{-1} &= P_{h12} P_{h22}^{-1} P_{h12}^T \\ P_{v11} - R_{v11}^{-1} &= P_{v12} P_{v22}^{-1} P_{v12}^T \end{aligned}$$

   where $P_{h22} = I$ and $P_{v22} = I$
2. Construct $P_h > 0$ and $P_v > 0$ and then the matrix $P = \text{diag}\{P_h, P_v\}$.
3. Compute the matrices M, N and Ψ.
4. Solve the LMI of (8.28) (where Θ is the unknown matrix) and hence obtain the controller state-space model matrices.

Also the attenuation level γ can be minimized using the following optimization procedure

$$\min_{P_{11}>0, R_{11}>0} \mu$$
$$\text{subject to (8.23) – (8.25) with } \mu = \gamma^2$$

Example 8.2.1. Consider the special case of the model of (8.5) defined by

$$A = \begin{bmatrix} 0.4 & 0.2 \\ 1.1 & 0.1 \end{bmatrix},\ B_0 = \begin{bmatrix} 0.3 & 0.3 \\ 0.4 & 0.9 \end{bmatrix},\ C = \begin{bmatrix} 0.7 & 0.4 \\ 0.2 & 0.1 \end{bmatrix},\ D_0 = \begin{bmatrix} 0.6 & 0.6 \\ 0.9 & 0.9 \end{bmatrix}$$

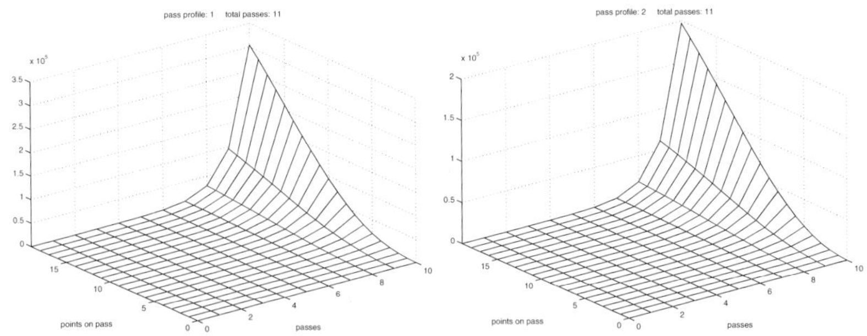

Fig. 8.1. Pass profile vector sequence of the uncontrolled process

$$B = \begin{bmatrix} 1.2 \\ 1.1 \end{bmatrix}, \; D = \begin{bmatrix} 3.0 \\ 1.7 \end{bmatrix}, \; B_{11} = \begin{bmatrix} 0.1 \\ 0.2 \end{bmatrix}, \; B_{12} = \begin{bmatrix} 0.2 \\ 0.3 \end{bmatrix}$$

This example is unstable along the pass – as confirmed by the simulation results of Fig. 8.1, where the left-hand plot corresponds to the first entry in the pass profile vector and that on the right the second, with the following boundary conditions

$$x_{k+1}(0) = \begin{bmatrix} 1 \\ 1 \end{bmatrix}, \; y_0(p) = \begin{bmatrix} 1 \\ 1 \end{bmatrix}, \; 0 \leq p \leq 20$$

Using the procedure of (8.12) gives the H_∞ disturbance attenuation level $\gamma = 2.2153$ and the matrices

$$P_h = 10^3 \times \begin{bmatrix} 1.7653 & -1.0543 & 0.0420 & 0 \\ -1.0543 & 0.6343 & -0.0251 & 0.0019 \\ 0.0420 & -0.0251 & 0.0010 & 0 \\ 0 & 0.0019 & 0 & 0.0010 \end{bmatrix}$$

$$P_v = \begin{bmatrix} 4.5436 & -7.4567 & 1.8072 & 0 \\ -7.4567 & 19.6458 & -4.2781 & 0.0837 \\ 1.8072 & -4.2781 & 1.0000 & 0 \\ 0 & 0.0837 & 0 & 1.0000 \end{bmatrix}$$

Hence the dynamic pass profile controller in this case is given by

$$A_c = \begin{bmatrix} -0.3693 & -3.0691 & 0.0420 & -0.0084 \\ 0.0305 & 0.3044 & 0.0074 & -0.0014 \\ 0.0145 & 0.4187 & 0.0222 & -0.0044 \\ -0.0003 & -0.0081 & -0.0008 & 0.0001 \end{bmatrix}$$

$$B_c = \begin{bmatrix} 4.1204 & 16.8363 \\ -0.1022 & -1.0908 \\ 2.3073 & 2.1374 \\ -0.0357 & -0.0283 \end{bmatrix}$$

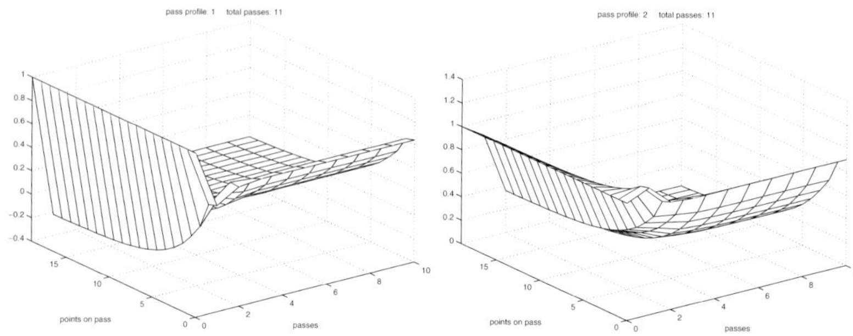

Fig. 8.2. Pass profile sequence of the controlled process

$$C_c = \begin{bmatrix} 0.0050 & 0.1238 & -0.0014 & 0.0003 \end{bmatrix}, \; D_c = \begin{bmatrix} -0.2938 & -0.3040 \end{bmatrix}$$

Figure 8.2, confirms that the resulting controlled process is stable along the pass (first entry in the pass profile vector on the left, second on the right).

Suppose also that the interest is in the level of disturbance rejection. Then one means of studying this is to examine the 2D frequency response between the disturbance and pass profile vector for the controlled process using the 2D transfer-function matrix description. The resulting 2D frequency responses are shown in Fig. 8.3 (first entry on the left and the second on the right). The maximum values are 1.4746 and 1.2013 respectively, which are both below the computed H_∞ attenuation level.

Consider again the state-space model of Example 1.2.10 (or (8.5) with the disturbance terms removed) and suppose that the defining matrices are subject to additive uncertainty. Then to design a dynamic pass profile controller for this case the starting point is to write the state-space model in the following form

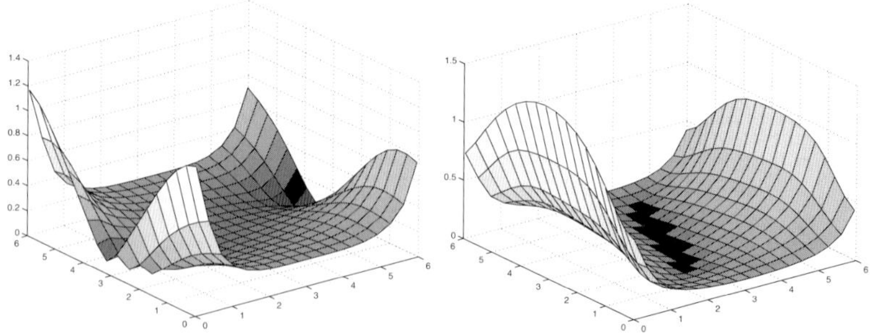

Fig. 8.3. The frequency responses for the controlled process

$$\begin{aligned}\begin{bmatrix} x_{k+1}(p+1) \\ y_{k+1}(p) \end{bmatrix} &= \left(\begin{bmatrix} A & B_0 \\ C & D_0 \end{bmatrix} + \begin{bmatrix} \Delta A & \Delta B_0 \\ \Delta C & \Delta D_0 \end{bmatrix} \right) \begin{bmatrix} x_{k+1}(p) \\ y_k(p) \end{bmatrix} \\ &+ \left(\begin{bmatrix} B \\ D \end{bmatrix} + \begin{bmatrix} \Delta B \\ \Delta D \end{bmatrix} \right) u_{k+1}(p) \end{aligned} \tag{8.30}$$

Also assume that the uncertainty here satisfies (first introduced in Sect. 8.1.1)

$$\begin{bmatrix} \Delta A & \Delta B_0 & \Delta B \\ \Delta C & \Delta D_0 & \Delta D \end{bmatrix} = \begin{bmatrix} H_1 \\ H_2 \end{bmatrix} \gamma^{-1} \mathcal{F} \begin{bmatrix} E_1 & E_2 & E_3 \end{bmatrix} \tag{8.31}$$

where H_1, H_2, E_1, E_2, E_3 are known and constant matrices, the scalar $\gamma > 0$ is given, and the matrix $\mathcal{F}$ satisfies (8.15).

Application of the dynamic pass profile controller gives a controlled process state-space model which can be written as

$$\begin{aligned}\begin{bmatrix} \overline{x}_{k+1}(p+1) \\ \overline{y}_{k+1}(p) \end{bmatrix} &= (\overline{A}_c + \overline{\Delta A}_c) \begin{bmatrix} \overline{x}_{k+1}(p) \\ \overline{y}_k(p) \end{bmatrix} \\ y_{k+1}(p) &= \overline{C}_c \begin{bmatrix} \overline{x}_{k+1}(p) \\ \overline{y}_k(p) \end{bmatrix} \end{aligned} \tag{8.32}$$

Here $\overline{x}_{k+1}(p)$ and $\overline{y}_k(p)$ (and Π used below) are as in the previous theorem and

$$\begin{aligned}\overline{A}_c + \overline{\Delta A}_c &:= \Pi \begin{bmatrix} \Phi + B_2 D_c C_2 & B_2 C_c \\ B_c C_2 & A_c \end{bmatrix} \Pi \\ &+ \Pi \begin{bmatrix} \Delta\Phi + \Delta\Psi_{ic} D_c C_2 & \Delta\Psi_{ic} C_c \\ 0 & 0 \end{bmatrix} \Pi \\ &= \Pi \begin{bmatrix} \Phi + B_2 D_c C_2 & B_2 C_c \\ B_c C_2 & A_c \end{bmatrix} \Pi \\ &+ \Pi \begin{bmatrix} \gamma^{-1} H \\ 0 \end{bmatrix} \mathcal{F} \begin{bmatrix} E & \\ +E_3 D_c C_2 & E_3 C_c \end{bmatrix} \Pi \\ &=: \overline{A}_c + \overline{H}\mathcal{F}\overline{E} \end{aligned}$$

where B_2 and C_2 are defined as in Theorem 8.2.1, and (as in Sect. 8.1.1)

$$H := \begin{bmatrix} H_1 \\ H_2 \end{bmatrix}, \quad E := \begin{bmatrix} E_1 & E_2 \end{bmatrix}$$

Theorem 8.2.6. *Consider a discrete unit memory linear repetitive process of the form given by (8.30) with uncertainty structure satisfying (8.31) and (8.15). Suppose also that a dynamic pass profile controller defined by (8.20) is applied to this process with resulting state-space model (8.32). Then this last process is stable along the pass if ∃ matrices $P_{11} > 0$, ($P_{11} := \text{diag}\{P_{h11}, P_{v11}\}$) and $R_{11} > 0$, ($R_{11} := \text{diag}\{R_{h11}, R_{v11}\}$) such that the following LMIs hold*

$$\begin{bmatrix} \mathcal{N}_1 & 0 & 0 \\ 0 & I & 0 \\ 0 & 0 & I \end{bmatrix}^T \begin{bmatrix} \Phi^T P_{11}\Phi - P_{11} & \Phi^T P_{11} H & E^T \\ H^T P_{11}\Phi & H^T P_{11} H - \gamma^2 I & 0 \\ E & 0 & -I \end{bmatrix} \begin{bmatrix} \mathcal{N}_1 & 0 & 0 \\ 0 & I & 0 \\ 0 & 0 & I \end{bmatrix} < 0 \tag{8.33}$$

$$\begin{bmatrix} \mathcal{N}_2 & 0 & 0 \\ 0 & I & 0 \\ 0 & 0 & I \end{bmatrix}^T \begin{bmatrix} \Phi R_{11}\Phi^T - R_{11} & \Phi R_{11} E^T & H \\ E R_{11}\Phi^T & -I + E R_{11} E^T & 0 \\ H^T & 0 & -\gamma^2 I \end{bmatrix} \begin{bmatrix} \mathcal{N}_2 & 0 & 0 \\ 0 & I & 0 \\ 0 & 0 & I \end{bmatrix} < 0 \tag{8.34}$$

$$\begin{bmatrix} P_{h11} & I \\ I & R_{h11} \end{bmatrix} \geq 0, \quad \begin{bmatrix} P_{v11} & I \\ I & R_{v11} \end{bmatrix} \geq 0 \tag{8.35}$$

where $\mathcal{N}_1$ *and* $\mathcal{N}_2$ *are full column rank matrices whose images satisfy*

$$\mathrm{Im}\mathcal{N}_1 = \ker(C_2^T), \quad \mathrm{Im}\mathcal{N}_2 = \ker([B_2^T \, E_3^T])$$

Proof. Based on the proof of Theorem 8.2.4 it can be shown that the stability along the pass condition in this case has the form

$$\begin{bmatrix} -\overline{P} & \overline{P}\,\overline{A}_c & \overline{P}\,\overline{H} & 0 \\ \overline{A}_c^T\overline{P} & -\overline{P} & 0 & \overline{E}^T \\ \overline{H}^T\overline{P} & 0 & -\gamma^2 I & 0 \\ 0 & \overline{E} & 0 & -I \end{bmatrix} < 0$$

Finally, apply similar transformations to those used in the proof of the previous theorem to obtain (8.33)–(8.35). □

To increase robustness, the term γ in the LMIs of (8.33)–(8.34) has to be minimized. This can be achieved by using the linear objective minimization procedure

$$\min_{P_{11}>0, R_{11}>0} \quad \mu$$
$$\text{subject to (8.33)} - \text{(8.35) with } \mu = \gamma^2$$

8.3 Guaranteed Cost Control of Discrete Processes

Many applications will require a control law (or controller) which not only guarantees stability along the pass in the presence of uncertainty but also ensures that the resulting controlled process satisfies specified performance criteria. Here we give a comprehensive treatment of one aspect of this general problem for discrete unit memory linear repetitive processes and, in particular, those described by (8.30) and associated uncertainty structure.

The problem is to obtain a control law which results in a controlled process which is stable along the pass with an associated cost function

$$J := \sum_{k=0}^{\infty}\sum_{p=0}^{\infty}\left(u_{k+1}^T(p)\Psi u_{k+1}(p)\right) + \sum_{k=0}^{\infty}\sum_{p=0}^{\infty}\left(\begin{bmatrix} x_{k+1}(p) \\ y_k(p) \end{bmatrix}^T \begin{bmatrix} Q_1 & 0 \\ 0 & Q_2 \end{bmatrix} \begin{bmatrix} x_{k+1}(p) \\ y_k(p) \end{bmatrix}\right) \tag{8.36}$$

where $\Psi > 0$, $Q_1 > 0$ and $Q_2 > 0$ are design matrices to be specified, which is bounded for all admissible uncertainties. In physical terms this cost function can be interpreted as the sum of quadratic costs on the input, state and pass profile vectors on each pass. (See also the last section of Chap. 7.)

Note 8.3.1. Repetitive processes are defined over the finite pass length α, and in practice the process will only complete a finite number of passes, say, N. Hence the corresponding cost function should be evaluated as

$$J' := \sum_{k=0}^{N}\sum_{p=0}^{\alpha}\left(u_{k+1}^T(p)\Psi u_{k+1}(p)\right) + \sum_{k=0}^{N}\sum_{p=0}^{\alpha}\left(\begin{bmatrix} x_{k+1}(p) \\ y_k(p) \end{bmatrix}^T \begin{bmatrix} Q_1 & 0 \\ 0 & Q_2 \end{bmatrix} \begin{bmatrix} x_{k+1}(p) \\ y_k(p) \end{bmatrix}\right)$$

However, as discussed in Chap. 1 it is routine to argue that the signals involved can be extended from $[0, \alpha]$ to the infinite interval in such a way that projection of the infinite interval solution onto the finite interval is possible. Likewise from the infinite set to $[0, N]$, and hence we will work with (8.36). The analysis which follows is from [133].

We start by developing the LMI condition which guarantees that the unforced (the control input terms are deleted) process is stable along the pass with an associated cost function which is bounded for all admissible uncertainties. These results are then extended to design a guaranteed cost control law.

8.3.1 Guaranteed Cost Bound

The process state-space model is that of (8.30) with the input terms deleted, i.e.

$$\begin{bmatrix} x_{k+1}(p+1) \\ y_{k+1}(p) \end{bmatrix} = \left(\begin{bmatrix} A & B_0 \\ C & D_0 \end{bmatrix} + \begin{bmatrix} \Delta A & \Delta B_0 \\ \Delta C & \Delta D_0 \end{bmatrix}\right) \begin{bmatrix} x_{k+1}(p) \\ y_k(p) \end{bmatrix} \tag{8.37}$$

and the associated cost function is

$$J_0 := \sum_{k=0}^{\infty}\sum_{p=0}^{\infty}\left(\begin{bmatrix} x_{k+1}(p) \\ y_k(p) \end{bmatrix}^T \begin{bmatrix} Q_1 & 0 \\ 0 & Q_2 \end{bmatrix} \begin{bmatrix} x_{k+1}(p) \\ y_k(p) \end{bmatrix}\right) \tag{8.38}$$

Now we have the following result which gives a sufficient condition for stability along the pass with guaranteed cost.

Theorem 8.3.1. *An unforced discrete unit memory linear repetitive process described by (8.37) with uncertainty structure satisfying (8.31) and (8.15) is stable along the pass if $\exists$ a scalar $\epsilon > 0$ and matrices $P_1 > 0$ and $P_2 > 0$ such that the following LMI holds*

$$\begin{bmatrix} -P_1 & 0 & P_1 A & P_1 B_0 & P_1 H_1 & P_1 H_1 \\ 0 & -P_2 & P_2 C & P_2 D_0 & P_2 H_2 & P_2 H_2 \\ A^T P_1 & C^T P_2 & Q_1 - P_1 + \epsilon E_1^T E_1 & 0 & 0 & 0 \\ B_0^T P_1 & D_0^T P_2 & 0 & Q_2 - P_2 + \epsilon E_2^T E_2 & 0 & 0 \\ H_1^T P_1 & H_2^T P_2 & 0 & 0 & -\epsilon I & 0 \\ H_1^T P_1 & H_2^T P_2 & 0 & 0 & 0 & -\epsilon I \end{bmatrix} < 0 \tag{8.39}$$

If this condition holds, the cost function (8.38) satisfies the upper bound

$$J_0 < \sum_{k=1}^{\infty} x_{k+1}(0) P_1 x_{k+1}(0) + \sum_{p=0}^{\infty} y_0^T(p) P_2 y_0(p) \tag{8.40}$$

Proof. Recall the definitions of the matrices $\widehat{A}_1$ and $\widehat{A}_2$ (first given in (2.66)) and $\widehat{\Delta A_1}$ and $\widehat{\Delta A_2}$ (see (8.1)) and also (first used in the proof of Theorem 8.2.1)

$$\zeta_k(p) = \begin{bmatrix} x_{k+1}^T(p) & y_k^T(p) \end{bmatrix}^T$$

Now rewrite (8.37) as

$$\begin{bmatrix} x_{k+1}(p+1) \\ y_{k+1}(p) \end{bmatrix} = \left((\widehat{A}_1 + \widehat{\Delta A_1}) + (\widehat{A}_2 + \widehat{\Delta A_2}) \right) \zeta_k(p)$$

and evaluate the Lyapunov function of (8.7) for the process state-space model considered here to obtain

$$\begin{aligned} \Delta V(k,t) &= \zeta_k^T(p) \Big((\widehat{A}_1 + \widehat{\Delta A_1})^T P (\widehat{A}_1 + \widehat{\Delta A_1}) \\ &+ (\widehat{A}_2 + \widehat{\Delta A_2}) P (\widehat{A}_2 + \widehat{\Delta A_2}) - P \Big) \zeta_k(p) \end{aligned}$$

where $P := \mathrm{diag}\{P_1, P_2\}$. Then stability along the pass holds if $\Delta V(k,p) < 0$ and also the inequality

$$\Delta V(k,p) + \zeta_k^T(p) Q \zeta_k(p) < 0$$

implies that (8.37) is stable along the pass where $Q := \mathrm{diag}\{Q_1, Q_2\} > 0$. Hence the following condition must hold

$$(\widehat{A}_1 + \widehat{\Delta A_1})^T P (\widehat{A}_1 + \widehat{\Delta A_1}) + (\widehat{A}_2 + \widehat{\Delta A_2}) P (\widehat{A}_2 + \widehat{\Delta A_2}) - P + Q < 0 \tag{8.41}$$

Next, an obvious application of, in turn, the Schur's complement formula and Lemma 5.4.2 to (8.41) yields

$$\begin{bmatrix} -P_1 & 0 & P_1A & P_1B_0 \\ 0 & -P_2 & P_2C & P_2D_0 \\ A^TP_1 & C^TP_2 & Q_1 - P_1 + \epsilon E_1^TE_1 & 0 \\ B_0^TP_1 & D_0^TP_2 & 0 & Q_2 - P_2 + \epsilon E_2^TE_2 \end{bmatrix}$$

$$+\epsilon^{-1}\begin{bmatrix} 0 & 0 & P_1H_1 & P_1H_1 \\ 0 & 0 & P_2H_2 & P_2H_2 \\ 0 & 0 & 0 & 0 \\ 0 & 0 & 0 & 0 \end{bmatrix}\begin{bmatrix} 0 & 0 & 0 & 0 \\ 0 & 0 & 0 & 0 \\ H_1^TP_1 & H_2^TP_2 & 0 & 0 \\ H_1^TP_1 & H_2^TP_2 & 0 & 0 \end{bmatrix} < 0$$

and use of the Schur's complement formula now gives (8.39). Finally, introduce

$$\Upsilon_{sd} := \sum_{k=0}^{\infty}\sum_{p=0}^{\infty}\zeta_k^T(p)Q\zeta_k(p)$$

then

$$\begin{aligned} \Upsilon_{sd} &< -\sum_{k=0}^{\infty}\left(\sum_{p=0}^{\infty}x_{k+1}(p+1)^TP_1x_{k+1}(p+1) - x_{k+1}^T(p)P_1x_{k+1}(p)\right) \\ &- \sum_{k=0}^{\infty}\left(\sum_{p=0}^{\infty}y_{k+1}^T(p)P_2y_{k+1}(p) - y_k^T(p)P_2y_k(p)\right) \\ &= \sum_{k=0}^{\infty}x_{k+1}^T(0)P_1x_{k+1}(0) + \sum_{p=0}^{\infty}y_0^T(p)P_2y_0(p) \end{aligned}$$

which ensures that (8.40) holds and the proof is complete. □

Note that it is possible to minimize the upper bound on the cost function (8.40) using the following optimization procedure

$$\begin{aligned} &\min_{P_1>0,P_2>0}\left[\sum_{k=0}^{\infty}x_{k+1}^T(0)P_1x_{k+1}(0) + \sum_{p=0}^{\infty}y_0^T(p)P_2y_0(p)\right] \\ =&\min_{P_1>0,P_2>0}\left[\sum_{k=0}^{\infty}\operatorname{trace}\left(P_1x_{k+1}(0)x_{k+1}^T(0)\right) + \operatorname{trace}\left(P_2\sum_{p=0}^{\infty}y_0(p)y_0^T(p)\right)\right] \end{aligned}$$
subject to (8.39)

8.3.2 Guaranteed Cost Control

Here it is assumed that all elements in the current pass state vector can be measured and hence a control law of the form (8.3) can be used. Consider again a discrete unit memory linear repetitive process described by (8.30) where the uncertainty in the state-space model is assumed to satisfy

$$\begin{bmatrix} \Delta A & \Delta B_0 & \Delta B \\ \Delta C & \Delta D_0 & \Delta D \end{bmatrix} = \begin{bmatrix} H_1 \\ H_2 \end{bmatrix}\mathcal{F}\begin{bmatrix} E_1 & E_2 & E_3 \end{bmatrix} \tag{8.42}$$

where $\mathcal{F}$ satisfies (8.15), i.e. (8.14) with the terms arising from the disturbance vector deleted from both sides of this equation. Then the associated cost function for the process resulting from the application of (8.3) to (8.30) is

$$J_{cp} := \sum_{k=0}^{\infty}\sum_{p=0}^{\infty}\left(\begin{bmatrix} x_{k+1}(p) \\ y_k(p) \end{bmatrix}^T \begin{bmatrix} Q_1 + K_1^T \Psi K_1 & K_1^T \Psi K_2 \\ K_2^T \Psi K_1 & Q_2 + K_2^T \Psi K_2 \end{bmatrix} \begin{bmatrix} x_{k+1}(p) \\ y_k(p) \end{bmatrix}\right) \tag{8.43}$$

which is of the form of that in Theorem 8.3.1 and we have the following result.

Theorem 8.3.2. *Suppose that a control law of the form (8.3) is applied to a discrete unit memory linear repetitive process described by (8.30) with uncertainty structure satisfying (8.42) and (8.15). Then the resulting process is stable along the pass if* $\exists$ *a scalar* $\epsilon > 0$ *and matrices* $W_1 > 0$, $W_2 > 0$, N_1 *and* N_2 *such that the following LMI holds*

$$\begin{bmatrix} -W_1 + 2\epsilon H_1 H_1^T & (\star) & (\star) & (\star) & (\star) & (\star) & (\star) & (\star) & (\star) \\ 2\epsilon H_1 H_2^T & -W_2 + 2\epsilon H_2 H_2^T & (\star) & (\star) & (\star) & (\star) & (\star) & (\star) & (\star) \\ W_1 A^T + N_1^T B^T & W_1 C^T + N_1^T D^T & -W_1 & (\star) & (\star) & (\star) & (\star) & (\star) & (\star) \\ W_2 B_0^T + N_2^T B^T & W_2 D_0^T + N_2 D^T & 0 & -W_2 & (\star) & (\star) & (\star) & (\star) & (\star) \\ 0 & 0 & E_1 W_1 + E_3 N_1 & 0 & -\epsilon I & (\star) & (\star) & (\star) & (\star) \\ 0 & 0 & 0 & E_2 W_2 + E_3 N_2 & 0 & -\epsilon I & (\star) & (\star) & (\star) \\ 0 & 0 & N_1 & N_2 & 0 & 0 & -\Psi^{-1} & (\star) & (\star) \\ 0 & 0 & W_1 & 0 & 0 & 0 & 0 & -Q_1^{-1} & (\star) \\ 0 & 0 & 0 & W_2 & 0 & 0 & 0 & 0 & -Q_2^{-1} \end{bmatrix} < 0 \tag{8.44}$$

If this condition holds, stabilizing control law matrices K_1, K_2 *are again given by (8.11) and the cost function (8.43) satisfies the following upper bound*

$$J_{cp} < \sum_{k=0}^{\infty} x_{k+1}^T(0) W_1^{-1} x_{k+1}(0) + \sum_{p=0}^{\infty} y_0^T(p) W_2^{-1} y_0(p) \tag{8.45}$$

Proof. Based on (8.39) we conclude that the controlled process is stable along the pass if the following matrix inequality is satisfied

$$\left[\begin{array}{cccc} -P_1 & 0 & P_1A+P_1BK_1 & P_1B_0+P_1BK_2 \\ 0 & -P_2 & P_2C+P_2DK_1 & P_2D_0+P_2DK_2 \\ A^TP_1+K_1^TB^TP_1 & C^TP_2+K_1^TD^TP_2 & Q_1-P_1+K_1^T\Psi K_1 & K_1^T\Psi K_2 \\ B_0^TP_1+K_2^TB^TP_1 & D_0^TP_2+K_2D^TP_1 & K_2^T\Psi K_1 & Q_2-P_2+K_2^T\Psi K_2 \end{array}\right]$$

$$+\begin{bmatrix} 0 & 0 & 0 & 0 \\ 0 & 0 & 0 & 0 \\ 0 & 0 & E_1^T+K_1^TE_3^T & 0 \\ 0 & 0 & 0 & E_2^T+K_2^TE_3^T \end{bmatrix}\begin{bmatrix} \mathcal{F}^T & 0 & 0 & 0 \\ 0 & \mathcal{F}^T & 0 & 0 \\ 0 & 0 & \mathcal{F}^T & 0 \\ 0 & 0 & 0 & \mathcal{F}^T \end{bmatrix}\begin{bmatrix} 0 & 0 & 0 & 0 \\ 0 & 0 & 0 & 0 \\ H_1^TP_1 & H_2^TP_2 & 0 & 0 \\ H_1^TP_1 & H_2^TP_2 & 0 & 0 \end{bmatrix}$$

$$+\begin{bmatrix} 0 & 0 & P_1H_1 & P_1H_1 \\ 0 & 0 & P_2H_2 & P_2H_2 \\ 0 & 0 & 0 & 0 \\ 0 & 0 & 0 & 0 \end{bmatrix}\begin{bmatrix} \mathcal{F} & 0 & 0 & 0 \\ 0 & \mathcal{F} & 0 & 0 \\ 0 & 0 & \mathcal{F} & 0 \\ 0 & 0 & 0 & \mathcal{F} \end{bmatrix}\begin{bmatrix} 0 & 0 & 0 & 0 \\ 0 & 0 & 0 & 0 \\ 0 & 0 & E_1+E_3K_1 & 0 \\ 0 & 0 & 0 & E_2+E_3K_2 \end{bmatrix}<0$$

Now set $W_1 = P_1^{-1}$, $W_2 = P_2^{-1}$, $U_1 = W_1Q_1W_1$ and $U_2 = W_2Q_2W_2$ and then pre- and post- multiply both sides of this last inequality by $\mathrm{diag}\{W_1, W_2, W_1, W_2\}$. Next, by an obvious application of the result of Lemma 5.4.2 we obtain

$$\left[\begin{array}{cccc} -W_1+2\epsilon H_1H_1^T & 2\epsilon H_2H_1^T & AW_1+BN_1 & B_0W_2+BN_2 \\ 2\epsilon H_1H_2^T & -W_2+2\epsilon H_2H_2^T & CW_1+DN_1 & D_0W_2+DN_2 \\ W_1A^T+N_1^TB^T & W_1C^T+N_1^TD^T & U_1-W_1+N_1^T\Psi N_1 & N_1^T\Psi N_2 \\ W_2B_0^T+N_2^TB^T & W_2D_0^T+N_2D^T & N_2^T\Psi N_1 & U_2-W_2+N_2^T\Psi N_2 \end{array}\right]+\epsilon^{-1}\begin{bmatrix} 0 & 0 & 0 & 0 \\ 0 & 0 & 0 & 0 \\ 0 & 0 & W_1E_1^T+N_1^TE_3^T & 0 \\ 0 & 0 & 0 & W_2E_2^T+N_2^TE_3^T \end{bmatrix}$$

$$\begin{bmatrix} 0 & 0 & 0 & 0 \\ 0 & 0 & 0 & 0 \\ 0 & 0 & E_1W_1+E_3N_1 & 0 \\ 0 & 0 & 0 & E_2W_2+E_3N_2 \end{bmatrix}<0$$

where $N_1 = K_1W_1$ and $N_2 = K_2W_2$. Making an obvious application of the Schur's complement formula now gives (8.44). Finally, the bound on the cost function can be established in an identical manner to that for its counterpart in the previous theorem and the proof is complete. □

Note 8.3.2. The presence of the nonlinear terms W_1^{-1} and W_2^{-1} in (8.45) means that it is not possible to apply a linear objective minimization procedure to minimize this cost function. However, a control law which minimizes the guaranteed cost can be achieved as follows. First note that

$$\sum_{k=0}^{N} x_{k+1}^T(0) W_1^{-1} x_{k+1}(0) = \sum_{k=0}^{N} \text{trace}\left(x_{k+1}^T(0) W_1^{-1} x_{k+1}(0)\right)$$
$$= \sum_{k=0}^{N} \text{trace}\left(W_1^{-1} x_{k+1}(0) x_{k+1}^T(0)\right)$$

and

$$\sum_{p=0}^{\alpha} y_0^T(p) W_2^{-1} y_0(p) = \sum_{p=0}^{\alpha} \text{trace}\left(y_0^T(p) W_2^{-1} y_0(p)\right)$$
$$= \sum_{p=0}^{\alpha} \text{trace}\left(W_2^{-1} y_0(p) y_0^T(p)\right)$$

Next, recall that the eigenvalue decomposition of a matrix $\widetilde{M} \geq 0$ can be written as $\widetilde{M} = V \Xi V^T$, where V is some unitary matrix and Ξ is a diagonal matrix with nonnegative diagonal entries. Therefore, the matrix square root of $\widetilde{M}$ can be defined as $\widetilde{M}^{\frac{1}{2}} = V \Xi^{\frac{1}{2}} V^T$ and computed in a well conditioned manner [76]. Based on this, the matrices $\Sigma_1^{\frac{1}{2}}$ and $\Sigma_2^{\frac{1}{2}}$ can be obtained as

$$\Sigma_1 = \Sigma_1^{\frac{1}{2}} \Sigma_1^{\frac{1}{2}} = \sum_{k=0}^{N} x_{k+1}^T(0) x_{k+1}(0), \quad \Sigma_2 = \Sigma_2^{\frac{1}{2}} \Sigma_2^{\frac{1}{2}} = \sum_{p=0}^{\alpha} y_0^T(p) y_0(p)$$

Also, introduce the symmetric matrices Ω_1 and Ω_2 which satisfy

$$\text{trace}\,(\Sigma_1^{\frac{1}{2}} W_1^{-1} \Sigma_1^{\frac{1}{2}}) < \text{trace}\,\Omega_1 \;\text{ and }\; \text{trace}\,(\Sigma_2^{\frac{1}{2}} W_2^{-1} \Sigma_2^{\frac{1}{2}}) < \text{trace}\,\Omega_2$$

respectively and hence we can write

$$\Sigma_1^{\frac{1}{2}} W_1^{-1} \Sigma_1^{\frac{1}{2}} < \Omega_1, \quad \Sigma_2^{\frac{1}{2}} W_2^{-1} \Sigma_2^{\frac{1}{2}} < \Omega_2$$

Application of the Schur's complement formula now gives

$$\begin{bmatrix} -\Omega_1 & \Sigma_1^{\frac{1}{2}} \\ \Sigma_1^{\frac{1}{2}} & -W_1 \end{bmatrix} < 0 \;\text{ and }\; \begin{bmatrix} -\Omega_2 & \Sigma_2^{\frac{1}{2}} \\ \Sigma_2^{\frac{1}{2}} & -W_2 \end{bmatrix} < 0 \tag{8.46}$$

Finally, the following minimization problem can be formulated

$$\min_{W_1>0, W_2>0, N_1, N_2} \quad \text{trace}(\Omega_1 + \Omega_2)$$
$$\text{subject to (8.44) and (8.46)}$$

which gives a control law for which the cost function is minimized.

8.4 H_∞ Control of Differential Processes

We will use the following signal space.

Definition 8.4.1. *Consider the sequence* $\{w_j(t)\}$ *whose elements are* $q \times 1$ *vectors* $w_j(t)$ *defined over the real interval* $0 \le t \le \infty$ *and the nonnegative integers* $0 \le j \le \infty$, *which is written as* $\{[0,\infty],[0,\infty]\}$ *Then the* L_2 *norm of this vector sequence is given by*

$$\|w\|_2 := \sqrt{\sum_{j=0}^{\infty} \int_0^{\infty} w_j^T(t) w_j(t)\, dt}$$

and this sequence is said to be a member of $L_2^q\{[0,\infty],[0,\infty]\}$, *or* L_2^q *for short, if* $||w||_2 < \infty$.

The process state-space model considered has the following form over $0 \le t \le \alpha,\ k \ge 0$

$$\begin{aligned} \dot{x}_{k+1}(t) &= Ax_{k+1}(t) + Bu_{k+1}(t) + B_0 y_k(t) + B_{11} w_{k+1}(t) \\ y_{k+1}(t) &= Cx_{k+1}(t) + Du_{k+1}(t) + D_0 y_k(t) + B_{12} w_{k+1}(t) \end{aligned} \tag{8.47}$$

where $w_{k+1}(t)$ is a $q \times 1$ disturbance vector which belongs to L_2^q i.e. the model of Example 1.2.3 with disturbance terms added to the state and pass profile vectors. As per the discrete case of Sect. 8.2, it is easy to see that stability along the pass holds under the same conditions as that for the disturbance free case.

Disturbance attenuation is defined as follows.

Definition 8.4.2. *A differential unit memory linear repetitive process described by (8.47) is said to have* H_∞ *disturbance attenuation* $\gamma > 0$ *if it is stable along the pass and the induced* L_2 *norm from* w *to* y *is bounded by* γ *i.e.*

$$\sup_{0 \ne w \in L_2^q} \frac{\|y\|_2}{\|w\|_2} < \gamma \tag{8.48}$$

Here we will also use the sufficient condition for stability along the pass of Lemma 8.4.1 below which is based on the following Lyapunov function

$$V(k,t) := V_1(k,t) + V_2(k,t) = x_{k+1}^T(t) P_1 x_{k+1}(t) + y_k^T(t) P_2 y_k(t) \tag{8.49}$$

where $P_1 > 0$ and $P_2 > 0$. The associated increment is

$$\Delta V(k,t) = \dot{V}_1(k,t) + \Delta V_2(k,t) \tag{8.50}$$

where

$$\begin{aligned} \dot{V}_1(k,t) &= \dot{x}_{k+1}^T(t) P_1 x_{k+1}(t) + x_{k+1}^T P_1 \dot{x}_{k+1}(t) \\ \Delta V_2(k,t) &= y_{k+1}^T(t) P_2 y_{k+1}(t) - y_k^T(t) P_2 y_k(t) \end{aligned}$$

Then we have the following result (the counterpart of Lemma 8.2.1).

Lemma 8.4.1. *A differential unit memory linear repetitive process described by (8.47) is stable along the pass if*

$$\Delta V(k,t) < 0 \tag{8.51}$$

Proof. Substituting from (8.47) (with $w_{k+1}(t) = 0$) into (8.50) gives

$$\begin{aligned} \Delta V(k,t) &= \xi_k^T(t)\left(\widehat{A}_1^T P + P\widehat{A}_1 + \widehat{A}_2^T R\widehat{A}_2 - R\right)\xi_k(t) \\ &=: \xi_k^T(t) S_{2D}\xi_k(t) \end{aligned} \tag{8.52}$$

where $P := \text{diag}\{P_1, 0\}$, $R := \text{diag}\{0, P_2\}$, $\xi_k(t) := \begin{bmatrix} x_{k+1}^T(t) & y_k^T(t) \end{bmatrix}^T$ and $\widehat{A}_1$ and $\widehat{A}_2$ are defined as in the discrete case (see also (2.66)). It is now routine to conclude (see [75]) that stability along the pass holds if $\Delta V(k,t) < 0$. (This is based on the fact the matrix S_{2D} in (8.52) is the 2D Lyapunov equation (first introduced in Sect. 4.2) for these processes and stability along the pass holds if $S_{2D} < 0$.)

The following result gives an H_∞ condition for stability along the pass in terms of an LMI.

Theorem 8.4.1. *A differential unit memory linear repetitive process described by (8.47) is stable along the pass and has H_∞ disturbance attenuation $\gamma > 0$ if $\exists$ matrices $P_1 > 0$ and $P_2 > 0$ such that the following LMI holds*

$$\begin{bmatrix} -P_2 & P_2 C & P_2 D_0 & P_2 B_{12} \\ C^T P_2 & A^T P_1 + P_1 A & P_1 B_0 & P_1 B_{11} \\ D_0^T P_2 & B_0^T P_1 & -P_2 + I & 0 \\ B_{12}^T P_2 & B_{11}^T P_1 & 0 & -\gamma^2 I \end{bmatrix} < 0 \tag{8.53}$$

Proof. It is easily shown that the H_∞ disturbance attenuation γ holds if the associated Hamiltonian

$$\begin{aligned} H(k,t) &:= \Delta V(k,t) + y_{k+1}^T(t) y_{k+1}(t) - \gamma^2 w_{k+1}^T(t) w_{k+1}(t) \\ &= \Delta V(k,t) + \xi_k^T(t)\overline{L}^T\overline{L}\xi_k(t) - \gamma^2 w_{k+1}^T(t) w_{k+1}(t) \\ &=: \begin{bmatrix} \xi_k^T(t) & w_{k+1}^T(t) \end{bmatrix} \Theta \begin{bmatrix} \xi_k(t) \\ w_{k+1}(t) \end{bmatrix} \end{aligned}$$

satisfies

$$H(k,t) < 0$$

Here $\Delta V(k,t)$ is given by (8.52) and

$$\Theta := \begin{bmatrix} \widehat{A}_1^T P + P\widehat{A}_1 + \widehat{A}_2^T S\widehat{A}_2 + \overline{L}^T\overline{L} - R & P\widehat{B}_1 + \widehat{A}_2^T \overline{S}\widehat{D}_1 \\ \widehat{B}_1^T P + \widehat{D}_1^T \overline{S}\widehat{A}_2 & \widehat{D}_1^T \overline{S}\widehat{D}_1 - \gamma^2 I \end{bmatrix}$$

where P is defined in the proof of Lemma 8.4.1, $\overline{S} := \text{diag}\{P_3, P_2\}$, and

$$\widehat{B}_1 := \begin{bmatrix} B_{11} \\ 0 \end{bmatrix}, \ \widehat{D}_1 := \begin{bmatrix} 0 \\ B_{12} \end{bmatrix}, \ \overline{L} := \begin{bmatrix} 0 & I \end{bmatrix}$$

Finally, an obvious application of the Schur's complement formula gives (8.53) and the proof is complete. □

In many cases of practical interest it will be necessary to compute the minimum disturbance rejection level γ. This can be obtained using the following linear objective minimization procedure

$$\min_{P_1>0, P_2>0} \quad \mu$$
$$\text{subject to (8.53) with } \mu = \gamma^2$$

Next we begin the development of an H_∞ based approach to the design of control laws for both stabilization and performance, where this last result provides the computational basis.

8.4.1 H_∞ Control with a Static Control Law

The result below is from [132] and solves the H_∞ control problem using a control law of the form (7.8), i.e.

$$u_{k+1}(t) = K_1 x_{k+1}(t) + K_2 y_k(t) =: \widetilde{K} \begin{bmatrix} x_{k+1}(t) \\ y_k(t) \end{bmatrix} \tag{8.54}$$

Theorem 8.4.2. *Suppose that a control law of the form (8.54) is applied to a differential unit memory linear repetitive process described by (8.47). Then the resulting process is stable along the pass and has prescribed H_∞ disturbance attenuation $\gamma > 0$ if $\exists$ matrices $W_1 > 0$, $W_2 > 0$, N_1 and N_2 such that the following LMI holds*

$$\begin{bmatrix} -W_2 & (\star) & (\star) & (\star) & (\star) \\ W_1 C^T + N_1^T D^T & \Omega_1 & (\star) & (\star) & (\star) \\ W_2 D_0^T + N_2^T D^T & W_2 B_0^T + N_2^T B^T & -W_2 & (\star) & (\star) \\ B_{12}^T & B_{11}^T & 0 & -\gamma^2 I & (\star) \\ 0 & 0 & W_2 & 0 & -I \end{bmatrix} < 0 \tag{8.55}$$

where P and R are defined in the proof of Lemma 8.4.1 and

$$\Omega_1 := W_1 A^T + N_1^T B^T + A W_1 + B N_1$$

If this condition holds, stabilizing control law matrices are given by $K_1 = N_1 W_1^{-1}$ and $K_2 = N_2 W_2^{-1}$ respectively.

Proof. Interpreting Theorem 8.4.1 in terms of the controlled process state-space model gives stability along the pass with prescribed H_∞ disturbance attenuation $\gamma > 0$ if

$$\begin{bmatrix} -\overline{S} & \overline{S}\,\overline{A}_2 & \overline{S}\widehat{D}_1 \\ \overline{A}_2^T\overline{S} & \overline{A}_1^T P + P\overline{A}_1 + \overline{L}^T\overline{L} - R & P\widehat{B}_1 \\ \widehat{D}_1\overline{S} & \widehat{B}_1^T P & -\gamma^2 I \end{bmatrix} < 0 \tag{8.56}$$

and

$$\overline{A}_1 := \begin{bmatrix} A + BK_1 & B_0 + BK_2 \\ 0 & 0 \end{bmatrix}, \quad \overline{A}_2 := \begin{bmatrix} 0 & 0 \\ C + DK_1 & D_0 + DK_2 \end{bmatrix}$$

Now make an obvious application of the Schur's complement formula to yield

$$\begin{bmatrix} -\overline{S} & \overline{S}\,\overline{A}_2 & \overline{S}\widehat{D}_1 & 0 \\ \overline{A}_2^T\overline{S} & \overline{A}_1^T P + P\overline{A}_1 - R & P\widehat{B}_1 & \overline{L}^T \\ \widehat{D}_1^T\overline{S} & \widehat{B}_1^T P & -\gamma^2 I & 0 \\ 0 & \overline{L} & 0 & -I \end{bmatrix} < 0$$

Next, substitute the formulas given above for $\overline{A}_1$ and $\overline{A}_2$ into this last expression, pre- and post-multiply the result by $\text{diag}\{P_3^{-1}, P_2^{-1}, P_1^{-1}, P_2^{-1}, I, I\}$ and then set $W_1 = P_1^{-1}$, $W_2 = P_2^{-1}$, $W_3 = P_3^{-1}$, $N_1 = K_1P_1^{-1}$, $N_2 = K_2P_2^{-1}$ Finally, noting that the result does not depend on matrix W_3, we obtain (8.55) and the proof is complete. □

8.4.2 H_∞ Control of Uncertain Differential Processes

Consider a differential linear repetitive processes of the form defined by Example 1.2.10 with uncertainty modelled as additive perturbations to the nominal state-space model matrices and written as

$$\begin{aligned} \begin{bmatrix} \dot{x}_{k+1}(t) \\ y_{k+1}(t) \end{bmatrix} &= \left(\begin{bmatrix} A & B_0 \\ C & D_0 \end{bmatrix} + \begin{bmatrix} \Delta A & \Delta B_0 \\ \Delta C & \Delta D_0 \end{bmatrix} \right) \begin{bmatrix} x_{k+1}(t) \\ y_k(t) \end{bmatrix} \\ &+ \left(\begin{bmatrix} B \\ D \end{bmatrix} + \begin{bmatrix} \Delta B \\ \Delta D \end{bmatrix} \right) u_{k+1}(t) \\ &+ \left(\begin{bmatrix} B_{11} \\ B_{12} \end{bmatrix} + \begin{bmatrix} \Delta B_{11} \\ \Delta B_{12} \end{bmatrix} \right) w_{k+1}(t) \end{aligned} \tag{8.57}$$

where the matrices describing the admissible uncertainties are assumed to satisfy

$$\begin{bmatrix} \Delta A & \Delta B_0 & \Delta B & \Delta B_{11} \\ \Delta C & \Delta D_0 & \Delta D & \Delta B_{12} \end{bmatrix} = \begin{bmatrix} H_1 \\ H_2 \end{bmatrix} \mathcal{F} \begin{bmatrix} E_1 & E_2 & E_3 & E_4 \end{bmatrix} \tag{8.58}$$

Here H_1, H_2, E_1, E_2, E_3, E_4 are known constant matrices of compatible dimensions and $\mathcal{F}$ is an unknown constant matrix which is assumed to satisfy (8.15).

Now we have the following result.

Theorem 8.4.3. *Suppose that a control law of the form (8.54) is applied to a differential unit memory linear repetitive process described by (8.57) with uncertainty structure satisfying (8.58) and (8.15). Then the resulting process is stable along the pass and has H_∞ disturbance attenuation $\gamma > 0$ if $\exists$ a scalar $\epsilon > 0$ and matrices $W_1 > 0$, $W_2 > 0$, N_1 and N_2 such that the following LMI holds*

$$\begin{bmatrix} -W_2 + 3\epsilon H_2 H_2^T & (\star) & (\star) & (\star) & (\star) & (\star) & (\star) & (\star) \\ W_1 C^T + N_1^T D^T + 3\epsilon H_1 H_2^T & \Omega_2 & (\star) & (\star) & (\star) & (\star) & (\star) & (\star) \\ W_2 D_0^T + N_2^T D^T & W_2 B_0^T + N_2^T B^T & -W_2 & (\star) & (\star) & (\star) & (\star) & (\star) \\ B_{12}^T & B_{11}^T & 0 & -\gamma^2 I & (\star) & (\star) & (\star) & (\star) \\ 0 & 0 & W_2 & 0 & -I & (\star) & (\star) & (\star) \\ 0 & E_1 W_1 + E_3 N_1 & 0 & 0 & 0 & -\epsilon I & (\star) & (\star) \\ 0 & 0 & E_2 W_2 + E_3 N_2 & 0 & 0 & 0 & -\epsilon I & (\star) \\ 0 & 0 & 0 & E_4 & 0 & 0 & 0 & -\epsilon I \end{bmatrix} < 0 \tag{8.59}$$

where

$$\Omega_2 := W_1 A^T + N_1^T B^T + A W_1 + B N_1 + 3\epsilon H_1 H_1^T$$

If this condition holds, the control law matrices K_1 and K_2 are given by $N_1 W_1^{-1}$ and $N_2 W_2^{-1}$ respectively.

Proof. First, incorporate the norm-bounded uncertainties into (8.55) to obtain

$$\begin{bmatrix} 0 & (\star) & (\star) & (\star) & (\star) \\ W_1 \Delta C^T + N_1^T \Delta D^T & \Omega_3 & (\star) & (\star) & (\star) \\ W_2 \Delta D_0^T + N_2^T \Delta D^T & W_2 \Delta B_0^T + N_2^T \Delta B^T & 0 & (\star) & (\star) \\ \Delta B_{12}^T & \Delta B_{11}^T & 0 & 0 & (\star) \\ 0 & 0 & 0 & 0 & 0 \end{bmatrix}$$
$$+ \begin{bmatrix} -W_2 & (\star) & (\star) & (\star) & (\star) \\ W_1 C^T + N_1^T D^T & \Omega_4 & (\star) & (\star) & (\star) \\ W_2 D_0^T + N_2^T D^T & W_2 B_0^T + N_2^T B^T & -W_2 & (\star) & (\star) \\ B_{12}^T & B_{11}^T & 0 & -\gamma^2 I & (\star) \\ 0 & 0 & W_2 & 0 & -I \end{bmatrix} < 0$$

where

$$\begin{aligned} \Omega_3 &:= W_1 \Delta A^T + N_1^T \Delta B^T + \Delta A W_1 + \Delta B N_1 \\ \Omega_4 &:= W_1 A^T + N_1^T B^T + A W_1 + B N_1 \end{aligned}$$

The first term in the above inequality can be rewritten as

$$\overline{H}_d \, \overline{\mathcal{F}} \, \overline{E}_d + \overline{E}_d^T \overline{\mathcal{F}}^T \overline{H}_d^T$$

where

$$\begin{aligned} \overline{H}_d &:= \begin{bmatrix} 0 & H_2 & H_2 & H_2 & 0 \\ 0 & H_1 & H_1 & H_1 & 0 \\ 0 & 0 & 0 & 0 & 0 \\ 0 & 0 & 0 & 0 & 0 \\ 0 & 0 & 0 & 0 & 0 \end{bmatrix}, \quad \overline{\mathcal{F}} := \mathrm{diag}\{\mathcal{F}, \mathcal{F}, \mathcal{F}, \mathcal{F}, \mathcal{F}\} \\ \overline{E}_d &:= \mathrm{diag}\{0, E_1 W_1 + E_3 N_1, E_2 W_2 + E_3 N_2, E_4, 0\} \end{aligned}$$

Finally, an obvious application of the result of Lemma 5.4.2 followed by application of the Schur's complement formula yields (8.59). □

Example 8.4.1. Consider the case when

$$A = \begin{bmatrix} -0.0050 & -5.8077 \\ 1 & -0.0050 \end{bmatrix}, \; B_0 = \begin{bmatrix} 0 \\ 0.0494 \end{bmatrix}, \; C = \begin{bmatrix} 1 & 0 \end{bmatrix}, D_0 = 0.7692 \tag{8.60}$$

and

$$B_{11} = \begin{bmatrix} 1.3827 \\ 1.1073 \end{bmatrix}, \quad B_{12} = 0.2644$$

Then for the particular choice of $\gamma = 1.4$, (8.59) holds with

$$\begin{aligned} W_1 &= \begin{bmatrix} 16.9993 & 0.9563 \\ 0.9563 & 1.2216 \end{bmatrix}, \quad W_2 = 18.1714 \\ N_1 &= \begin{bmatrix} -11.0732 - 6.5747 \end{bmatrix}, \quad N_2 = -5.7055 \end{aligned}$$

and the corresponding control law matrices are

$$K_1 = \begin{bmatrix} -0.3647 - 5.0965 \end{bmatrix}, \; K_2 = -0.3140$$

This control law guarantees stability along the pass and ensures that the H_∞ norm bound is never greater than 1.4.

8.4.3 H_∞ Control with a Dynamic Pass Profile Controller

In the control law considered previously in this section, full access to the current state vector has been assumed. Here (recall the discrete case) we consider the problem of using a controller which is activated only by the current pass profile vector to guarantee stability along the pass of the controlled process together with a guaranteed H_∞ disturbance attenuation level.

The controller used here (the differential counterpart of (7.13) for the discrete case) has the following state-space model

$$\begin{aligned}\begin{bmatrix} \dot{x}^c_{k+1}(t) \\ y^c_{k+1}(t) \end{bmatrix} &= \begin{bmatrix} A_{c11} & A_{c12} \\ A_{c21} & A_{c22} \end{bmatrix} \begin{bmatrix} x^c_{k+1}(t) \\ y^c_k(t) \end{bmatrix} + \begin{bmatrix} B_{c1} \\ B_{c2} \end{bmatrix} y_{k+1}(t) \\ u_{k+1}(t) &= \begin{bmatrix} C_{c1} & C_{c2} \end{bmatrix} \begin{bmatrix} x^c_{k+1}(t) \\ y^c_k(t) \end{bmatrix} + D_c y_{k+1}(t)\end{aligned} \tag{8.61}$$

where $x^c_{k+1}(t)$ and $y^c_k(t)$ are state vectors for the controller.

Suppose therefore that this controller is applied to a process described by (8.47). Recall also the definition of the augmented plant matrix Φ (for the state-space model of Example 1.2.3)) and introduce the notation

$$\begin{aligned} B_1 &:= \begin{bmatrix} B_{11} \\ B_{12} \end{bmatrix}, \; B_2 := \begin{bmatrix} B \\ D \end{bmatrix}, \; C_2 := \begin{bmatrix} 0 & I \end{bmatrix} \\ A_c &:= \begin{bmatrix} A_{c11} & A_{c12} \\ A_{c21} & A_{c22} \end{bmatrix}, \; B_c := \begin{bmatrix} B_{c1} \\ B_{c2} \end{bmatrix}, \; C_c := \begin{bmatrix} C_{c1} & C_{c2} \end{bmatrix} \end{aligned}$$

and define so-called augmented state and pass profile vectors as

$$\dot{\overline{x}}_{k+1}(t) := \begin{bmatrix} \dot{x}_{k+1}(t) \\ \dot{x}^c_{k+1}(t) \end{bmatrix}, \; \overline{y}_k(t) := \begin{bmatrix} y_k(t) \\ y^c_k(t) \end{bmatrix}$$

Also recall the matrix Π from the discrete process analysis and introduce

$$\Pi_1 := \begin{bmatrix} I & 0 & 0 & 0 \\ 0 & 0 & I & 0 \\ 0 & 0 & 0 & 0 \\ 0 & 0 & 0 & 0 \end{bmatrix}, \; \Pi_2 := \begin{bmatrix} 0 & 0 & 0 & 0 \\ 0 & 0 & 0 & 0 \\ 0 & I & 0 & 0 \\ 0 & 0 & 0 & I \end{bmatrix}$$

Then the resulting process state-space model can be written as

$$\begin{aligned} \begin{bmatrix} \dot{\overline{x}}_{k+1}(t) \\ \overline{y}_{k+1}(t) \end{bmatrix} &= \overline{A} \begin{bmatrix} \overline{x}_{k+1}(t) \\ \overline{y}_k(t) \end{bmatrix} + \overline{B} w_{k+1}(t) \\ y_{k+1}(t) &= \overline{C} \begin{bmatrix} \overline{x}_{k+1}(t) \\ \overline{y}_{k+1}(t) \end{bmatrix} \end{aligned} \tag{8.62}$$

where

$$\begin{aligned} \overline{A} &:= \Pi_1 \begin{bmatrix} \Phi + B_2 D_c C_2 & B_2 C_c \\ B_c C_2 & A_c \end{bmatrix} \Pi \\ &+ \Pi_2 \begin{bmatrix} \Phi + B_2 D_c C_2 & B_2 C_c \\ B_c C_2 & A_c \end{bmatrix} \Pi \\ &=: \overline{A}_1 + \overline{A}_2 \end{aligned}$$

$$\overline{B} := \Pi_1 \begin{bmatrix} B_1 \\ 0 \end{bmatrix} + \Pi_2 \begin{bmatrix} B_1 \\ 0 \end{bmatrix} =: \overline{B}_1 + \overline{B}_2, \; \overline{C} := \begin{bmatrix} C_2 & 0 \end{bmatrix} \Pi$$

Introduce the matrix of controller data (recall the discrete process case) as

$$\Theta := \begin{bmatrix} D_c & C_c \\ B_c & A_c \end{bmatrix}$$

and also

$$\begin{aligned}
\mathcal{A}_1 :=& \Pi_1 \begin{bmatrix} \Phi & 0 \\ 0 & 0 \end{bmatrix} \Pi, \ \mathcal{A}_2 := \Pi_2 \begin{bmatrix} \Phi & 0 \\ 0 & 0 \end{bmatrix} \Pi, \ \Gamma_1 := \Pi_1 \begin{bmatrix} B_2 & 0 \\ 0 & I \end{bmatrix} \\
\Gamma_2 :=& \Pi_2 \begin{bmatrix} B_2 & 0 \\ 0 & I \end{bmatrix}, \ \mathcal{C}_2 := \begin{bmatrix} C_2 & 0 \\ 0 & I \end{bmatrix} \Pi, \ \mathcal{C} := \begin{bmatrix} C_2 & 0 \end{bmatrix} \Pi, \ \mathcal{D}_2 := \begin{bmatrix} I \\ 0 \end{bmatrix} \\
\mathcal{B}_1 :=& \Pi_1 \begin{bmatrix} B_1 \\ 0 \end{bmatrix}, \ \mathcal{B}_2 := \Pi_2 \begin{bmatrix} B_1 \\ 0 \end{bmatrix}
\end{aligned}$$

Then the resulting process state-space model matrices can be written in the following form (i.e. affine in the controller data matrix Θ) which is central to the subsequent analysis

$$\begin{aligned}
\overline{A}_1 =& \mathcal{A}_1 + \Gamma_1 \Theta \mathcal{C}_2, \ \overline{A}_2 = \mathcal{A}_2 + \Gamma_2 \Theta \mathcal{C}_2, \ \overline{B}_1 = \mathcal{B}_1 + \Gamma_1 \Theta \mathcal{D}_2 \\
\overline{B}_2 =& \mathcal{B}_2 + \Gamma_2 \Theta \mathcal{D}_2, \ \overline{C} = \mathcal{C}
\end{aligned} \tag{8.63}$$

Theorem 8.4.1 can now be applied to the resulting state-space model to obtain the following result which serves as an existence condition for the controller considered here.

Theorem 8.4.4. *Suppose that a dynamic pass profile controller defined by (8.61) is applied to a differential unit memory linear repetitive process described by (8.47) with resulting state-space model (8.62). Then this process is stable along the pass with H_∞ disturbance attenuation $\gamma > 0$ if $\exists$ matrices $\mathcal{S}_h > 0$, $\mathcal{S}_v > 0$, and $P_h > 0$ such that*

$$\begin{bmatrix}
-\mathcal{S} & \mathcal{S}\overline{A}_2 & \mathcal{S}\overline{B}_2 & 0 \\
\overline{A}_2^T \mathcal{S} & \overline{A}_1^T P + P\overline{A}_1 - R & P\overline{B}_1 & \overline{C}^T \\
\overline{B}_2^T \mathcal{S} & \overline{B}_1^T P & -\gamma^2 I & 0 \\
0 & \overline{C} & 0 & -I
\end{bmatrix} < 0 \tag{8.64}$$

where $S := \text{diag}\{I, \mathcal{S}_v\}$, $P := \text{diag}\{P_h, I\}$, *and* $R := \text{diag}\{0, \mathcal{S}_v\}$.

The following result extends this last theorem to give a controller design algorithm.

Theorem 8.4.5. *Consider the controlled process of Theorem 8.4.4. Then this process is stable along the pass and has prescribed H_∞ disturbance attenuation $\gamma > 0$ if $\exists$ matrices $P_{h_{11}} > 0$, $U_{h_{11}} > 0$, $\mathcal{S}_{v_{11}} > 0$, and $T_{v_{11}} > 0$ such that the LMIs defined by (8.65)–(8.67) below hold*

$$\begin{bmatrix} \mathcal{N}_c^T & 0 & 0 \\ 0 & I & 0 \\ 0 & 0 & I \end{bmatrix} \begin{bmatrix} -T_{v_{11}} + D_0 T_{v_{11}} D_0^T & CU_{h_{11}} + D_0 T_{v_{11}} B_0^T & B_{12} & D_0 T_{v_{11}} \\ U_{h_{11}} C^T + B_0 T_{v_{11}} D_0^T & U_{h_{11}} A^T + AU_{h_{11}} + B_0 T_{v_{11}} B_0^T & B_{11} & B_0 T_{v_{11}} \\ B_{12}^T & B_{11}^T & -\gamma^2 I & 0 \\ T_{v_{11}} D_0^T & T_{v_{11}} B_0^T & 0 & -I \end{bmatrix} \begin{bmatrix} \mathcal{N}_c & 0 & 0 \\ 0 & I & 0 \\ 0 & 0 & I \end{bmatrix} < 0 \tag{8.65}$$

$$\begin{bmatrix} A^T P_{h_{11}} + P_{h_{11}} A + C^T \mathcal{S}_{v_{11}} C & (\star) \\ B_{11}^T P_{h_{11}} + B_{12}^T \mathcal{S}_{v_{11}} C & B_{12}^T \mathcal{S}_{v_{11}} B_{12} - \gamma^2 I \end{bmatrix} < 0 \tag{8.66}$$

$$\begin{bmatrix} P_{h11} & I \\ I & U_{h11} \end{bmatrix} \geq 0, \quad \begin{bmatrix} \mathcal{S}_{v11} & I \\ I & T_{v11} \end{bmatrix} \geq 0 \tag{8.67}$$

where $\mathcal{N}_c$ is a full column rank matrix whose image satisfies

$$\mathrm{Im}\mathcal{N}_c = \ker\left(\begin{bmatrix} D^T \\ B^T \end{bmatrix}\right)$$

Proof. Using the definitions of (8.63), (8.64) can be rewritten as

$$\begin{bmatrix} -\mathcal{S} & \mathcal{S}\mathcal{A}_2 + \mathcal{S}\Gamma_2\Theta\mathcal{C}_2 & \mathcal{S}\mathcal{B}_2 & 0 \\ \mathcal{A}_2^T\mathcal{S} + \mathcal{C}_2^T\Theta^T\Gamma_2^T\mathcal{S} & \mathcal{V} & P\mathcal{B}_1 & \mathcal{C}^T \\ \mathcal{B}_2^T\mathcal{S} & \mathcal{B}_1^T P & -\gamma^2 I & 0 \\ 0 & \mathcal{C} & 0 & -I \end{bmatrix}$$
$$= \begin{bmatrix} -\mathcal{S} & \mathcal{S}\mathcal{A}_2 & \mathcal{S}\mathcal{B}_2 & 0 \\ \mathcal{A}_2^T\mathcal{S} & \mathcal{A}_1^T P + P\mathcal{A}_1 - R & P\mathcal{B}_1 & \mathcal{C}^T \\ \mathcal{B}_2^T\mathcal{S} & \mathcal{B}_1^T P & -\gamma^2 I & 0 \\ 0 & \mathcal{C} & 0 & -I \end{bmatrix} + \begin{bmatrix} 0 & \mathcal{S}\Gamma_2\Theta\mathcal{C}_2 & 0 & 0 \\ 0 & P\Gamma_1\Theta\mathcal{C}_2 & 0 & 0 \\ 0 & 0 & 0 & 0 \\ 0 & 0 & 0 & 0 \end{bmatrix}$$
$$+ \begin{bmatrix} 0 & 0 & 0 & 0 \\ \mathcal{C}_2^T\Theta^T\Gamma_2^T\mathcal{S} & \mathcal{C}_2^T\Theta^T\Gamma_1^T P & 0 & 0 \\ 0 & 0 & 0 & 0 \\ 0 & 0 & 0 & 0 \end{bmatrix} \tag{8.68}$$

where

$$\mathcal{V} := \mathcal{A}_1^T P + \mathcal{C}_2^T\Theta\Gamma_1^T P + P\mathcal{A}_1 + P_1\Gamma_1^T\Theta\mathcal{C}_2 - R$$

At this stage (8.68) can be rewritten as

$$\begin{bmatrix} -\mathcal{S} & \mathcal{S}\mathcal{A}_2 & \mathcal{S}\mathcal{B}_2 & 0 \\ \mathcal{A}_2^T\mathcal{S} & \mathcal{A}_1^TP+P\mathcal{A}_1-R & P\mathcal{B}_1 & \mathcal{C}^T \\ \mathcal{B}_2^T\mathcal{S} & \mathcal{B}_1^TP & -\gamma^2I & 0 \\ 0 & \mathcal{C} & 0 & -I \end{bmatrix} + \begin{bmatrix} \mathcal{S}\Gamma_2 \\ P\Gamma_1 \\ 0 \\ 0 \end{bmatrix} \Theta \begin{bmatrix} 0 & \mathcal{C}_2 & 0 & 0 \end{bmatrix}$$
$$+ \begin{bmatrix} \Gamma_2^T\mathcal{S} & \Gamma_1^TP & 0 & 0 \end{bmatrix} \Theta^T \begin{bmatrix} 0 \\ \mathcal{C}_2^T \\ 0 \\ 0 \end{bmatrix} \tag{8.69}$$

Introducing the matrices

$$\Psi := \begin{bmatrix} -\mathcal{S} & \mathcal{S}\mathcal{A}_2 & \mathcal{S}\mathcal{B}_2 & 0 \\ \mathcal{A}_2^T\mathcal{S} & \mathcal{A}_1^TP+P\mathcal{A}_1-R & P\mathcal{B}_1 & \mathcal{C}^T \\ \mathcal{B}_2^T\mathcal{S} & \mathcal{B}_1^TP & -\gamma^2I & 0 \\ 0 & \mathcal{C} & 0 & -I \end{bmatrix}, \quad M^T := \begin{bmatrix} \mathcal{S}\Gamma_2 \\ P\Gamma_1 \\ 0 \\ 0 \end{bmatrix}$$
$$N := \begin{bmatrix} 0 & \mathcal{C}_2 & 0 & 0 \end{bmatrix}$$

now enables us to rewrite (8.69) in the form

$$\Psi + M^T\Theta N + N^T\Theta^T M < 0$$

Also we can apply Lemma 8.2.2 to eliminate Θ and obtain

$$W_M^T\Psi W_M < 0 \text{ and } W_N^T\Psi W_N < 0$$

where

$$W_M \in \ker(M), \quad W_N \in \ker(N) \tag{8.70}$$

Note now that

$$M = \begin{bmatrix} \Gamma_2^T\mathcal{S} & \Gamma_1^TP & 0 & 0 \end{bmatrix} = \begin{bmatrix} \Gamma_2^T & \Gamma_1^T & 0 & 0 \end{bmatrix} \begin{bmatrix} \mathcal{S} & 0 & 0 & 0 \\ 0 & P & 0 & 0 \\ 0 & 0 & I & 0 \\ 0 & 0 & 0 & I \end{bmatrix} = M_nV \tag{8.71}$$

where the matrices P and $\mathcal{S}$ are partitioned as

$$P = \begin{bmatrix} P_{h_{11}} & P_{h_{12}} & 0 & 0 \\ P_{h_{12}}^T & P_{h_{22}} & 0 & 0 \\ 0 & 0 & I & 0 \\ 0 & 0 & 0 & I \end{bmatrix}, \quad \mathcal{S} = \begin{bmatrix} I & 0 & 0 & 0 \\ 0 & I & 0 & 0 \\ 0 & 0 & \mathcal{S}_{v_{11}} & \mathcal{S}_{v_{12}} \\ 0 & 0 & \mathcal{S}_{v_{12}}^T & \mathcal{S}_{v_{22}} \end{bmatrix}$$

and

$$P_h = \begin{bmatrix} P_{h_{11}} & P_{h_{12}} \\ P_{h_{12}}^T & P_{h_{22}} \end{bmatrix}, \quad P_v = \begin{bmatrix} I & 0 \\ 0 & I \end{bmatrix}$$

$$\mathcal{S}_h = \begin{bmatrix} I & 0 \\ 0 & I \end{bmatrix}, \quad \mathcal{S}_v = \begin{bmatrix} \mathcal{S}_{v_{11}} & \mathcal{S}_{v_{12}} \\ \mathcal{S}_{v_{12}}^T & \mathcal{S}_{v_{22}} \end{bmatrix}$$

It is easy to see that the solution of (8.64) is independent of the matrices P_v and S_h but they are necessary to compute V^{-1} (The matrix V is defined in (8.71)). Also

$$\ker(M_n V) = V^{-1}\ker(M_n)$$

and using (8.70) yields that

$$W_M = V^{-1} W_{M_n}$$

Hence

$$W_{M_n}^T V^{-T} \Psi V^{-1} W_{M_n} < 0$$

and also

$$\widetilde{\Phi} := V^{-T}\Psi V^{-1} = \begin{bmatrix} -S^{-1} & \mathcal{A}_2 P^{-1} & \mathcal{B}_2 & 0 \\ P^{-1}\mathcal{A}_2^T & P^{-1}\mathcal{A}_1^T + \mathcal{A}_1 P^{-1} - R & \mathcal{B}_1 & P^{-1}\mathcal{C}^T \\ \mathcal{B}_2^T & \mathcal{B}_1^T & -\gamma^2 I & 0 \\ 0 & \mathcal{C}P^{-1} & 0 & -I \end{bmatrix}$$

At this stage, we have established the following inequalities which are not LMIs

$$W_{M_n}^T \Phi W_{M_n} < 0 \tag{8.72}$$

and

$$W_N^T \Psi W_N < 0 \tag{8.73}$$

respectively. Now write out the matrices M_n and N as

$$M_n = \begin{bmatrix} \Gamma_2^T & \Gamma_1^T & 0 & 0 \end{bmatrix} = \left[\begin{array}{cccc|cccc|c|c} 0 & 0 & 0 & 0 & B^T & 0 & 0 & 0 & 0 & 0 \\ 0 & 0 & D^T & 0 & 0 & I & 0 & 0 & 0 & 0 \\ 0 & 0 & 0 & I & 0 & 0 & 0 & 0 & 0 & 0 \end{array}\right]$$

and

$$N = \left[\begin{array}{c|c|c|c} 0 & \mathcal{C}_2 & 0 & 0 \end{array}\right] = \left[\begin{array}{cccc|cccc|c|c} 0 & 0 & 0 & 0 & 0 & 0 & I & 0 & 0 & 0 \\ 0 & 0 & 0 & 0 & 0 & I & 0 & 0 & 0 & 0 \\ 0 & 0 & 0 & 0 & 0 & 0 & 0 & I & 0 & 0 \end{array}\right]$$

respectively. Then it follows immediately that the kernels of M_n and N are the images of W_{M_n} and W_N respectively as detailed next

$$W_{M_n} = \begin{bmatrix} I & 0 & 0 & 0 & 0 & 0 & 0 \\ 0 & I & 0 & 0 & 0 & 0 & 0 \\ 0 & 0 & \mathcal{N}_D & 0 & 0 & 0 & 0 \\ 0 & 0 & 0 & 0 & 0 & 0 & 0 \\ 0 & 0 & \mathcal{N}_B & 0 & 0 & 0 & 0 \\ 0 & 0 & 0 & 0 & 0 & 0 & 0 \\ 0 & 0 & 0 & I & 0 & 0 & 0 \\ 0 & 0 & 0 & 0 & I & 0 & 0 \\ 0 & 0 & 0 & 0 & 0 & I & 0 \\ 0 & 0 & 0 & 0 & 0 & 0 & I \end{bmatrix}, \quad W_N = \begin{bmatrix} I & 0 & 0 & 0 & 0 & 0 & 0 \\ 0 & I & 0 & 0 & 0 & 0 & 0 \\ 0 & 0 & I & 0 & 0 & 0 & 0 \\ 0 & 0 & 0 & I & 0 & 0 & 0 \\ 0 & 0 & 0 & 0 & I & 0 & 0 \\ 0 & 0 & 0 & 0 & 0 & 0 & 0 \\ 0 & 0 & 0 & 0 & 0 & 0 & 0 \\ 0 & 0 & 0 & 0 & 0 & 0 & 0 \\ 0 & 0 & 0 & 0 & 0 & I & 0 \\ 0 & 0 & 0 & 0 & 0 & 0 & I \end{bmatrix}$$

where

$$\mathcal{N}_B \in \ker(B^T), \quad \mathcal{N}_D \in \ker(D^T)$$

Also the matrices P^{-1}, $\mathcal{S}^{-1}$ and R can be partitioned as follows

$$P^{-1} = \begin{bmatrix} U_{h_{11}} & U_{h_{12}} & 0 & 0 \\ U_{h_{12}}^T & U_{h_{22}} & 0 & 0 \\ 0 & 0 & I & 0 \\ 0 & 0 & 0 & I \end{bmatrix}, \ \mathcal{S}^{-1} = \begin{bmatrix} I & 0 & 0 & 0 \\ 0 & I & 0 & 0 \\ 0 & 0 & T_{v_{11}} & T_{v_{12}} \\ 0 & 0 & T_{v_{12}}^T & T_{v_{22}} \end{bmatrix}$$

$$R = \begin{bmatrix} 0 & 0 & 0 & 0 \\ 0 & 0 & 0 & 0 \\ 0 & 0 & \mathcal{S}_{v_{11}} & \mathcal{S}_{v_{12}} \\ 0 & 0 & \mathcal{S}_{v_{12}}^T & \mathcal{S}_{v_{22}} \end{bmatrix}$$

and hence the matrix $\widetilde{\Phi}$ can be rewritten as

$$\widetilde{\Phi} = \begin{bmatrix} -I & 0 & 0 & 0 & 0 & 0 & 0 & 0 & 0 & 0 \\ 0 & -I & 0 & 0 & 0 & 0 & 0 & 0 & 0 & 0 \\ 0 & 0 & -T_{v_{11}} & -T_{v_{12}} & CU_{h_{11}} & CU_{h_{12}} & D_0 & 0 & B_{12} & 0 \\ 0 & 0 & -T_{v_{12}}^T & -T_{v_{22}} & 0 & 0 & 0 & 0 & 0 & 0 \\ 0 & 0 & U_{h_{11}}C^T & 0 & U_{h_{11}}A^T + AU_{h_{11}} & AU_{h_{12}} & B_0 & 0 & B_{11} & 0 \\ 0 & 0 & U_{h_{12}}^T C^T & 0 & U_{h_{12}}^T A^T & 0 & 0 & 0 & 0 & 0 \\ 0 & 0 & D_0^T & 0 & B_0^T & 0 & -\mathcal{S}_{v_{11}} & -\mathcal{S}_{v_{12}} & 0 & I \\ 0 & 0 & 0 & 0 & 0 & 0 & -\mathcal{S}_{v_{12}}^T & -\mathcal{S}_{v_{22}} & 0 & 0 \\ 0 & 0 & B_{12}^T & 0 & B_{11}^T & 0 & 0 & 0 & -\gamma^2 I & 0 \\ 0 & 0 & 0 & 0 & 0 & 0 & I & 0 & 0 & -I \end{bmatrix}$$

Noting that the fourth and sixth rows of W_{M_n} are zero, the inequality (8.72) can be rewritten as

$$\begin{bmatrix} \mathcal{N}_D^T & \mathcal{N}_B^T & 0 & 0 & 0 & 0 \\ 0 & 0 & I & 0 & 0 & 0 \\ 0 & 0 & 0 & I & 0 & 0 \\ 0 & 0 & 0 & 0 & I & 0 \\ 0 & 0 & 0 & 0 & 0 & I \end{bmatrix} \begin{bmatrix} -T_{v_{11}} & CU_{h_{11}} & D_0 & 0 & B_{12} & 0 \\ U_{h_{11}}C^T & U_{h_{11}}A^T + AU_{h_{11}} & B_0 & 0 & B_{11} & 0 \\ D_0^T & B_0^T & -\mathcal{S}_{v_{11}} & -\mathcal{S}_{v_{12}} & 0 & I \\ 0 & 0 & -\mathcal{S}_{v_{12}}^T & -\mathcal{S}_{v_{22}} & 0 & 0 \\ B_{12}^T & B_{11}^T & 0 & 0 & -\gamma^2 I & 0 \\ 0 & 0 & I & 0 & 0 & -I \end{bmatrix}$$

$$\times \begin{bmatrix} \mathcal{N}_D & 0 & 0 & 0 & 0 \\ \mathcal{N}_B & 0 & 0 & 0 & 0 \\ 0 & I & 0 & 0 & 0 \\ 0 & 0 & I & 0 & 0 \\ 0 & 0 & 0 & I & 0 \\ 0 & 0 & 0 & 0 & I \end{bmatrix} < 0$$

and application of the Schur's complement formula yields (8.65).

In order to obtain (8.66), routine manipulations show that (8.73) reduces to

$$\begin{bmatrix} -\mathcal{S}_{v11} & -\mathcal{S}_{v12} & \mathcal{S}_{v11}C & \mathcal{S}_{v11}B_{12} & 0 \\ -\mathcal{S}_{v12}^T & -\mathcal{S}_{v22} & \mathcal{S}_{v12}^TC & \mathcal{S}_{v12}^TB_{12} & 0 \\ C^T\mathcal{S}_{v11} & C^T\mathcal{S}_{v12} & A^TP_{h11}+P_{h11}A & P_{h11}B_{11} & 0 \\ B_{12}^T\mathcal{S}_{v11} & B_{12}^T\mathcal{S}_{v12} & B_{11}^TP_{h11} & -\gamma^2 I & 0 \\ 0 & 0 & 0 & 0 & -I \end{bmatrix} < 0$$

Finally, application of the Schur's complement formula gives (8.66). Also application of Lemma 8.2.3 requires the conditions given in (8.67) and the proof is complete. □

Suppose now that this last result holds. Then the following is a systematic procedure for obtaining the corresponding controller state-space matrices.

(1) Compute the matrices S_{v12}, P_{v12} using

$$\mathcal{S}_{v11} - T_{v11}^{-1} = \mathcal{S}_{v12}\mathcal{S}_{v12}^T$$
$$P_{h11} - U_{h11}^{-1} = P_{h12}P_{h12}^T$$

where $P_{h22} = I$ and $\mathcal{S}_{v22} = I$

(b) Construct the matrices $P_h > 0$ and $S_v > 0$ as

$$P_h = \begin{bmatrix} P_{h11} & P_{h12}^T \\ P_{h12} & I \end{bmatrix}, \quad \mathcal{S}_v = \begin{bmatrix} \mathcal{S}_{v11} & \mathcal{S}_{v12}^T \\ \mathcal{S}_{v12} & I \end{bmatrix}$$

and hence the matrices $\mathcal{S} = \text{diag}\{I, \mathcal{S}_v\}$, $P = \text{diag}\{P_h, I\}$, $R = \text{diag}\{0, \mathcal{S}_v\}$

(c) Compute the matrices M, N and Ψ where

$$\Psi := \begin{bmatrix} -\mathcal{S} & \mathcal{S}\mathcal{A}_2 & \mathcal{S}\mathcal{B}_2 & 0 \\ \mathcal{A}_2^T\mathcal{S} & \mathcal{A}_1^TP + P\mathcal{A}_1 - R & P\mathcal{B}_1 & \mathcal{C}^T \\ \mathcal{B}_2^T\mathcal{S} & \mathcal{B}_1^TP & -\gamma^2 I & 0 \\ 0 & \mathcal{C} & 0 & -I \end{bmatrix}$$
$$M := \begin{bmatrix} \Gamma_2^T S & \Gamma_1^T P & 0 & 0 \end{bmatrix}, \quad N := \begin{bmatrix} 0 & \mathcal{C}_2 & 0 & 0 \end{bmatrix}$$

(d) Solve the following LMI

$$\Psi + M^T\Theta N + N^T\Theta^T M < 0$$

to obtain

$$\Theta = \begin{bmatrix} D_c & C_c \\ B_c & A_c \end{bmatrix}$$

i.e. the matrices which define the controller state-space model.

8.5 Guaranteed Cost Control of Differential Processes

As for the discrete case, many applications will require a control law which not only guarantees stability along the pass but also meets specified performance criteria. Here we give a comprehensive treatment of one aspect of this problem for differential linear repetitive processes described by (8.57) with the disturbance terms deleted, i.e.

$$\begin{aligned}\begin{bmatrix} \dot{x}_{k+1}(t) \\ y_{k+1}(t) \end{bmatrix} &= \left(\begin{bmatrix} A & B_0 \\ C & D_0 \end{bmatrix} + \begin{bmatrix} \Delta A & \Delta B_0 \\ \Delta C & \Delta D_0 \end{bmatrix} \right) \begin{bmatrix} x_{k+1}(t) \\ y_k(t) \end{bmatrix} \\ &+ \left(\begin{bmatrix} B \\ D \end{bmatrix} + \begin{bmatrix} \Delta B \\ \Delta D \end{bmatrix} \right) u_{k+1}(t) \end{aligned} \tag{8.74}$$

and associated uncertainty structure. The problem solved here is the same as in the discrete case treated earlier in this chapter, i.e. design a control law which ensures stability along the pass together with a guaranteed bound on the associated cost function

$$\begin{aligned} J &:= \sum_{k=0}^{\infty} \int_0^{\infty} \left(u_{k+1}^T(t) \Psi u_{k+1}(t) \right) dt \\ &+ \sum_{k=0}^{\infty} \int_0^{\infty} \left(\begin{bmatrix} x_{k+1}(t) \\ y_k(t) \end{bmatrix}^T \begin{bmatrix} Q_1 & 0 \\ 0 & Q_2 \end{bmatrix} \begin{bmatrix} x_{k+1}(t) \\ y_k(t) \end{bmatrix} \right) dt \end{aligned} \tag{8.75}$$

where $\Psi > 0$, $Q_1 > 0$ and $Q_2 > 0$ are given matrices, is bounded for all admissible uncertainties. The results in this section are from [131].

Note 8.5.1. Repetitive processes are defined over the finite pass length α and, in practice, only a finite number of passes, say N, will actually be completed. Hence the cost function (8.75) should be replaced by

$$\begin{aligned} J' &:= \sum_{k=0}^{N} \int_0^{\alpha} \left(u_{k+1}^T(t) \Psi u_{k+1}(t) \right) dt \\ &+ \sum_{k=0}^{N} \int_0^{\alpha} \left(\begin{bmatrix} x_{k+1}(t) \\ y_k(t) \end{bmatrix}^T \begin{bmatrix} Q_1 & 0 \\ 0 & Q_2 \end{bmatrix} \begin{bmatrix} x_{k+1}(t) \\ y_k(t) \end{bmatrix} \right) dt \end{aligned}$$

However, it is routine to argue that the signals involved can be extended from $[0, \alpha]$ to the infinite interval in such a way that projection of the infinite interval solution onto the finite interval is possible. The same is true for the pass-to-pass direction and hence we will work with (8.75).

We start by developing the LMI condition which guarantees that the unforced (the control input terms are deleted) process is stable along the pass and also the associated cost function is bounded for all admissible uncertainties. These results are then extended to design a guaranteed cost controller.

8.5.1 Guaranteed Cost Bound

In this sub-section we are interested in finding an upper bound for corresponding cost function of the unforced process ($u_{k+1}(t) = 0$ in (8.74)) with associated cost function

$$J_0 := \sum_{k=0}^{\infty} \int_0^{\infty} \left(\begin{bmatrix} x_{k+1}(t) \\ y_k(t) \end{bmatrix}^T \begin{bmatrix} Q_1 & 0 \\ 0 & Q_2 \end{bmatrix} \begin{bmatrix} x_{k+1}(t) \\ y_k(t) \end{bmatrix} \right) dt \tag{8.76}$$

The following theorem gives a sufficient condition for stability along the pass with guaranteed cost.

Theorem 8.5.1. *A differential linear repetitive process described by the unforced version of (8.74) is stable along the pass if exists a scalar $\epsilon > 0$ and matrices $P_1 > 0$ and $P_2 > 0$ such that the following LMI holds*

$$\begin{bmatrix} -P_2 & P_2C & P_2D_0 & P_2H_2 & P_2H_2 \\ C^TP_2 & A^TP_1 + P_1A + Q_1 + \epsilon E_1^TE_1 & P_1B_0 & P_1H_1 & P_1H_1 \\ D_0^TP_2 & B_0^TP_1 & -P_2 + Q_2 + \epsilon E_2^TE_2 & 0 & 0 \\ H_2^TP_2 & H_1^TP_1 & 0 & -\epsilon I & 0 \\ H_2^TP_2 & H_1^TP_1 & 0 & 0 & -\epsilon I \end{bmatrix} < 0 \tag{8.77}$$

If this condition holds, the cost function (8.76) satisfies the upper bound

$$J_0 < \sum_{k=0}^{\infty} x_{k+1}^T(0)P_1x_{k+1}(0) + \int_0^{\infty} y_0^T(t)P_2y_0(t)dt \tag{8.78}$$

Proof. Recall the definition of the matrices $\widehat{A}_1$, $\widehat{A}_2$, define $\widehat{\Delta A}_1$, and $\widehat{\Delta A}_2$ as in the discrete case (see (8.1)) and rewrite (8.74) as

$$\begin{bmatrix} \dot{x}_{k+1}(t) \\ y_{k+1}(t) \end{bmatrix} = \left((\widehat{A}_1 + \widehat{\Delta A}_1) + (\widehat{A}_2 + \widehat{\Delta A}_2) \right) \xi_k(t)$$

(where $\xi_k(t)$ is defined in (8.52)). Recall now the Lyapunov function of (8.49) and interpret it in terms of the current case. Then (see (8.52)) stability along the pass holds if

$$(\widehat{A}_1 + \widehat{\Delta A}_1)P + P(\widehat{A}_1 + \widehat{\Delta A}_1) + (\widehat{A}_2 + \widehat{\Delta A}_2)R(\widehat{A}_2 + \widehat{\Delta A}_2) - R < 0$$

where

$$P := \begin{bmatrix} P_1 & 0 \\ 0 & 0 \end{bmatrix}, \quad R := \begin{bmatrix} 0 & 0 \\ 0 & P_2 \end{bmatrix} \tag{8.79}$$

Next, it follows immediately that the inequality

$$\Delta V(k,t) + \xi_k^T(t) \begin{bmatrix} Q_1 & 0 \\ 0 & Q_2 \end{bmatrix} \xi_k(t) < 0$$

implies that the unforced process is stable along the pass. Also

$$\Upsilon_{bs} := \sum_{k=0}^{\infty} \int_0^{\infty} \left(\xi_k^T(t) \begin{bmatrix} Q_1 & 0 \\ 0 & Q_2 \end{bmatrix} \xi_k(t) \right) dt$$

and, since the process is stable along the pass, we now have that

$$\begin{aligned}
\Upsilon_{bs} &< -\sum_{k=0}^{\infty} \int_0^{\infty} (\dot{V}_1(k,t) + \Delta V_2(k,t))dt \\
&= -\sum_{k=0}^{\infty} x_{k+1}^T(t) P_1 x_{k+1}(t) \Big|_0^{\infty} \\
&\quad - \int_0^{\infty} \left(\sum_{k=0}^{\infty} (y_{k+1}^T(t) P_2 y_{k+1}(t) - y_k^T(t) P_2 y_k(t)) \right) dt \\
&= \sum_{k=0}^{\infty} x_{k+1}^T(0) P_1 x_{k+1}(0) - \int_0^{\infty} \left(y_{\infty}^T(t) P_2 y_{\infty}(t) - y_0^T(t) P_2 y_0(t) \right) dt \\
&= \sum_{k=0}^{\infty} x_{k+1}^T(0) P_1 x_{k+1}(0) + \int_0^{\infty} y_0^T(t) P_2 y_0(t) dt
\end{aligned}$$

and hence (8.78) holds.

At this stage we have that stability along the pass with the cost bound exists if

$$(\widehat{A}_1 + \widehat{\Delta A}_1)^T P + P(\widehat{A}_1 + \widehat{\Delta A}_1) + (\widehat{A}_2 + \widehat{\Delta A}_2)^T S(\widehat{A}_2 + \widehat{\Delta A}_2) - R + Q < 0$$

where $Q := \text{diag}\{Q_1, Q_2\}$, $S := \text{diag}\{P_3, P_2\}$, and the matrix $P_3 > 0$ has compatible dimensions. Next, an obvious application of the Schur's complement formula yields

$$\begin{bmatrix} -P_3 & (\star) & (\star) & (\star) \\ 0 & -P_2 & (\star) & (\star) \\ 0 & C^T P_2 + \Delta C^T P_2 & \Lambda & (\star) \\ 0 & D_0^T P_2 + \Delta D_0^T P_2 & B_0^T P_1 + \Delta B_0^T P_1 & Q_2 - P_2 \end{bmatrix} < 0 \tag{8.80}$$

where

$$\Lambda := A^T P_1 + \Delta A^T P_1 + P_1 A + P_1 \Delta A + Q_1$$

On removing the block $-P_3$ which is always negative definite, (8.80) gives the equivalent condition

$$\begin{bmatrix} -P_2 & P_2 C & P_2 D_0 \\ C^T P_2 & A^T P_1 + P_1 A + Q_1 & P_1 B_0 \\ D_0^T P_2 & B_0^T P_1 & Q_2 - P_2 \end{bmatrix}$$
$$+ \begin{bmatrix} 0 & 0 & 0 \\ 0 & E_1^T & 0 \\ 0 & 0 & E_2^T \end{bmatrix} \begin{bmatrix} \mathcal{F}^T & 0 & 0 \\ 0 & \mathcal{F}^T & 0 \\ 0 & 0 & \mathcal{F}^T \end{bmatrix} \begin{bmatrix} 0 & 0 & 0 \\ H_2^T P_2 & H_1^T P_1 & 0 \\ H_2^T P_2 & H_1^T P_1 & 0 \end{bmatrix}$$
$$+ \begin{bmatrix} 0 & P_2 H_2 & P_2 H_2 \\ 0 & P_1 H_1 & P_1 H_1 \\ 0 & 0 & 0 \end{bmatrix} \begin{bmatrix} \mathcal{F} & 0 & 0 \\ 0 & \mathcal{F} & 0 \\ 0 & 0 & \mathcal{F} \end{bmatrix} \begin{bmatrix} 0 & 0 & 0 \\ 0 & E_1 & 0 \\ 0 & 0 & E_2 \end{bmatrix} < 0$$

An obvious application of the result of Lemma 5.4.2 to this last expression now gives

$$\begin{bmatrix} -P_2 & P_2 C & P_2 D_0 \\ C^T P_2 & \Theta_{pm} & P_1 B_0 \\ D_0^T P_2 & B_0^T P_1 & Q_2 - P_2 + \epsilon E_2^T E_2 \end{bmatrix}$$
$$+ \epsilon^{-1} \begin{bmatrix} 0 & P_2 H_2 & P_2 H_2 \\ 0 & P_1 H_1 & P_1 H_1 \\ 0 & 0 & 0 \end{bmatrix} \begin{bmatrix} 0 & 0 & 0 \\ H_2^T P_2 & H_1^T P_1 & 0 \\ H_2^T P_2 & H_1^T P_1 & 0 \end{bmatrix} < 0$$

where

$$\Theta_{pm} := A^T P_1 + P_1 A + \epsilon E_1^T E_1 + Q_1$$

Finally, an obvious application of the Schur's complement formula gives (8.77) and the proof is complete. □

Note 8.5.2. Note that it is possible to minimize the upper bound on the cost function (8.78) using the following optimization procedure

$$\min \left[\sum_{k=0}^{\infty} x_{k+1}^T(0) P_1 x_{k+1}(0) + \int_0^{\infty} y_0^T(t) P_2 y_0(t) dt \right]$$
$$= \min \left[\sum_{k=0}^{\infty} \operatorname{trace}\left(P_1 x_{k+1}(0) x_{k+1}^T(0)\right) + \operatorname{trace}\left(P_2 \int_0^{\infty} y_0(t) y_0^T(t) dt\right) \right]$$
$$\text{subject to } (8.77)$$

8.5.2 Guaranteed Cost Control

The control law used here is again given by (8.54) and the uncertainty associated with the state-space model is assumed to satisfy

$$\begin{bmatrix} \Delta A & \Delta B_0 & \Delta B \\ \Delta C & \Delta D_0 & \Delta D \end{bmatrix} = \begin{bmatrix} H_1 \\ H_2 \end{bmatrix} \mathcal{F} \begin{bmatrix} E_1 & E_2 & E_3 \end{bmatrix} \tag{8.81}$$

i.e. (8.58) with the terms arising from the disturbance vector deleted from both sides of this equation, and $\mathcal{F}$ satisfies (8.15),. The associated cost function for the controlled process is given by

$$J_{cp} := \sum_{k=0}^{\infty} \int_0^{\infty} \left(\begin{bmatrix} x_{k+1}(t) \\ y_k(t) \end{bmatrix}^T \begin{bmatrix} Q_1 + K_1^T \Psi K_1 & K_1^T \Psi K_2 \\ K_2^T \Psi K_1 & Q_2 + K_2^T \Psi K_2 \end{bmatrix} \begin{bmatrix} x_{k+1}(t) \\ y_k(t) \end{bmatrix} \right) dt \tag{8.82}$$

which is of the form of that in Theorem 8.5.1 and we have the following result.

Theorem 8.5.2. *Suppose that a control law of the form(8.54) is applied to a differential linear repetitive process described by (8.74) with uncertainty structure satisfying (8.81) and (8.15). Then the resulting process is stable along the pass if $\exists$ a scalar $\epsilon > 0$ and matrices $W_1 > 0$, $W_2 > 0$, N_1 and N_2 such that the following LMI holds*

$$\begin{bmatrix} -W_2 + 2\epsilon H_2 H_2^T & (\star) & (\star) & (\star) & (\star) & (\star) & (\star) & (\star) \\ W_1 C^T + N_1^T D^T + 2\epsilon H_1 H_2^T & \widetilde{\Phi} & (\star) & (\star) & (\star) & (\star) & (\star) & (\star) \\ W_2 D_0^T + N_2^T D^T & W_2 B_0^T + N_2^T B^T & -W_2 & (\star) & (\star) & (\star) & (\star) & (\star) \\ 0 & E_1 W_1 + E_3 N_1 & 0 & -\epsilon I & (\star) & (\star) & (\star) & (\star) \\ 0 & 0 & E_2 W_2 + E_3 N_2 & 0 & -\epsilon I & (\star) & (\star) & (\star) \\ 0 & N_1 & N_2 & 0 & 0 & -\Psi^{-1} & (\star) & (\star) \\ 0 & W_1 & 0 & 0 & 0 & 0 & -Q_1^{-1} & (\star) \\ 0 & 0 & W_2 & 0 & 0 & 0 & 0 & -Q_2^{-1} \end{bmatrix} < 0 \tag{8.83}$$

where $\widetilde{\Phi} := W_1 A^T + A W_1 + N_1^T B^T + B N_1 + 2\epsilon H_1 H_1^T$ *and* $\Psi > 0$, $Q_1 > 0$ *and* $Q_2 > 0$ *are the given matrices for the cost function (8.75). If this condition holds, stabilizing control law matrices* K_1, K_2 *are given by*

$$K_1 = N_1 W_1^{-1}, \; K_2 = N_2 W_2^{-1} \tag{8.84}$$

and the cost function (8.82) satisfies the following upper bound

$$J_{cp} < \sum_{k=0}^{\infty} x_{k+1}^T(0) W_1^{-1} x_{k+1}(0) + \int_0^{\infty} y_0^T(t) W_2^{-1} y_0(t) dt \tag{8.85}$$

Proof. Based on (8.77) we conclude that the controlled process is stable along the pass if the following matrix inequality holds

$$\begin{bmatrix} -P_2 & (\star) & (\star) \\ C^T P_2 + K_1^T D^T P_2 & \Psi_1 & (\star) \\ D_0^T P_2 + K_2^T D^T P_2 & B_0^T P_1 + K_2^T B^T P_1 + K_2^T \Psi K_1 & \Psi_2 \end{bmatrix} + \begin{bmatrix} 0 & (\star) & (\star) \\ \Delta C^T P_2 + K_1^T \Delta D^T P_2 & \Psi_3 & (\star) \\ \Delta D_0^T P_2 + K_2^T \Delta D^T P_2 & \Delta B_0^T P_1 + K_2^T \Delta B^T P_1 & 0 \end{bmatrix} < 0$$

where

$$\begin{aligned} \Psi_1 :=& A^T P_1 + P_1 A + K_1^T B^T P_1 + P_1 B K_1 + Q_1 + K_1^T \Psi K_1 \\ \Psi_2 :=& -P_2 + Q_2 + K_2^T \Psi K_2 \\ \Psi_3 :=& \Delta A^T P_1 + P_1 \Delta A + K_1^T \Delta B^T P_1 + P_1 \Delta B K_1 \end{aligned}$$

Now set $W_1 = P_1^{-1}$, $W_2 = P_2^{-1}$, $U_1 = W_1 Q_1 W_1$ and $U_2 = W_2 Q_2 W_2$ and then pre- and post- multiply both sides of this last inequality by $\text{diag}\{W_2, W_1, W_2\}$ to obtain

$$\begin{bmatrix} -W_2 & (\star) & (\star) \\ W_1 C^T + W_1 K_1^T D^T & \Psi_4 & (\star) \\ W_2 D_0^T + W_2 K_2^T D^T & W_2 B_0^T + W_2 K_2^T B^T + W_2 K_2^T \Psi K_1 W_1 & \Psi_5 \end{bmatrix} + \begin{bmatrix} 0 & (\star) & (\star) \\ W_1 \Delta C^T + W_1 K_1^T \Delta D^T & \Psi_6 & (\star) \\ W_2 \Delta D_0^T + W_2 K_2^T \Delta D^T & W_2 \Delta B_0^T + W_2 K_2^T \Delta B^T & 0 \end{bmatrix} < 0$$

where

$$\begin{aligned} \Psi_4 :=& W_1 A^T + A W_1 + W_1 K_1^T B^T + B K_1 W_1 + U_1 + W_1 K_1^T \Psi K_1 W_1 \\ \Psi_5 :=& -W_2 + U_2 + W_2 K_2^T \Psi K_2 W_2 \\ \Psi_6 :=& \Delta W_1 A^T + \Delta A W_1 + W_1 K_1^T \Delta B^T + \Delta B K_1 W_1 \end{aligned}$$

This last inequality can be rewritten as

$$\begin{bmatrix} -W_2 & (\star) & (\star) \\ W_1C^T + W_1K_1^TD^T & \Psi_7 & (\star) \\ W_2D_0^T + W_2K_2^TD^T & W_2B_0^T + W_2K_2^TB^T + W_2K_2^T\Psi K_1W_1 & \Psi_8 \end{bmatrix}$$
$$+ \begin{bmatrix} 0 & H_2 & H_2 \\ 0 & H_1 & H_1 \\ 0 & 0 & 0 \end{bmatrix} \begin{bmatrix} \mathcal{F} & 0 & 0 \\ 0 & \mathcal{F} & 0 \\ 0 & 0 & \mathcal{F} \end{bmatrix}$$
$$\times \begin{bmatrix} 0 & 0 & 0 \\ 0 & E_1W_1 + E_3K_1W_1 & 0 \\ 0 & 0 & E_2W_2 + E_3K_2W_2 \end{bmatrix}$$
$$+ \begin{bmatrix} 0 & 0 & 0 \\ 0 & W_1K_1^TE_3^T + W_2E_1^T & 0 \\ 0 & 0 & W_2K_2^TE_3^T + W_2E_2^T \end{bmatrix}$$
$$\times \begin{bmatrix} \mathcal{F}^T & 0 & 0 \\ 0 & \mathcal{F}^T & 0 \\ 0 & 0 & \mathcal{F}^T \end{bmatrix} \begin{bmatrix} 0 & 0 & 0 \\ H_2^T & H_1^T & 0 \\ H_2^T & H_1^T & 0 \end{bmatrix} < 0$$

where

$$\Psi_7 := W_1A^T + AW_1 + W_1K_1^TB^T + BK_1W_1 + U_1 + W_1K_1^T\Psi K_1W_1$$
$$\Psi_8 := -W_2 + U_2 + W_2K_2^T\Psi K_2W_2$$

and by an obvious application of the result of Lemma 5.4.2 we now obtain

$$\begin{bmatrix} -W_2 & (\star) & (\star) \\ W_1C^T + N_1^TD^T & \Psi_9 & (\star) \\ W_2D_0^T + N_2^TD^T & W_2B_0^T + N_2^TB^T + N_2^T\Psi N_1 & \Psi_{10} \end{bmatrix}$$
$$+ \epsilon \begin{bmatrix} 0 & H_2 & H_2 \\ 0 & H_1 & H_1 \\ 0 & 0 & 0 \end{bmatrix} \begin{bmatrix} 0 & 0 & 0 \\ H_2^T & H_1^T & 0 \\ H_2^T & H_1^T & 0 \end{bmatrix}$$
$$+ \epsilon^{-1} \begin{bmatrix} 0 & 0 & 0 \\ 0 & N_1^TE_3^T + W_2E_1^T & 0 \\ 0 & 0 & N_2^TE_3^T + W_2E_2^T \end{bmatrix}$$
$$\times \begin{bmatrix} 0 & 0 & 0 \\ 0 & E_1W_1 + E_3N_1 & 0 \\ 0 & 0 & E_2W_2 + E_3N_2 \end{bmatrix} < 0$$

where

$$\begin{aligned} \Psi_9 &= W_1A^T + AW_1 + N_1^TB^T + BN_1 + U_1 + N_1^T\Psi N_1 \\ \Psi_{10} &= -W_2 + U_2 + N_2^T\Psi N_2 \end{aligned}$$

and $N_1 = K_1W_1$ and $N_2 = K_2W_2$. Finally, making an obvious application of the Schur's complement formula to this last expression gives (8.83) and the proof is complete. □

Note 8.5.3. Note that it is possible to minimize the upper bound on the cost function (8.82) using the following optimization procedure

$$\min \left[\sum_{k=0}^{\infty} x_{k+1}^T(0) W_1^{-1} x_{k+1}(0) + \int_0^{\infty} y_0^T(t) W_2^{-1} y_0(t) dt \right]$$
$$= \min \left[\sum_{k=0}^{\infty} \text{trace}\left(W_1^{-1} x_{k+1}(0) x_{k+1}^T(0)\right) + \text{trace}\left(W_2^{-1} \int_0^{\infty} y_0(t) y_0^T(t) dt \right) \right]$$
subject to (8.83).

The convex optimization algorithm detailed in Note 8.5.1 for the discrete case cannot be applied here because of the nonlinear terms W_1^{-1} and W_2^{-1}. However, a control law which minimizes the guaranteed cost here can be obtained as follows.

First note that (using an elementary property of the trace of the product of two matrices)

$$\begin{aligned} \sum_{k=0}^{\bar{k}} x_{k+1}^T(0) W_1^{-1} x_{k+1}(0) &= \sum_{k=0}^{\bar{k}} \text{trace}\left(x_{k+1}^T(0) W_1^{-1} x_{k+1}(0)\right) \\ &= \sum_{k=0}^{\bar{k}} \text{trace}\left(W_1^{-1} x_{k+1}(0) x_{k+1}^T(0)\right) \end{aligned}$$

and

$$\begin{aligned} \int_0^{\infty} y_0^T(t) W_2^{-1} y_0(t) dt &= \int_0^{\infty} \text{trace}\left(y_0^T(t) W_2^{-1} y_0(t)\right) dt \\ &= \int_0^{\infty} \text{trace}\left(W_2^{-1} y_0(t) y_0^T(t)\right) dt \end{aligned}$$

where $\bar{k}$ denotes the number of passes completed. Also we can write

$$\begin{aligned} \Upsilon &= \Upsilon^{\frac{1}{2}} \Upsilon^{\frac{1}{2}} = \sum_{k=0}^{\bar{k}} x_{k+1}(0) x_{k+1}^T(0) \\ \Sigma &= \Sigma^{\frac{1}{2}} \Sigma^{\frac{1}{2}} = \int_0^{\infty} y_0(t) y_0^T(t) dt \end{aligned}$$

and if we introduce symmetric matrices Ξ, and Ω which satisfy

$$\begin{aligned} \text{trace}(\Upsilon^{\frac{1}{2}} W_1^{-1} \Upsilon^{\frac{1}{2}}) &< \text{trace}\ \Xi \\ \text{trace}(\Sigma^{\frac{1}{2}} W_2^{-1} \Sigma^{\frac{1}{2}}) &< \text{trace}\ \Omega \end{aligned}$$

then we can write

$$\begin{aligned} \Upsilon^{\frac{1}{2}} W_1^{-1} \Upsilon^{\frac{1}{2}} &< \Xi \\ \Sigma^{\frac{1}{2}} W_2^{-1} \Sigma^{\frac{1}{2}} &< \Omega \end{aligned}$$

An obvious application of the Schur's complement formula to this last expression now yields

$$\begin{bmatrix} -\Xi & \Upsilon^{\frac{1}{2}} \\ \Upsilon^{\frac{1}{2}} & -W_1 \end{bmatrix} < 0 \quad \text{and} \quad \begin{bmatrix} -\Omega & \Sigma^{\frac{1}{2}} \\ \Sigma^{\frac{1}{2}} & -W_2 \end{bmatrix} < 0 \tag{8.86}$$

respectively, and the optimization problem can be stated as

$$\begin{aligned} \min_{W_1>0, W_2>0, N_1, N_2} \quad & (\text{trace } \Xi + \text{trace } \Omega) \\ \text{subject to}: \quad & (8.83) \text{ and } (8.86) \end{aligned} \tag{8.87}$$

The solution for the control law matrices here now guarantees that the cost function is minimized over the finite pass length in the case when only a finite number of passes is actually completed. Since (8.87) is a convex optimization problem it poses no computational or accuracy problems.

To illustrate the control law design procedure just developed, we now give the following example.

Example 8.5.1. Consider the special case when the state-space model matrices A, B_0, C and D_0 are given by (8.60) and the boundary conditions are

$$x_{k+1}(0) = \begin{bmatrix} 0 \\ 0 \end{bmatrix}, \quad y_0(t) = 1, \quad 0 \le t \le \alpha$$

Suppose also that the matrices B, D, H_1, H_2, E_1, E_2 and E_3 are given by

$$\begin{aligned} B &= \begin{bmatrix} 1.2 \\ 0.8 \end{bmatrix}, \; D = 1.2, \; H_1 = \begin{bmatrix} 0.2 \\ 0.4 \end{bmatrix}, \; H_2 = 0.1, \\ E_1 &= 1.0 \cdot 10^{-3} \begin{bmatrix} 0.3 & 0.2 \end{bmatrix}, \; E_2 = 0.2, \; E_3 = 0.5 \end{aligned}$$

Assume also that

$$Q_1 = \begin{bmatrix} 10 & 0 \\ 0 & 10 \end{bmatrix}, \; Q_2 = 20, \; \Psi = 10$$

Then application of the procedure of (8.87) for 20 passes (i.e. $\overline{k} = 20$) and $\alpha = 10$ gives the solution matrices

$$\begin{aligned} W_1 &= \begin{bmatrix} 0.0256 & -0.0017 \\ -0.0017 & 0.0022 \end{bmatrix}, \; W_2 = 0.02249, \; \Sigma = 444.5218, \\ N_1 &= \begin{bmatrix} -0.0199 & -0.0025 \end{bmatrix}, \; N_2 = -0.0050, \quad \epsilon = 0.0069 \end{aligned}$$

The corresponding control law matrices are

$$K_1 = \begin{bmatrix} -0.8965 & -1.8275 \end{bmatrix}, \; K_2 = -0.2201$$

This control law guarantees stability along the pass and ensures that the cost bound (8.85) holds which has value 444.5218 (i.e. the value of trace Σ). Conversely, it follows from (8.85) that

$$\int_0^{10} y_0^T(t) W_2^{-1} y_0(t) dt = 10 W_2^{-1} = 444.642$$

8.6 H_2 and Mixed H_2/H_∞ Control of Differential Processes

8.6.1 The H_2 Norm and Stability Along the Pass

We consider again a differential unit memory linear repetitive process of the form defined by Example 1.2.3 and (in the case of zero boundary conditions) 2D transfer-function matrix $G(s, z_2)$ and 2D characteristic polynomial $\mathcal{C}(s, z_2)$ given by (1.53) (or any equivalent form) and (1.55) respectively.

Using the 1D case as motivation, consider a single input stable along the pass process (note again that this can be analyzed mathematically by letting the pass length $\alpha \to \infty$) and let the $m \times 1$ vector $g_k(t)$ denote the response to an impulse, denoted by $\delta_k(t)$, applied at $t = 0$ on pass k. Then, by invoking Parseval's theorem in the along pass direction on each pass and summing over the pass index, the H_2 norm is given by

$$\|G\|_2 := \sqrt{\|g_{k+1}(t)\|_2^2} = \sqrt{\sum_{k=0}^{\infty} \int_0^{\infty} g_{k+1}^T(t) g_{k+1}(t) dt}$$

To extend this definition to vector-valued inputs, introduce

$$u_k^h(t) := \delta_k(t) e^h$$

where e^h is the $r \times 1$ vector whose entries are zero except for a unit entry in position $h, 1 \leq h \leq r$. Then we have that

$$\|G\|_2 = \sqrt{\sum_{h=1}^{r} \sum_{k=0}^{\infty} \int_0^{\infty} g_{k+1}^{hT}(t) g_{k+1}^h(t) dt} \tag{8.88}$$

To determine $g_{k+1}^h(t)$, first introduce

$$\xi_{k+1}^h(t) := \begin{bmatrix} \dot{x}_{k+1}^h(t) \\ y_{k+1}^h(t) \end{bmatrix}, \; \zeta_{k+1}^h(t) := \begin{bmatrix} x_{k+1}^h(t) \\ y_k^h(t) \end{bmatrix}$$

and

$$\Omega := \begin{bmatrix} B \\ D \end{bmatrix}, \ \Psi := \begin{bmatrix} C & D_0 \end{bmatrix}$$

Then (recall again the definition of the matrices $\widehat{A}_1$ and $\widehat{A}_2$)

$$\xi_{k+1}^h(t) = (\widehat{A}_1 + \widehat{A}_2)\zeta_{k+1}^h(t) + \Omega\delta_{k+1}(t)e^h = \begin{cases} \widehat{\Omega}_h, & \text{for } k = 0 \\ (\widehat{A}_1 + \widehat{A}_2)\zeta_{k+1}^h(t), & \text{for } k > 0 \\ 0, & \text{otherwise} \end{cases}$$

and

$$g_{k+1}^h(t) = \Psi\zeta_{k+1}^h(t) + D\delta_{k+1}(t)e^h = \begin{cases} \widehat{D}_h, & \text{for } k = 0,\ 0 \leq t \leq \alpha \\ \Psi\zeta_{k+1}^h(t), & \text{for } k > 0,\ 0 \leq t \leq \alpha \\ 0, & \text{otherwise} \end{cases} \tag{8.89}$$

where $\widehat{D}_h$ and $\widehat{\Omega}_h$ denote column h of the matrices D and Ω respectively.

The following is the first major result in this general area and gives a sufficient condition for stability along the pass together with an upper bound on the H_2 norm of the 2D transfer-function matrix.

Theorem 8.6.1. *A differential unit memory linear repetitive process of the form defined by Example 1.2.3 is stable along the pass and has H_2 norm bound $\gamma_2 > 0$, i.e. $\|G\|_2 < \gamma_2$, if there exist matrices $P_1 > 0$ and $P_2 > 0$ such that the following LMIs hold*

$$\begin{bmatrix} -P_2 & P_2C & P_2D_0 \\ C^TP_2 & A^TP_1 + P_1A + C^TC & P_1B_0 + C^TD_0 \\ D_0^TP_2 & B_0^TP_1 + D_0^TC & -P_2 + D_0^TD_0 \end{bmatrix} < 0 \tag{8.90}$$

and

$$\text{trace}(D^TD + B^TP_1B + D^TP_2D) - \gamma_2^2 < 0 \tag{8.91}$$

Proof. Clearly, (8.90) can be rewritten as

$$\begin{bmatrix} -P_2 & P_2C & P_2D_0 \\ C^TP_2 & A^TP_1 + P_1A & P_1B_0 \\ D_0^TP_2 & B_0^TP_1 & -P_2 \end{bmatrix} + \begin{bmatrix} 0 \\ C^T \\ D_0^T \end{bmatrix} \begin{bmatrix} 0 & C & D_0 \end{bmatrix} < 0 \tag{8.92}$$

Also since the second term on the left hand side of (8.92) is nonnegative definite, it follows immediately that

$$\begin{bmatrix} -P_2 & P_2C & P_2D_0 \\ C^TP_2 & A^TP_1 + P_1A & P_1B_0 \\ D_0^TP_2 & B_0^TP_1 & -P_2 \end{bmatrix} < 0$$

which is easily shown to be equivalent to the 2D Lyapunov equation of (5.68) applied to this case and hence stability along the pass holds. Next, we will show that the prescribed H_2 performance bound γ_2 holds. To proceed, we use the Lyapunov function approach and introduce

$$\begin{aligned}\Delta V^h(k,t) &:= \dot{x}_{k+1}^{hT}(t)P_1x_{k+1}^h(t) + x_{k+1}^{hT}(t)P_1\dot{x}_{k+1}^h(t)\\ &+ y_{k+1}^{hT}(t)P_2y_{k+1}^h(t) - y_k^{hT}(t)P_2y_k^h(t)\end{aligned}$$

Also

$$\Delta V(k,t) = \sum_{h=1}^{r}\Delta V^h(k,t)$$

where

$$\Delta V^h(k,t) = \zeta_{k+1}^{hT}(t)\left(\widehat{A}_1^TP + P\widehat{A}_1 + \widehat{A}_2^TR\widehat{A}_2 - R\right)\zeta_{k+1}^h(t)$$

with P and R defined in (8.79). If stability along the pass holds then

$$\sum_{k=0}^{\infty}\int_0^{\infty}\Delta V^h(k,t)dt = \sum_{k=0}^{\infty}\int_0^{\alpha}\Delta V^h(k,t)dt = 0 \tag{8.93}$$

and this property also gives that $\lim_{t\to\infty} x_{k+1}(t) = 0$ and $\lim_{k\to\infty} y_k(t) = 0$. The result follows immediately on explicitly evaluating the expression defining this term. Further, using (8.88) and (8.89),

$$\begin{aligned}\|G\|_2^2 &= \sum_{h=1}^{r}\sum_{k=0}^{\infty}\int_0^{\infty} g_{k+1}^{hT}(t)g_{k+1}^h(t)dt\\ &= \sum_{h=1}^{r}\left(\widehat{D}_h^T\widehat{D}_h + \sum_{k=0}^{\infty}\int_0^{\infty}\zeta_{k+1}^{hT}(t)\Psi^T\Psi\zeta_{k+1}^h(t)dt\right)\end{aligned}$$

Obviously, (using (8.93)) we have that

$$\|G\|_2^2 = \sum_{h=1}^{r}\left(\widehat{D}_h^T\widehat{D}_h + \sum_{k=0}^{\infty}\int_0^{\infty}\zeta_{k+1}^{hT}(t)\Psi^T\Psi\zeta_{k+1}^h(t)dt + \sum_{k=0}^{\infty}\int_0^{\infty}\Delta V^h(k,t)dt\right) \tag{8.94}$$

and routine manipulations now show that (8.94) is equivalent to

$$\begin{aligned}\|G\|_2^2 &= \sum_{h=1}^{r}\Bigg(\widehat{D}_h^T\widehat{D}_h + \widehat{\Omega}_h^TW\widehat{\Omega}_h\\ &+ \sum_{k=0}^{\infty}\int_0^{\infty}\zeta_{k+1}^{T}(t)\left(\Psi^T\Psi + \widehat{A}_1^TP + P\widehat{A}_1 + \widehat{A}_2^TS\widehat{A}_2 - R\right)\zeta_{k+1}(t)dt\Bigg)\\ &= \mathrm{trace}(D^TD + \Omega^TW\Omega)\\ &+ \sum_{k=0}^{\infty}\int_0^{\infty}\zeta_{k+1}^{T}(t)\left(\Psi^T\Psi + \widehat{A}_1^TP + P\widehat{A}_1 + \widehat{A}_2^TS\widehat{A}_2 - R\right)\zeta_{k+1}(t)dt\\ &=: \chi_1 + \chi_2\end{aligned}$$

where $W := \text{diag}\{P_1, P_2\}$, $S := \text{diag}\{P_3, P_2\}$, and $P_3 > 0$ are any given matrices of the required dimensions. Also it follows immediately from (8.91) that $\chi_1 < \gamma_2^2$ and an obvious application of the Schur's complement formula to (the right-hand side of) χ_2 yields (8.90) (where we have also made use of the fact that $-P_3 < 0$). Also if (8.90) holds then $\chi_2 < 0$ and these facts together imply that $\chi_1 + \chi_2 < \gamma_2^2$. However, $\gamma_2 > 0$ by assumption and hence $||G||_2 < \gamma_2$ and the proof is complete.

□

Note 8.6.1. The H_2 norm bound here can be minimized using the following linear objective minimization procedure

$$\min_{P_1>0, P_2>0} \quad \mu$$

$$\text{subject to (8.90) and (8.91) with } \mu = \gamma^2$$

8.6.2 Static H_2 Control Law Design

As noted previously in this chapter, some applications areas will clearly require the design of control laws which guarantee stability along the pass and also have the maximum possible disturbance attenuation. Here we show how to address this question in an H_2 setting, for which the following is the relevant background.

The process state-space model considered has the following form over $0 \le t \le \alpha$, $k \ge 0$

$$\begin{aligned} \dot{x}_{k+1}(t) &= Ax_{k+1}(t) + Bu_{k+1}(t) + B_0 y_k(t) + B_{11} w_{k+1}(t) \\ y_{k+1}(t) &= Cx_{k+1}(t) + Du_{k+1}(t) + D_0 y_k(t) + B_{12} w_{k+1}(t) \end{aligned} \tag{8.95}$$

where $w_{k+1}(t)$ is an $q \times 1$ disturbance vector which belongs to L_2^q.

The control law employed is again given by (8.54) and the problem considered is as follows: for a given $\gamma_2 > 0$, find a control law of the form (8.54) for the process (8.95) such that the controlled process is stable along the pass and the H_2 norm of the 2D transfer-function matrix between the disturbance and pass profile vectors, denoted here by $G_d(s, z)$, is bounded by γ_2, i.e. $||G_d||_2 < \gamma_2$ – also termed the H_2 disturbance rejection attenuation (or bound). This transfer-function matrix is computed by replacing $\begin{bmatrix} B^T & D^T \end{bmatrix}^T$ with $\begin{bmatrix} B_{11}^T & B_{12}^T \end{bmatrix}^T$) in the equation defining the 2D transfer-function matrix between the input and pass profile vectors.

The following result gives a solution to this problem with an algorithm for designing the control law.

Theorem 8.6.2. *Suppose that a control law of the form (8.54) is applied to differential unit memory linear repetitive process described by (8.95). Then* *the resulting process is stable along the pass and has H_2 disturbance attenuation $\gamma_2 > 0$ if $\exists$ matrices $W_1 > 0$, $W_2 > 0$, N_1 and N_2 such that the following LMIs hold*

$$\begin{bmatrix} -W_2 & (\star) & (\star) & (\star) \\ N_1^T D^T + W_1 C^T & W_1 A^T + A W_1 + N_1^T B^T + B N_1 & (\star) & (\star) \\ N_2^T D^T + W_2 D_0^T & W_2 B_0^T + N_2^T B^T & -W_2 & (\star) \\ 0 & C W_1 + D N_1 & D_0 W_2 + D N_2 & -I \end{bmatrix} < 0 \tag{8.96}$$

and

$$\begin{aligned} &\operatorname{trace}(X) < \gamma_2^2 - \operatorname{trace}(B_{12}^T B_{12}) \\ &\begin{bmatrix} X & B_{11}^T & B_{12}^T \\ B_{11} & W_1 & 0 \\ B_{12} & 0 & W_2 \end{bmatrix} > 0 \end{aligned} \tag{8.97}$$

where X is a symmetric matrix of compatible dimensions. If these conditions hold, the control law matrices K_1 and K_2 are given by

$$K_1 = N_1 W_1^{-1}, \quad K_2 = N_2 W_2^{-1} \tag{8.98}$$

Proof. Applying Theorem 8.6.1 to the controlled process state-space model yields the following sufficient condition for stability along the pass

$$\begin{bmatrix} -P_2 & (\star) & (\star) & (\star) \\ K_1^T D^T P_2 + C^T P_2 & \Lambda_1 & (\star) & (\star) \\ K_2^T D^T P_2 + D_0^T P_2 & B_0^T P_1 + K_2^T B^T P_1 & -P_2 & (\star) \\ 0 & C + D K_1 & D_0 + D K_2 & -I \end{bmatrix} < 0$$

where $\Lambda_1 := A^T P_1 + P_1 A + K_1^T B^T P_1 + P_1 B K_1$. Now set $W_1 = P_1^{-1}$, $W_2 = P_2^{-1}$, and pre- and post-multiply both sides of this last inequality by $\operatorname{diag}\{W_2, W_1, W_2, I\}$ to obtain

$$\begin{bmatrix} -W_2 & (\star) & (\star) & (\star) \\ W_1 K_1^T D^T + W_1 C^T & \Lambda_2 & (\star) & (\star) \\ W_2 K_2^T D^T + W_2 D_0^T & W_2 B_0^T + W_2 K_2^T B^T & -W_2 & (\star) \\ 0 & C W_1 + D K_1 W_1 & D_0 W_2 + D K_2 W_2 & -I \end{bmatrix} < 0$$

where $\Lambda_2 := W_1 A^T + A W_1 + W_1 K_1^T B^T + B K_1 W_1$. Next, setting $N_1 = K_1 W_1$ and $N_2 = K_2 W_2$ now yields (8.96). Also in this case the inequality (8.91) becomes

$$\operatorname{trace}(B_{12}^T B_{12}) + \operatorname{trace}\left(\begin{bmatrix} B_{11}^T & B_{12}^T \end{bmatrix} \begin{bmatrix} W_1 & 0 \\ 0 & W_2 \end{bmatrix}^{-1} \begin{bmatrix} B_{11} \\ B_{12} \end{bmatrix} \right) < \gamma_2^2 \tag{8.99}$$

which is equivalent to (8.97). To see this, first note that this last expression is equivalent to

$$\begin{bmatrix} I & -B_{11}^T W_1^{-1} & -B_{12}^T W_2^{-1} \\ 0 & I & 0 \\ 0 & 0 & I \end{bmatrix} \begin{bmatrix} X & B_{11}^T & B_{12}^T \\ B_{11} & W_1 & 0 \\ B_{12} & 0 & W_2 \end{bmatrix} \begin{bmatrix} I & 0 & 0 \\ -W_1^{-1}B_{11} & I & 0 \\ -W_2^{-1}B_{12} & 0 & I \end{bmatrix}$$
$$= \begin{bmatrix} X - B_{11}^T W_1^{-1} B_{11} - B_{12}^T W_2^{-1} B_{12} & 0 & 0 \\ 0 & W_1 & 0 \\ 0 & 0 & W_2 \end{bmatrix}$$

and also

$$X > B_{11}^T W_1^{-1} B_{11} + B_{12}^T W_2^{-1} B_{12}$$

This guarantees that the constraint (8.97) is satisfied when (8.99) holds and the proof is complete. □

Note 8.6.2. The H_2 attenuation bound γ_2 in the LMI of (8.97) can be minimized using the linear objective minimization procedure

$$\min_{W_1>0, W_2>0, X, N_1, N_2} \quad \mu$$
$$\text{subject to (8.96) and (8.97) with } \mu = \gamma_2^2$$

8.6.3 The Mixed H_2/H_∞ Control Problem

Consider a differential unit memory linear repetitive process described by the following state-space model over $0 \le t \le \alpha$, $k \ge 0$,

$$\begin{aligned} \dot{x}_{k+1}(t) &= Ax_{k+1}(t) + Bu_{k+1}(t) + B_0 y_k(t) + B_{11} w_{k+1}(t) + B_{21} \nu_{k+1}(t) \\ y_{k+1}(t) &= Cx_{k+1}(t) + Du_{k+1}(t) + D_0 y_k(t) + B_{12} w_{k+1}(t) + B_{22} \nu_{k+1}(t) \end{aligned} \tag{8.100}$$

where $w_{k+1}(t)$ and $\nu_{k+1}(t)$ are disturbance vectors.

Now we address the question of when does there exist a control law of the form (8.54) which minimizes the H_2 norm from w to y, denoted here by $||G_{d2}||_2$, and keeps the H_∞ norm from ν to y, denoted here by $||G_{d1}||_\infty$, below some prescribed level. Note also that if only w is present then this problem reduces to the H_2 control problem already solved in this chapter. Similarly, if only ν is present then we obtain the H_∞ control problem also solved previously in this chapter.

Based on these last two facts, it is clear that there exists a control law of the form (8.54) which (for the resulting controlled process) minimizes the H_2 norm from w to y and keeps the H_∞ norm from ν to y below γ_∞ if the inequalities (8.96), (8.97) and

$$\begin{bmatrix} -R_2 & CR_1 + DK_1R_1 & D_0R_2 + DK_2R_2 & B_{22} & 0 \\ R_1C^T + R_1K_1^TD^T & R_1A^T + R_1K_1^TB^T + AR_1 + BK_1R_1 & B_0R_2 + BK_2R_2 & B_{21} & 0 \\ R_2D_0^T + R_2K_2^TD^T & R_2B_0^T + R_2K_2^TB^T & -R_2 & 0 & R_2 \\ B_{22}^T & B_{21}^T & 0 & -\gamma_\infty^2 I & 0 \\ 0 & 0 & R_2 & 0 & -I \end{bmatrix} < 0$$

hold for some $P_1 > 0$, $P_2 > 0$, $R_1 > 0$ and $R_2 > 0$.

The main problem now is that we cannot linearize simultaneously the terms K_1R_1, K_1P_1 and K_2R_2, K_2P_2 in this last expression. This can be overcome by enforcing $P_1 = R_1$ and $P_2 = R_2$ (at the possible expense of increased conservativeness). Under these assumptions, the following result provides the LMI condition for mixed H_2/H_∞ control law design.

Theorem 8.6.3. *Suppose that a control law of the form (8.54) is applied to a differential linear repetitive process described by (8.100). Then the resulting process is stable along the pass and has prescribed H_2 and H_∞ attenuation bounds $\gamma_2 > 0$ and $\gamma_\infty > 0$ respectively if $\exists$ matrices $W_1 > 0$, $W_2 > 0$, N_1 and N_2 such that the LMIs (8.96), (8.97) and*

$$\begin{bmatrix} -W_2 & (\star) & (\star) & (\star) & (\star) \\ W_1C^T + N_1^TD^T & W_1A^T + N_1^TB^T + AW_1 + BN_1 & (\star) & (\star) & (\star) \\ W_2D_0^T + N_2^TD^T & W_2B_0^T + N_2^TB^T & -W_2 & (\star) & (\star) \\ B_{22}^T & B_{21}^T & 0 & -\gamma_\infty^2 I & (\star) \\ 0 & 0 & W_2 & 0 & -I \end{bmatrix} < 0 \quad (8.101)$$

hold. If this is the case, then the H_2/H_∞ control law matrices K_1 and K_2 are given by (8.98).

Proof. This follows immediately from the results given or referenced above and hence the details are omitted. □

Note 8.6.3. A control law which minimizes the H_2 norm of the 2D transfer-function matrix from ν to y, and keeps the H_∞ norm of that from w to y below some prescribed level γ_∞, can be computed using the following convex optimization procedure

$$\begin{aligned} &\min_{W_1>0, W_2>0, X, N_1, N_2} \quad \mu \\ &\text{subject to} \;\; (8.96),\; (8.97) \text{ and } (8.101) \;\text{ with } \mu = \gamma_2^2 \end{aligned} \quad (8.102)$$

Note 8.6.4. This result is based on choosing a single Lyapunov function for both the H_2 and H_∞ criteria. In the 1D systems case this is a well known procedure termed the 'Lyapunov shaping paradigm' in the literature [162].

Recall that in 1D system theory the H_∞ norm is used as a measure of system robustness and here this last result can be interpreted as follows; keeping the H_∞ norm of the 2D transfer-function matrix from ν to y less than γ_∞ guarantees that the example considered is robust to unstructured perturbations of the form

$$\nu = \mathbf{\Delta} y, \quad \|\mathbf{\Delta}\|_\infty \leq \gamma_\infty$$

and, simultaneously, the performance cost (in the H_2 norm sense) is minimized. This means that choosing a lower value of γ_∞ reduces the process robustness and vice versa.

Note 8.6.5. Note that by adjusting γ_∞ we can trade-off between H_∞ and H_2 performance. Hence, a trade-off curve can be constructed for a given example which allows the designer to choose the control law which satisfies the compromise between robustness (measured with the H_∞ norm) and performance (measured with the H_2 norm).

8.6.4 H_2 Control of Uncertain Processes

In the analysis just completed, we have seen that the solution to the H_2/H_∞ control problem allows us to cope with uncertainty under an H_2 performance constraint. Hence, it is clear that we can immediately proceed to solve the H_2 control problem for differential linear repetitive processes in the case when there is norm bounded (or additive) uncertainty in the matrices A, B_0, C and D_0.

Consider now a differential unit memory linear repetitive process with the uncertainty modelled as additive perturbations to the nominal model matrices, resulting in the state-space model over $0 \leq t \leq \alpha$, $k \geq 0$

$$\begin{aligned} \dot{x}_{k+1}(t) &= (A + \Delta A)x_{k+1}(t) + Bu_{k+1}(t) + (B_0 + \Delta B_0)y_k(t) \\ &+ B_{11}w_{k+1}(t) \\ y_{k+1}(t) &= (C + \Delta C)x_{k+1}(t) + Du_{k+1}(t) + (D_0 + \Delta D_0)y_k(t) \\ &+ B_{12}w_{k+1}(t) \end{aligned} \tag{8.103}$$

where ΔA, ΔB_0, ΔC, ΔD_0 represent admissible uncertainties. We also assume that these uncertainty matrices satisfy

$$\begin{bmatrix} \Delta A & \Delta B_0 \\ \Delta C & \Delta D_0 \end{bmatrix} = \begin{bmatrix} H_1 \\ H_2 \end{bmatrix} \sigma^{-1} \mathcal{F} \begin{bmatrix} E_1 & E_2 \end{bmatrix} \tag{8.104}$$

Here H_1, H_2, E_1, E_2 are known constant matrices and $\mathcal{F}$ is an unknown matrix with constant entries which satisfies (8.15). The design parameter σ can be considered as a term available for use to attenuate the effects of the uncertainty.

It is well known fact in the 1D linear systems case that linear fractional transformations (LFT) provide a general framework for control law design in the presence of (certain types of) uncertainty. The basic use of an LFT (see, for example, [182]) is to isolate the uncertainty and represent it as interconnection of the nominal process with uncertainty block – see Fig. 8.4 (where LRP denotes an uncertainty free differential linear repetitive process of the form defined by Example 1.2.3).

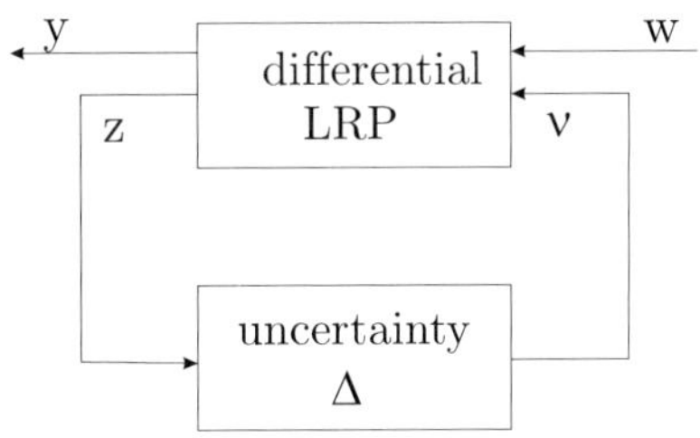

Fig. 8.4. Schematic of LFT representation for a differential process

Application of this technique to the process model of (8.103) (with the current pass input terms deleted as they do not have uncertainty associated with them) gives

$$\begin{aligned}\dot{x}_{k+1}(t) =& Ax_{k+1}(t) + B_0 y_k(t) + B_{11} w_{k+1}(t) + H_1 \nu_{k+1}(t)\\ y_{k+1}(t) =& Cx_{k+1}(t) + D_0 y_k(t) + B_{12} w_{k+1}(t) + H_2 \nu_{k+1}(t)\\ z_{k+1}(t) =& E_1 x_{k+1}(t) + E_2 y_k(t) + H_3 \nu_{k+1}(t)\\ \nu_{k+1}(t) =& \sigma^{-1}\mathcal{F} z_{k+1}(t)\end{aligned}$$

Now substituting the 3rd equation into the last one in the above set of four and solving for $\nu_{k+1}(t)$ gives

$$\nu_{k+1}(t) = \left(I - \sigma^{-1}\mathcal{F}H_3\right)^{-1}\sigma^{-1}\mathcal{F}\left(E_1 x_{k+1}(t) + E_2 y_k(t)\right)$$

The process model (8.103) can now be rewritten in the form of (8.100) as

$$\begin{aligned}\dot{x}_{k+1}(t) &= (A + H_1 \Upsilon E_1)x_{k+1}(t) + Bu_{k+1}(t)\\ &+ (B_0 + H_1 \Upsilon E_2)y_k(t) + B_{11} w_{k+1}(t)\\ y_{k+1}(t) &= (C + H_2 \Upsilon E_1)x_{k+1}(t) + Du_{k+1}(t)\\ &+ (D_0 + H_2 \Upsilon E_2)y_k(t) + B_{12} w_{k+1}(t)\end{aligned} \tag{8.105}$$

where

$$\Upsilon := \left(I - \sigma^{-1}\mathcal{F}H_3\right)^{-1}\sigma^{-1}\mathcal{F}$$

and clearly $\det(I - \sigma^{-1}\mathcal{F}H_3) \neq 0$ is required to give a well-posed feedback interconnection. In particular, setting $H_3 = 0$ in (8.105) gives this property and hence

$$\begin{aligned}
\dot{x}_{k+1}(t) &= (A + \sigma^{-1}H_1\mathcal{F}E_1)x_{k+1}(t) + Bu_{k+1}(t) \\
&+ (B_0 + \sigma^{-1}H_1\mathcal{F}E_2)y_k(t) + +B_{11}w_{k+1}(t) \\
y_{k+1}(t) &= (C + \sigma^{-1}H_2\mathcal{F}E_1)x_{k+1}(t) + Du_{k+1}(t) \\
&+ (D_0 + \sigma^{-1}H_2\mathcal{F}E_2)y_k(t) + B_{12}w_{k+1}(t)
\end{aligned}$$

which is of the form (8.103) and we immediately have the following result.

Theorem 8.6.4. *Suppose that a control law of the form (8.54) is applied to a differential unit memory linear repetitive process described by (8.103) with uncertainty structure satisfying (8.104) and (8.15). Then the resulting process is stable along the pass with H_2 norm bound $\gamma_2 > 0$ if $\exists$ matrices $W_1 > 0$, $W_2 > 0$, N_1 and N_2 such that the LMIs (8.96) and (8.97) of Theorem 8.6.2 hold together with*

$$\begin{bmatrix}
-W_2 & (\star) & (\star) & (\star) & (\star) \\
W_1C^T + N_1^TD^T & W_1A^T + AW_1 + BN_1 + N_1^TB^T & (\star) & (\star) & (\star) \\
W_2D_0^T + N_2^TD^T & W_2B_0^T + N_2^TB^T & -W_2 & (\star) & (\star) \\
H_2^T & H_1^T & 0 & -\sigma^2 I & (\star) \\
0 & E_1W_1 & E_2W_2 & 0 & -I
\end{bmatrix} < 0 \tag{8.106}$$

If these conditions hold, the stabilizing control law matrices K_1 and K_2 are given by (8.98).

Proof. Introduce the matrices

$$\begin{aligned}
\overline{A}_1 &:= \begin{bmatrix} A + BK_1 & B_0 + BK_2 \\ 0 & 0 \end{bmatrix}, \quad \overline{A}_2 := \begin{bmatrix} 0 & 0 \\ C + DK_1 & D_0 + DK_2 \end{bmatrix} \\
\overline{H}_1 &:= \begin{bmatrix} H_1 \\ 0 \end{bmatrix}, \quad \overline{H}_2 := \begin{bmatrix} 0 \\ H_2 \end{bmatrix}, \quad E := \begin{bmatrix} E_1 & E_2 \end{bmatrix}
\end{aligned}$$

Then, from the associated Lyapunov function, it follows immediately that a process described by (8.105) is stable along the pass if $\exists$ matrices $P_1 > 0$, $P_2 > 0$ and $P_3 > 0$ such that the following LMI holds

$$\left(\overline{A}_1 + \overline{H}_1\sigma^{-1}\mathcal{F}E\right)^T P + P\left(\overline{A}_1 + \overline{H}_1\gamma^{-1}\mathcal{F}E\right)$$

$$+ \left(\overline{A}_2 + \overline{H}_2\sigma^{-1}\mathcal{F}E\right)^T \mathcal{S}\left(\overline{A}_2 + \overline{H}_2\gamma^{-1}\mathcal{F}E\right) - R < 0$$

where $P := \operatorname{diag}\{P_1, 0\}$, $\mathcal{S} := \operatorname{diag}\{P_3, P_2\}$, $R := \operatorname{diag}\{0, P_2\}$. An obvious application of the Schur's complement formula to this last expression now yields

$$\begin{bmatrix} -P_2 & (\star) & (\star) \\ C^T P_2 + K_1^T D^T P_2 & A^T P_1 + P_1 A + P_1 B K_1 + K_1^T B^T P_1 & (\star) \\ D_0^T P_2 + K_2^T D^T P_2 & B_0^T P_1 + K_2^T B^T P_1 & -P_2 \end{bmatrix}$$

$$+ \begin{bmatrix} 0 \\ E_1^T \\ E_2^T \end{bmatrix} \mathcal{F}^T \begin{bmatrix} \sigma^{-1} H_2^T P_2 & \sigma^{-1} H_1^T P_1 & 0 \end{bmatrix} + \begin{bmatrix} \sigma^{-1} P_2 H_2 \\ \sigma^{-1} P_1 H_1 \\ 0 \end{bmatrix} \mathcal{F} \begin{bmatrix} 0 & E_1 & E_2 \end{bmatrix} < 0$$

Now make an obvious application of Lemma 5.4.2, and pre- and post-multiply the result by $\mathrm{diag}(\epsilon^{-\frac{1}{2}} I, \epsilon^{-\frac{1}{2}} I, \epsilon^{-\frac{1}{2}} I)$. Also introduce the notation $\overline{P}_1 := \epsilon^{-1} P_1$, $\overline{P}_2 := \epsilon^{-1} P_2$ and then an obvious application of the Schur's complement formula gives

$$\begin{bmatrix} -\overline{P}_2 & (\star) & (\star) & (\star) & (\star) \\ C^T \overline{P}_2 + K_1^T D^T \overline{P}_2 & A^T \overline{P}_1 + \overline{P}_1 A + \overline{P}_1 B K_1 + K_1^T B^T \overline{P}_1 & (\star) & (\star) & (\star) \\ D_0^T \overline{P}_2 + K_2^T D^T \overline{P}_2 & B_0^T \overline{P}_1 + K_2^T B^T \overline{P}_1 & -\overline{P}_2 & (\star) & (\star) \\ H_2^T \overline{P}_2 & H_1^T \overline{P}_1 & 0 & -\sigma^2 I & (\star) \\ 0 & E_1 & E_2 & 0 & -I \end{bmatrix} < 0$$

Note that this last condition is not linear in $\overline{P}_1$, $\overline{P}_2$, $\overline{P}_3$, K_1 and K_1. However, this difficulty can be avoided by employing the following transformations. First, pre- and post-multiply the last expression by $\mathrm{diag}(\overline{P}_2^{-1}, \overline{P}_1^{-1}, \overline{P}_2^{-1}, I, I)$ and finally set $W_1 = P_1^{-1}$, $W_2 = \overline{P}_2^{-1}$, $N_1 = K_1 \overline{P}_1^{-1}$, $N_2 = K_2 \overline{P}_2^{-1}$ to obtain (8.106). □

Note 8.6.6. A control law which minimizes σ and keeps the H_2 norm of the 2D transfer-function matrix from ν to y below some prescribed level γ_2, can be computed using the following minimization procedure

$$\min_{W_1 > 0, W_2 > 0, X, N_1, N_2} \mu$$
$$\text{subject to (8.96), (8.97) and (8.106) with } \mu = \sigma^2$$

which, due to the presence of the term σ^{-1} in the uncertainty model of (8.104), provides the essential advantage of allowing us to extend the uncertainty boundaries, i.e. increase the robustness.

The last result (i.e. the LMI (8.106)) shows that there exists a link between mixed H_2/H_∞ and robust H_2 control problems. This means that the same control law solves the mixed H_2/H_∞ and robust H_2 control problems. To see this, assume that $\sigma = \gamma_\infty$ and apply the same transformation used to obtain (8.101) from (8.56) where the matrices $\widehat{B}_1$, $\widehat{D}_1$ and L are now given by

$$\widehat{B}_1 := \begin{bmatrix} H_1 \\ 0 \end{bmatrix}, \ \widehat{D}_1 := \begin{bmatrix} 0 \\ H_2 \end{bmatrix}, \ L := \begin{bmatrix} E_1 & E_2 \end{bmatrix}$$

This link can also be established using the Lyapunov stability condition of Lemma 8.4.1. In particular from (8.15)

$$\sigma^{-2}\mathcal{F}^T\mathcal{F} \leq \sigma^{-2}I$$

and hence, assuming $\sigma = \gamma_\infty$, we have

$$\begin{aligned}\nu_{k+1}^T(t)\nu_{k+1}(t) &= \gamma_\infty^{-2} z_{k+1}^T(t)\mathcal{F}^T\mathcal{F}z_{k+1}(t)\\ &\leq \gamma_\infty^{-2} z_{k+1}^T(t)z_{k+1}(t)\end{aligned}$$

which can be rewritten as

$$\gamma_\infty^2\nu_{k+1}^T(t)\nu_{k+1}(t) - z_{k+1}^T(t)z_{k+1}(t) \leq 0$$

Hence the inequality

$$\Delta V(k,t) + z_{k+1}^T(t)z_{k+1}(t) - \gamma_\infty^2\nu_{k+1}^T(t)\nu_{k+1}(t) < 0 \tag{8.107}$$

can hold only if the term $\Delta V(k,t) < 0$, i.e. stability along the pass holds. Also this inequality can be regarded as arising from the associated Hamiltonian for differential linear repetitive processes – see [132] for further details (in addition to its use in the H_∞ analysis in this chapter). Moreover, if (8.107) holds then the process is stable along the pass and the H_∞ norm from ν to z is kept below the prescribed level $\gamma_\infty > 0$. Finally, routine manipulations establish the link between robust H_2 control and H_2/H_∞ control detailed above.

Example 8.6.1. Consider the following special case of Example 1.2.3

$$A = \begin{bmatrix} -0.0050 & -5.8077 \\ 1 & -0.0050 \end{bmatrix},\ B_0 = \begin{bmatrix} 0 \\ 0.0494 \end{bmatrix},\ C = \begin{bmatrix} 1 & 0 \end{bmatrix}, D_0 = 0.7692$$

which is easily shown to be unstable along the pass. The boundary conditions are

$$x_{k+1}(0) = \begin{bmatrix} 1 \\ 1 \end{bmatrix},\ y_0(t) = 1,\ 0 \leq k \leq 10, 0 \leq t \leq 2$$

Suppose also that the matrices B, D, B_{11}, B_{12} are given by

$$B = \begin{bmatrix} 0.9 \\ 0.2 \end{bmatrix},\ D = 0.6,\ B_{11} = \begin{bmatrix} 0.9 \\ 1.3 \end{bmatrix},\ B_{12} = 1.2$$

and the uncertainty is modelled with

$$H_1 = \begin{bmatrix} 0.9 \\ 0.8 \end{bmatrix},\ H_2 = 1.3,\ E_1 = \begin{bmatrix} -0.8 & -0.7 \end{bmatrix},\ E_2 = 1.3$$

Then executing the design procedure of Theorem (8.6.4) in Matlab gives the solution as $\sigma = 2.8$ and

$$\begin{aligned} W_1 &= \begin{bmatrix} 1.3567 & -0.0541 \\ -0.0541 & 0.2257 \end{bmatrix},\ W_2 = 0.2509 \\ N_1 &= \begin{bmatrix} -2.5479 & -0.1611 \end{bmatrix},\ N_2 = -0.3204 \end{aligned}$$

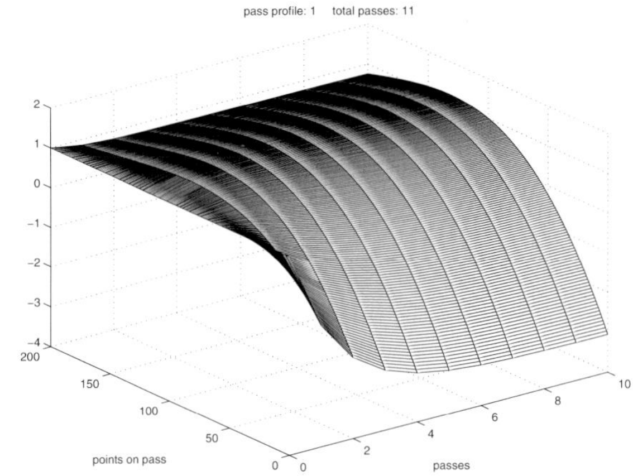

Fig. 8.5. Controlled process pass profile sequence

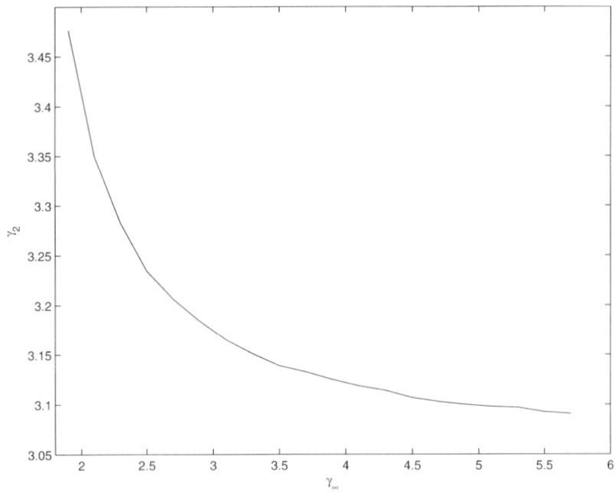

Fig. 8.6. Trade-off curve.

The corresponding control law matrices are

$$K_1 = \begin{bmatrix} -1.9248 & -1.1749 \end{bmatrix}, \; K_2 = -1.2772$$

This controller guarantees a stable along the pass process for all admissible uncertainties and ensures that the $\mathcal{H}_2$ norm bound is never greater than 3.9730.

To illustrate the application of the designed control law, Fig. 8.5 shows the pass profile sequence generated in the case when $\mathcal{F} = 0.8$.

Suppose also that the $\mathcal{H}_2$ disturbance rejection bound is set as $\beta\gamma_2$, where β is a given positive scalar. Then, by adjusting β we can trade off the $\mathcal{H}_2$ performance and the level of uncertainty attenuation σ. This leads to a trade-off curve between σ and γ_2 as shown in Fig. 8.6.

9 Application to Iterative Learning Control

Summary. In this chapter, the application of repetitive process theory to iterative learning control is considered. The first part uses the stability theory of Chap. 2 to establish a fundamental link between convergence of a powerful class of such schemes and the resulting error dynamics. Following this, the norm optimal approach is developed and an experimental verification on a chain conveyor system described. Finally, some highly promising results on robust control are obtained and experimentally verified on a gantry robot system.

9.1 Stability and Convergence of ILC Schemes

The state-space model of the plant to be controlled by an ILC scheme is of the form given in Eample 1.2.4, i.e.

$$\begin{aligned} \dot{x}_k(t) &= Ax_k(t) + Bu_k(t),\ 0 \leq t \leq T \\ y_k(t) &= Cx_k(t) \end{aligned} \tag{9.1}$$

where on trial k, $x_k(t)$ is the $n \times 1$ state vector, $y_k(t)$ is the $m \times 1$ output vector, $u_k(t)$ is the $r \times 1$ vector of control inputs, and the trial length $T < \infty$. If the signal to be tracked is denoted by $r(t)$ then $e_k(t) = r(t) - y_k(t)$ is the error on trial k. Also without loss of generality in this section (except where stated) we set $x_k(0) = 0,\ k \geq 0$.

The class of ILC schemes considered in this section are those defined by Example 1.2.4, where on trial $k+1$ the control input is calculated using

$$u_{k+1}(t) = \sum_{j=1}^{M} \alpha_j u_{k+1-j}(t) + \sum_{j=1}^{M} (K_j e_{k+1-j})(t) + (K_0 e_{k+1})(t) \tag{9.2}$$

In addition to the 'memory' M, the design parameters in this control law are the static scalars α_j, $1 \leq j \leq M$, the linear operator K_0 which describes the current trial error contribution, and the linear operator K_j, $1 \leq j \leq M$, which describes the contribution from the error on trial $k+1-j$.

It has been shown in Example 1.2.4 that the closed-loop (or controlled) process in this case can be expressed as a special case of the abstract model S (Definition 1.2.1) in the form

$$\hat{e}_{k+1} = L_T \hat{e}_k + b \tag{9.3}$$

where

$$\hat{e}_k(t) = \left[\begin{array}{ccc} e^T_{k+1-M}(t) & \cdots & e^T_k(t) \end{array} \right]^T$$

is the so-called error super-vector.

It is now routine to argue that convergence of a closed-loop ILC scheme of the form considered here as $k \to \infty$ is equivalent to asymptotic stability of its linear repetitive process interpretation. Hence by direct application of Definition 2.1.1 and Theorem 2.1.1 we have the following result.

Theorem 9.1.1. *An ILC scheme of the form defined by (9.1) and (9.2) is asymptotically stable and hence converges if, and only if, all roots of*

$$z^M - \alpha_1 z^{M-1} - \cdots - \alpha_{M-1} z - \alpha_M = 0 \tag{9.4}$$

have modulus strictly less than unity.

Recall from the analysis of Example 1.2.4 that the error dynamics on trial $k+1$ here can be written in convolution form as

$$e_{k+1}(t) = r(t) - (Gu_{k+1})(t), \ 0 \leq t \leq T$$

Suppose also that the condition of Theorem 9.1.1 holds. Then the closed-loop error dynamics converge (use the limit profile concept starting from Definition 2.1.3 and Theorems 2.1.1 and 2.1.2) in the norm topology of $L_p[0,T]$ to

$$e_\infty = (I + GK_{\text{eff}})^{-1} r \tag{9.5}$$

where the so-called effective controller K_{eff} is given by

$$K_{\text{eff}} := \frac{K}{1-\beta}$$

and

$$\beta := \sum_{i=1}^{M} \alpha_i, \quad K := \sum_{i=0}^{M} K_i$$

(The simplest way to obtain (9.5) is to replace all variables by their strong limits and then re-arrange.)

The following result, whose proof follows by direct application of repetitive process stability theory (effectively a special case of Lemma 2.2.1) and is hence omitted here, gives a bound on the error sequence under Theorem 9.1.1.

Theorem 9.1.2. *Suppose that the condition of Theorem 9.1.1 holds. Then the resulting error sequence is bounded by an expression of the form*

$$||\hat{e}_k - \hat{e}_\infty|| \leq M_1 \{\max(||e_0||, \cdots, ||e_{N-1}||) + M_2\} \lambda_e^k$$

where M_1 and M_2 are positive real scalars, and $\lambda_e \in (\max|\mu_i|, 1)$ where μ_i, $1 \leq i \leq M$, is a solution of (9.4).

As in all repetitive process theory, the result of Theorem 9.1.1 is counter-intuitive in the sense that stability is largely independent of the plant and the controllers used. This is a direct result of the fact that the trial duration T is finite and over such an interval a linear system can only produce a bounded output irrespective of its stability properties and in this definition unstable outputs of this kind are still acceptable. Hence even if the error sequence generated is guaranteed to converge to a limit, this terminal error may be unstable and/or possibly worse than the first trial error, i.e. the use of ILC has produced no improvement in performance. To guarantee an acceptable (i.e. stable (as the most basic requirement)) limit error the stronger concept of stability along the trial (see below) has to be used.

Theorems 9.1.1 and 9.1.2 can be used to derive important results on parameter selection. In particular, we have the following (see [126]).

(a) Convergence is predicted to be 'rapid' if λ_e is small and will be geometric in form, converging approximately with λ_e^k. The limit error is nonzero but is usefully described by a (1D linear systems) unity negative feedback system with effective controller K_{eff} defined above. If $\max_i(|\mu_i|) \to 0+$ then the limit error is essentially the first learning iterate, i.e. use of ILC has little benefit and will simply lead to the normal large errors encountered in simple feedback loops. There is hence pressure to let $\max_i|\mu_i|$ be close to unity when K_{eff} is a high gain controller which will lead (roughly speaking) to small limit errors.

(b) Zero limit error can only be achieved if $b = 0$ in (9.3) which, in turn, requires that $\sum_{i=1}^{M} \alpha_i = 1$ which is not possible if $r(L_T) < 1$ (but is possible for the case of $r(L_T) = 1$). This situation is reminiscent of classical control where the inclusion of an integrator (on the stability boundary) into the controller results in zero steady state (limit) error in response to constant reference signals.

There is a conflict in the above conclusions which has implications on the systems and control structure from both the theoretical and practical points of view. In particular, consider for ease of presentation the case when $K_i = 0,\ 1 \leq i \leq M$. Then small learning errors will require high effective gain yet GK_0 should be stable under such gains.

High gain feedback systems are described in essential detail by the system root-locus. The general multiple-input multiple-output case requires application of the theory of multivariable root-loci (see, for example [122]), which is somewhat complex and does not permit the useful connection between the frequency and time domain properties of the system as in the SISO case. Fortunately, however, the two cases are structurally similar and hence we only consider the latter here.

In the SISO case, the results can be summarized as the requirement that GK_0 is minimum phase and relative degree one or two. This leads to the heuristics given next.

(a) Convergence is rapid for well-conditioned relative degree one minimum phase systems, particularly if GK_0 has interlaced poles and zeros and has positive high frequency gain. (This result is proved by a detailed pole-zero and residue analysis of the limit error dynamics.)
(b) Convergence is guaranteed for well-conditioned relative degree two systems which are minimum phase and have a positive high frequency gain with a negative intercept/pivot of the root locus of GK_0. Convergence in this case is, however, almost always slower than in the relative degree one case, is non-uniform in the derivative, and theoretically leads to limit errors of amplitude of the order of (considering $M = 2$ for ease of presentation) $1 - \alpha_1 - \alpha_2$ and in the form of an exponentially decaying oscillation with a high frequency of the order of $(1 - \alpha_1 - \alpha_2)^{-\frac{1}{2}}$.

In what follows we consider the application of stability along the trial to ILC schemes which can be written in the abstract form (9.3). The starting point is to note that there are two essential problems with asymptotic stability applied to ILC schemes.

The first of these is that only statements about the situation after an infinite number of trials have occurred are possible and little information is available concerning performance from trial-to-trial. The second problem is (as noted above) that although a limit error is guaranteed to exist, it could well have unacceptable dynamic characteristics. In particular, exponentially growing (unstable) signals can be accepted because of the finite trial length and, in practice, this is clearly undesirable. In the next section, we will introduce the concept of norm optimal control to deal with the first problem and below we address the second using the concept of stability along the trial, (Definition 2.2.1 and Theorem 2.2.1).

To illustrate the application of the stability along the trial property, it is instructive (see, for example, [6] and [128]) to consider the uncontrolled case where the trial-to-trial error dynamics are defined by the integral operator $L : e_k \rightarrow e_{k+1}$

$$e_{k+1}(t) = L(e_k(\cdot))(t) = De_k(t) + C \int_0^t e^{A(t-\tau)} B\, e_k(\tau)\, d\tau, \ \ 0 \leq t \leq T \quad (9.6)$$

with $E_T = C([0,T]; \mathbb{R}^m)$ the space of bounded continuous real-valued functions over the interval $0 \leq t \leq T$ on which the norm is defined as $||e(t)|| = \sup_{0 \leq t \leq T} ||e(t)||_m$, where $|| \cdot ||_m$ is any convenient norm in $\mathbb{R}^m$, e.g. $||y||_m = \max_{1 \leq i \leq m} |y_i|$. In this case, asymptotic stability is easily shown to be equivalent to $r(D) < 1$; to provide a physical explanation of this property suppose that $r(D) \leq ||D|| < 1$ (a sufficient condition) holds. Then $e_{k+1}(0) = De_k(0)$ is reduced from trial-to-trial, i.e. the sequence of initial errors is reduced from trial-to-trial (since it only depends on D). By continuity of e_{k+1}, this also occurs for t slightly greater than zero. In this way, the matrix D 'squeezes' the error to zero, starting from $t = 0$ and working to $t = T$. Unfortunately, depending on the state-space triple $\{A, B, C\}$, it could be that for $t \gg 0$ the error actually increases over the first few trials and it takes a large number of trials before the error is small everywhere. This is shown by first noting that asymptotic stability guarantees (see Lemma 2.2.1) the existence of real scalars $M_T > 0$ and $\lambda_T \in (0, 1)$ such that

$$||e_k - e_\infty|| \leq M_T \lambda_T^k \left\{ ||e_0|| + \frac{||b||}{1 - \lambda_T} \right\}, \; k \geq 0 \tag{9.7}$$

The term involving λ_T in this last equation relates to the error reduction due to D and the term M_T relates to, and depends on, the system structure defined by the state-space triple $\{A, B, C\}$. This whole process can be visualized as squeezing something out of a tube, e.g. toothpaste, where, when the end is already flat, a bulge develops in the middle and after 'squeezing' long enough everything drops out at the end.

The analysis of this case can be usefully extended by adding a disturbance term $b_{k+1}(t)$, $k \geq 0$, to the right-hand side of (9.6) arising, for example, from non-zero state initial conditions on each trial of the form $x_{k+1}(0) = d_{k+1}$, $k \geq 0$, when

$$b_{k+1}(t) = Ce^{At} d_{k+1}, \; k \geq 0 \tag{9.8}$$

In this case if the limit disturbance $b_\infty(t)$ is non-zero then (as noted earlier) the limit error is given by $e_\infty = (I - L_T)^{-1} b_\infty$ or, in state-space form,

$$\begin{aligned} \dot{x}_\infty(t) &= (A + B(I_m - D)^{-1}C)x_\infty(t) + B(I_m - D)^{-1}b_\infty(t), \; x_\infty(0) = d_\infty \\ e_\infty(t) &= (I_m - D)^{-1}Cx_\infty(t) + (I_m - D)^{-1}b_\infty(t) \end{aligned}$$

where d_∞ is the strong limit of the sequence $\{d_k\}_{k \geq 1}$.

In this last equation, $(I_m - D)^{-1}$ exists since $r(D) < 1$ but the system is unstable in the standard, or 1D, sense unless all eigenvalues of $A + B(I_m - D)^{-1}C$ have strictly negative real parts. If this is not the case then an exponentially growing limit error with an unacceptable magnitude for most practical applications could arise. Such cases require the stronger concept of stability along the trial. In particular, with $E_{T_e} = C([0, T_e]; \mathbb{R}^m)$ where T_e is arbitrary and not necessarily finite, we obtain the following result for stability along the trial in the case when the error dynamics are of the form (9.6). (The proof of this result is again omitted since it is a direct application of repetitive process stability theory.)

Theorem 9.1.3. *Suppose that the pair $\{A, B\}$ is controllable and the pair $\{C, A\}$ is observable. Then the ILC process (9.6) with $T_e \geq T$ is stable along the trial if, and only if,*

(a) $r(D) < 1$, $\det(sI - A) \neq 0$, $\mathrm{Re}(s) \geq 0$; *and*
(b) all eigenvalues of the transfer-function matrix

$$L(s) := C(sI_n - A)^{-1}B + D$$

have modulus strictly less than unity for $s = \imath\,\omega,\ \omega \geq 0$, *where* $L(s)$ *is termed the learning transmission operator.*

This last theorem has a well defined physical interpretation. In particular, first note that since stability along the trial is essentially an infinite time concept, the use of the Laplace transform is possible (and is actually used in the theorem!). Taking the Laplace transform of (9.6) with the disturbance term $b_{k+1}(t)$ added leads to the following recursive equation in the Laplace domain

$$e_{k+1}(s) = L(s)e_k(s) + b_{k+1}(s)$$

Theorem 9.1.3 then requires that $L(s)$ is a stable transfer-function matrix with a minimal realization which satisfies

$$r(L(\imath\,\omega)) < 1,\ \forall\,\omega \geq 0 \tag{9.9}$$

i.e. the error transmission operator must satisfy the spectral radius condition at all frequencies. In contrast, asymptotic stability alone requires this at only one point, i.e. infinity (since $D = \lim_{\omega\to\infty} L(\imath\,\omega)$)

Consider now the SISO case when (9.9) reduces to

$$|L(\imath\,\omega)| < 1,\ \forall\,\omega \geq 0$$

Also if $b_k = 0$, the error $|e_k(\imath\,\omega)|$ is pointwise bounded by

$$|e_k(\imath\,\omega)| \leq |L(\imath\,\omega)|^k |e_0(\imath\,\omega)|$$

Hence the error in the frequency domain is monotonically reduced at the geometric rate α_∞ where

$$\alpha_\infty := \sup_{\omega\geq 0} |L(\imath\,\omega)|$$

This connects Theorem 9.1.3 in this case to Definition 2.2.1 with $M_\infty = 1$ and $\lambda_\infty = \alpha_\infty$ easily computable.

It is of interest to compare the above stability analysis with alternatives for linear ILC schemes. The original work [13] contained a convergence analysis adapted to a specific algorithm for a continuous-time plant. The definition of convergence is similar to that considered here, i.e. the system is convergent if the error with $b_{k+1}(t) = 0$, $k \geq 0$, goes to zero as $k \to \infty$ with respect to the supremum-norm in $[0, T]$. Their condition (expressed in the notation used here) states that the system is convergent if $||D|| < 1$.

In contrast to Definition 2.2.1 for the general case adopted here, the analysis of [13] has no perturbation γ on the error transmission model and the disturbance $b_k = 0$, for all k. Also the proof of the main result in this work used the so-called λ-norm

$$||e(t)||_\lambda := \sup_{0 \le t \le T} e^{-\lambda t}|e(t)|, \ \lambda > 0$$

The proof proceeds by showing that if $||D|| < 1$ then there exists a $\lambda > 0$ such that

$$||e_{k+1}||_\lambda \le \rho ||e_k||_\lambda \tag{9.10}$$

holds with $\rho < 1$. Hence for $\lambda > 0$

$$e^{-\lambda t}||e(t)|| \le ||e(t)||_\lambda \le ||e(t)|| \tag{9.11}$$

and therefore if (9.10) holds convergence also occurs in the supremum-norm since from (9.10) and (9.11) it follows that

$$\begin{aligned} \lim_{k\to\infty} ||e_k(t)|| &\le \lim_{k\to\infty} e^{\lambda T}||e_k(t)||_\lambda \le e^{\lambda T}\rho^k \lim_{k\to\infty} ||e_0||_\lambda \\ &= e^{\lambda T}||e_0||_\lambda \lim_{k\to\infty} \rho^k = 0 \end{aligned}$$

This last result provides an interesting connection (missing from the many other ILC stability results derived by this route) to (9.7). In particular, the constants M_T and λ_T in this case are given by

$$M_T \le e^{\lambda T}, \ \lambda_T = \rho$$

Note also that M_T could be 'very large' (for 'large' λ and T) and hence it could take a 'very large' number of trials before the error is actually reduced in terms of the usual supremum-norm.

The work in [130] produces a convergence condition for the continuous-time case by frequency domain methods. Use of the Laplace transform requires that all signals are given in the time interval $[0, \infty)$ instead of the usual ILC finite range of $[0, T]$. This was assumed in [130] with the justification that the work was aimed at mathematical rigor and unification. In this case, the system is said to be L_2-convergent if $||e_k(t)||_{L_2[0,\infty)} \to 0$ as $k \to \infty$. The formal result is as follows.

Theorem 9.1.4. *Suppose that the pair $\{A, B\}$ is controllable and the pair $\{C, A\}$ is observable. Then the ILC system (9.6) with (9.8) added is L_2-convergent if the initial error is bounded, i.e. $e_0(t) \in L_2[0, \infty)$, the disturbance vector b_k is independent of k, i.e. $b_k(t) = b_\infty(t) \in L_2[0, \infty)$, and*

$$||C(sI_n - A)^{-1}B + D||_\infty \le 1$$

and the set of ω such that $||C(\imath\omega I_n - A)^{-1}B + D||_\infty = 1$ is of measure zero. Here $||\cdot||_\infty$ is the standard H_∞ norm, i.e. $||L(s)||_\infty := \sup_{0\le\omega<\infty} ||L(\imath\omega)||$

where $||L(\imath\omega)||$ *denotes any matrix norm of* $L(\imath\omega)$ *at a given* ω*, e.g. the spectral norm* $\overline{\sigma}(L(\imath\omega))$*, where again* $\overline{\sigma}(\cdot)$ *denotes the largest singular value of its matrix argument.*

This result gives a sufficient condition for stability as opposed to the necessary and sufficient conditions of Theorem 9.1.3. Note, however, that the definitions of stability are slightly different in that the repetitive process stability theory from which Theorem 9.1.3 is derived includes a degree of robustness (the model perturbation γ in the basic theory) and requires for this reason a strict inequality but the condition developed in [130] has no implicit robustness and may allow that $||G(\imath\omega)||_{\infty} = 1$. The difference is important since it often happens in SISO linear ILC schemes that the learning transmission operator L is proper with $D = 1$. Hence (noting that $\lim_{\omega \to \infty} L(\imath\omega) = D$) such a system is never convergent in terms of Theorem 9.1.3 but it could be in terms of Theorem 9.1.4. The difference only arises due to the introduction of a perturbation in the definition of convergence and this makes Theorem 9.1.3 robust in applications terms.

It has long been known that systems theory for 2D discrete and (more recently) continuous-discrete 2D linear systems described by well known and extensively studied state-space models can be applied to iterative learning control. In the discrete case, the 2D systems state-space model is that due to Roesser but alternatives (such as the Fornasini-Marchesini class) could also be used. For example, [100] reported work in this area for ILC in the linear discrete time case, and for linear continuous-time systems [31] (and others) have reported designs using the theory of so-called 2D continuous-discrete linear systems – see, for example, [92].

Focusing on the latter case, recall that differential linear repetitive processes are distinct from the classes of 2D continuous-discrete linear systems studied by [92] (and others) in that information propagation in one of the two separate directions only occurs over a finite duration – the trial length. This has a more natural link with ILC and hence it is to be expected that the linear repetitive process theory will be more appropriate to the analysis and design of the relevant class of ILC schemes. In what follows, we demonstrate that the results in [31] are a special case of the repetitive process based ILC theory given in this section with the added advantage of making available computable information on performance. A similar conclusion holds for the corresponding discrete case but the details are omitted here for brevity.

In [31], the plant to be controlled has the form of (9.1) but (for consistency) we will follow notation in this paper and write it in the form

$$\begin{aligned} \frac{\partial x(t,k)}{\partial t} &= Ax(t,k) + Bu(t,k) \\ y(t,k) &= Cx(t,k) \end{aligned} \tag{9.12}$$

Then in this setting a general learning law is defined as

$$u(t,k+1) = u(t,k) + \Delta u(t,k) \tag{9.13}$$

where $\Delta u(t,k)$ denotes a modification of the control input. Convergence is defined as follows.

Definition 9.1.1. *The learning law (9.13) is said to be convergent if, and only if, for any $x(0,k) = x_0$, where x_0 is a given $n \times 1$ vector with known constant entries, and any initial control sequence $\{u(t,0)\}$ it generates a control sequence $\{u(t,k)\}$ for the system (9.12)–(9.13) such that $y(t,k) \to r(t)$, $u(k,t) \to u_\infty(t)$, $0 \leq t \leq T$, $k \to \infty$.*

The learning law actually used in [31] is obtained by first writing $e(t,k) = r(t) - y(t,k)$ and introducing

$$\eta(t,k) := \int_0^t [x(t,k+1) - x(t,k)]\, dt$$

Then $\eta(0,k) = 0$, and, after some routine analysis,

$$\frac{\partial \eta(t,k)}{\partial t} = A\eta(t,k) + B\int_0^t \Delta u(\tau,k)\, d\tau \tag{9.14}$$

Also

$$e(t,k+1) - e(t,k) = -CA\eta(t,k) - CB\int_0^t \Delta u(\tau,k)\, d\tau \tag{9.15}$$

and if $r(t)$ is differentiable the learning law can be expressed as

$$\Delta u(t,k) = K\frac{\partial e(t,k)}{\partial t}$$

Using (9.14) and (9.15) the so-called control error system can be written as a 2D continuous-discrete Roesser type model of the form

$$\begin{bmatrix} \frac{\partial \eta(t,k)}{\partial t} \\ e(t,k+1) \end{bmatrix} = \begin{bmatrix} A & BK \\ -CA & I - CBK \end{bmatrix} \begin{bmatrix} \eta(t,k) \\ e(t,k) \end{bmatrix} + \begin{bmatrix} -BK \\ CBK \end{bmatrix} e(0,k)$$

with boundary conditions

$$\begin{aligned} &\eta(0,k) = 0,\ k \geq 0 \\ &e(t,0) = r(t) - Ce^{At}x_0 - C\textstyle\int_0^t e^{A(t-\tau)}Bu(\tau,0)\, d\tau,\ t > 0 \end{aligned}$$

Also we have that $e(0,k) = r(0) - Cx_0 = 0$, and hence

$$\begin{bmatrix} \frac{\partial \eta(t,k)}{\partial t} \\ e(t,k+1) \end{bmatrix} = \begin{bmatrix} A & BK \\ -CA & I - CBK \end{bmatrix} \begin{bmatrix} \eta(t,k) \\ e(t,k) \end{bmatrix} \tag{9.16}$$

Note 9.1.1. If $e(0,k) = r(0) - Cx_0 \neq 0$, then the transformation $\tilde{e}(t,k) = e(t,k) - r(0) + Cx_0$ can be used to write the control error system in the form of (9.16).

To analyze the convergence of learning, the route now is to exploit the transition matrix (or fundamental matrix sequence) for the 2D continuous-discrete Roesser model and hence establish conditions on the description of (9.16). This yields the following result.

Theorem 9.1.5. *Suppose that the reference signal $r(t)$ is differentiable. Then for the ILC law*

$$u(t,k+1) = u(t,k) + K\left[\frac{dr(t)}{dt} - \frac{\partial y(t,k)}{\partial t}\right] \tag{9.17}$$

applied to (9.12), the tracking error $e(t,k)$ converges asymptotically to zero if, and only if, $r(I - CBK) < 1$.

It is immediate that a matrix K exists such that $r(I-CBK) < 1$ if, and only if, the matrix CB has full row rank and hence we have the following result.

Theorem 9.1.6. *Suppose that the reference trajectory $r(t)$ is differentiable. Then there exists a convergent learning law (9.17) if, and only if,* $\text{rank}(CB) = \text{rank}(C)$.

A detailed analysis of the consequences of Theorems 9.1.5 and 9.1.6 can be undertaken but it is also clear that learning laws of the form of (9.17) are special cases of (9.2). Moreover, it can be argued that the route via repetitive process theory is much more powerful than alternatives. Note also that this route first develops a general theory which can then be specialized to differential or discrete processes respectively. Other work in the same setting can, for example, be found in [117].

9.2 Norm Optimal ILC

It has been argued that the area of ILC can be enhanced by the analysis of benchmark problems that answer questions such as (i) are there globally convergent algorithms that do not rely on detailed systems properties? and (ii) what insights do these problems give on issues that affect convergence rates? In this section, a convergent ILC approach is developed which provides an answer to these questions with a realization available in terms of current trial feedback mechanisms combined with feedforward of previous trial data.

The approach is based on splitting the 2D dynamics into two separate sets of 1D dynamics. This is done by introducing a performance criterion as the basis of specifying the control input to be used on each trial. The algorithm uses the criterion to evaluate the performance of the system on a given trial by 'averaging over time' and hence removing the dimension of time from the analysis. The performance criterion is then used to construct and solve an optimization problem whose solution is the proposed control input for the new trial.

The optimization problem is solved firstly at the abstract level using operator theory. These results are then converted in an illustrative and important case of practical interest into a well-known optimal tracking problem which can be solved by Riccati methods. Mathematically, this results in what appears to be a non-causal representation of the solution. In actual fact, however, it is causal in the ILC context as it can be represented by a causal Riccati feedback of current trial data plus a feedforward component obtained from previous trial records. The feedback component of the solution representation opens up the possibility of enhancing robustness of the algorithm to plant modelling errors.

In what follows, the ILC algorithm is formulated in the general form using operator methods from functional analysis in a Hilbert space setting. This approach offers great benefits as it demonstrates the generality of the principles and makes it easy to derive algorithms for a wide range of system types using standard mathematical methods. Then the convergence of the algorithm is established and in a subsequent section experimental verification on a chain conveyor system is detailed.

The mathematical definition of ILC used in the following analysis has the following general form.

Definition 9.2.1. *Consider a dynamic system with input u and output y. Let $\mathcal{Y}$ and $\mathcal{U}$ be the output and input function spaces respectively and let $r \in \mathcal{Y}$ be a desired reference trajectory for the system. An ILC algorithm is successful if, and only if, it constructs a sequence of control inputs $\{u_k\}_{k\geq 0}$ which, when applied to the system (under identical experimental conditions), produces an output sequence $\{y_k\}_{k\geq 0}$ with the following properties of convergent learning*

$$\lim_{k\to\infty} y_k = r, \quad \lim_{k\to\infty} u_k = u_\infty$$

Here convergence is interpreted in terms of the topologies assumed in $\mathcal{Y}$ and $\mathcal{U}$ respectively.

Note that this general description of the problem allows a simultaneous description of linear and nonlinear dynamics, continuous or discrete plant with either time-invariant or time varying dynamics.

Let the space of output signals $\mathcal{Y}$ be a real Hilbert space and $\mathcal{U}$ also be a real (and possibly distinct) Hilbert space of input signals. The respective inner products (denoted by $\langle \cdot, \cdot \rangle$) and norms $\|\cdot\|^2 = \langle \cdot, \cdot \rangle$ are indexed in a way that reflects the space if it is appropriate to the discussion e.g. $\|x\|_{\mathcal{Y}}$ denotes the norm of $x \in \mathcal{Y}$. The Hilbert space structure induced by the inner product is essential in what follows but is not restrictive, as specific choices of this structure enables the analysis of, for example, continuous or discrete-time systems.

The dynamics of the systems considered here are assumed to be linear and represented in operator form as

$$y = Gu + z_0$$

where $G : \mathcal{U} \to \mathcal{Y}$ is the system input/output operator (assumed to be bounded and typically a convolution operator) and z_0 represents the effects of system initial conditions. If $r \in \mathcal{Y}$ is the reference trajectory or desired output then the tracking error is defined as

$$e = r - y = r - Gu - z_0 = (r - z_0) - Gu$$

Hence without loss of generality, it is possible to replace r by $r - z_0$ and consequently assume that $z_0 = 0$.

It is clear that the ILC procedure, if convergent, solves the problem $r = Gu_\infty$ for u_∞. If G is invertible, then the formal solution is just $u_\infty = G^{-1}r$. A basic assumption of the ILC paradigm is that direct inversion of G is not acceptable since, for example, this would require exact knowledge of the plant and involve derivatives of the reference trajectory. This high-frequency gain characteristic would make the approach sensitive to noise and other disturbances. Also inversion of the whole plant G is unnecessary as the solution only requires finding the pre-image of the reference trajectory r under G.

The problem considered here can easily be seen to be equivalent to finding the minimizing input u_∞ for the optimization problem

$$\min_u \{\|e\|^2 : \ e = r - y, \quad y = Gu\}$$

The optimization problem here can be interpreted as a singular optimal control problem [64] that, by its very nature, needs an iterative solution. This iterative solution is traditionally seen as a numerical problem in numerical analysis but, in the context of this chapter, it is seen as an experimental procedure. The difference between the two viewpoints is the fact that an experimental procedure has an implicit causality structure that is not naturally there in numerical computation. The development which now follows is taken in the main from [7].

There are an infinity of potential iterative procedures to solve the optimization problem considered in this section. There is a clear advantage in the use of descent algorithms of a suitable type [65]. The gradient based ILC algorithm class generates the control input to be used on trial $k+1$ from the relation

$$u_{k+1} = u_k + \epsilon_{k+1} G^* e_k$$

where $G^* : \mathcal{Y}^* \to \mathcal{U}^*$ is the adjoint operator of G and ϵ_{k+1} is a step length to be chosen at each trial. This approach suffers from the need to choose a step length and the feedforward structure of the trial which takes no account of current trial effects including disturbances and plant modelling errors.

The improved approach considered here results in an algorithm with the following two important properties.

(a) Automatic choice of step size.
(b) Potential for improved robustness through the use of causal feedback of current trial data and feedforward of data from previous trials.

The algorithm considered here, on completion of trial k, calculates the control input on trial $k+1$ as the solution of the minimum norm optimization problem

$$u_{k+1} = \arg\min_{u_{k+1}}\{J_{k+1}(u_{k+1}) : \; e_{k+1} = r - y_{k+1}, \; y_{k+1} = Gu_{k+1}\}$$

where the performance index, or optimality criterion, used is defined to be

$$J_{k+1}(u_{k+1}) := \|e_{k+1}\|^2_{\mathcal{Y}} + \|u_{k+1} - u_k\|^2_{\mathcal{U}}$$

The initial control $u_0 \in \mathcal{U}$ can be arbitrary in theory but, in practice, will be a good first guess at the solution of the problem.

This problem can be interpreted as the determination of the control input on trial $k+1$ with the properties that: (i) the tracking error is reduced in an optimal way; and (ii) this new control input does not deviate too much from the control input used on trial k. The relative weighting of these two objectives can be absorbed into the definitions of the norms in $\mathcal{Y}$ and $\mathcal{U}$ in a manner that will become more apparent in what follows.

The benefits of this approach are immediate from the simple interlacing result

$$\|e_{k+1}\|^2 \leq J_{k+1}(u_{k+1}) \leq \|e_k\|^2, \; \forall k \geq 0 \tag{9.18}$$

which follows from optimality and the fact that the (non-optimal) choice of $u_{k+1} = u_k$ would lead to the relation $J_{k+1}(u_k) = \|e_k\|^2$. The result states that the algorithm is a descent algorithm as the norm of the error is monotonically non-increasing in k. Also, equality holds if, and only if, $u_{k+1} = u_k$, i.e. when the algorithm has converged and no more input-updating takes place. There is an implicit choice of step size here which avoids the choice of this parameter in contrast to steepest-descent methods.

The controller on trial $k+1$ is obtained from the stationarity condition, necessary for a minimum, by Fréchet differentiation of (9.18) with respect to u_{k+1} as

$$u_{k+1} = u_k + G^* e_{k+1}, \; \forall k \geq 0$$

This equation is the formal update relation for the class of ILC algorithms considered in this section.

Using $e = r - Gu$ now gives the tracking error update relation

$$e_{k+1} = (I + GG^*)^{-1} e_k, \; \forall k \geq 0$$

and the recursive relation for the input evolution

$$u_{k+1} = (I + G^*G)^{-1}(u_k + G^* r), \; \forall k \geq 0$$

The algorithm has a number of other useful properties. For example, monotonicity immediately shows that the following limits exist

$$\lim_{k\to\infty} \|e_k\|^2 = \lim_{k\to\infty} J_k(u_k) =: J_\infty \geq 0$$

Also inductive argument and the inequality $\|y\| \leq \|G\|\|u\|$ yields the relations

$$\sum_{k\geq 0} \|u_{k+1} - u_k\|^2 < \|e_0\|^2 - J_\infty < \infty$$

$$\sum_{k\geq 0} \|e_{k+1} - e_k\|^2 < \|G\|^2(\|e_0\|^2 - J_\infty) < \infty \tag{9.19}$$

and hence

$$\lim_{k\to\infty} \|u_{k+1} - u_k\|^2 = 0, \quad \lim_{k\to\infty} \|e_{k+1} - e_k\|^2 = 0 \tag{9.20}$$

The properties given in (9.20) show that the algorithm has an implicit choice of step size as the incremental input converges to zero. This asymptotic slow variation is a prerequisite for convergence. Also the summation of the energy costs from the first to the last trial is bounded, as shown by (9.19). This implicitly contains information on convergence rates.

The result following result from [7] is a proof of convergent learning, where the notation $\mathcal{R}(\cdot)$ is used to denote the range of an operator.

Theorem 9.2.1. *If either $r \in \mathcal{R}(G)$ or $\mathcal{R}(G)$ is dense in $\mathcal{Y}$, then the ILC tracking error sequence $\{e_k\}_k$ converges in norm to zero in $\mathcal{Y}$, i.e. the ILC algorithm has guaranteed convergence of learning.*

The guaranteed convergence together with the monotonicity of the tracking error sequence represent powerful properties of the algorithm. Note again that the abstract proof using the techniques of functional analysis enables the wide applicability of the ILC algorithm to both continuous and discrete-time systems. The realization of this potential will finally rely on the conversion of the abstract results into a causal ILC algorithm, where 'causal' means that it can be realized in the form of a sequence of experiments.

This last requirement is not immediately obvious here as the relation $u_{k+1} = u_k + G^* e_{k+1}$, although apparently of a feedback form, suggests that the relationship is not causal. For example, if G is the convolution operator in $L_2^m[0,T]$ (endowed with the inner product $\langle w, v\rangle_{L_2^m[0,T]} = \int_0^T w^T(t)v(t)dt$) described by the relation $(Gu)(t) := \int_0^t K(t-\tau)u(\tau)d\tau$, then $(G^*e)(t) = \int_t^T K^T(\tau - t)e(\tau)d\tau$. This means that evaluation of $G^* e_{k+1}$ requires knowledge of future values of the tracking errors. Such data is not, of course, available in practice. The special causality structure of ILC allows, however, the transformation of the algorithm into a causal procedure, as detailed for one case of particular interest below.

The general analysis given above has to be converted into computational procedures that will depend in their detail on the form of the dynamics of the systems to be controlled. The essential aspect of this conversion is that the procedure is causal in the ILC sense, as defined next.

Definition 9.2.2. *An ILC algorithm is causal if, and only if, the value of the input $u_{k+1}(t)$ at time t on trial $k+1$ is computed only from data that is available from trial $k+1$ in the time interval $0 \leq t \leq t$ and from previous trials on the whole of the time interval $0 \leq t \leq T$.*

Note 9.2.1. Such an algorithm is not causal in the classical (or 1D) sense as data from times $t' > t$ can be used, but only from previous trials.

Now these calculations are detailed for one case of practical interest, namely the choice of $L_2[0,T]$ input and output spaces. The problem then reduces to a form of a familiar linear quadratic tracking problem.

Suppose that the plant has m outputs and r inputs with dynamics described by a linear, possibly time-varying, state-space model. The input-output map $G : \mathcal{U} \rightarrow \mathcal{Y}$ and, in particular, the relations $y_k = Gu_k$ and $e_k = r - y_k$ take the form

$$\begin{aligned}
\dot{x}_k(t) &= A(t)\,x_k(t) + B(t)\,u_k(t), \;\; x_k(0) = 0 \\
y_k(t) &= C(t)x_k(t) \\
e_k(t) &= r(t) - C(t)\,x_k(t), \; 0 \leq t \leq T, \; k \geq 0
\end{aligned}$$

The choice of input and output spaces is as follows

$$\mathcal{U} = L_2^r[0,T], \;\; \mathcal{Y} = L_2^m[0,T] \times \mathbb{R}^m$$

The unusual choice of output space as the Cartesian product of a familiar L_2 space with $\mathbb{R}^m$ is required for generality but, more importantly, for the avoidance of numerical convergence problems in the final moments of the trials.

The inner products on $\mathcal{Y}$ and $\mathcal{U}$ are defined as:

$$\begin{aligned}
\langle (y_1,z_1),(y_2,z_2)\rangle_{\mathcal{Y}} &:= \frac{1}{2}\int_{t=0}^{T} y_1^T(t)Qy_2(t)\,dt + \frac{1}{2}z_1^T F z_2 \\
\langle u_1(t),u_2(t)\rangle_{\mathcal{U}} &:= \frac{1}{2}\int_{t=0}^{T} u_1^T(t)Ru_2(t)\,dt
\end{aligned}$$

where $Q > 0$, $R > 0$ and $F \geq 0$. (Formally, we should take $F > 0$ but, if F has a full rank decomposition $F = VV^T$, a simple redefinition of $\mathcal{Y}$ as $L_2^m[0,T] \times V^T\mathbb{R}^m$ regains the Hilbert space structure. The details are omitted for brevity.)

The initial conditions are taken to be homogeneous without loss of generality because the plant response due to non-zero initial conditions can be absorbed into $r(t)$, as discussed before. The index J_{k+1} with the specified norms in $\mathcal{Y}$ and $\mathcal{U}$ becomes a familiar linear quadratic performance criterion, i.e.

$$\begin{aligned} J_{k+1}(u_{k+1}) &= \frac{1}{2}\int_0^T \Big\{ e_{k+1}^T(t)\, Q e_{k+1}(t) + (u_{k+1}(t) - u_k(t))^T \\ &\times \; R\,(u_{k+1}(t) - u_k(t)) \Big\} dt + \frac{1}{2} e_{k+1}^T(T)\, F\, e_{k+1}(T) \end{aligned}$$

In particular, it is a combination of the optimal tracking (tracking of $r(t)$) and the disturbance accommodation problem (regarding $u_k(t)$ as a known disturbance on trial $k+1$).

Note 9.2.2. In the cost function used here it is $u_{k+1} - u_k$, i.e. the difference between the control inputs on any two successive trials which is penalized. This should help to avoid unacceptably (from and implementations standpoint) changes in the magnitude of the elements in the control input vector from one trial to the next.)

The optimal solution u_{k+1} is $u_{k+1} = u_k + G^* e_{k+1}$ and the abstract definition of the adjoint operator can be transformed into a more firm description using the definitions of G^* and the inner products. In particular, in this case the equation $u_{k+1} - u_k = G^* e_{k+1}$ containing G^* becomes the well known costate system

$$\begin{aligned} \dot{\psi}_{k+1}(t) &= -A^T(t)\,\psi_{k+1}(t) - C^T(t) Q\, e_{k+1}(t) \\ \psi_{k+1}(T) &= C^T(T) F e_{k+1}(T) \\ u_{k+1}(t) &= u_k(t) + R^{-1} B^T(t)\,\psi_{k+1}(t),\; T \geq t \geq 0 \end{aligned}$$

This system has a terminal condition (at $t = T$) instead of an initial condition, making it (as expected) as an anti-causal representation of the solution. It cannot therefore be implemented in this form. This problem is removed by the derivation of an alternative, but equivalent, causal representation as detailed next.

The non-causal representation is transformed into a causal algorithm using a state feedback representation. In particular, the optimal control is transformed by writing for the costate $\psi_{k+1}(t) = -K(t)(x_{k+1}(t) - x_k(t)) + \xi_{k+1}(t)$, where $\xi_{k+1}(t)$ is the so-called predictive or feedforward term defined below, and hence

$$u_{k+1}(t) = u_k(t) + R^{-1} B^T(t)\,[-K(t)\,(x_{k+1}(t) - x_k(t)) + \xi_{k+1}(t)]$$

Standard techniques then yield the matrix gain $K(t)$ as the solution of the familiar matrix Riccati differential equation on the interval $t \in [0, T]$

$$\dot{K}(t) = -A^T K(t) - K(t) A + K(t) B R^{-1} B^T K(t) - C^T Q C,\;\; K(T) = C^T F C$$

This equation is independent of the inputs, states and outputs of the system. In contrast, the predictive or feedforward term $\xi_{k+1}(t)$ is generated by

$$\dot{\xi}_{k+1}(t) = -\left(A - BR^{-1}B^T K(t)\right)^T \xi_{k+1}(t) - \zeta\, C^T Q\, e_k(t)$$

with terminal boundary condition

$$\xi_{k+1}(T) = C^T F \zeta e_k(T)$$

The predictive term is hence driven by a combination of the tracking error and the input on the previous trial k and also the reference signal. This is therefore a causal ILC algorithm consisting of current trial full state feedback combined with feedforward from the previous trial output tracking error data. This representation of the solution is causal in the ILC sense because the costate system can be solved off-line, between trials, by reverse time simulation using available previous trial data. The differential matrix Riccati equation for the feedback matrix $K(t)$ needs to be solved only once before the sequence of trials begin.

The ILC algorithm developed above can be implemented in practice if full state feedback is available. For an implementation, the free terms Q, R and F must be chosen appropriately. With the objective of minimizing the error norm in mind, intuitive guidelines for the choice of these parameters are provided below. The convergence properties are assumed to be described by the sequence $\{J_k\}_{k\geq 0}$ which simultaneously represents the behavior of the error sequence and the rate of change of the input signals.

Changing the parameters affects the speed of decrease of the entries in $\{J_k\}_{k\geq 0}$. The parameter Q is related to the size of the error, the parameter R to the size of the change of the input, and the parameter F to the size of the error at the end of the trial. An increase in the relative magnitudes of Q and R represents less caution in learning and hence greater convergence speeds. This is seen in numerical studies (see, for example, [6]).

The last parameter F is not easily related to the overall decrease of $\|e_k\|$ and 'rules of thumb' for its choice need more intuition. It is suggested that it is advantageous to choose F, which appears in the terminal condition in the defining Riccati equation, such that $K(t)$ is as close to being constant as possible.

A natural extension of norm-optimal ILC, termed predictive norm-optimal ILC, can also be derived. The intuition that motivated this work [8] is that predictive control contains the key to improved performance. The proposed form of 'norm optimal predictive iterative learning control' extends the performance criterion to take future predicted error signals into account. The extended criterion for computing the input u_{k+1} on trial $k+1$ is

$$J_{k+1}(u_{k+1,N}) := \sum_{i=1}^{N} \lambda^{i-1} \left(\|e_{k+i}\|^2 + \|u_{k+i} - u_{k+i-1}\|^2\right)$$

This criterion includes the error not only of the next trial, but of the next N trials, together with the corresponding changes in input. The weight $\lambda > 0$

determines the importance of more distant (future) errors and incremental inputs compared to the present ones.

The actual ILC algorithm follows uniquely from minimization of the proposed cost criterion. It only remains to compute the minimizing input. This is done by dynamic programming. Once the input is found, a recursive formulation for the evolution of the error (and input) is computed. All interesting properties and characteristics of the norm-optimal predictive ILC algorithm can, as in the norm optimal case, be obtained by examining the properties of the operators appearing in the recursive formulation.

9.3 Norm Optimal ILC Applied to Chain Conveyor Systems

Chain conveyor systems are prevalent in a number of industries, particularly food manufacturing. This highly competitive industry needs to reduce manufacturing costs by obtaining greater production rates. However, the combination of simple mechanical components and standard three term (PID) controllers can make accurate position control difficult as the throughput is increased. Improvements in production rate by increasing the speed of operation are limited as the acceleration of the product quickly becomes unacceptable. Further improvements must be obtained by minimizing the dead-time required to allow the system to settle. This can only be achieved by reducing following errors. To obtain this desired improvement in control, it is necessary to look at methods beyond that of the traditional PID controller.

The type of conveyor systems found in the food industry are constructed from components that contain a number of non-linearities. Generally, motion is provided by an induction motor which has a non-linear characteristic, and encoder feedback produces quantization of the position signal. Simple mechanical components are used which lead to significant Coulomb friction, and there is variation in friction and stiffness values as the position of the conveyor changes. Analysis of control methods based entirely on simulation may therefore suggest solutions that are not valid experimentally. To allow experimental analysis of controller algorithms, a section of a typical conveyor and dispenser system has been constructed, and linear and non-linear (see, for example, [18] and the relevant cited references in this work for further details) models obtained for design purposes.

Due to operational constraints there is typically no position sensor on the conveyor or dispenser. Feedback is only available at the motor shaft, placing the conveyor and dispenser outside the control loop. Closing this loop, without adding to the sensor requirements, has been considered. The conveyor and dispenser form a two axis system that is required to remain synchronized. Methods of controlling the relative position of the two axes rather than the absolute position of each individual axis have also been investigated [18].

As the level of factory automation increases, the requirement to move product swiftly, accurately and reliably around the production site becomes more critical. Recent advances in technology have seen the introduction of automatic remote guided vehicles (ARGV's) to move product, but the conveyor still dominates in the majority of applications. It's fundamental simplicity, combined with the advances in power electronics and motor design make it the ideal material handling solution from automation of warehouse storage to bulk handling at coal mines, quarries and power stations. The very different applications to which conveyors are suited also exhibit very different requirements. Bulk handling applications generally require constant speed operation, capable of moving product immense distances. The principal difficulties with these form of conveyors is starting and the safety of personnel working near them, especially important in the confines of a mine and they are frequently based around belt conveyors.

A wide range of conveyors are used in the manufacturing industry and are described as either overhead, above floor or below floor, where the above floor type is the most common. These are then divided into three groups based on their construction; belt, roller or chain. Each type of conveyor has advantages that make it suitable to a particular application.

Belt conveyors are ideal where a workpiece is to be removed and replaced on the conveyor, or where the workpiece is of a complex shape. Usually they are quieter than other types and require little maintenance. However, as there is only a friction connection between the conveying surface and the drive they tend to be restricted to applications where acceleration is limited.

Roller conveyors consist of a large number of consecutive rollers. To work satisfactorily the items placed on the conveyor must have an underside that is flat and sufficiently large to extend over a minimum of three rollers, otherwise some form of carrying tray is required. If the application requires a drop in height then unpowered rollers provide a simple solution. A roller conveyor can be powered by using a belt to connect the rollers to a shaft placed beneath them.

A recent development in roller conveyor technology is to construct the roller as a small induction motor. The outside of the roller forms the rotor with power fed to the stator via fixed points at the ends. Some applications, especially in flexible manufacturing systems, require that the products on the conveyor are delayed before reaching the next process. Roller conveyors achieve this by slipping or using spring loaded clutches that disengage from the drive when the load becomes too high. Queues can be created by blocking the progress of one unit causing the rollers under the load to slip or disengage. As the conveyor upstream still operates further product will be brought up behind the first, thereby creating an accumulator.

Where the conveyor is required to follow a stop/start profile, a positive connection between the drive and the conveying surface is needed. Chain conveyors fulfil this requirement and are used in a number of industries where either high loads or high accelerations are demanded. Their robust construction makes them eminently suitable for systems dealing with corrosive chemicals or where hygiene requirements mean regular cleaning. Simplicity of construction allows straightforward and inexpensive design of conveyors suited to a specific task, and chain manufacturers have developed a wide range of attachments to expedite this.

Recent developments have considered the use of plastics in the construction of the chain and associated components. Sprockets made from nylon appear to provide an increase in lifetime over their steel counterparts. Plastic components are obviously of interest where there may be concerns about chain lubricant contaminating the product. The small amounts of lubricant on most chains makes this concern unnecessary unless there is direct contact between the chain and the product.

The control of conveyors that require a repeatable index can be accomplished either electronically or mechanically. Systems based on cams or a hydraulic walking beam can offer a solution that achieves an accurate and repeatable index at minimal complexity. Cam systems do provide a very high speed, accurate index and are used in lightweight, high volume systems such as printed circuit board (PCB) assembly. However, they are susceptible to wear and require regular maintenance. Alteration of the indexing profile can only be achieved by replacement of the cam.

Detailed reviews of conveyor types and their selection have been published in, for example, [88]. The majority of published material concerning conveyors deals with statistical modelling for determining the most efficient layout of a manufacturing plant, see, for example, [79].

The type of system considered here is a chain conveyor of a type frequently found in the food industry. Typically they are constructed from very simple mechanical components, the chain and sprocket being based on standard roller chain, driven from either a helical gearbox or via a belt drive. Originally, motion was provided by a d.c. servo system, but the requirement of a high level of hygiene necessitates the use of high pressure water or steam hoses in cleaning. A d.c. motor capable of withstanding such environmental demands is expensive and requires a long period of downtime for maintenance. With the advent of inexpensive power electronics the use of induction motors became viable. Position feedback is provided by an optical encoder placed on the motor shaft with control provided by a PID controller.

This approach has a number of limitations. The tuning of the controller is achieved manually which is unlikely to produce an optimum solution. Alteration of system dynamics over time will cause a decrease in performance unless the controller is manually re-tuned. These perform adequately but are unable to account directly for higher order effects that may be introduced by non-linearities. The use of modern drives could be used to overcome some of these problems. Flux vector control of induction motors claims to provide a performance that is comparable to a d.c. servo system, and the supply of inverters based around this philosophy is becoming more widespread. Replacement of existing drives with flux vector versions is often undesirable due to the expense and re-training involved.

Feedback for the controller is provided by an encoder mounted on the motor shaft. This leaves the mechanical components of the system outside the control loop which causes a number of control difficulties. As the chain ages, the bushings wear causing the chain to effectively become longer. Teeth in the gearbox and the sprockets suffer wear, and drive belts stretch introducing backlash into the system. The net effect is to decrease the accuracy of the indexing motion and this can be a reason for not using chain conveyors.

Currently the effect of open-loop disturbances is overcome by regular manual resetting or automatic homing of the system once every revolution of the conveyor. Both methods are undesirable as while the conveyor is being homed, production is delayed and throughput decreased. Automatic compensation for position disturbances, commonly referred to as drift, would increase the production efficiency of the system but must be achieved without adding significantly to the sensing requirement. Any sensing must be non-invasive and impervious to contamination by environmental factors such as water, steam or electromagnetic interference. It should also not require long runs of expensive multi-core screened cable. An encoder placed on a mechanical output shaft is unacceptable for cabling reasons and the difficulty in achieving accurate mounting. A system based on optics is also unacceptable as it may easily become obscured by spilt product.

The objective now considered is how to overcome some of these problems to produce an improvement in the performance of the system. Performance is measured by the number of units completed per hour. Improvement could be achieved by increasing the speed of operation, however this is limited by acceleration. Accelerations that are too rapid make it difficult to maintain the position of the product on the conveyor, though slight increases in speed are possible if position accuracy is not degraded. To obtain the most significant increase in throughput, the accuracy of the position trajectory following should be improved.

Currently it is necessary to introduce a time delay to allow the conveyor to settle. For example, if the time taken to dispense the product is 0.3 secs, a delay of 0.1 secs may be added before starting dispensing to ensure that the system is in the correct position. If the accuracy with which the conveyor followed the demand profile was improved then this delay time could be reduced. For each individual unit the reduction may only be a fraction of a second, but over the course of a 15 hour day this would accumulate to a significant number of additional units completed. The control problem is therefore one of accurate position tracking and the application of more recently developed control algorithms should provide an increase in performance.

The experimental apparatus used in this study is representative of manufacturing systems frequently found in the food industry – see Fig. 9.1. It

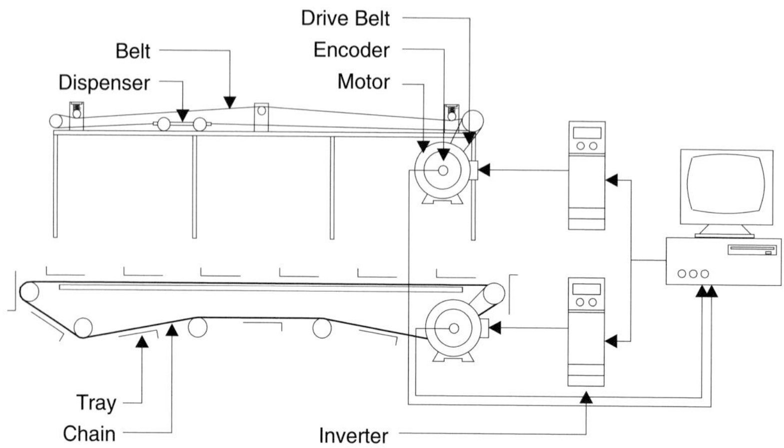

Fig. 9.1. Experimental Apparatus

consists of a short section of chain conveyor operating in conjunction with a product dispenser. Such systems are designed to have two modes of operation, either indexing or synchronizing. In indexing, the conveyor positions items placed upon it consecutively under the product dispenser in a stop/start motion. Product is only dispensed while the conveyor is stationary. Synchronizing requires the conveyor to move at constant velocity with the dispenser accelerating to match, product being dispensed when the velocities are synchronized. The dispenser then rapidly returns to the origin in readiness for the next unit. Figure 9.2 shows the demands for both synchronizing and indexing profiles. In both operating modes the requirement is for accurate position control, especially during the dispensing time. Typical specifications are for ± 1mm which equates to approximately ± 0.05 rad at the motor shaft with a dispensing time of 0.4 secs.

In summary, the design of the system allows implementation and testing of various controllers to be undertaken. The two axes can be operated as a single unit or individually. Under the individual mode of operation a spare encoder counter is available for additional feedback or measurement. The design of the system also allows the recording of time, demand and feedback data for subsequent analysis.

Modelling of the system can be undertaken using two separate approaches. The first method is based on writing equations to describe the behavior of the individual system components which, after verification, can be assembled into a state-space model of the complete system. Here we use linear models identified from frequency domain tests as the basics for controller design. How these models were obtained is described next.

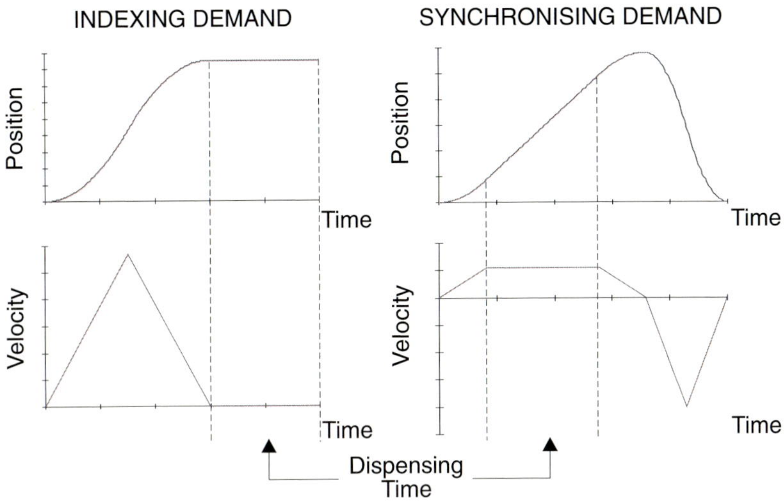

Fig. 9.2. Indexing and Synchronizing

Frequency response analysis provides linear models and information on effects that are not readily apparent from time response methods, such as pure time delays. For the purposes of experimentation the system excitation is a sinusoidal voltage, with speed at the relevant shaft as the output.

The frequency response of the drive system with a sinusoidal excitation of 1 V amplitude and 5V d.c. offset is linear to around 30 Hz. Above this, the motor shaft frequency falls as the demand frequency increases suggesting a sampler at around 60 Hz on the inverter input (Fig. 9.3). A sampler placed on the input of the theoretical model produces excessive oscillation of the motor shaft. Repositioning the sampler on the frequency demand output from the inverter control block restores a satisfactory step response.

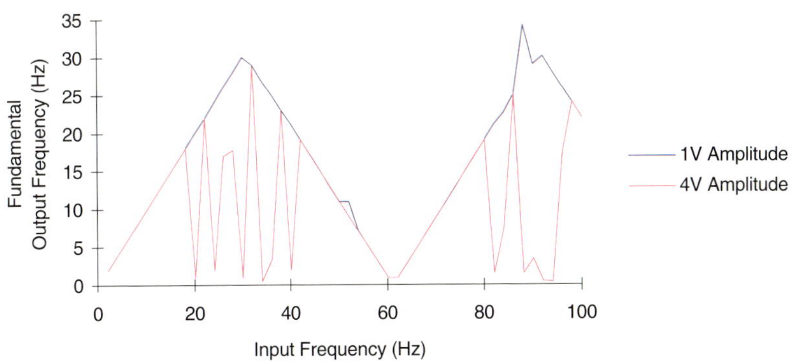

Fig. 9.3. Input-output frequency characteristics of the invertor drive

Increasing the excitation amplitude introduces a frequency quenching effect into the response (Fig. 9.3) particularly noticeable around 30 Hz. To generate sinusoidal motion at the motor shaft the output of the inverter is required to be both a frequency modulated (FM) and an amplitude modulated (AM) signal. For an ideal VVVF inverter the output signal will have the form

$$v(t) = \tilde{\beta}x(t)\sin(\beta x(t)t)$$

where $x(t) = \sin(\omega t)$ is the inverter input and $\tilde{\beta}$ and β are constants to provide the correct relationship between input voltage, output voltage and output frequency. Figure 9.4 shows the voltage spectrum for a sinusoidal input with amplitude 4 V and frequency 30 Hz, with a d.c. offset of 5 V. A demand of 10 V corresponds to full speed at 50 Hz, the 5 V offset therefore demands half speed which produces the component at 25 Hz. The main peaks in the spectrum are separated by a frequency equal to that of the demand frequency. If the inverter is incapable of producing the higher frequencies, or producing a demand with the correct frequency spectrum, then the output will not be as required. If attenuation was the only effect then the output for 4 V peak demand would be expected to be a sinusoid of the correct frequency and speed offset but of reduced amplitude. Instead the frequency at 25 Hz is found to dominate causing the output to be at a constant speed. As the demand amplitude is increased, the magnitude of the frequency components other than the offset increase and the inverter becomes less able to create the full range of required frequencies.

A d.c. servo has been used to excite the mechanical components of each axis. By recording the speed of the motor and output shafts and subtracting the two frequency responses obtained, the frequency response of the mechanical components was derived.

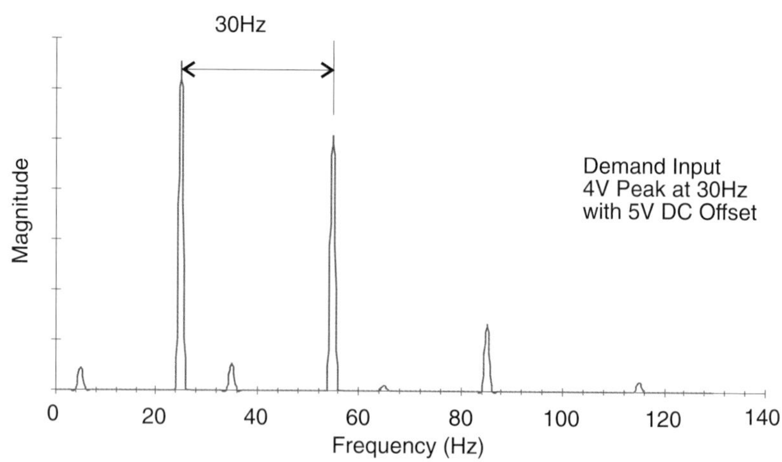

Fig. 9.4. Ideal VVVF inverter voltage spectrum

To observe the effect of non-linear friction on the frequency response of the conveyor, the experiment was performed twice. First with zero d.c. offset in the input excitation and then again with sufficient d.c. offset to ensure that the conveyor velocity never fell to zero. As expected, the results indicate that when the velocity does not pass through zero the system exhibits less damping (Fig. 9.5). The transfer-functions obtained from both sets of experiments are given next

$$G_{con}(s) = \frac{20.3(s+200)}{s^2 + 100s + 100^2}$$

$$G_{con(offset)}(s) = \frac{46(s+170)}{s^2 + 84s + 140^2}$$

The pure time delay apparent from the frequency response is caused by gaps in the chain links. As the system is excited the chain links start moving consecutively, creating a time delay between input and output shafts. To a first approximation, however, this can be ignored (and the transfer-functions given above used in subsequent analysis).

In the case of the conveyor, a time delay was observed from the frequency response for the dispenser but again this can be ignored to a first approximation– Fig. 9.5. The dispenser is therefore also represented by a second-order model

$$G_{dis}(s) = \frac{126}{s^2 + 440s + 550^2}$$

For controller design purposes a linear model of the complete system is required. This has been achieved by using the PC as a signal generator and determining the frequency response of the complete system up to 30 Hz

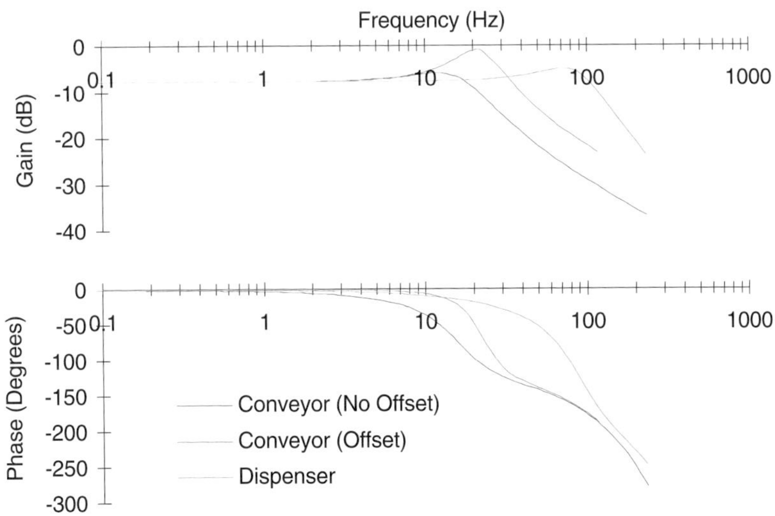

Fig. 9.5. Conveyor and dispenser frequency responses

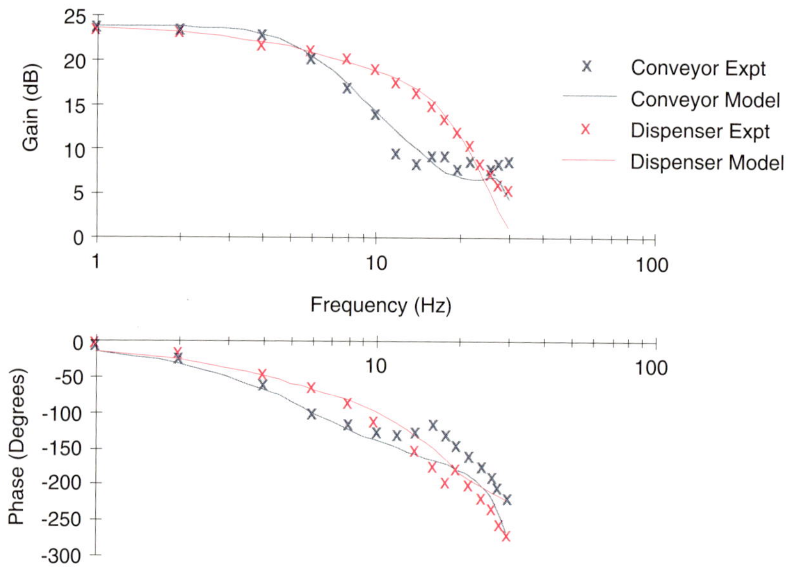

Fig. 9.6. Complete frequency response

(Fig. 9.6). For the conveyor, a fourth order model was obtained with a third order approximation being sufficient to model the dispenser dynamics

$$G_{conveyor}(s) = \frac{615.06 \times 10^6}{(s^2 + 49s + 35^2)(s^2 + 54s + 180^2)}$$

$$G_{dispenser}(s) = \frac{6.47 \times 10^6}{(s + 35)(s^2 + 99s + 110^2)}$$

9.3.1 Experimental Results

Previous research has applied conventional (i.e. 1D in the terminology used in this monograph) control laws designed using a variety of techniques to chain conveyor systems. In order to be able to compare the performance of practical controllers, it has been necessary to define a series of experiments and evaluate the results using a performance index. The results given here are from [4, 5, 19] (and we mainly follow the style of presentation in these references).

For all controller experiments, each axis was started from the same point and continuously run for 40 units, the position of the last 20 units being recorded at 100 Hz and averaged. Each test was analyzed to provide a measure of the mean square error (MSE) for the entire profile, MSE during the dispensing time (MSE(D_t)) and the settling time. Settling time is measured from the start of the dispensing time to the point where the system reaches and maintains the position specification.

To obtain a measure of the performance of the controller, each measured parameter is averaged across a range of unit rates, 30–60 units per minute (UPM). The values for the three average parameters are then combined using a cost function to obtain a single value indicating overall controller performance. The cost function, J, is defined as

$$J := \gamma_1 \bar{e_t^2} + \gamma_2 T_{st} + \gamma_3 \bar{e_{dt}^2}$$

where $\bar{e_t^2}$, T_{st} and $\bar{e_{dt}^2}$ are the MSE for the total profile, the MSE during the dispensing time and the settling time respectively. The weights, γ_1, γ_2 and γ_3 are 1, 2 and 3 respectively, though the MSE during the dispensing time had previously been multiplied by 1000 to bring it to the same order of magnitude as the other variables. It is considered essential for the controller to bring the system to within specification during the dispensing time, hence the highest weighting. A short settling time is also highly desirable, while the accuracy over the entire profile is more a consequence of good results for the previous two measures. Values for the weighting parameters that are greater then those indicated tended place too much emphasis on a single parameter, swamping the effect of other parameters. Smaller weighting did not adequately emphasize those parameters that are considered important. Results obtained were normalized such that the results for the PID controller were 1.

A variety of 'classical and modern' 1D linear control schemes have been tested. These are PID, Lag/Lead compensation, PID with velocity feedforward, velocity estimation (exploiting a Kalman filter), multiple loop PID control with a Kalman estimator, flux vector invertors (invertor open-loop and closed-loop) respectively. In summary, we can state the following concerning the performance of these controllers.

The performance of the PID controller has been shown to require improvement, especially as the unit rate is increased. Different implementations of the basic PID structure have been investigated. The addition of a velocity feedforward term produces a significant performance improvement but the performance continues to degrade as the unit rate is increased. It is also apparent that the mechanical load is open-loop, and though the motor may be within specification, the load may not be. The performance of the conveyor appears to be worse than that of the dispenser, even though the conveyor follows the simpler demand. Once the conveyor velocity has fallen to zero no further improvement is possible, whereas the dispenser error can be reduced throughout the profile.

Comparison of the original inverter with a modern constant V/f inverter has been shown to provide some improvement. Conversion of the modern drive to closed-loop mode utilizing flux vector current decoupling, dramatically improves performance but requires tuning of up to six gains. This method can not be retro-fitted to existing systems without replacement of components, and also has an increase in complexity for the commissioning engineer.

As the system is required to follow the same position profile for each unit, it is ideally suited to ILC where first this has been constructed around the existing well tuned PID loop. The PID controller is required to provide disturbance rejection and the learning controller slowly removes position error as the number of completed trials increases. A memory is used to store the cumulative effect of some proportion of the tracking error over the previous trials, which is then used to update the demand for the current trial. This can be represented by

$$u_{k+1}(t) = u_k(t) + L(t)e_{k+1}(t)$$

where $L(t)$ is the learning controller. Substituting for $u_k(t)$ gives

$$u_{k+1}(t) = [u_{k-1}(t) + L(t)e_k(t)] + L(t)e_{k+1}(t)$$

This can be generalized to

$$u_{k+1}(t) = u_{k-j}(t) + L(t) \sum_{i=k-j+1}^{k+1} e_i(t)$$

If $j = k$ then this reduces to

$$u_{k+1}(t) = u_0(t) + L(t) \sum_{i=1}^{k+1} e_i(t)$$

where $e_{k+1}(t) = r(t) - y_{k+1}(t)$ and

$$u_0(t) = 0, \;\; v(t) = u(t) + r(t)$$

This is implemented as indicated in Fig. 9.7 where N is the number of samples in one trial where in the experimental results reported here only a proportional gain (K) has been used. When the iterative learning controller is implemented in the manner of Fig. 9.7, the mean square error reduces by a small amount before growing, eventually leading to instability (Fig. 9.8).

For a step input applied at time t, the output will not reach the desired value until some time later, $t + \tau$. If the response at t is required to be at the desired value then it will be necessary to apply the step t' seconds earlier. For the iterative learning controller this means that the learning demand is

$$u_{k+1}(t) = r(t) + L(t) \sum_{i=1}^{k+1} e_i(t - t')$$

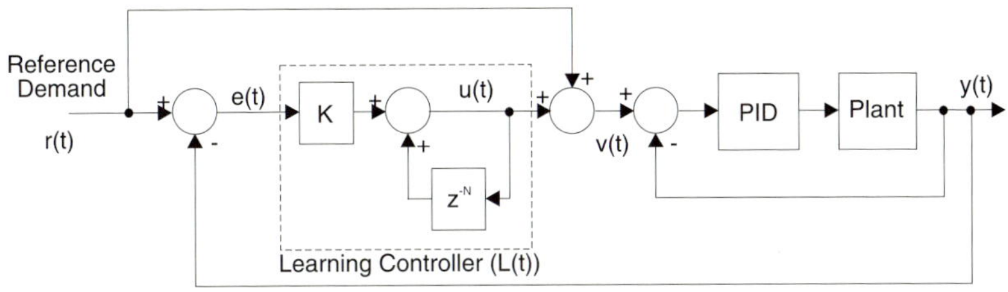

Fig. 9.7. Learning controller block diagram

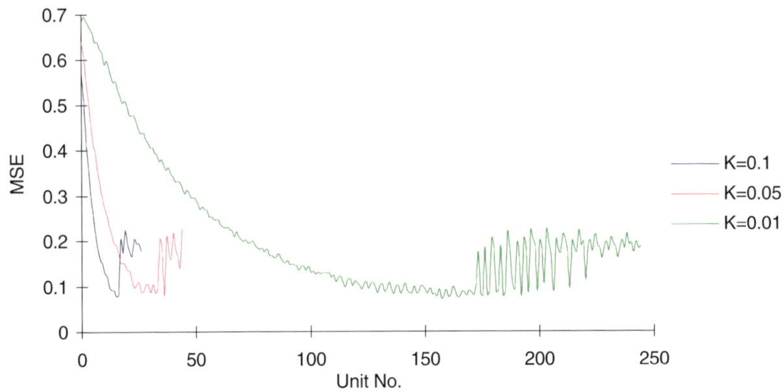

Fig. 9.8. Error results

This is very simply represented in the z-transform domain, with the delay $z^{-(n+\tau)}$ where $\tau = t' \times f_s$.

A gain value of $K = 0.05$ was used for all tests. This produces a superior control effort profile than the choice of $K = 0.1$. There is little difference between control efforts for $K = 0.05$ and $K = 0.01$ but the higher value converges considerably quicker.

The cost function results for this controller are 0.26 and 0.01 for the conveyor and dispenser axes respectively compared to the normalized PID controller cost function. The dispenser is improved well beyond that of the conveyor, performance of the conveyor being limited by the requirement for it to come to rest. As the dispenser velocity never falls to zero continual improvement is possible. The performance of this controller on both axes is significantly improved over that of a standard PID controller, and is achieved at very little computational expense or increase in controller complexity. Moreover, it is also very consistent as the unit rate is increased. Only the flux vector system has performance indices that are similar and this requires new equipment and tuning of up to six gain terms. It is also incapable of reacting to changes in dynamics without manual intervention.

Fig. 9.9 shows results obtained for the dispenser with a PID controller and the alternative controller after it has learnt. This latter system follows the

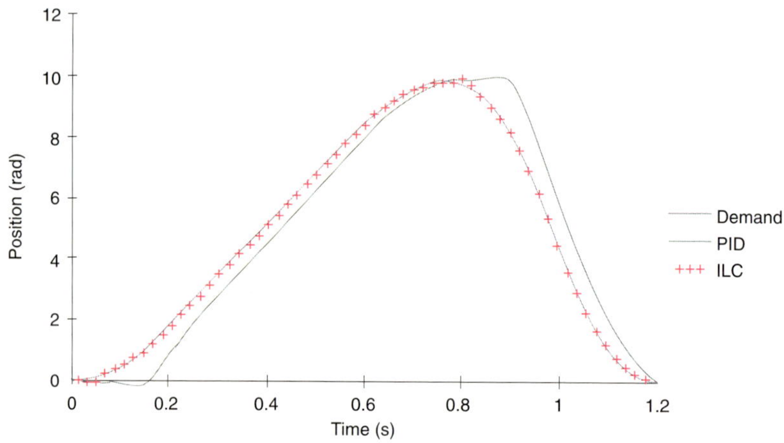

Fig. 9.9. Output performance over the 100th trial at 50 UPM

demand with considerably greater accuracy than the equivalent PID system. The integral term in both controllers has been de-tuned and this demonstrates the ability of the alternative controller to improve poorly tuned systems.

When the dispenser is in synchronizing mode, the precise position of the dispenser is less important than the relative position between the two axes. Fig. 9.10 shows the relative error between the conveyor and the dispenser after 100 trials. The first system has a PID controller on both axes and the second has a PID controller on the conveyor axis and a learning controller on the dispenser axis. The performance of the PID controller for the constant speed requirement of the conveyor in this mode of operation has been deemed to be acceptable. The learning system maintains the relative position within specification for the majority of the profile. The PID system is only just within specification by the start of the dispensing time. As the unit rate is increased, the PID controller cannot meet the specification by the start of the dispensing time.

The effect of the iterative learning controller on shaping the demand profile is shown in Fig. 9.11 where a dispenser demand signal is presented. The lag in the system causes the updated demand to appear in advance of the reference signal. However, closer inspection reveals that the learnt demand is not simply a time displaced copy of the reference. Firstly, the time difference between the two signals is not constant. Secondly, the effect of non-linearities in the system can be seen, especially where the velocity demand passes through zero. To overcome the effect of the inverter dead-zone at this point the learnt signal differs considerably from that of the reference.

The repeatability of a system is commonly quoted as a range around the mean equivalent to three standard deviations [84]. If the error is considered to have a normal distribution then three standard deviations from the mean will include 99.8% of all units. The results averaged across the unit rate

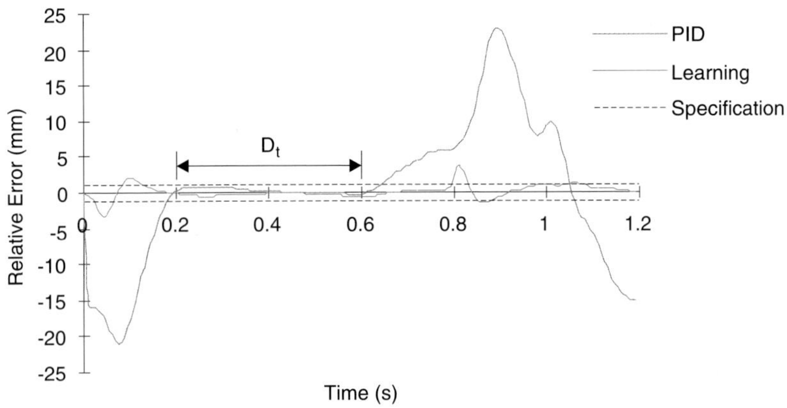

Fig. 9.10. Relative error for learning controller (50UPM)

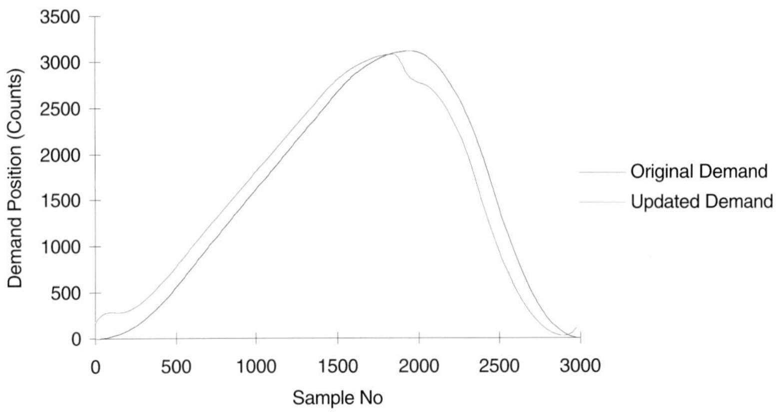

Fig. 9.11. Demand signal shaping for the dispenser axis

range are reproduced in Table 9.1. Neither controller succeeds in operating the conveyor in such a way that the mechanical components are consistently within the position specification of 53 mrad though the motor is. The poor mechanics make repetition very difficult compared to the motor, hence the increase in standard deviation. Also the control is hindered by the mechanical components being outside the control loop.

The dispenser PID controller achieves the repeatability requirement but the error causes the system to be outside the position specification. The learning controller has a considerably superior error value with a similar repeatability, ensuring that the motor is within specification at almost all unit rates (Fig. 9.12).

For both axes the repeatability of the two controllers is very similar. This is to be expected as the disturbance rejection in the learning controller is provided by the PID loop.

Table 9.1. Average error and repeatability results during D_t

	Conveyor Error (mrad)		Dispenser Error (mrad)	
Location	Learning	PID	Learning	PID
Motor	-3 ± 47	-13 ± 44	19 ± 24	127 ± 33
Mechanical	-33 ± 140	-39 ± 140	20 ± 21	137 ± 28

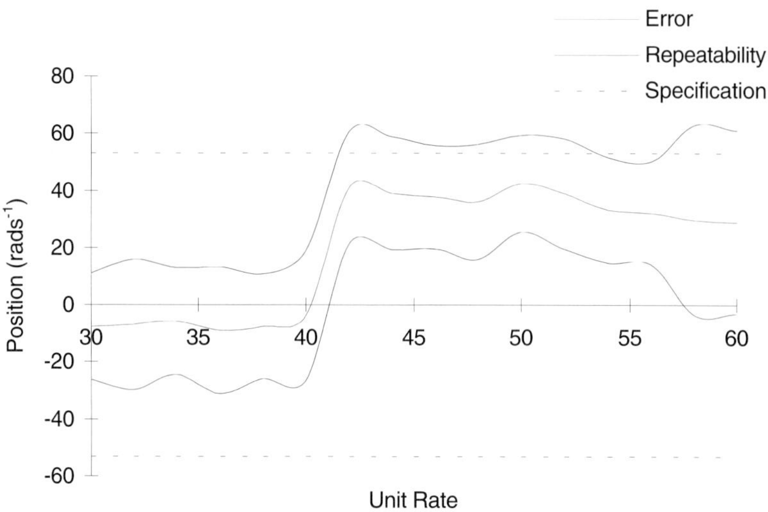

Fig. 9.12. Dispenser repeatability under learning control

Clearly, the error profile should decay. This is confirmed in Fig. 9.13 where the improvement in the actual error as the unit number increases is clearly visible. From this plot it would appear that the learning controller is unable to compensate for all of the dynamics. However, if the system could be left to operate for long enough without becoming unstable, the error would fall to zero across the whole profile.

Although the ILC algorithm described here outperforms the PID controller, it does have a limitation; namely for good performance system time lags need to be determined to ensure stable operation. In particular, a low value of the learning gain must be used, leading to a slow rate of convergence. Additionally, the ILC must be switched off if the MSE exceeds a threshold value.

A norm-optimal ILC algorithm has also been implemented (with the state vector in the cost function replaced by the trial profile (and hence Q in this case is also a scalar parameter). The PID settings were identical to those

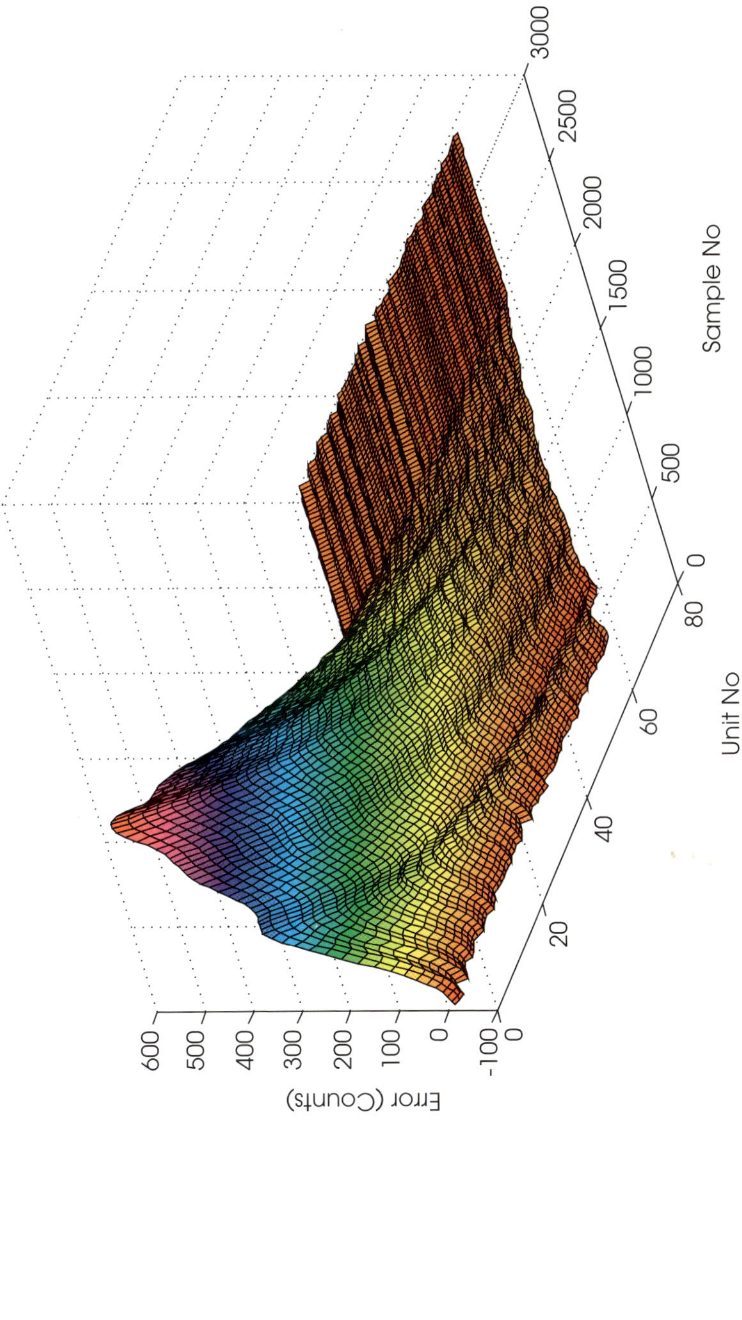

Fig. 9.13. Learning Error Profile at 50 UPM

used above and it remains to consider the selection of the weighting matrices in the norm optimal cost function. The experimental results discussed below were obtained with $Q = 1$, $R = 14$, and $F = 0.1$.

In order to compare the norm-optimal controller's performance with the results obtained using the well-tuned PID controller only, measurements have been obtained over the range of unit rates from $35 - 60$ UPM for the dispenser operating in a synchronized mode. Fig. 9.14 shows the performance of the controller in terms of the tracking MSE over the first 100 trials. As with the simple structure ILC algorithm the plant delay, τ, has been included. Compared to the previously used controllers, overall performance has been improved. At 50 UPM the rate of convergence is greater for the same value of learning gain. For all unit rates the MSE by the 20th trial is satisfactory. An analysis of the 20th trial in terms of the total MSE over the trial, the MSE during dispensing and the settling time respectively has been completed and the results are given in Fig. 9.15. Compared to those obtained for the PID controller alone, the results are very impressive and the overall normalized controller performance cost function value is 0.009, indicating an improvement in performance when compared to the simple ILC implementation after 100 trials and a 100 fold improvement in performance when compared to a well-tuned PID controller.

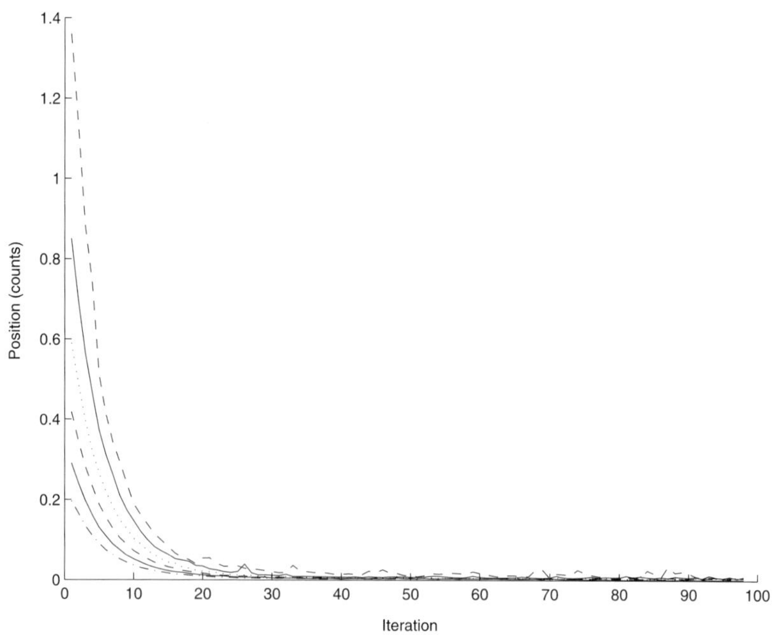

Fig. 9.14. The norm optimal behavior in terms of MSE

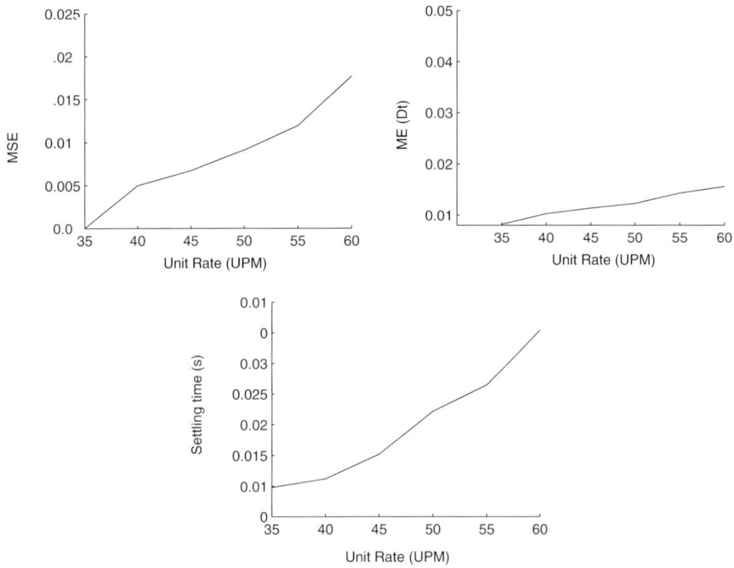

Fig. 9.15. Dispenser results under norm optimal ILC

This last point can also be illustrated by observation of the output response of the dispenser as the number of trials increases. Figure 9.16 shows the system response at 50 UPM and by trial 20 there is almost perfect tracking and only a slight position error when there is a change in direction demand. The relative error between the dispenser and conveyor at 50 UPM, is detailed in Fig. 9.17, by the 20th trial the relative error is within specification during the dispensing time but is out of specification when the dispenser if flying back to meet the next item on the conveyor. By the 40th trial the performance of the norm-optimal controller is superior to that achieved by the simple structure ILC after 100 trials. In practical terms this is an important conclusion because it indicates that using an advanced control strategy can reduce the system set up time. As with the simple structure ILC hysteresis switching has been implemented to ensure long term stability. In order to observe the performance of the system having obtained good tracking the proportional term of the controller was de-tuned to half of its original value. In this case after 75 trials (Fig. 9.18) the learning loop is reactivated and the dispenser output is again within specification by the 95th trial. In comparison, it can be seen that the norm-optimal controller's improved response results in far less trials where the dispenser output is out of specification. Finally, note that these results could vary somewhat if the weighting matrices are changed but we have found by extensive tests that the trends in the results given here always hold.

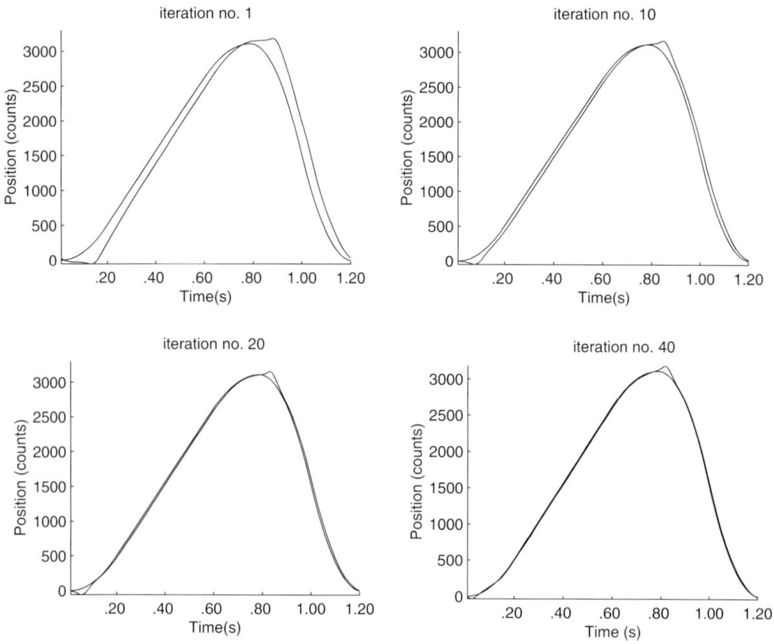

Fig. 9.16. Norm-optimal ILC output response against reference signal at speed 50 UPM

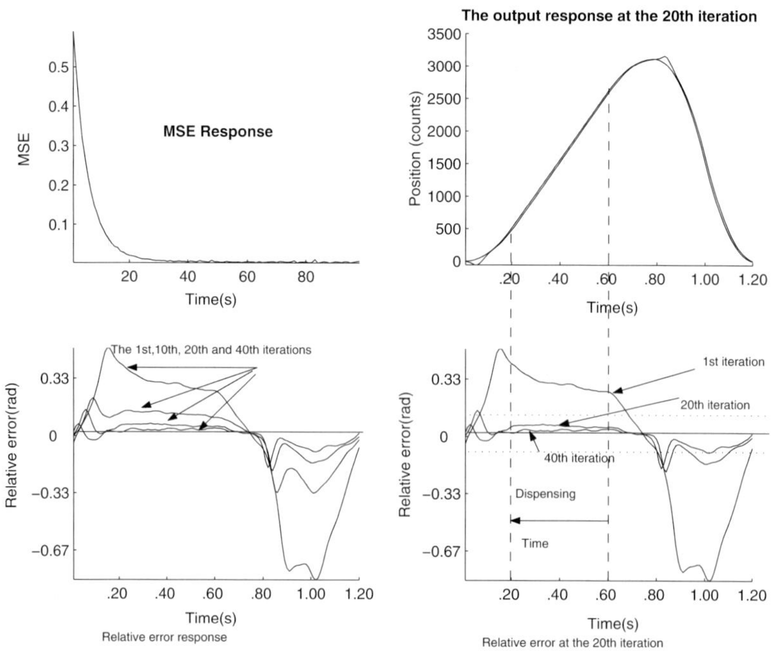

Fig. 9.17. MSE and relative error at speed 50 UPM

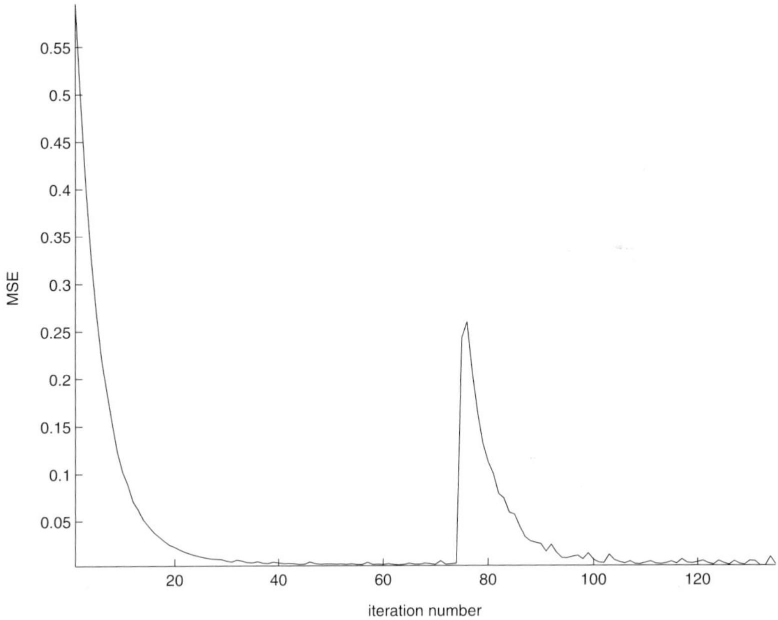

Fig. 9.18. The norm-optimal ILC behavior in the presence of a disturbance

9.4 Robust ILC

The problem of what is meant by 'robustness' of an ILC control scheme is more involved than in the 1D case. For example, there is robustness to unmodelled dynamics and also to the effects of the process not resetting to the same value before the start of each new trial (this is usually known as robustness to initial conditions). In this section we develop new results in the former area for SISO 1D discrete linear systems for a steepest-descent ILC law. The resulting algorithm is then verified experimentally on a gantry robot as described in the next section.

The starting point is a controllable and observable SISO discrete linear system defined over the trial length $T < \infty$ by the state-space model

$$\begin{aligned} x(p+1) &= \Phi x(p) + \Gamma u(p), \;\; x(0) = x_o \\ y(p) &= Cx(p) \end{aligned} \tag{9.21}$$

where $x(p)$ is the $n \times 1$ state vector. Also without loss of generality we assume that $C\Gamma > 0$, and $r(p)$ denotes the signal to be tracked.

Since (9.21) is defined over a finite time-interval, it can be equivalently described by a matrix equation $\hat{y}_k = \widetilde{G}_e \hat{u}_k$, where

$$\widetilde{G}_e := \begin{bmatrix} 0 & 0 & 0 & \dots & 0 \\ C\Gamma & 0 & 0 & \dots & 0 \\ C\Phi\Gamma & C\Gamma & 0 & \dots & 0 \\ \vdots & \vdots & \vdots & \ddots & \vdots \\ C\Phi^{N-1}\Gamma & C\Phi^{N-2}\Gamma & C\Phi^{N-3}\Gamma & \dots & 0 \end{bmatrix}$$

$\hat{u}_k := [\ u_k(0) \quad u_k(1) \quad \dots \quad u_k(T)\ ]^T, \hat{y}_k := [\ y_k(0) \quad y_k(2) \quad \dots \quad y_k(T)\ ]^T$, and the elements $C\Phi^j\Gamma$ of G_e are the Markov parameters of the plant (9.21). We also make the standing assumption that the reference signal $r(t)$ satisfies $r(0) = Cx_o$. Then it can be shown, see [85], that for analysis it is sufficient to analyze the so-called 'lifted' plant equation $y_k = G_e u_k$ where $u_k := [u_k(0)\ u_k(1)\ \dots\ u_k(T-1)]^T$, $y_k = [y_k(1)\ y_k(2)\ \dots\ y_k(T)]^T$ and

$$G_e := \begin{bmatrix} C\Gamma & 0 & 0 & \dots & 0 \\ C\Phi\Gamma & C\Gamma & 0 & \dots & 0 \\ C\Phi^2\Gamma & C\Phi\Gamma & C\Gamma & \dots & 0 \\ \vdots & \vdots & \vdots & \ddots & \vdots \\ C\Phi^{N-1}\Gamma & C\Phi^{N-2}\Gamma & C\Phi^{N-3}\Gamma & \dots & C\Gamma \end{bmatrix}$$

This lifted model forms the basis for the analysis which now follows. (For other uses of lifting in ILC (and repetitive) control systems analysis and design see, for example, [41, 116].)

The ILC steepest-descent algorithm [65] has been extensively studied over the years and is a form of optimal control where the cost function has the structure

$$J(u_{k+1}) := \|e_{k+1}\|^2, \ e_{k+1} = r - G_e u_{k+1}$$

during each repetition. Suppose now that we select the input on trial $k+1$ as

$$u_{k+1} = u_k + \epsilon_{k+1}\delta_{k+1}$$

where ϵ_{k+1} is a scaling factor and δ_{k+1} is the vector that determines the direction of the update (vector). Then on trial $k+1$

$$\begin{aligned} J(u_{k+1}) &= J(u_k + \epsilon_{k+1}\delta_{k+1}) = \|e_{k+1}\|^2 \\ &= \|e_k\|^2 - 2\epsilon_{k+1}\delta_{k+1}^T G_e^T e_k + \epsilon_{k+1}^2 \delta_{k+1}^T G_e^T G_e \delta_{k+1} \end{aligned}$$

and hence

$$\|e_{k+1}\|^2 - \|e_k\|^2 = -2\epsilon_{k+1}\delta_{k+1}^T G_e^T e_k + \epsilon_{k+1}^2 \delta_{k+1}^T G_e^T G_e \delta_{k+1} \tag{9.22}$$

Monotonic convergence clearly requires that the right-hand side of (9.22) is negative and one way to achieve this is to set $\delta_{k+1} = G_e^T e_k$, resulting in the control law

$$u_{k+1} = u_k + \epsilon_{k+1} G_e^T e_k \tag{9.23}$$

The difference $\|e_{k+1}\|^2 - \|e_k\|^2$ is then given by

$$\|e_{k+1}\|^2 - \|e_k\|^2 = -2\epsilon_{k+1}\|G_e^T e_k\|^2 + \epsilon_{k+1}^2\|G_e G_e^T e_k\|^2 \tag{9.24}$$

where the term $-\epsilon_{k+1}\|G_e^T e_k\|^2$ is of $o(\epsilon)$ and the term $\epsilon_{k+1}^2\|G_e G_e^T e_k\|^2$ is of $o(\epsilon^2)$. Hence by selecting a sufficiently small positive ϵ_{k+1} the right-hand side of (9.24) can be made negative (note that G_e was assumed to be invertible), resulting in monotonic convergence.

In order to automate the selection process for ϵ_{k+1}, it was proposed in [65] that ϵ_{k+1} be taken as the solution of the optimization problem

$$\arg \min_{\epsilon_{k+1} \in \mathbb{R}} J(u_k + \epsilon_{k+1} G_e^T e_k)$$

Given this choice, it is a routine exercise to show that the optimal ϵ_{k+1}, denoted here by $\epsilon_{k+1}^{\text{opt}}$, is given by

$$\epsilon_{k+1}^{\text{opt}} = \frac{\|G_e^T e_k\|^2}{\|G_e G_e^T e_k\|^2}$$

which results in

$$\|e_{k+1}\|^2 - \|e_k\|^2 = -\frac{\|G_e^T e_k\|^4}{\|G_e G_e^T e_k\|^2}$$

where the right-hand side is obviously negative, implying monotonic convergence to zero tracking error.

Note 9.4.1. If the original plant is positive-definite in the sense that $G_e + G_e^T > 0$, then it follows immediately from the above analysis that the simple control law $u_{k+1}(p) = u_k(p) + \gamma e_k(p+1)$ will result in convergence if γ is selected to be sufficiently small. In order to automate the selection of γ, [129] has developed a so-called parameter-optimal ILC algorithm that results in monotonic convergence for 'positive-definite plants'. More results in this general area can be found in, for example, [127].

In practical applications modelling uncertainty is always present, in particular at high frequencies, due, for example, to measurement noise and nonlinearities in the real plant. Hence what is really required in ILC is a robust control theory as for 1D linear systems. Next, we develop some new results in this still very much open research area where, as also noted above, the problem is more involved than in the 1D case. The results developed are then experimentally verified on a gantry robot (in the next section).

The starting point here is the assumption that the true plant contains multiplicative uncertainty, which is modelled as $G_e = G_o U$, where G_o is the nominal model (i.e. an estimate of the true plant), and U reflects the multiplicative uncertainty (i.e. modelling errors). Also G_o is used instead of G_e in the update law, i.e. $u_{k+1} = u_k + \epsilon_{k+1} G_o^T e_k$, and then $\|e_{k+1}\|^2 - \|e_k\|^2$ becomes

$$\|e_{k+1}\|^2 - \|e_k\|^2 = -2\epsilon_{k+1} e_k^T G_o U G_o^T e_k + \epsilon_{k+1}^2 e_k^T G_o G_e^T G_e G_o^T e_k \tag{9.25}$$

and we have the following result for this case.

Lemma 9.4.1. *Suppose that $U + U^T > 0$. Then if $\|e_k\| \neq 0$ $\exists$ an $\epsilon_{k+1} > 0$ such that $\|e_{k+1}\|^2 - \|e_k\|^2 < 0$.*

Proof. By assumption $U+U^T > 0$ and hence the term $-2\epsilon_{k+1}e_k^T G_o U G_o^T e_k = -2\epsilon_{k+1}e_k^T G_o(\frac{U+U^T}{2})G_o^T e_k$ in (9.25) is strictly negative for an arbitrary non-zero e_k and $\epsilon_{k+1} > 0$, and the second term $\epsilon_{k+1}^2 e_k^T G_o G_e^T G_e G_o^T e_k$ in this same equation is strictly positive for an arbitrary non-zero e_k. Also, the first term is of $o(\epsilon_{k+1})$ whereas the second term is of $o(\epsilon_{k+1}^2)$. Hence a sufficiently small non-nonzero ϵ_{k+1} will result in monotonic convergence. □

In the standard steepest-descent algorithm, ϵ_{k+1} is given by

$$\epsilon_{k+1} = \frac{\|G_o^T e_k\|^2}{\|G_o G_o^T e_k\|^2}$$

and hence there is no clear mechanism for choosing ϵ_{k+1} sufficiently small enough to satisfy Lemma 9.4.1. To remove this difficulty, we now develop a new modified steepest-descent algorithm that will result in monotonic convergence for plants with multiplicative uncertainty U, where $U + U^T > 0$.

Consider again the algorithm

$$u_{k+1} = u_k + \epsilon_{k+1} G_e^T e_k$$

where ϵ_{k+1} is selected as the solution of the optimization problem

$$\min_{\epsilon_{k+1}\in\mathbb{R}} J(\epsilon_{k+1}), \quad J(\epsilon_{k+1}) := \|e_{k+1}\|^2 + w\epsilon_{k+1}^2 \tag{9.26}$$

where the real number $w > 0$ and this choice of $J(\epsilon_{k+1})$ reflects two design objectives. In particular, the first term reflects the objective that the tracking error should be small during each trial whereas the second attempts to keep the magnitude of ϵ_{k+1} small, possibly resulting in a more cautious and robust algorithm when compared to the standard steepest-descent one.

It is easy to show that the optimal solution of (9.26) is given by

$$\epsilon_{k+1} = \frac{\|G_e^T e_k\|^2}{w + \|G_e G_e^T e_k\|^2} \tag{9.27}$$

and the following result on the convergence properties of this algorithm can now be established.

Lemma 9.4.2. *If $w > 0$ then $\|e_{k+1}\| \leq \|e_k\|$, where equality holds if, and only if, $\epsilon_{k+1} = 0$. Also*

$$\lim_{k\to\infty} \|e_k\| = 0 \quad \text{and} \quad \lim_{k\to\infty} \epsilon_k = 0$$

and hence monotonic convergence to zero tracking error.

Proof. Selecting a sub-optimal choice $\epsilon_{k+1} = 0$ in the cost function (9.26) yields $J(0) = \|e_k\|^2$ and hence

$$\|e_k\|^2 \geq \|e_{k+1}\|^2 + w\epsilon_{k+1}^2 \geq \|e_{k+1}\|^2 \tag{9.28}$$

which establishes monotonic convergence. This inequality also shows that $\|e_{k+1}\| = \|e_k\|$ if, and only if, $\epsilon_{k+1} = 0$. Using (9.28) we have that

$$0 \leq \|e_{k+1}\|^2 \leq \|e_k\|^2 - w\epsilon_{k+1}^2$$

and by induction on $\|e_k\|^2$

$$0 \leq \|e_0\|^2 - w\sum_{i=1}^{k+1} \epsilon_i^2$$

Hence, since k is arbitrary, $\sum_{i=1}^{\infty} \epsilon_i^2 < \infty$, and therefore $\lim_{k\to\infty} \epsilon_k = 0$ and we then have

$$0 = \lim_{k\to\infty} \epsilon_{k+1} = \lim_{k\to\infty} \frac{\|G_e^T e_k\|^2}{w + \|G_e G_e^T e_k\|^2}$$

This is only possible if $\lim_{k\to\infty} e_k = 0$ since G_e^{-1} exists (by assumption). □

Using the nominal plant G_o in the update law (9.23) gives

$$u_{k+1} = u_k + \epsilon_{k+1} G_o^T e_k$$

and

$$\epsilon_{k+1} = \frac{\|G_o^T e_k\|^2}{w + \|G_o G_o^T e_k\|^2}$$

Also substituting the optimal ϵ_{k+1} into (9.25) (which defines $\|e_{k+1}\|^2 - \|e_k\|^2$ when G_e contains multiplicative uncertainty) gives

$$\begin{aligned}\|e_{k+1}\|^2 - \|e_k\|^2 &= -2\frac{\|G_o^T e_k\|^2}{w + \|G_o G_o^T e_k\|^2} e_k^T G_o\left(\frac{U + U^T}{2}\right) G_o^T e_k \\ &+ \frac{\|G_o^T e_k\|^4}{(w + \|G_o G_o^T e_k\|^2)^2} \|G_e G_o^T e_k\|^2 \end{aligned} \tag{9.29}$$

and for monotonic convergence $\|e_{k+1}\|^2 - \|e_k\|^2 < 0$ must hold for a non-zero e_k. The next result shows how this can be achieved by taking w to be a sufficiently large positive number.

Lemma 9.4.3. *Assume that $U + U^T > 0$ and w is chosen such that*

$$w > \frac{1}{2} \frac{\|G_o^T\|^2 \|G_e G_o^T\|^2 \|e_0\|^2}{\underline{\sigma}(G_o \frac{(U+U^T)}{2} G_o^T)}$$

(where again $\underline{\sigma}(\cdot)$ denotes the smallest singular value of its matrix argument). Then the tracking error sequence satisfies $\|e_{k+1}\| < \|e_k\|$ when $e_k \neq 0$.

Proof. Routine manipulations applied to (9.29) show that for $\|e_{k+1}\| < \|e_k\|$, $e_k \neq 0$, it is required that

$$w + \|G_o G_o^T e_k\|^2 > \frac{1}{2} \frac{\|G_o^T e_k\|^2 \|G_e G_o^T e_k\|^2}{e_k^T G_o(\frac{U+U^T}{2}) G_o^T e_k}$$

but it is not clear how to select w such that this last condition holds. If, however, the term on the right-hand side is estimated as

$$\frac{\|G_o^T e_k\|^2 \|G_e G_o^T e_k\|^2}{e_k^T G_o(\frac{U+U^T}{2}) G_o^T e_k} \leq \frac{\|G_o^T\|^2 \|G_e G_o^T\|^2 \|e_k\|^2}{\underline{\sigma}(G_o(\frac{U+U^T}{2}) G_o^T)}$$

then a sufficient condition for strictly monotonic convergence is

$$w > \frac{1}{2} \frac{\|G_o^T\|^2 \|G_e G_o^T\|^2 \|e_k\|^2}{\underline{\sigma}(G_o(\frac{U+U^T}{2}) G_o^T)} \tag{9.30}$$

Further, because the initial guess u_0 results in a bounded tracking error e_0, w can be selected such that (9.30) holds for $\|e_0\|^2$. This results in $\|e_1\|^2 < \|e_0\|^2$, $e_0 \neq 0$, and consequently this inequality also holds with this particular choice of w for $\|e_1\|^2$. Using an induction argument, the inequality holds for an arbitrary trial k, implying strictly monotonic convergence. □

Note 9.4.2. The estimate for w in this last result can be very conservative, since $e_k^T G_o(\frac{U+U^T}{2}) G_o^T e_k$ is estimated in terms of the smallest singular value of $G_o(\frac{U+U^T}{2}) G_o^T$. Also if w is selected to be excessively large in magnitude, this can have an undesirable effect on the convergence speed, because such a choice will result in a small ϵ_{k+1}, implying that $u_{k+1} \approx u_k$. Consequently this result should be understood as an existence condition, and in practice w can be selected by resorting to a trial and improvement approach.

Note 9.4.3. Using (9.30) the required (sufficient) value of w decreases as $\|e_k\|^2$ decreases. Hence the potential to reduce w with each successive trial k with the added benefit that such a reduction could result in an increase in the convergence speed. A natural choice to exploit this would be $w = w_1\|e_k\|^2$. This could, however, lead, as $\|e_k\|^2 \to 0$, to 'dangerously high inputs' being applied to the plant. A simple remedy would be to choose $w = w_0 + w_1\|e_k\|^2$.

The next result shows that in addition to monotonic convergence, $U^T + U > 0$ also implies that $\lim_{k\to\infty} e_k = 0$ if w is selected to be sufficiently large.

Lemma 9.4.4. *Under the assumptions of Lemma 9.4.3 the algorithm converges monotonically to zero tracking error.*

Proof. Since $\|e_k\| \geq \|e_{k+1}\|$, $\lim_{k\to\infty} \|e_k\|$ exists, i.e. $\lim_{k\to\infty} \|e_k\| = E$, $E \geq 0$. Assume now that $E > 0$. Then there exists a subsequence $\{e_{k_j}\}$ of $\{e_k\}$

such that $\lim_{k_j \to \infty} e_{k_j} = e_\infty$ where e_∞ satisfies $\|e_\infty\| = E$, and $e_{k_j} \neq 0$ for all k_j. Consider now the sequence $\{e_{k_j+1}\}$ generated by

$$e_{k_j+1} = (I - \epsilon_{k_j+1} G_e G_o^T) e_{k_j}$$

where

$$\epsilon_{k_j+1} = \frac{\|G_o^T e_{k_j}\|^2}{w + \|G_o G_o^T e_{k_j}\|^2}$$

Then by Lemma 9.4.3, the sequence $\{e_{k_j+1}\}$ satisfies the inequality

$$\lim_{k_j \to \infty} \|e_{k_j+1}\| < \lim_{k_j \to \infty} \|e_{k_j}\| = E \tag{9.31}$$

However, the sequence $\{e_{k_j+1}\}$ is obviously a subsequence of $\{e_k\}$, and therefore $\lim_{k_j \to \infty} \|e_{k_j+1}\| \geq E$. This contradicts (9.31) and the proof is complete. □

Note 9.4.4. In the above analysis, monotonic convergence to zero tracking error requires that the multiplicative uncertainty U has to be positive in the sense that $U^T + U > 0$. However, in most applications the nominal 'matrix' model G_o is obtained by truncating the 'lifted' transfer-function $zG_o(z)$, where $G_o(z)$ is the transfer-function of the nominal plant model when time-axis is infinite. Hence it is important to understand this positivity result in terms of these transfer-function models. In which context, note that using transfer-function descriptions of the plant, the uncertainty model becomes $zG(z) = U(z)G_o(z)$, or $U(z) = G(z)^{-1}G_o(z)$. Further, it can be shown that if $U(z)$ is a positive-real system (or equivalently, its Nyquist diagram lies strictly in the right-half plane, see, for example, [42]), the truncated system (matrix) U is positive, i.e. $U^T + U > 0$ (see [57]). In Nyquist-diagram terms, this is equivalent to the condition that phase of the uncertainty model $U(z)$ has to lie inside $\pm 90°$, demonstrating a good degree of robustness.

In summary, the modified steepest-descent algorithm will converge monotonically to zero tracking error if the multiplicative uncertainty U satisfies the positivity condition $U^T + U > 0$ and w is chosen to be sufficiently large. Note that in the standard algorithm $w = 0$ and hence the introduction of w has resulted in a straightforward mechanism to find a balance between convergence speed and robustness.

9.5 Experimental Verification of Robust ILC on a Gantry Robot

The robust optimal algorithm developed in this previous section has been experimentally validated on an industrial gantry robot. As background to presenting a representative set of the results obtained, we first give a brief description of the gantry robot test facility and the procedure used to obtain frequency domain models for each component of the robot. A detailed treatment of the design and construction of this facility is given in [137] together with other tests both for the algorithm of the previous section and others not considered here.

The gantry robot of Fig. 9.19 is representative of a typical tracking control problem. It is mounted above a chain conveyor system and is designed to collect items from a dispenser and place them on the moving conveyor. This is a fairly complex task as the gantry is required to accurately synchronize speed and position with the conveyor before releasing the item. This type of operation can be found in many different industrial systems such as processed food canning, bottle filling, automotive assembly and electronic circuit assembly.

Fig. 9.19. The gantry robot

The gantry robot is constructed from two types of linear motion device and the lowest horizontal axis, X, consists of one brushless linear dc motor and a free running slide parallel to the motor. The next horizontal axis, Y, is perpendicular to the X-axis and has one end attached to the linear motor and the other end to the slide. The Y-axis is a single brushless linear dc motor. The X and Y-axes are 1.02 m and 0.91 m long respectively. Finally, the vertical axis, Z, is a short 0.10 m travel linear ball-screw stage driven by a rotary brushless dc motor. All axes are powered by matched brushless motor dc amplifiers and axis motion is detected and recorded with appropriate optical encoder systems.

Each axis of the gantry robot has been modelled individually in the velocity control mode of operation. The axes dynamics have been determined by performing a series of open-loop frequency response tests. Note that in velocity control mode, the amplifier receives the encoder data as well as the computer, therefore providing an inner closed-loop and an integrating action to the system. From the frequency response measurements linear approximations of the transfer-function for each axis were determined. To refine each axis model further, a Least-Mean-Squares optimization technique was applied to the approximations to fine-tune the transfer-function. The frequency response obtained for the X-axis is shown in Fig. 9.20 (the responses of the other axes were similar). To accurately model the complicated high-frequency behavior of each axis, a relatively high-order model is needed to describe their dynamics. The optimal fit for the X-axis was found to be $21st$ order, for the Y-axis $7th$ order, and for the Z-axis $4th$ order. For each axis a state-space model of the form (9.21) was obtained using a sampling rate of 100 Hz (where the choice of this value is justified below).

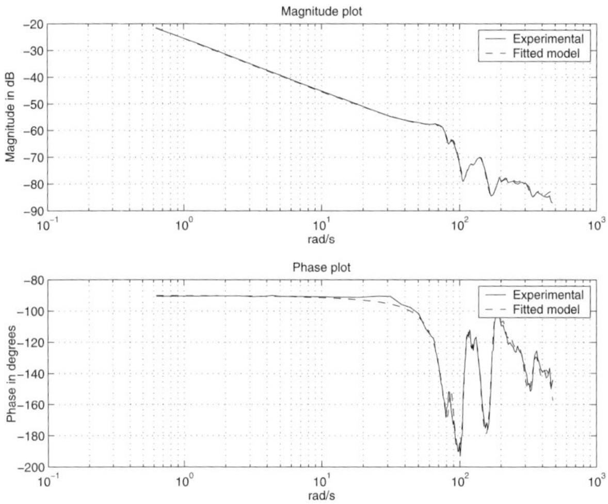

Fig. 9.20. X-axis Bode plot, experimental data and fitted model

All tests were performed in ILC format (this facility can also be configured to run in repetitive control mode – see, for example, [138]), implying that between each trial there is an undefined stoppage time which allows computation of the input sequence for the following trial. Before each new trial, the axes of the gantry robot are also returned to within ±30 microns of a homing position defined by the reference trajectories. During each trial, test data is saved to memory and is then transferred to disk in the stoppage time between trials. A sample instant from one trial therefore occurs at exactly the same time point within the trial duration as the corresponding sample instant of any other trial.

The control task for the ILC algorithm is of the trajectory tracking type. The trajectories for each axis are shown in Fig. 9.21. With all axes operating simultaneously, the trajectories generate a three dimensional synchronizing

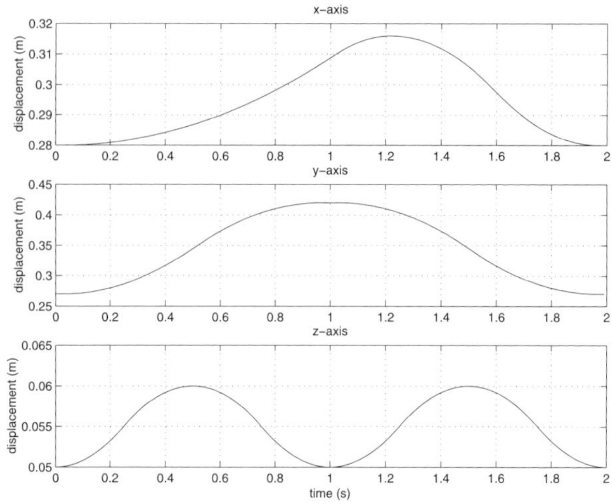

Fig. 9.21. Reference trajectories

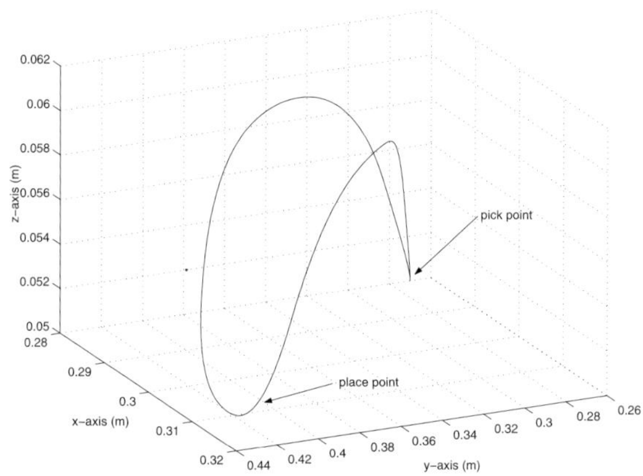

Fig. 9.22. 3D reference signal, 'pick and place'

'pick and place' action shown in Fig. 9.22. The trajectories are designed to produce an operation speed of 30 units per minute (UPM), equivalent to a trial duration of 2 seconds.

The size of the G_o matrices is determined by the sampling frequency and the time period of one trial. For the robust optimal algorithm, the selected sampling frequency is low at only 100 Hz and this has several advantages A low sample frequency minimizes the sizes of the G_o matrices implying that less memory and computation time are required to generate the next input to the plant. Using a low sample frequency also acts as a simple low pass filter by aliasing out high frequency noise. This is useful for achieving good long term performance in ILC systems [104] (see also below). The gantry robot axes are positioned using a velocity control mode where the linear motor amplifiers operate with their own closed-loop control. The input to the amplifiers is therefore simply a setpoint adjustment and is not directly required to achieve a stable system. Therefore the chosen frequency of 100 Hz is sufficient to accurately control the plant.

Data recorded during testing consists of the control input voltage supplied to the motor amplifiers, the axis position data and the tracking error relative to the reference trajectory. These variables are recorded at each sample instant. The MSE (mean squared error) for each trial is also calculated. Comparing the MSE of different trials shows whether the tracking performance is generally improving or degrading.

In ILC research the three factors which are generally used to compare the performance of different algorithms are convergence speed (rate of tracking error reduction), minimization of final error bound, and long-term performance. Also ILC research/algorithm development has, in the main, focused on maximizing the convergence speed so that the algorithm achieves minimum error in the least number of iterations. For practical application, this is also a valid goal however convergence speed although desirable, is not always critical. It is equally important for the algorithm to perform within acceptable error bounds for (in theory) all trials, i.e. as $k \to \infty$. If this cannot be achieved practically then it will be necessary to reset the system periodically, which is again wasteful of both time and resources. To summarize, in practical ILC research, emphasis must be placed on all three performance criteria, rather than focussing on one alone.

As a first set of experiments, the performance of the modified steepest descent algorithm developed in the previous section has been compared against two other algorithms. The reason for this is to establish that this new algorithm is capable of providing improved performance and the results can be used to assess robustness issues using the modified robust optimal algorithm. The two other algorithms used in this set of experiments are a standard three term (PID) feedback controller (acting on the current trial error alone with K_p, K_i and K_d denoting the proportional, integral and derivative gains respectively) and a simple proportional or P-type anticipatory ILC algorithm

$$u_{k+1}(p) = u_k(p) + \gamma e_k(p+1) \tag{9.32}$$

where γ is a scalar learning gain which is chosen by the operator. The anticipatory component (the $(p+1)$ term) here means that the error value of the update is shifted forward by one sample instant. This aims to capture the trend of the error data rather than treating each sampling instant individually.

Figure 9.23 shows the MSE data obtained for the three types of controller, where the performance of the PID controller does not change as it has no ability to learn. To allow fair comparison of the controllers, the first trial of the P-type and robust ILC are controlled by the PID controller. The PID

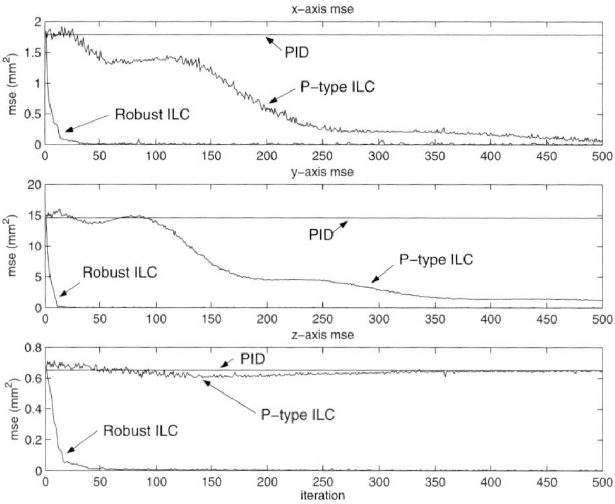

Fig. 9.23. Comparison of PID, P-type and robust ILC

controller is then switched out and the ILC controller operates alone. This automatically sets the MSE for the first trial at a uniform value. For the data shown, the value of γ was set equal to 0.1 for all axes, and the value of w was set to 0.

From the data it is immediately evident that both the P-type and robust controllers have the ability to improve the tracking performance beyond that of the PID controller. It is also clear that the performance of the robust optimal algorithm is superior to that of the P-type, as the convergence rate and level of error reduction are far greater. The performance of the P-type algorithm also appears to be non-monotonic as there is a tendency for the MSE to grow first before reducing. In an attempt to match the performance of the robust algorithm, the learning gain γ of the P-type algorithm was set at $0.01, 0.1$ and 1 respectively. At 0.01, the learning rate was very slow or unnoticeable. With $\gamma = 1$ the learning rate was much faster, but still did not match the robust controller. With this higher learning gain value the MSE also began steadily increasing again after approximately 95 trials. This was accompanied by a noticeable vibration of the gantry structure which became increasingly violent. On reaching 200 trials, the system was switched off to prevent damaging the mechanical components.

The increase in MSE and the mechanical vibration are clear signs of instability. This is most likely caused by high frequency noise which is amplified at each cycle of the trial loop (see [104]). It is important to note that the P-type ILC is operated at a sampling frequency of 1 KHz, i.e. ten times faster than the robust algorithm. Hence it is possible for much higher frequencies to be added to the control signal at each trial. High frequency noise is not a repeatable disturbance and so the ILC is unable to reduce it.

Having completed these experiments and demonstrated the relative performance of the modified steepest decent algorithm, it is now possible to investigate the properties of the robust optimal algorithm. To apply this algorithm, it is necessary to determine G_o and U ($G_e = G_oU$) for the particular application under consideration. Clearly there is a very wide range of admissible G_o and U and here we seek to demonstrate the capabilities of this uncertainty structure by selecting these parameters to closely resemble those likely to be made by a practicing engineer. In particular, we consider in turn the cases when (i) there is an error only in the zero frequency gain of the transfer-function built from measured frequency responses, (ii) a model is produced based only on the low frequency characteristics, and (iii) the axis dynamics can be represented by a pure integrator with a tunable gain.

In the case of (i) we can investigate the consequences of this by constructing G_o starting from the measured Bode plots in the continuous domain and then investigating what happens when the positive real scalar U is varied. For (ii), a standard form of differential linear systems model for a linear motor drive is $\frac{k}{s(s+a)}$ where k and a are positive scalars. In physical terms, this amounts to assuming that there is no interaction between the individual axes of the robot and all higher frequency dynamics are ignored. The tuned models actually used for the three axes are

$$G_{ox}(s) = \frac{11}{s(s+200)}$$

$$G_{oy}(s) = \frac{17}{s(s+300)}$$

$$G_{oz}(s) = \frac{10}{s(s+400)}$$

respectively. Finally, in (iii) no account of the individual axis frequency responses is taken when obtaining G_o.

9.5.1 Model uncertainty – Case (i)

Results for Case (i) with $\omega = 0$ are shown in Fig. 9.24 where U is selected in the range between 0.75 and 1.25. It follows immediately from (9.27) that if $U < 1$ then this will result in an increase in the learning gain and consequently performance will be severely degraded because the algorithm will over compensate for the error at each iteration. For example, Fig. 9.25 shows the poor tracking performance that results from a zero frequency gain error of 0.5.

As described in Note 9.4.3 one option to restore a high convergence rate, while maintaining stability, is to allow the adaptation of the magnitude of ω as a function of the current level of tracking error, i.e. to use

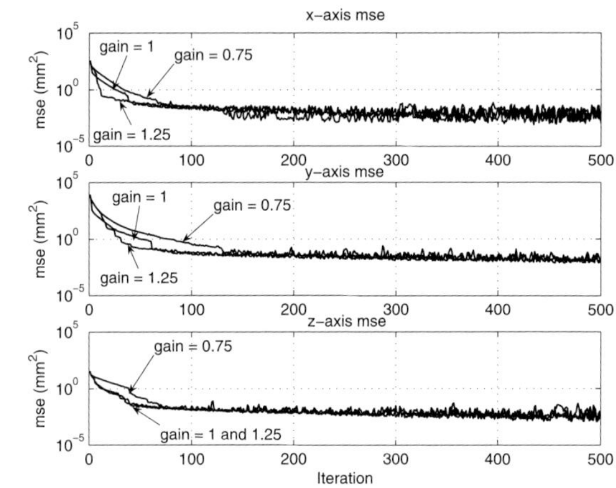

Fig. 9.24. mse for different scalar gains – Case (i)

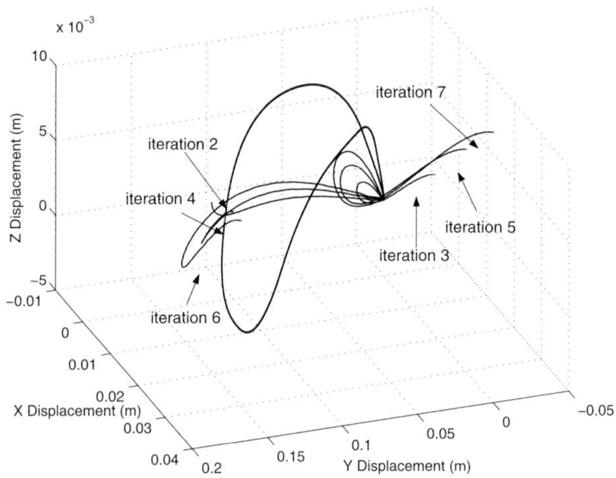

Fig. 9.25. 3D tracking progression $U = 0.5$ – Case (i)

$$\omega_{k+1} = \omega_0 + \omega_1 ||\mathbf{e}||^2 \tag{9.33}$$

where ω_0 and ω_1 are two new tuning parameters, which must be appropriately selected to match the operation of the control system. Parameter ω_0 can be used to guarantee stability by setting a baseline magnitude for ω_{k+1}, while $\omega_1 ||\mathbf{e}||^2$ adapts ω_{k+1} to match the change in tracking error.

Figure 9.26 gives the X-axis displacement profiles for iterations 490−−500 with $U = 0.5$ and $\omega_{k+1} = 2 \times 10^{-7} ||\mathbf{e}||^2$. The adaptive variant of the gain tuning parameter has clearly improved the convergence rate and minimum tracking error. Table 9.2 gives a measure of the relative performance of each axis over the first 100 iterations for various choices of U and the tuning

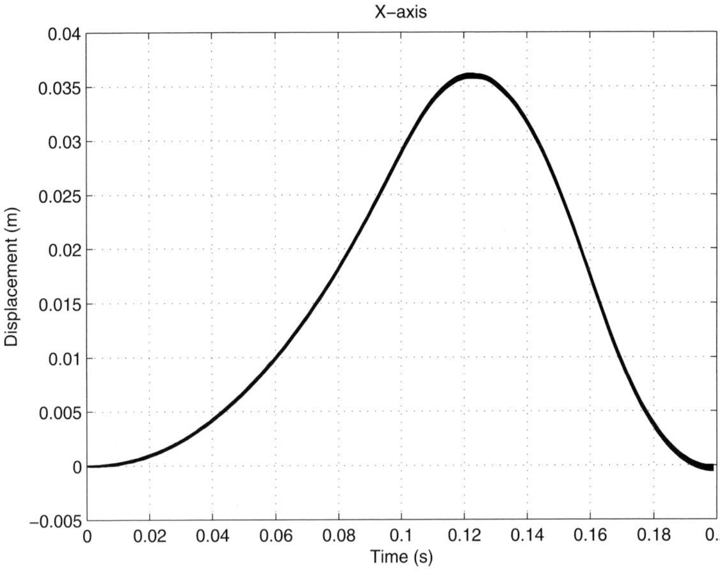

Fig. 9.26. X-axis tracking performance iterations 490-500 ($\omega_{k+1} = 2 \times 10^{-7}||\mathbf{e}||^2$ – Case (i), $U = 0.5$)

Table 9.2. Gain modelling error, PI_{100} values – Case (i)

Scalar gain	ω	X-axis	Y-axis	Z-axis
1.5	0	2.182499	1.982648	4.650866
1.25	0	2.088590	2.315949	3.045411
0.75	0	3.035958	2.454343	6.282407
0.5	0	100.000000	100.000000	100.000000
0.5	5×10^{-8}	21.685481	11.737366	69.209067
0.5	$2 \times 10^{-7}\|\|\mathbf{e}\|\|^2$	5.640472	3.309684	39.097583
0.5	$\omega_z = 0$	–	–	11.132384

parameters. The relative performance is measured through the use of

$$PI_{100} = \sum_{n=1}^{100} \overline{e^2}_n \tag{9.34}$$

This reveals that the Z-axis performance remains poor in comparison to the other axes. However, this can be corrected by setting $\omega_{k+1} = 0$. The required Z-axis travel is sufficiently small, that the initial overshoot caused by the increased learning gain does not exceed axis travel limits. By iteration 100, the output displacement accurately follows the reference trajectory to high accuracy.

9.5.2 Model uncertainty – Cases (ii) and (iii)

For Case (ii), the representative log mse results in Fig. 9.27 demonstrate that the robustness structure assumed in this work can be used to design a controller based on a reduced order model whose comparative performance is very close to that achieved with a full order model. As expected, the difference is most noticeable for the X-axis where the performance of the low order model is noticeably worse that of the high order model. The X-axis has the most significant high frequency dynamics and so is most affected by the simplification process.

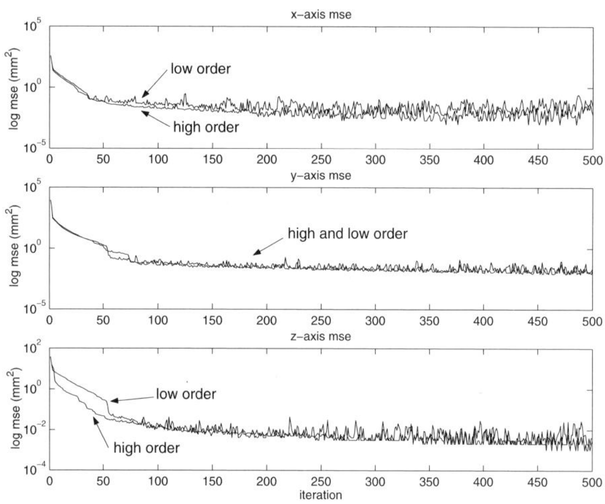

Fig. 9.27. Comparison of high and low order plant models, $w = 0$ – Case (ii)

The stability theory (see Note 9.4.4) allows G_o to be constructed from a (1D differential linear systems) model of the form β/s, where β is the tunable gain, as in Case (iii). (This represents the effects of choosing any other transfer-function whose phase lies between 0 and $-180°$.) Using this transfer-function for controller design, a range of values of β from 0.01 to 1.0 have been considered in detail to investigate the effect of the model gain on stability and convergence. Figure 9.28 compares the data for β equal to 1.00, 0.05 and 0.03. This shows that with a large value for β the rate of learning is slower. As β is reduced, the learning steadily becomes faster until an optimum β is reached. Increasing β any further then rapidly reduces the learning rate and very quickly the system performance degrades. The best values of β for each axis were found to be 0.05 for the X and Y-axes and 0.03 for the Z-axis. In all three cases the zero frequency gain of the resulting simplified model is a close match to that of the higher order model

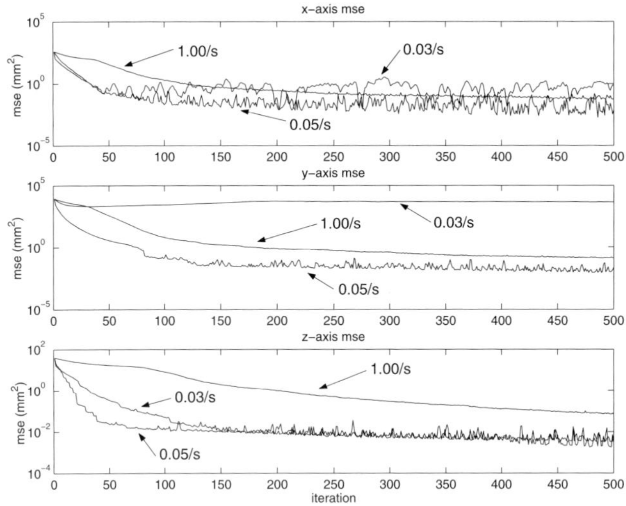

Fig. 9.28. Comparison of convergence rate for Case (iii) (β/s)

When β is too large, the controller will assume that a small change in controller effort will have a larger effect at the output. This effectively reduces the convergence rate. If β is too small, then the resulting controller assumes that large changes in controller effort are required to effect the output. Consequently there will be over-compensation for a given tracking error and performance will be severely degraded. This suggests a very simple tuning rule for the controller in this case.

1. Set β to a large value (if any form of plant model is available ensure that β generates a gain which is greater than the gain of the model at low frequency).
2. Operate the system and ensure that stability is achieved.
3. Reduce β.
4. Continue steps 2 and 3 until optimal convergence is achieved or until the system begins to severely degrade.
5. If this occurs, increase β slightly.
6. β is set.

To investigate the $\pm 90°$ degree stability boundary, a plant model of $1/s^2$ has been used on each axis (this generates a phase shift of $-180°$). At low frequency, each axis of the plant has a phase shift of $-90°$ and the system is therefore just on the theoretical stability boundary. The gain of $1/s^2$ generates a conservative learning system for each axis. Figure 9.29 displays the trajectory described by the X-axis from iteration 1 where the trajectory is zero to iteration 201, in intervals of 20 iterations, where the thicker line represents the desired reference trajectory. By iteration 121, the axis initially

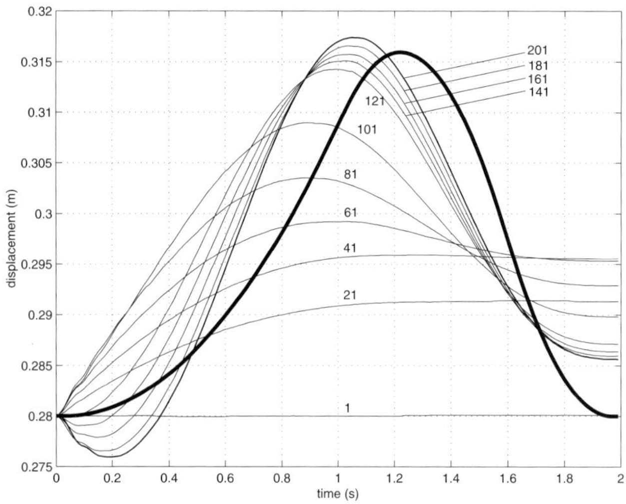

Fig. 9.29. X-axis performance using model $1/s^2$

travels in the opposite direction to what is required. This effect becomes steadily worse and by iteration 201, the rapid change of direction at the start of the iteration is severe. The test was in fact terminated at 203 iterations to prevent damaging the gantry robot. In addition, there is a growing oscillatory component of the displacement waveform which occurs at time 0.1 seconds. This initially appears at iteration 41 and is clearly visible at iteration 201. Small oscillations such as this can usually be seen in other ILC systems as their performance begins to degrade. These observations suggest that use of a $1/s^2$ model will produce a poor controller as it introduces an additional phase shift of $90°$ at low frequency when compared to the plant (see also Note 9.4.4).

The results above have only considered the case when $w = 0$, but the theory predicts that a non-zero w could increase robustness of the controller. The approximate model β/s with β selected too small is an example of such a situation and we now investigate this case in more detail.

Focussing on the Y-axis, consider first the case when $\beta = 0.2$. This is well below the best choice of 0.5 and causes the controller to severely degrade within the first few iterations. With this model, a range of values of w were applied. When w is too small, the system performance does not improve but as w is increased, a point is reached at which it does but at the cost of a slow convergence rate. Increasing w further simply causes the convergence rate to slow even further.

Figure 9.30 shows the displacement recorded every 50 iterations from 1 to 451 for the Y-axis with $w = 2.5 \times 10^{-7}$. If w is reduced to 2.0×10^{-7} the gantry robot cannot operate for more than 3 iterations. Therefore the

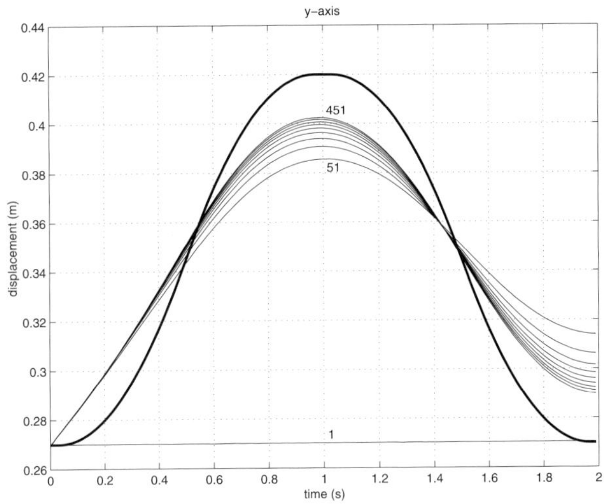

Fig. 9.30. Y-axis, model $= 0.2/s$, $w = 2.5 \times 10^{-7}$

first choice here gives the better convergence rate without degrading performance but at the cost of a poor final tracking error bound. This suggests that w becomes increasingly conservative as the error reduces and gradually prevents any learning from occurring. To prevent this, w needs to be reduced proportionally with the error.

An alternative selection method has already been proposed here in the form of (9.33). If w_1 is non-zero, then w is linked to the magnitude of the error, which is exactly what is required. Figure 9.31 shows the same test as before, but with $w_0 = 0$ and $w_1 = 2 \times 10^{-7}$. Clearly, appropriate adaptation of w can greatly improve the convergence rate for the unstable plant model.

9.5.3 Robustness to initial state error

The ILC algorithms developed here make the assumption that the plant is reset to an identical known starting location before each iteration begins. In practise, it is impossible to home the system to exactly the correct position, therefore initial state error always occurs. There has been some research on generating ILC algorithms which are tolerant to, or can limit, initial state error. As the robust optimal algorithm has proved tolerant to plant modelling errors, it is a natural progression to investigate its performance with respect to initial state error. Here we give some experimental evidence on the potential performance of the algorithm class considered by introducing a pseudo-random offset to the reset position before the start of each new iteration.

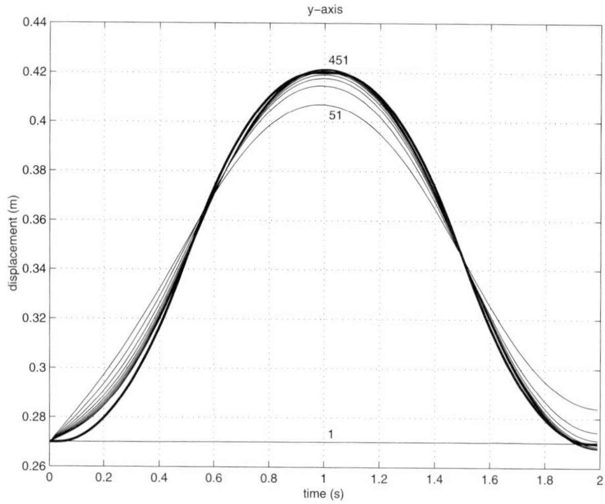

Fig. 9.31. Y-axis, model $= 0.2/s$, $w = 0 + 2 \times 10^{-7}\|e_k\|^2$

In describing the experimental facility here it has been stated that homing of the gantry axes is performed to within ±30 microns of the desired position. This initial state error is very small compared to the axes motion over a whole trajectory. An initial state error has been artificially created by adding a pseudo-random position error to the homing routine of each iteration. Tests were performed for different error bounds over the range 0.2 through to $2mm$.

Figure 9.32 gives the log mse results with initial position errors of $0.2mm$ and $2.0mm$ respectively. Clearly, the robust optimal algorithm maintains good performance over the 500 iterations shown. Even for the Z-axis where the initial error reaches 20% of the total trajectory motion, the algorithm still converges. The initial error does however have an effect on the final tracking error bound. A larger initial error results in a larger final error bound. This is to be expected since the input adjustment calculated after the previous iteration is less likely to match the shifted position of the next iteration.

9.5.4 Long-term performance

Some ILC algorithms have been found to exhibit poor transient performance after a relatively large number of iterations. In practice, no ILC algorithm will operate for ever. Conversely, there will clearly be many applications where good transient performance will be required over as large a number of iterations as possible. Here we have investigated this experimentally by running the experiment for 4000 iterations.

Figure 9.33 shows the log mse data for the robust optimal controller designed using plant models $0.5/s$ for the X and Y-axes and $0.3/s$ for the Z-axis. The final steady state error is reached by 2000 iterations on all three

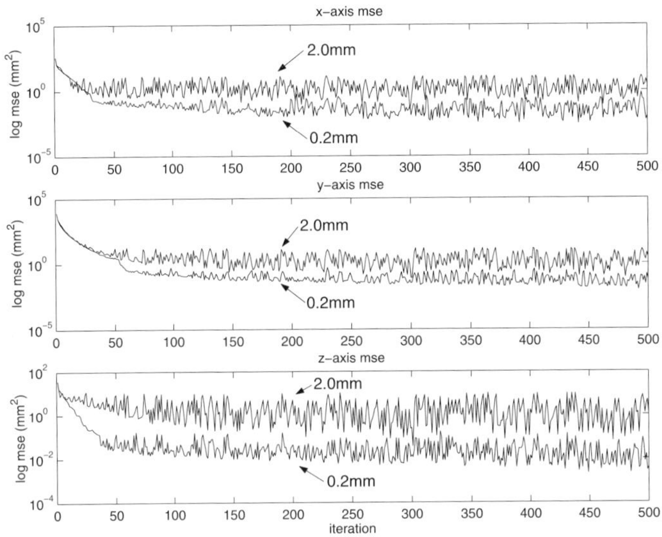

Fig. 9.32. Initial state error bounds 0.2mm and 2.0mm

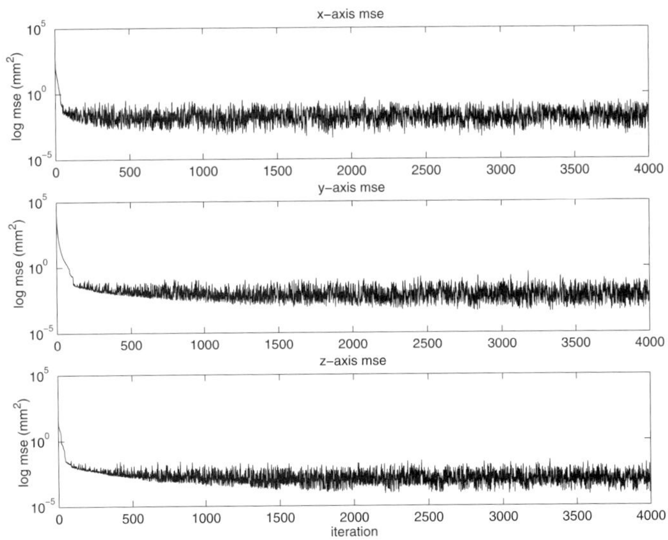

Fig. 9.33. β/s long-term performance, 4000 iterations

axes. A further 2000 iterations were completed with no indication of an increasing mse. Poor performing ILC systems tend to exhibit an initial decrease in mse while the dominant low frequency transients are reduced, followed by an immediate increase in mse as the high frequency transients begin to grow. Here, there is no sign of high frequency transient growth. The high frequency

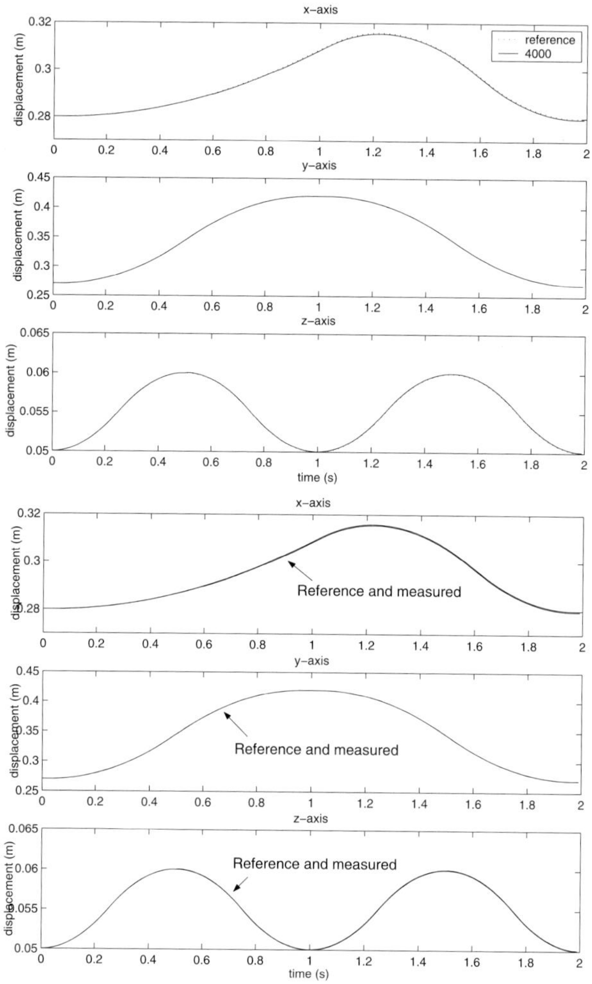

Fig. 9.34. Iteration 4000, reference and measured displacement

transients also tend to show during the iteration as small oscillations superimposed on the main trajectory. Figure 9.34 shows the position data for iteration 4000. The tracking remains excellent and there are no signs of high frequency oscillation.

10 Conclusions and Further Work

Summary. In this chapter we first review progress in terms of the results given in this monograph. Then we proceed to discuss areas for further work. These include further development/extension of the results given here and new applications areas.

10.1 Conclusions

In Chap. 1 the two long standing industrial examples, i.e. long-wall coal cutting and metal rolling, have been used to introduce the novel features of repetitive processes and, in particular, the unique control problem. This has been followed by a review of the abstract model representation together with the special cases considered in subsequent chapters. These include the first two so-called algorithmic applications, i.e. ILC (iterative learning control) and iterative solution algorithms for nonlinear optimal control problems based on the maximum principle. The first of these has been treated in Chap. 9 (see also below) and in the case of the second it is important note that, although not treated here beyond showing how the problem can be formulated in repetitive process terms and the abstract model based stability theory applied, much further development work, including software, has been undertaken as part of ongoing research in this area, e.g.[144]. Recently there have also appeared work (at various stages of development) on other physical and algorithmic applications where repetitive process theory could have an important role in onward developments, e.g. [47, 115].

The remainder of Chap. 1 first explores, at the model structure level, the links with other classes of 2D linear systems. In the case of discrete linear repetitive processes, an equivalent Roesser or Fornasini-Marchesini 2D discrete linear systems state-space model interpretation of the underlying dynamics has been developed which moves this area beyond the previously known fact that there is an equivalence in terms of stability theory between these two classes of linear systems. These models have then formed the basis for (part of) the controllability analysis undertaken in Chap. 6. Also a 1D equivalent state-space model for these processes has been developed which has proved of major use in onward analysis (as discussed further below). A unique feature of this model is that the vectors and matrices which define it have constant dimensions as do the entries in the matrices themselves.

The last topic in Chap. 1 is the development of 2D transfer-function descriptions for the dynamics of differential and discrete linear repetitive processes respectively which have also been used in subsequent analysis. One immediate use for these has been the development of block diagram representations of the dynamics which clearly highlight the critical role of the previous pass dynamics. Some related results concerning matrix primeness etc have also been introduced. The key point again is that the situation is much more complex than in the 1D case.

Chapter 2 presents the previously reported abstract model based stability theory and applies it to the processes considered in this monograph. One key fact which emerges is the critical dependence on the structure of the boundary conditions and, in particular, the pass state initial vector sequence. In particular, the structure of this sequence alone is enough to make an example unstable – contrary to what happens in the 1D linear systems case. Hence in any possible application due care must be given to the modelling of this sequence.

Previous work had already reported the development of computationally feasible stability tests which reduce to testing the locations of the eigenvalues of constant matrices and, in essence, computing (in the SISO case for simplicity) the Nyquist diagram generated by the (1D) transfer-function describing the contribution of the previous pass dynamics to those of the current. A new result here shows that this also extends to the case of discrete linear repetitive processes in the presence of so-called dynamic boundary conditions, i.e. the pass state initial vector is an explicit function of points along the previous pass profile. This case for differential processes is still an open problem.

From this last analysis, 2D characteristic polynomials for the differential and discrete cases respectively (only with non-dynamic (or static) boundary conditions) have been developed which play exactly the same role as those for their 1D linear systems counterparts and they are used (Chap. 6) to develop the concept of a pole for these processes which has an exponential trajectory (as opposed to a purely algebraic) interpretation and also as a starting point for control law design (Chaps. 7 and 9).

The final two sections of Chap. 2 have focused on the development of stability tests which can be implemented without the need to compute with matrices which have complex entries. In the first of these, previously reported approaches based on the availability of step response data and some elementary properties of nonnegative matrices have been presented. These are sufficient but not necessary conditions but do yield computable information regarding the rate of approach to the limit profile and the error on any intermediate pass. Moreover, they extend (unlike all other currently available tests) to the case of inter-pass smoothing. The final section of this chapter has continued this theme using (in relative terms) more advanced tools centered round the numerical radius of a matrix and some extensions particular to 2D discrete linear systems/discrete linear repetitive processes. These results are preliminary in nature and much further work remains to be done in this area.

Chapters 3 and 4 deal with Lyapunov equations for the stability analysis of discrete and differential linear repetitive processes respectively. Here there ar two types of such equations, termed 1D and 2D respectively. In the former case, the defining matrices are functions of a complex variable and in the latter they have constant entries.

Consider first the discrete case. Then here it has been shown that the 1D Lyapunov equation is necessary and sufficient for stability along the pass and if it holds then the sequence of pass profiles generated approaches the resulting limit profile at a computable geometric rate. Moreover, by exploiting properties of the matrix Kronecker product necessary and sufficient conditions for stability along the pass have been obtained which can be computed using matrices with constant entries. More work is required on the numerical properties of these tests, e.g. the matrices involved could be of 'very large' dimensions.

The basic form of 2D Lyapunov equation in the discrete case is the same as that reported previously for 2D discrete linear systems and a similar route, i.e. a counter-example, has been used to show that it is a sufficient but not condition (but necessary and sufficient in special cases, e.g. SISO examples). Some refinements, based on results first reported for the Fornasini-Marchesini 2D state-space model, have also been given together with candidate solution algorithms. The main onward roles for the 2D Lyapunov equation are for robustness analysis (Chap. 5) and as the essential basis for stability analysis and control law design based on LMI tools. Together these form one of the major and most significant blocks of new results in this monograph (Chaps. 7 and 8). In particular, the 2D Lyapunov equation is instrumental in the definition of Lyapunov functions for these processes.

The 1D Lyapunov equation for differential processes developed in the first part of Chap. 4 mirrors that for the discrete processes of Chap. 3 (with the unit circle in the complex plane replaced by the imaginary axis). Again, the 2D Lyapunov equation for differential processes is sufficient but not necessary for stability along the pass (but is in special cases, e.g. SISO processes) and is very closely linked to that for certain classes of linear delay differential systems. Moreover, the main onward role in this monograph is as stated above for the discrete case. In the final section of Chap. 4 the first results (in either the differential or discrete cases) on a Lyapunov/Riccati equation approach to analysis for a special case of dynamic boundary conditions have been developed. The special case considered establishes the link between the stability of differential linear repetitive processes and that for classes of linear delay differential systems. There is much more productive research to be done in this area.

Robustness without the application of control laws is the subject of Chap. 5. In the discrete case, three starting points have been used. These are the 2D Lyapunov equation, the theory of nonnegative matrices (following on from the stability tests in Chap. 2) and LMIs respectively. The results for the first two of these have close parallels in the theory of 2D discrete linear systems described by the Roesser and Fornasini-Marchesini state-space models. It is, however, only the LMI approach which can currently be followed through to control law design, including the case when there is uncertainty in the model and the same is true for the differential case.

Chapter 6 starts out by developing a transition matrix for discrete unit memory linear processes using as a starting point the 2D singular Fornasini-Marchesini equivalent model for the underlying dynamics developed in Chap. 1. This has subtle differences with that for the 2D discrete linear systems and is then used to develop a substantial body of results on local controllabiltiy/reachability, i.e. steering the process to produce a specified vector at some point along some pass. The resulting conditions are expressed in terms of matrix rank based tests (the matrices have constant entries) but further research is needed on reliable numerical means of testing them for a given example.

As for other classes of 2D linear systems there is more than one definition of controllability for discrete linear repetitive processes. In particular, the property of pass profile controllability, i.e. require the process to produce a pre-specified pass profile vector either on a given pass or on some pass during its evolution, has obvious appeal in terms of applications. Here this property has been defined and characterized in terms of matrix rank based tests using the 1D equivalent model. Other forms of controllability (and observability) for discrete processes can be found in, for example, [48].

Controllability of discrete processes with dynamic boundary conditions has also been analyzed and the key message here is again the importance of these in terms of the process dynamics. In the case of differential processes, a substantial batch of results have been given but these are much less well developed relative to the discrete case and this is an obvious area for further research. Other results for such processes can be found in, for example, [49].

In summary, substantial progress has been made on controllability and observability of differential and discrete linear repetitive processes but much remains as areas for further research. Some areas have already been alluded to above but there is also one other major area which has not yet seen any real progress. This is the general question of how does, for example, a given controllability property relate to the structure and design of control laws? For example, is it meaningful to expect a result which says that a controllability property is necessary and sufficient for the existence of a stabilizing control law? (i.e. a parallel of the pole placement result in 1D linear systems theory).

Chapter 6 also contains a section which gives results on systems equivalence as defined in terms of the 2D transfer-function matrix descriptions developed in Chap. 1. These results are the first and much further work is required on this area. A more general question is: can a comprehensive Rosenbrock system matrix theory for these processes be developed?

In the final part of Chap. 6 the behavioral approach to systems theory has been used to begin the development of the theory of poles and zeros for these processes. In the case of poles it has been shown that these can be defined as in the 1D case, i.e. in terms of the 2D characteristic polynomial introduced in Chap. 1 and also they provide an exponential trajectory interpretation of stability along the pass. The zeros also have an exponential trajectory interpretation but the overall picture is even more complicated than that for poles and requires much further development.

Chapter 7 undertakes a detailed treatment of key aspects of control law design for differential and discrete linear repetitive processes (including dynamic boundary conditions in the discrete case). First, the potential for control using only current pass information is considered. The main conclusion is that, although relatively simple in structure (and hence implementation) such control laws are very limited in what they can actually achieve.

A detailed treatment of control law design to achieve the most basic requirement of stability along the pass under the action of control laws which are composed of two terms, one activated by current pass information, state or pass profile, and the second by the previous pass profile has also been undertaken in Chap. 7. The design route in both cases is via an LMI and numerical examples are given. Of course, the control law which uses current and previous pass profile information only is weaker than the one which uses current pass state and previous pas profile information. As noted above, these results extend in a natural manner to the case of discrete processes with dynamic boundary conditions – the differential counterpart is still an open research question.

Having obtained results which enable the design of control laws to ensure stability along the pass, the next step is to design for this property plus performance objectives. In the latter respect, one key question is how to force a stable along the pass example to produce a pre-specified limit profile. Here the first substantial results on this problem have been obtained for both differential and discrete processes. The design algorithms again exploit LMIs and, in the discrete case, the 1D equivalent model. A repetitive process generalization of the familiar proportional plus integral, or PI, control law is also developed for both cases, where the 'integral term' is defined as a summation across the passes.

After a brief treatment of the direct application of delay differential theory to the differential case, the remainder of this chapter has focused on optimal control. This considered the problem where the cost function is the sum over the passes of a standard quadratic cost function along each pass. Here it has been shown, using ideas from dynamic programming, that this problem has a causal solution in the form of feedback of the current pass state vector.

There is clearly much further work to be done in this area, e.g. computational aspects – see also below.

Chapter 8 deals with the control of both discrete and differential repetitive processes in the presence of uncertainty, including the case where stability long the pass is required together with a guaranteed level of performance. First we use an H_∞ setting to solve the problem of stabilizing a discrete repetitive process and, simultaneously, minimizing the influence of an additive disturbance (on both the state and pass profile updating equations). The starting point is the use of a static control law activated by both current and previous pass information. Next these results are extended to the case of a so-called dynamic pass profile controller, whose main advantage is that it does not need the use of a state observer.

The next part of this chapter focuses on the guaranteed cost control of discrete processes, where we additionally introduce a cost/perfomance index. This has been followed by extension of all these results to the case of differential processes. The key point to note here is that this is by no means a routine task.

The last part of Chap. 8, reports the first significant results for the case when the influence of disturbances is modelled in the H_2 setting. Here the design of a static control law is developed and this technique is applied to enhance the robust control scheme. Finally the mixed H_2/H_∞ control problem, where some part of the disturbance influence is modelled by use of the H_2 norm and the remaining by the H_∞ norm, is considered. These results are given for differential processes.

The results in this chapter, although substantial, are by no means a complete treatment. Areas for future research include the extension of μ analysis. Moreover, many control problems either cannot be expressed in LMI form or such a representation leads to very conservative results. Hence research is required into the use of bilinear or more general polynomial matrix inequality forms in this general area.

In Chap. 9, the role of repetitive process theory in the ILC area has been considered. The first part shows how a direct application of asymptotic stability leads immediately to a major trade-off, i.e. it is possible to converge trial-to-trial to a limit error which has unacceptable along the trial dynamics – a fact which is overlooked in many papers in this area. Moreover, it is shown that previously reported results derived using the Roesser model are, in fact, available as special cases of the repetitive process analysis.

The rest of this chapter focuses on norm optimal and robust ILC. The cost function used in the norm optimal approach and the resulting solution have a clear repetitive process flavor and one feature which could well arise in other applications is the ability to penalize large increases in the control effort required from pass-to-pass with consequent effects on the health of the actuators employed. The experimental results demonstrate that control algorithms for repetitive processes (ILC related or otherwise) beyond those with a simple SISO structure can indeed be implemented to good effect. This general area will continue to develop with many opportunities for interchange of ideas/algorithms/proofs between them.

10.2 Further Research

Aside from the topics raised in the context of the chapter-by-chapter review, there are many others which require further research. In general terms, these can be split into the following: (i) a detailed treatment and improvement of the currently available computational algorithms, (ii) enhancing the currently available analysis tools, and (iii) extending the abstract model setting currently used to include cases not currently covered but for which there are potential applications areas.

Next we give one area where extension is possible and some results are already available.

10.2.1 Repetitive Processes with Switching Dynamics

Consider again the metal rolling example of Chap. 1. Then the particular cases of the dynamics treated are by no means all that can occur. In particular, a number of passes may be completed under one regime and then the dynamics change to allow further processing to take place. One way of modelling such a case is by switching the dynamics from one state-space model to an alternative (or alternatives). More generally, there are (at least) two distinct possibilities for switching dynamics to occur in repetitive processes; either from pass-to-pass or along a pass respectively and recently the development of tools for the analysis of these two cases has been started. Both of these scenarios are practically motivated, e.g. switching from pass-to-pass can occur as summarized above in metal rolling or, as an alternative, when considering multiple operation robot arms. Along the pass switching can arise in the analysis of ILC applied to processes with periodic dynamics.

Consider the case of a discrete unit memory linear repetitive process of the form defined by Example 1.2.10 whose dynamics switch in the pass-to-pass direction as described by the following state-space model over $0 \le p \le \alpha - 1,\ f = 0, 1, \cdots$

$$
\begin{aligned}
x_{f+1}(p+1) &= \begin{cases} A_1 x_{f+1}(p) + B_1 u_{f+1}(p) + B_{01} y_f(p), & f = 0, 2, \ldots \\ A_2 x_{f+1}(p) + B_2 u_{f+1}(p) + B_{02} y_f(p), & f = 1, 3, \cdots \end{cases} \\
y_{f+1}(p) &= \begin{cases} C_1 x_{f+1}(p) + D_1 u_{f+1}(p) + D_{01} y_f(p), & f = 0, 2, \cdots \\ C_2 x_{f+1}(p) + D_2 u_{f+1}(p) + D_{02} y_f(p), & f = 1, 3, \cdots \end{cases}
\end{aligned}
\tag{10.1}
$$

This model assumes that the dynamics switch on completion of each pass profile and is clearly not the most general case but, given the absence of any previous results in this area, it will act as a starting point with the possibility that the experience gained will lead to straightforward generalizations to other cases.

Another possibility is a process where switching occurs in the along the pass direction as described by the state-space model

$$x_{k+1}(h+1) = \begin{cases} A_1 x_{k+1}(h) + B_1 u_{k+1}(h) + B_{01} y_k(h), & h = 0, 2, \cdots \\ A_2 x_{k+1}(h) + B_2 u_{k+1}(h) + B_{02} y_k(h), & h = 1, 3, \cdots \end{cases}$$

$$y_{k+1}(h) = \begin{cases} C_1 x_{k+1}(h) + D_1 u_{k+1}(h) + D_{01} y_k(h), & h = 0, 2, \cdots \\ C_2 x_{k+1}(h) + D_2 u_{k+1}(h) + D_{02} y_k(h), & h = 1, 3, \cdots \end{cases}$$

Both these models can be easily transformed to the model structure of Example 1.2.10, and hence the associated theory can be applied, by use of simple substitutions which doubles the dimensions of the matrices and vectors involved. This, however, comes at the possible price of losing (potentially) useful structure in, say, some of the original model matrices due to their 'embedding' in the structure of the resulting model. Also, when considering the case of uncertainty in the model, the overall problem becomes non-convex and hence more difficult to solve.

Overall, this indicates that developing a comprehensive control theory for repetitive processes with switched dynamics will be a non-trivial task and, in general, will require results and techniques beyond those currently available from applying this transformation. To illustrate the results to-date, consider the first model above and a switched control law of the form

$$u_{f+1}(p) = \begin{cases} K_1^1 x_{f+1}(p) + K_2^1 y_f(p), & f = 0, 2, \ldots \\ K_1^2 x_{f+1}(p) + K_2^2 y_f(p), & f = 1, 3, \cdots \end{cases} \tag{10.2}$$

Then we have the following result [23].

Theorem 10.2.1. *Suppose that a control law of the form (10.2) is applied to a discrete linear repetitive process described by (10.1). Then the resulting process is stable along the pass, if* $\exists$ *matrices* $X_i > 0$, $Y > 0$, *square matrices* V_i, $i = 1, 2, 3$ *and* Z, *and rectangular matrices* L_i, $i = 1, 2, 3$ *and* N *such that*

$$\left[\begin{array}{cccc} -X + Y & 0 & \bar{A}_1 \bar{V} + \bar{B}_1 \bar{L} & 0 \\ 0 & -X - Y & 0 & \bar{A}_2^1 \bar{Z} + \bar{B}_2^1 \bar{N} \\ \bar{V}^T \bar{A}_1^T + \bar{L}^T \bar{B}_1^T & 0 & -\bar{V} - \bar{V}^T & 0 \\ 0 & \bar{Z}^T \bar{A}_2^{1T} + \bar{N}^T \bar{B}_2^{1T} & 0 & -\bar{Z} - \bar{Z}^T \\ 0 & 0 & X & \bar{V}^T \bar{A}_2^{2T} + \bar{L}^T \bar{B}_2^{2T} \\ 0 & 0 & 0 & \bar{V}^T \bar{A}_2^{2T} + \bar{L}^T \bar{B}_2^{2T} \end{array} \right.$$

$$\begin{array}{cc} 0 & 0 \\ 0 & 0 \\ X & 0 \\ \bar{A}_2^2\bar{V} + \bar{B}_2^2\bar{L} & \bar{A}_2^2\bar{V} + \bar{B}_2^2\bar{L} \\ -X & 0 \\ 0 & X - \bar{V} - \bar{V}^T \end{array} \left.\vphantom{\begin{array}{c}0\\0\\0\\0\\0\\0\end{array}}\right] < 0$$

where

$$X := \begin{bmatrix} X_1 & 0 \\ 0 & X_2 \end{bmatrix}, \quad \bar{V} := \begin{bmatrix} V_1 & 0 & 0 \\ 0 & V_2 & 0 \\ 0 & 0 & V_3 \end{bmatrix}, \quad \bar{Z} := \begin{bmatrix} Z & 0 & 0 \\ 0 & Z & 0 \\ 0 & 0 & Z \end{bmatrix}$$

$$\bar{L} := \begin{bmatrix} L_1 & 0 & 0 \\ 0 & L_2 & 0 \\ 0 & 0 & L_3 \end{bmatrix}, \quad \bar{N} := \begin{bmatrix} N & 0 & 0 \\ 0 & N & 0 \\ 0 & 0 & N \end{bmatrix}$$

and

$$\bar{A}_1 := \begin{bmatrix} A_1 & 0 & B_{01} \\ 0 & A_2 & 0 \\ 0 & C_2 & 0 \end{bmatrix}, \quad \bar{B}_1 := \begin{bmatrix} B_1 & 0 & B_1 \\ 0 & B_2 & 0 \\ 0 & D_2 & 0 \end{bmatrix}$$

$$\bar{K}_1 := \begin{bmatrix} K_1^1 & 0 & 0 \\ 0 & K_1^2 & 0 \\ 0 & 0 & K_2^1 \end{bmatrix}, \quad \bar{A}_2^1 := \begin{bmatrix} 0 & 0 & 0 \\ 0 & B_{02} & 0 \\ 0 & 0 & D_{02} \end{bmatrix}$$

$$\bar{B}_2^1 := \begin{bmatrix} 0 & 0 & 0 \\ 0 & B_2 & 0 \\ 0 & 0 & D_2 \end{bmatrix}, \quad \bar{K}_2 := \begin{bmatrix} K_2^2 & 0 & 0 \\ 0 & K_2^2 & 0 \\ 0 & 0 & K_2^2 \end{bmatrix}$$

$$\bar{A}_2^2 := \begin{bmatrix} 0 & 0 & 0 \\ C_1 & 0 & D_{01} \\ C_1 & 0 & D_{01} \end{bmatrix}, \quad \bar{B}_2^2 := \begin{bmatrix} 0 & 0 & 0 \\ D_1 & 0 & D_1 \\ D_1 & 0 & D_1 \end{bmatrix}$$

If this condition holds, the control law matrices are given by

$$\bar{K}_1 = \bar{L}\bar{V}^{-1} = \begin{bmatrix} K_1^1 & 0 & 0 \\ 0 & K_1^2 & 0 \\ 0 & 0 & K_2^1 \end{bmatrix}, \quad \bar{K}_2 = \bar{N}\bar{Z}^{-1} = \begin{bmatrix} K_2^2 & 0 & 0 \\ 0 & K_2^2 & 0 \\ 0 & 0 & K_2^2 \end{bmatrix}$$

A similar result is also available for the case of dynamics which switch along the pass and also for their differential process counterparts. Areas for further work here include processes with uncertainty in their model structure and the extension to the cases when (i) there is more than one switching operation, and (ii) there is a random number of passes between successive switching operations.

A major possible application area for these switched models is in the analysis and control of so-called bi-directional repetitive processes, see [52], where the pass profiles on successive passes are produced in opposite directions, i.e. over (in the discrete case) $[0, \alpha - 1]$ and $[\alpha - 1, 0]$ respectively. It is not at all clear how, if at all, such dynamics could be included in the models studied in the previous research but this is not true if switching is allowed. Further research is clearly required in this general area.

10.3 Spatially Interconnected Systems – a Role for Repetitive Processes?

Recently, motivated by the need to control ever increasing complex systems in various applications areas, there has been a considerable body of work reported on the development of tools for the analysis and control of spatio-temporal interconnected systems. Of this work the most relevant here is that on stability and feedback stabilization of various classes, see, for example, [37].

The intrinsic feature of this approach is that the spatial index-number of the subsystem plays the role of an independent variable, and hence the interconnection of the basic building blocks leads to a spatio-temporal multi-dimensional system, where the basic building block is illustrated in Fig. 10.1. By cascade interconnection it is then possible to consider an infinite system (Fig. 10.2), or a periodic system (Fig. 10.3). It is also possible to extend the

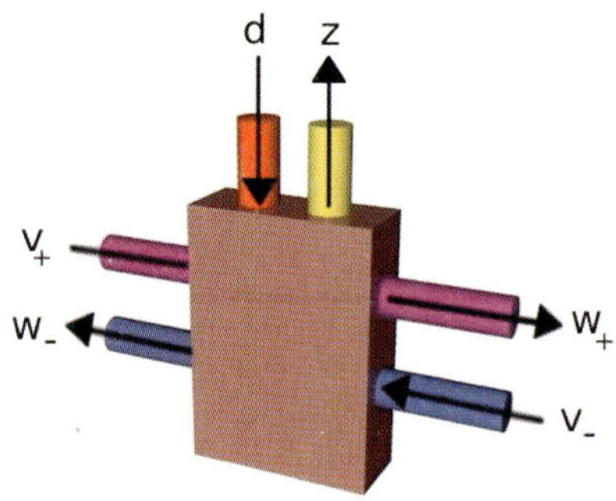

Fig. 10.1. Basic building block

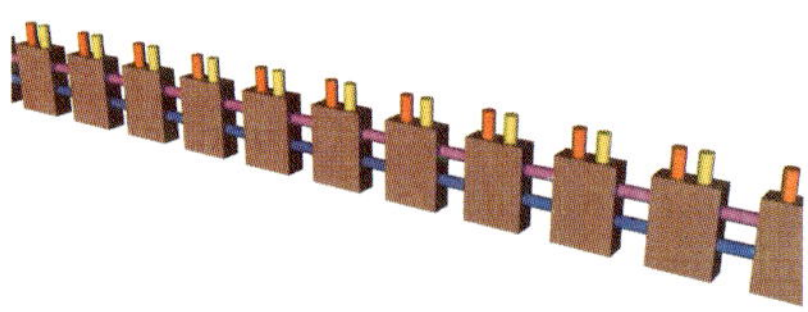

Fig. 10.2. Infinite interconnection

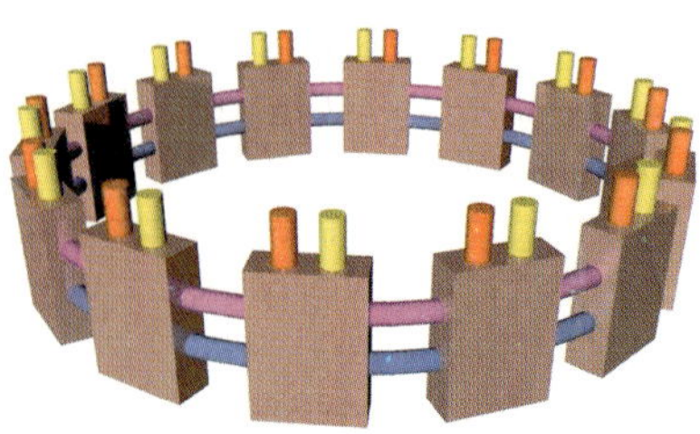

Fig. 10.3. Periodic interconnection

analysis in [37] to spatially variable systems [38], to interconnections over an arbitrary graph [101], and also over a discrete group [39]. This considerably extends the applicability of the overall approach.

It is apparent that the repetitive processes considered in this work can encompass some cases of interconnected systems and hence provide additional/enabling tools for the solution of control problems for them. To support this statement, assume in the state-space model of Example 1.2.10 that the variable k has the meaning of the discrete time for each of the subsystems and the variable p indexes α of them connected spatially. Then shift invariance requires each of them to be the same but the robustness analysis of Chap. 8 could allow variations around the 'nominal' to be considered.

The work discussed above is solely related, in repetitive process terms, to asymptotic stability and stabilization of the interconnected system and cannot guarantee that the amplitude of a signal does not increase in the spatial direction, i.e. from a given subsystem to the next one. Note, however, that if each subsystem is described by a model whose dimension is not equal to all others then LMI based computations cannot be undertaken. In such cases, the only currently available option is to use the equivalent 1D model, but then we can only consider asymptotic stability and this is clearly a weakness of the repetitive process approach advocated here.

Another weakness arises from the fact that the model of Example 1.2.10 is only causal in the positive quadrant (of the (k, p) plane) and hence, unlike the models in the papers cited above, it can only be used to study the influence of the preceding subsystem on the next one. There are, however, several possible ways of extending the state-space model of Example 1.2.10 to include the structure necessary to study spatially-interconnected systems as discussed next.

The discrete linear repetitive process state-space model considered in [69] (and termed there a model with acausal updating of the pass profile vector) has the following form over $0 \leq p \leq \alpha - 1,\ k \geq 0$

$$\begin{aligned} x_{k+1}(p+1) &= Ax_{k+1}(p) + Bu_{k+1}(p) + B_0 y_k(p) + B_1 y_k(p+1) \\ y_{k+1}(p) &= Cx_{k+1}(p) + Du_{k+1}(p) + D_0 y_k(p) + D_1 y_k(p+1) \end{aligned}$$

where the vectors x, y, and u are as in Example 1.2.10. The terms $B_1 y_k(p+1)$ and $D_1 y_k(p+1)$ mean that this model does not have the upper right quarter plane causality property of either the Roesser and Fornasini-Marchesini models or the model of Example 1.2.10 (to which it reduces on deleting these terms). This is, however, acceptable here due to the combined time and space characteristics of the indeterminates. To complete the process description in this case, it is necessary to augment the boundary conditions for the model of Example 1.2.10 with a 'stopping condition' at the end of each pass of the form

$$y_k(\alpha) = \gamma_k, \quad k \geq 1$$

where γ_k is an $n \times 1$ vector with known real entries.

In general, the analysis of this model is much more complex relative to that of Example 1.2.10, e.g. stability along the pass. Following the method given in Sect. 1.3 for constructing such models, introduce the substitutions $l := k+1,\ v_l := y_{l-1}$ and defined the global pass profile, state, and input super-vectors of dimensions $m\alpha \times 1$, $n\alpha \times 1$, and $r\alpha \times 1$ respectively as

$$V(l) := \begin{bmatrix} v_l(0) \\ v_l(1) \\ v_l(2) \\ \vdots \\ v_l(\alpha-1) \end{bmatrix} X(l) := \begin{bmatrix} x_l(1) \\ x_l(2) \\ x_l(3) \\ \vdots \\ x_l(\alpha) \end{bmatrix}, U(l) := \begin{bmatrix} u_l(0) \\ u_l(1) \\ u_l(2) \\ \vdots \\ u_l(\alpha-1) \end{bmatrix}$$

Then the repetitive dynamics in this case are equivalently described by the following 1D discrete linear systems state-space model given next where $l \geq 1$

$$\begin{aligned} V(l+1) &= \Delta V(l) + \Psi U(l) + \Theta d_l + \Xi v_l(\alpha) \\ X(l) &= \Phi V(l) + \Gamma U(l) + \Gamma_0 d_l + \Gamma_\alpha v_l(\alpha) \end{aligned}$$

and of the matrices involved it is only necessary to detail

$$\Delta := \begin{bmatrix} D_0 & D_1 & 0 & \dots & 0 \\ CB_0 & CB_1 + D_0 & D_1 & \dots & 0 \\ CAB_0 & C(\cdot) & CB_1 + D_0 & \dots & \vdots \\ \vdots & \vdots & \vdots & \ddots & D_1 \\ CA^{\alpha-2}B_0 & CA^{\alpha-3}(\cdot) & CA^{\alpha-4}(\cdot) & \dots & CB_1 + D_0 \end{bmatrix}$$

where $(\cdot) = (B_0 + AB_1)$. Asymptotic stability now holds if, and only if, $r(\Delta) < 1$ (as compared with $r(D_0) < 1$ in the case of the model of Example 1.2.10). The only possible difficulty here is numerical problems induced by the potentially very large dimensions of this matrix. It is also possible to extend the pass profile controllability analysis of Chap. 6 to this case and also the control law design to achieve asymptotic stability results from Chap. 7.

A more general form of the model above was introduced in [71] (where it was termed 'the generalized model') and has the following state-space description over $0 \leq p \leq \alpha - 1,\ k \geq 0$,

$$\begin{aligned} x_{k+1}(p+1) &= Ax_{k+1}(p) + Bu_{k+1}(p) \\ &+ B_0 y_k(p) + B_1 y_k(p+1) \\ y_{k+1}(p) &= Cx_{k+1}(p) + Du_{k+1}(p) \\ &+ \sum_{j=0}^{\alpha-1} D_j y_k(j) \end{aligned}$$

with the same boundary conditions as the previous model. The novel feature here arises from the fact that the pass profile dynamics at point p on any pass also (explicitly) depend on the dynamics at all points $p+1, \cdots, \alpha-1$ of the previous pass. The state dynamics are, however, still causal along the pass and hence (in interconnected systems terms) only the preceding $\alpha - 1$ subsystems can influence the state dynamics of a given subsystem.

As for the previous model, stability along the pass is still an open question but the 1D equivalent model can be used to characterize asymptotic stability and the design of control laws to ensure this property. The 1D equivalent model has the form (using the notation of the previous model)

$$V(l+1) = \Delta V(l) + \Psi U(l) + \Theta x_l(0)$$

Now we consider repetitive processes which have a 'wave' characteristic in the along the pass dynamics, i.e. the dynamics in this direction evolve in both right and left directions (in relative terms) [70]. This is a very important feature since it allows the inclusion of the bidirectional character of the subsystem interactions within a spatially interconnected system.

Such processes could be described in the normal singular systems framework but attempts to make progress with this approach encountered numerous analysis and numerical implementation difficulties. Here we propose the basis of a new approach to analysis using specific 2D and so-called global 1D state-space models with, in particular, highly promising initial results in the areas of stability analysis and related problems.

In the case of linear processes with discrete dynamics along the pass, the 2D model investigated here has the following form

$$\begin{aligned} x_{k+1}(p) &= A_{-1}x_{k+1}(p-1) + A_1 x_{k+1}(p+1) \\ &\quad + B_{-1}u_{k+1}(p-1) + B_0 u_{k+1}(p) \\ &\quad + B_1 u_{k+1}(p+1) + B_{0-1}y_k(p-1) \\ &\quad + B_{00}y_k(p) + B_{01}y_k(p+1) \\ y_{k+1}(p) &= Cx_{k+1}(p) + D_0 y_k(p) + Du_{k+1}(p) \end{aligned} \tag{10.3}$$

where $p = -\alpha+1, \ldots, \alpha-1$, and hence the pass length is equal to $2\alpha+1$. Here on pass k, $x_k(p)$ is the $n \times 1$ state vector, $y_k(p)$ is the $m \times 1$ pass profile vector, and $u_k(p)$ is the $r \times 1$ vector of control inputs.

To complete the process description, it is necessary to specify the initial, or boundary, conditions, where, in contrast to the previously considered cases, here we must specify the left and right boundary values for the state vector on each pass and the initial pass profile. These are taken to be of the following form (but others are possible)

$$\begin{aligned} x_{k+1}(-\alpha) &= d_{k+1}^-, \quad k \geq 0 \\ x_{k+1}(\alpha) &= d_{k+1}^+, \quad k \geq 0 \\ y_0 &= y(p), \quad -\alpha \leq p \leq \alpha \end{aligned} \tag{10.4}$$

where d_{k+1}^- and d_{k+1}^+ are $n \times 1$ vectors with known constant entries and the elements in the $m \times 1$ vector $y(p)$ are known functions of p. Also it is obviously necessary to assume that the sequences $\{d_{k+1}^-\}$ and $\{d_{k+1}^+\}$ are bounded.

To construct the 1D equivalent model in this case, again use the substitutions $l := k+1, \ v_l := y_{l-1}$ from Chap. 1 and define the so-called global pass state and input super-vectors respectively as follows

$$Y(l) := \begin{bmatrix} v_l(-\alpha) \\ v_l(-\alpha+1) \\ \vdots \\ v_l(\alpha-1) \\ v_l(\alpha) \end{bmatrix}, \ X(l) := \begin{bmatrix} x_l(-\alpha+1) \\ x_l(-\alpha+2) \\ \vdots \\ x_l(\alpha-2) \\ x_l(\alpha-1) \end{bmatrix}, \ U(l) := \begin{bmatrix} u_l(-\alpha) \\ u_l(-\alpha+1) \\ \vdots \\ u_l(\alpha-1) \\ u_l(\alpha) \end{bmatrix}$$

respectively. Also, define the vector of boundary conditions on a given pass as

$$\epsilon(l) := \begin{bmatrix} \epsilon(l)^- \\ \epsilon(l)^+ \end{bmatrix} \text{where} \begin{array}{c} \epsilon(l)^- := d_l^-(-\alpha) \\ \epsilon(l)^+ := d_l^+(\alpha) \end{array}$$

Then the 1D equivalent state-space model for processes described by (10.3) and (10.4) is given by

$$Y(l+1) = CX(l) + DU(l) + D_0Y(l) + \Lambda\epsilon(l) \tag{10.5}$$
$$EX(l) = BU(l) + B_0Y(l) + \Upsilon\epsilon(l) \tag{10.6}$$

where the matrices here have the following structure

$$E := \begin{bmatrix} I_n & -A_1 & & & 0 \\ -A_{-1} & I_n & -A_1 & 0 & \\ & \ddots & \ddots & \ddots & \\ & & -A_{-1} & I_n & -A_1 \\ 0 & & & -A_{-1} & I_n \end{bmatrix}$$

$$B := \begin{bmatrix} B_{-1} & B_0 & B_1 & & 0 \\ & \ddots & \ddots & \ddots & \\ 0 & & B_{-1} & B_0 & B_1 \end{bmatrix}$$

$$B_0 := \begin{bmatrix} B_{0-1} & B_{00} & B_{01} & & 0 \\ & \ddots & \ddots & \ddots & \\ 0 & & B_{0-1} & B_{00} & B_{01} \end{bmatrix}$$

$$\mathcal{C} := \begin{bmatrix} 0 \\ \mathrm{diag}_{2\alpha-1}\{C\} \\ 0 \end{bmatrix}, \; \mathcal{D} := \mathrm{diag}_{2\alpha+1}\{D\}, \; \mathcal{D}_0 := \mathrm{diag}_{2\alpha+1}\{D_0\}$$

$$\Lambda := \begin{bmatrix} C & 0 \\ 0 & \vdots \\ \vdots & 0 \\ 0 & C \end{bmatrix}, \quad \Upsilon := \begin{bmatrix} A_{-1} & 0 \\ 0 & \vdots \\ \vdots & 0 \\ 0 & A_1 \end{bmatrix}$$

This equivalent model is easily transformed to the familiar 1D form. In particular, from (10.6) we can write (if E is nonsingular)

$$X(l) = E^{-1}BU(l) + E^{-1}B_0Y(l) + E^{-1}\Upsilon\epsilon(l) \tag{10.7}$$

and substituting this last result into (10.5) yields

$$Y(l+1) = \Psi Y(l) + \Delta U(l) + \Theta\epsilon(l) \tag{10.8}$$

where $\Psi := \mathcal{D}_0 + \mathcal{C}E^{-1}B_0$, $\Delta := \mathcal{D} + \mathcal{C}E^{-1}B$, $\Theta := \Lambda + \mathcal{C}E^{-1}\Upsilon$.

One factor which could cause difficulties in constructing (10.7) and (10.8) is the numerics of inverting the matrix E, resulting from the fact that this matrix is generally very badly conditioned. Hence, the next step is to develop an alternative 1D equivalent model which does not involve matrix inversion.

To develop this so-called extended model, return to (10.6), write $l+1$ instead of l and substitute the result into (10.5) to obtain

$$\begin{aligned} \begin{bmatrix} E & 0 \\ 0 & I \end{bmatrix} \begin{bmatrix} X(l+1) \\ Y(l+1) \end{bmatrix} &= H \begin{bmatrix} X(l) \\ Y(l) \end{bmatrix} + \begin{bmatrix} B \\ 0 \end{bmatrix} U(l+1) \\ &+ \begin{bmatrix} B_0\mathcal{D} \\ \mathcal{D} \end{bmatrix} U(l) + \begin{bmatrix} \Upsilon \\ 0 \end{bmatrix} \epsilon_{l+1} \\ &+ \begin{bmatrix} B_0\Lambda \\ \Lambda \end{bmatrix} \epsilon(l) \end{aligned}$$

where

$$H := \begin{bmatrix} B_0\mathcal{C} & B_0\mathcal{D}_0 \\ \mathcal{C} & \mathcal{D}_0 \end{bmatrix}$$

Using this last model, a sufficient condition for asymptotic stability is that $\exists$ a matrix $P > 0$ such that the following LMI holds

$$\widetilde{\Delta}P\widetilde{\Delta}^T - \Xi P \Xi^T < 0$$

where

$$\widetilde{\Delta} := \begin{bmatrix} B_0\mathcal{C} & B_0\mathcal{D}_0 \\ \mathcal{C} & \mathcal{D}_0 \end{bmatrix}, \quad \Xi := \begin{bmatrix} E & 0 \\ 0 & I \end{bmatrix}$$

This condition does not involve the inversion of the matrix E. Hence, as in other cases (Chap. 8), we can apply the LMI/Lyapunov function based methods to design a control law which ensures asymptotic stability, including the case when there is uncertainty in the model structure, and performance. Control law design to ensure stability along the pass for this class of processes is still an open problem.

Following [32], an alternative to this last model is over $0 \le p \le \alpha - 1, k \ge 0$

$$\begin{aligned} x_{k+1}(p) &= A_1 x_k(p-1) + A_2 x_k(p) + A_3 x_k(p+1) \\ &+ B_1 u_k^1(p) + B_2 u_k^2(p) + B_3 u_k^3(p) \end{aligned} \tag{10.9}$$

with boundary conditions

$$\begin{aligned} x_0(p) &= g(p), \ 0 \le p \le \alpha \\ x_k(-1) &= 0 \\ x_k(\alpha) &= g_k, \ k > 0 \end{aligned} \tag{10.10}$$

where $g(p)$ is an $n \times 1$ vector whose entries are known functions of p and g_k is an $n \times 1$ vector with known constant entries and the sequence $\{g_k\}$ is bounded.

In comparison to that of Example 1.2.10, the updating structure in this last model is radically different and also its boundary conditions differ in that now it is the final value on each pass which is specified. The dynamics of (10.9) and, in particular, the updating structure can be visualized as in Fig. 10.4, and the difference with the process model of Example 1.2.10 is apparent.

This model is intrinsically non-causal in the right-upper quadrant sense, but physical motivation does not contradict this as the pass dynamics can posses a space, as opposed to a temporal, characteristic and hence right-upper quadrant causality is not required. Such a feature is especially important for the possible use in interconnected systems analysis as it allows us to encompass bi-directional interactions between subsystems. This and more general situations also occur in signal processing and are related to so-called semi-causal and minimum neighbor systems [102]. Also, similar models arise when discretizing some systems of partial differential equations describing, for example, transportation systems [47].

As in the previous cases, it is a simple task to obtain a condition for asymptotic stability via the 1D equivalent model. The only possible difficulty could be numerical problems due to the dimensions of the matrices involved. As detailed below, however, it is already possible to obtain LMI based conditions for so-called quadratic stability. This new property has a Lyapunov function interpretation in a similar manner to that for a process of the form defined by Example 1.2.10 and for the current model (in the interconnected systems context) allows the spatial axis (or the number of connected systems) to tend to infinity.

The basic idea (recall again the unique control problem for these processes and the Lyapunov function based analysis of Chap. 8) is to define a quadratic

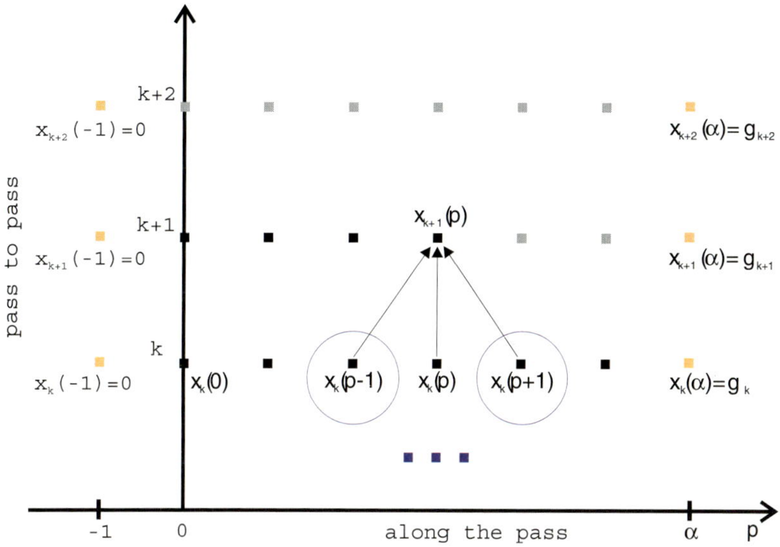

Fig. 10.4. Illustrating the updating structure of (10.9) and (10.10)

Lyapunov energy function for each point on a pass and then sum over $[0, \infty]$ to give the so-called total Lyapunov function for that pass. Quadratic stabilization then demands that the process dissipates this energy from pass-to-pass. Moreover, as shown below the methods developed for this case operate with matrices whose dimensions equal those of the distinct subsystems or small multiples of them. The so-called quadratic, or total, Lyapunov function is defined as

$$V(k) := \sum_{p=0}^{\infty} x_k^T(p)\widetilde{P}x_k(p) \tag{10.11}$$

where $\widetilde{P} > 0$, and

$$V(0) < \infty \tag{10.12}$$

and the following is the formal definition of quadratic stability.

Definition 10.3.1. *A discrete linear repetitive process described by* (10.9) *and (10.10) is said to be quadratically stable if, and only if,* $\exists$ *a matrix* $\widetilde{P} > 0$ *such that*

$$V(k+1) < V(k), \ \forall\, k \geq 0 \tag{10.13}$$

and (10.12) holds.

The following result gives an LMI based interpretation of this property which forms the basis of the analysis which follows. Note that this condition is sufficient but not necessary and hence there is a degree of conservativeness associated with its use.

Theorem 10.3.1. *A discrete linear repetitive process described by* (10.9) *and* (10.10) *is quadratically stable if* $\sum_{p=0}^{\infty} ||x_0(p)|| < \infty$ *and* $\exists$ *matrices* $P > 0$, $Q > 0$ *and* $Z > 0$ *such that the following LMI holds*

$$\begin{bmatrix} A_1^T(P+Q+Z)A_1 - P & A_1^T(P+Q+Z)A_2 & A_1^T(P+Q+Z)A_3 \\ A_2^T(P+Q+Z)A_1 & A_2^T(P+Q+Z)A_2 - Q & A_2^T(P+Q+Z)A_3 \\ A_3^T(P+Q+Z)A_1 & A_3^T(P+Q+Z)A_2 & A_3^T(P+Q+Z)A_3 - Z \end{bmatrix} < 0 \quad (10.14)$$

Proof. From (10.9) and (10.11) we have that

$$\begin{aligned} V(k+1) &= \sum_{p=0}^{\infty} x_{k+1}^T(p)\widetilde{P}x_{k+1}(p) \\ &= \sum_{p=0}^{\infty} \Big(A_1x_k(p-1) + A_2x_k(p) + A_3x_k(p+1)\Big)^T \widetilde{P} \\ &\times \Big(A_1x_k(p-1) + A_2x_k(p) + A_3x_k(p+1)\Big) \\ &= \sum_{p=0}^{\infty} \Big[x_k(p-1)^T A_1^T \widetilde{P} A_1 x_k(p-1) + x_k(p-1)^T A_1^T \widetilde{P} \\ &\times A_2x_k(p) + x_k(p-1)^T A_1^T \widetilde{P} A_3 x_k(p+1) + x_k(p)^T A_2^T \widetilde{P} \\ &\times A_1x_k(p-1) + x_k(p)^T A_2^T \widetilde{P} A_2 x_k(p) + x_k(p)^T A_2^T \widetilde{P} \\ &\times A_3x_k(p+1) + x_k(p+1)^T A_3^T \widetilde{P} A_1 x_k(p-1) \\ &+ x_k(p+1)^T A_3^T \widetilde{P} A_2 x_k(p) + x_k(p+1)^T A_3^T \widetilde{P} A_3 x_k(p+1)\Big] \end{aligned}$$

and introduce $\widetilde{P} := P+Q+Z$, $P > 0, Q > 0, Z > 0$. Then, since $x_k(-1) = 0$, (10.11) can be rewritten as

$$V(k) = x_k^T(0)Zx_k(0) + \widetilde{S}(k)$$

where

$$\begin{aligned} \widetilde{S}(k) &:= \sum_{p=0}^{\infty} \Big\{x_k^T(p-1)Px_k(p-1) \\ &+ x_k^T(p)Qx_k(p) + x_k^T(p+1)Zx_k(p+1)\Big\} \end{aligned}$$

It now follows immediately that the requirement of (10.14) is equivalent to $V(k+1) \leq \widetilde{S}(k)$. Finally, the assumption on Z shows that (10.13) holds and the proof is complete. □

In what follows, we produce the first result on the control of processes described by (10.9) and (10.10) under the action of the following control law defined over $0 \leq p \leq \alpha$, $k \geq 0$,

$$\begin{bmatrix} u_k^1(p) \\ u_k^2(p) \\ u_k^3(p) \end{bmatrix} = I_3 \otimes K \begin{bmatrix} x_k(p-1) \\ x_k(p) \\ x_k(p+1) \end{bmatrix} \tag{10.15}$$

Applying the control law and then the condition of Theorem 10.3.1 to the resulting state-space model gives quadratic stability if we can find a matrix K and matrices $P > 0$, $Q > 0$ and $Z > 0$ such that the following LMI holds

$$\begin{bmatrix} (A_1+B_1K)^T\widetilde{P}(A_1+B_1K)-P & (A_1+B_1K)^T\widetilde{P}(A_2+B_2K) & (A_1+B_1K)^T\widetilde{P}(A_3+B_3K) \\ (A_2+B_2K)^T\widetilde{P}(A_1+B_1K) & (A_2+B_2K)^T\widetilde{P}(A_2+B_2K)-Q & (A_2+B_2K)^T\widetilde{P}(A_3+B_3K) \\ (A_3+B_3K)^T\widetilde{P}(A_1+B_1K) & (A_3+B_3K)^T\widetilde{P}(A_2+B_2K) & (A_3+B_3K)^T\widetilde{P}(A_3+B_3K)-Z \end{bmatrix} < 0 \tag{10.16}$$

Theorem 10.3.2. *A discrete linear repetitive process described by* (10.9) *and* (10.10) *is quadratically stabilizable by a control law of the form (10.15) if* $\exists$ *matrices* N, $X > 0$, $Y > 0$, $J > 0$ *and* $M > 0$ *such that the following LMI holds*

$$\begin{bmatrix} -X & (*) & (*) & (*) \\ 0 & -Y & (*) & (*) \\ 0 & 0 & -J & (*) \\ A_1\widetilde{M}+B_1N & A_2\widetilde{M}+B_2N & A_3\widetilde{M}+B_3N & -\widetilde{M} \end{bmatrix} < 0 \tag{10.17}$$

and

$$\widetilde{M} := X + Y + J$$

If (10.17) holds, then a stabilizing K *in the control law (10.15) is given by*

$$K = N\widetilde{M}^{-1} \tag{10.18}$$

Proof. An obvious application of the Schur's complement formula to (10.16) yields

$$\begin{bmatrix} -P & 0 & 0 & (A_1+B_1K)^T \\ 0 & -Q & 0 & (A_2+B_2K)^T \\ 0 & 0 & -Z & (A_3+B_3K)^T \\ A_1+B_1K & A_2+B_2K & A_3+B_3K & -(P+Q+Z)^{-1} \end{bmatrix} < 0$$

Next, pre- and post-multiply this last expression by

$$\mathrm{diag}\{I, I, I, (P+Q+Z)\}$$

and then pre- and post-multiply the result by

$$\mathrm{diag}\{(P+Q+Z)^{-1}, (P+Q+Z)^{-1}, (P+Q+Z)^{-1}, (P+Q+Z)^{-1}\}$$

and follow this with the substitutions

$$\begin{aligned} X &:= (P+Q+Z)^{-1}P(P+Q+Z)^{-1} > 0 \\ Y &:= (P+Q+Z)^{-1}Q(P+Q+Z)^{-1} > 0 \\ J &:= (P+Q+Z)^{-1}Z(P+Q+Z)^{-1} > 0 \\ \widetilde{M} &:= (P+Q+Z)^{-1} > 0 \end{aligned} \tag{10.19}$$

to obtain

$$\begin{bmatrix} -X & (*) & (*) & (*) \\ 0 & -Y & (*) & (*) \\ 0 & 0 & -J & (*) \\ (A_1+B_1K)\widetilde{M} & (A_2+B_2K)\widetilde{M} & (A_3+B_3K)\widetilde{M} & -\widetilde{M} \end{bmatrix} < 0$$

and it is immediate from (10.19) that $\widetilde{M} = X + Y + J$. Finally, use of (10.18) completes the proof. □

Next we develop an alternative form of the results of Theorems 10.3.1 and 10.3.2 which can be directly applied to the case of uncertainty in the process state-space model. Moreover, this gives the unexpected result that the process considered here is equivalent from the stability point of view to some 3D systems described by the Roesser state-space model. The implications of this is an obvious area for further research.

Introduce the matrices

$$\begin{aligned} \widehat{A} &:= \begin{bmatrix} A_1 & A_2 & A_3 \\ A_1 & A_2 & A_3 \\ A_1 & A_2 & A_3 \end{bmatrix}, \quad \widehat{B} := \begin{bmatrix} B_1 & B_2 & B_3 \\ B_1 & B_2 & B_3 \\ B_1 & B_2 & B_3 \end{bmatrix} \\ \Theta &:= \begin{bmatrix} P & 0 & 0 \\ 0 & Q & 0 \\ 0 & 0 & Z \end{bmatrix}, \quad \widehat{K} := \begin{bmatrix} K & 0 & 0 \\ 0 & K & 0 \\ 0 & 0 & K \end{bmatrix} \end{aligned} \tag{10.20}$$

and hence we can rewrite (10.14) as

$$\widehat{A}^T \Theta \widehat{A} - \Theta < 0 \tag{10.21}$$

To analyze quadratic stability we proceed as follows (see also [36] and [74]).

Theorem 10.3.3. *A discrete linear repetitive process described by* (10.9) *and* (10.10) *is quadratically stable if* $\exists$ *matrices* $\widehat{S} := \mathrm{diag}\{S,S,S\}$ *and* $\widehat{G} := \mathrm{diag}\{G,G,G\}$ *such that the following LMI holds*

$$\begin{bmatrix} \widehat{G}+\widehat{G}^T-\widehat{S} & \widehat{G}^T\widehat{A}^T \\ \widehat{A}\widehat{G} & \widehat{S} \end{bmatrix} > 0 \tag{10.22}$$

Proof. Assume that (10.22) holds and then

$$\widehat{G}+\widehat{G}^T-\widehat{S} > 0$$

Also since $\widehat{G}$ is of full rank and $\widehat{S} > 0$ we have that

$$(\widehat{S}-\widehat{G})^T\widehat{S}^{-1}(\widehat{S}-\widehat{G}) \geq 0$$

which is equivalent to

$$\widehat{G}^T\widehat{S}^{-1}\widehat{G} \geq \widehat{G}^T+\widehat{G}-\widehat{S}$$

This, together with (10.22), leads to

$$\begin{bmatrix} \widehat{G}^T\widehat{S}^{-1}\widehat{G} & \widehat{G}^T\widehat{A}^T \\ \widehat{A}\widehat{G} & \widehat{S} \end{bmatrix} > 0$$

which, in turn, is equivalent to

$$\begin{bmatrix} \widehat{G}^T & 0 \\ 0 & \widehat{S} \end{bmatrix}\begin{bmatrix} \widehat{S}^{-1} & \widehat{A}^TS^{-1} \\ \widehat{S}^{-1}\widehat{A} & \widehat{S}^{-1} \end{bmatrix}\begin{bmatrix} \widehat{G} & 0 \\ 0 & \widehat{S} \end{bmatrix} > 0$$

Finally setting $\Theta = \widehat{S}^{-1}$ yields

$$\begin{bmatrix} \Theta & \widehat{A}^T\Theta \\ \Theta\widehat{A} & \Theta \end{bmatrix} > 0 \tag{10.23}$$

and the proof is complete since (by an obvious application of the Schur's complement formula) (10.23) is an equivalent LMI to that of (10.21). □

Now apply the control law (10.15) to (10.22) to obtain

$$\begin{bmatrix} \widehat{G}+\widehat{G}^T-\widehat{S} & \widehat{G}^T(\widehat{A}^T+\widehat{K}^T\widehat{B}^T) \\ (\widehat{A}+\widehat{B}\widehat{K})\widehat{G} & \widehat{S} \end{bmatrix} > 0 \tag{10.24}$$

where $\widehat{K}$ is defined in (10.20) and we have the following result.

Theorem 10.3.4. *Suppose that a control law of the form* (10.15) *is applied to a discrete linear repetitive process described by* (10.9) *and* (10.10). *Then the resulting process is quadratically stable if* $\exists$ *a matrix* $\widehat{S} := \text{diag}\{S, S, S\} > 0$ *and matrices* $\widehat{G} := \text{diag}\{G, G, G\}$, $\widehat{N} := \text{diag}\{N, N, N\}$ *such that the following LMI holds*

$$\begin{bmatrix} \widehat{G} + \widehat{G}^T - \widehat{S} & \widehat{G}^T\widehat{A}^T + \widehat{N}^T\widehat{B}^T \\ \widehat{A}\widehat{G} + \widehat{B}\widehat{N} & \widehat{S} \end{bmatrix} > 0 \tag{10.25}$$

If this condition holds, a stabilizing matrix K *in the control law* (10.15) *is given by*

$$\widehat{K} = \widehat{N}\widehat{G}^{-1} \tag{10.26}$$

Proof. Follows immediately from substituting for $\widehat{K}$ in (10.24). □

The control law considered above is constant in the sense that it is independent of both the current pass number and the position along the pass. In Theorem 10.3.3, this necessitated an additional constraint for an LMI solution at the cost of increased conservativeness. To reduce this we now consider the following control law over $0 \leq p \leq \alpha - 1$

$$\begin{bmatrix} u_k^1(p) \\ u_k^2(p) \\ u_k^3(p) \end{bmatrix} = \text{diag}\{K_1, K_2, K_3\} \begin{bmatrix} x_k(p-1) \\ x_k(p) \\ x_k(p+1) \end{bmatrix} \tag{10.27}$$

The resulting controlled process is given by

$$\begin{aligned} x_{k+1}(p-1) &= (A_1 + B_1K_1)x_k(p-2) \\ &+ (A_2 + B_2K_2)x_k(p-1) + (A_3 + B_3K_3)x_k(p) \\ x_{k+1}(p) &= (A_1 + B_1K_1)x_k(p-1) \\ &+ (A_2 + B_2K_2)x_k(p) + (A_3 + B_3K_3)x_k(p+1) \\ x_{k+1}(p+1) &= (A_1 + B_1K_1)x_k(p) \\ &+ (A_2 + B_2K_2)x_k(p+1) + (A_3 + B_3K_3)x_k(p+2) \end{aligned}$$

and we have the following result.

Theorem 10.3.5. *Suppose that a control law of the form (10.27) is applied to a discrete linear repetitive process described by (10.9) and (10.10). Then the resulting process is stable along the pass if* $\exists$ *matrices* $\widehat{P} := \text{diag}\{P, Q, Z\} > 0$ *and* $\widehat{N} := \text{diag}\{N_1, N_2, N_3\}$ *such that*

$$\begin{bmatrix} -\widehat{P} & \widehat{P}\widehat{A}^T + \widehat{N}^T\widehat{B}^T \\ \widehat{A}\widehat{P} + \widehat{B}\widehat{N} & -\widehat{P} \end{bmatrix} < 0 \tag{10.28}$$

If this condition holds, stabilizing K_1, K_2 and K_3 in the control law (10.27) are given by

$$\widehat{K} = \widehat{N}\widehat{P}^{-1} \tag{10.29}$$

where $\widehat{K} := \mathrm{diag}\{K_1, K_2, K_3\}$.

The proof of this result follows almost identical steps to that for the previous cases and hence the details are omitted here.

References

1. Agathoklis, P. and Foda, S. (1989) Stability and the matrix Lyapunov equation for delay differential systems, International Journal of Control, 49(2), pp. 417–432.
2. Agathoklis, P., Jury, E. I. and Mansour, M. (1989) The discrete-time strictly bounded-real lemma and the computation of positive definite solutions to the 2-D Lyapunov equation, IEEE Transactions on Circuits and Systems, 36(6), pp. 830–837.
3. Agathoklis, P., Jury, E. I. and Mansour, M. (1993) Algebraic necessary and sufficient conditions for the stability of 2-D discrete systems, IEEE Transactions on Circuits and Systems II: Analog and Digital Signal Processing, 40(4), pp. 251–258.
4. Al-Towlaim, T. (2004) Practical Implementation of Norm-optimal and Predictive Iterative Learning Control on a Chain Conveyor System, PhD thesis, University of Southampton, UK.
5. Al-Towlaim, T., Barton, A. D., Lewin, P. L., Rogers, E. and Owens, D. H. (2004) Iterative learning control – 2D systems from theory to application, International Journal of Control, 77(9), pp. 877–893.
6. Amann, N. (1996) Optimal Algorithms for Iterative Learning Control, PhD Thesis, University of Exeter, UK.
7. Amann, N., Owens, D. H. and Rogers, E. (1996) Iterative learning control using optimal feedback and feedforward actions, International Journal of Control, 65(2), pp. 277–293.
8. Amann, N., Owens, D. H. and Rogers, E. (1998) Predictive optimal iterative learning control, International Journal of Control, 69(2), pp. 203–226.
9. Anderson, B. D. O. and Moore, J. M. (1972) Linear Optimal Control, Prentice-Hall, Englewood Cliffs.
10. Anderson, B. D. O. and Vongpanitlerd, S. (1973) Network Analysis and Synthesis, A Modern Systems Theory Approach, Prentice-Hall, Englewood Cliffs.
11. Anderson, B. D. O., Agathoklis, P., Jury, E. I. and Mansour, M. (1986) Stability and the matrix Lyapunov equation for discrete 2-dimensional systems, IEEE Transactions on Circuits and Systems, CAS-33(3), pp. 261–266.
12. Aravena, J. L., Shafiee, M. and Porter, W. A. (1990) State models and stability for 2-D filters, IEEE Transactions on Circuits and Systems, 37(12), pp. 1509–1519.
13. Arimoto, S., Kawamura, S. and Miyazaki, F. (1984) Bettering operations of robots by learning, Journal of Robotic Systems, 1, pp. 123–140.
14. Aubin, J-B. (1979) Applied Functional Analysis, John Wiley, New York and London.

15. Bachelier, O., Bernussou, J., de Oliveria, M. C. and Geromel, J. C. (1999) Parameter dependent Lyapunov control design: numerical evaluation, Proceedings of 38th IEEE Conference on Decision and Control, pp. 293–297.
16. Barker, G. P. and Campbell, S. L. (1984) Internal stability of two-dimensional systems, Linear and Multilinear Algebra, 14, pp. 365–370.
17. Barnett, S. (1983) Polynomials and Linear Control Systems, Marcel Dekker, New York.
18. Barton, A. D., Lewin, P. L. and Brown, D. J. (1997) Analysis and control of a high speed indexing chain conveyor, Proceedings of the 8th International Conference on Electrical Machines and Drives, pp. 314–318.
19. Barton, A. D. and Lewin, P. L. (2000) Experimental Comparison of the Performance of Different Chain Conveyor Controllers, Proceedings of The Institution of Mechanical Engineers, Part I, Journal of Systems and Control Engineering, 214(5), pp. 361–369.
20. Beauchamp, G. (1990) Algorithms for Singular Systems, PhD Thesis, Georgia Institute of Technology, USA.
21. Benton, S. E. (2000) Analysis and Control of Linear Repetitive Processes, PhD Thesis, University of Southampton, UK.
22. Benton, S. E., Rogers, E. and Owens, D. H. (2001) Fast sampling control of a class of differential linear repetitive processes, Proceedings of The Institution of Mechanical Engineers, Part I, Journal of Systems and Control Engineering, 215(I), pp. 521–530.
23. Bochniak, J., Galkowski, K., Rogers, E., Mehdi, D., Bachelier, O. and Kummert, A. (2006) Stabilization of discrete linear repetitive processes with switched dynamics, Multidimensional Systems and Signal Processing, 17(1–2), pp. 271–295.
24. Boland, F. M. and Owens, D. H. (1979) Linear multipass processes – a two-dimensional interpretation, Proceedings of The Institution of Electrical Engineers, Part D, Control Theory and Applications, 127(5), pp. 189–193.
25. Boyd, S., Ghaoui, L. E., Feron, E. and Balakrishnan, V. (1994) Linear Matrix Inequalities In System and Control Theory, SIAM Studies in Applied Mathematics, Vol. 15, SIAM, Philadelphia.
26. Brierley, S. D., Chiasson, J. N., Lee, E. B. and Zak, S. H. (1982) On stability independent of delay for linear systems, IEEE Transactions on Automatic Control, AC-27(1), pp. 252–254.
27. Brockett, R. W. (1970) Finite-dimensional Linear Systems, John Wiley, New York.
28. Buchsbaum, D. A. and Eisenbud, D. (1995) Generic free resolutions and a family of generically perfect ideals, Advances in Mathematics, 18, pp. 245–301.
29. Byers, R. (1988) A bisection method for measuring the distance of a stable matrix to the unstable matrices, SIAM Journal on Scientific and Statistical Computing, 9(5), pp. 875–881.
30. Campbell, S. L., Chancelier, J-P. and Nikoukhah, R. (2006) Modelling and Simulation in Scilab/Scicos, Springer-Verlag, Berlin.
31. Chow, T. W. S. and Fang, Y. (1998) An iterative learning control method for continuous-time systems based on 2-D system theory, IEEE Transactions on Circuits and Systems I: Fundamental Theory and Applications, 45(4), pp. 683–689.

32. Cichy, B., Galkowski, K., Rogers, E. and Kummert, A. (2005) Control of a class of "wave" discrete linear repetitive processes, Proceedings of the Joint International Symposium on Intelligent Control and 13th Mediterranean Conference on Control and Automation, pp. 370–375.
33. Collins, W. D. (1981) Approximate controllability of multipass systems described by linear ordinary differential equations, Proceedings of the Third IMA Conference on Control Theory, Academic Press, London, pp. 685–707.
34. Collins, W. D. (1984) Controllability and canonical forms for multipass systems described by linear ordinary differential equations, IMA Journal of Mathematical Control and Information, 1, pp. 1–25.
35. Curtain, R. F. and Pritchard, A. J. (1978) Infinite Dimensional Linear Systems Theory, Springer-Verlag, Berlin.
36. Daafouz, J. and Bemussou, J. (2001) Parameter dependent Lyapunov functions for discrete time systems with time-varying parameters, Systems and Control Letters, 43, pp. 355–359.
37. D'Andrea, R. and Dullerud, G. E. (2003) Distributed control design for spatially interconnected systems, IEEE Transactions on Automatic Control, 48(9), pp. 1478–1495.
38. D'Andrea, R. and Dullerud, G. E. (2004) Distributed control of heterogeneous systems, IEEE Transactions on Automatic Control, 49(12), pp. 2113–2128.
39. D'Andrea, R. and Dullerud, G. E. (2004) Distributed control of systems over discrete groups, IEEE Transactions on Automatic Control, 49(9), pp. 1446–1452.
40. DeCarlo, R. E., Muarry, J. and Saeks, R. (1977) Mutivariable Nyquist theory, International Journal of Control, 25, pp. 657–675.
41. DeRoover, D., Bosgra, O. H. and Steinbuch, M. (2000) Internal model-based design of repetitive and iterative learning controllers for linear multivariable systems, International Journal of Control, 73(10), pp. 914–929.
42. Desoer, C. A. and Vidyasagar, M. (1975) Feedback Systems: Input-Output Properties, Academic Press, London.
43. Dieudonne, J. J. (1969) Foundations of Modern Analysis, Academic Press, London.
44. Dolecki, S. and Russell, D. L. (1977) A general theory of observation and control, SIAM Journal on Control and Optimization, 15, pp. 185–220.
45. Dullerud, G. E. and Paganini, F. (2000) A Course in Robust Control Theory: A Convex Approach, Texts in Applied Mathematics Series, Springer-Verlag, New York.
46. Du, C. and Xie, L. (2002) H_∞ Control and Filtering of Two-dimensional Systems, Lecture Notes in Control and Information Sciences, Vol. 278, Springer-Verlag, Berlin.
47. Dymkou, S. and Jank, G. (2005) 2-D control theory in gas transport network modelling, Proceedings of the fourth International Workshop on Multidimensional Systems (NDS 2005), pp. 241–246.
48. Dymkov, M., Gasihun, I., Rogers, E. and Owens, D. H. (2002) z-transform and Volterra operator based approaches to controllability and observability analysis for discrete linear repetitive processes, Multidimensional Systems and Signal Processing, 14, pp. 365–395.
49. Dymkov, M., Gasihun, I., Rogers, E. and Owens, D. H. (2004) Control theory for a class of 2D continuous-discrete linear systems, International Journal of Control, 77(9), pp. 847–860.

50. Edwards, J. B. (1974) Stability problems in the control of multipass processes, Proceedings of The Institution of Electrical Engineers, 121(11), pp. 1425–1432.
51. Edwards, J. B. and Greenberg, J. M. (1977) Longitudinal interaction in multipass processes, Proceedings of The Institution of Electrical Engineers, 124(4), pp. 385–392.
52. Edwards, J. B. and Owens, D. H. (1982) Analysis and Control of Multipass Processes, Research Studies Press (A Division of Wiley).
53. Eisenbud, D. (1995) Commutative Algebra with a view towards Algebraic Geometry, Springer-Verlag, Berlin.
54. Erdelyi, A. (1962) Operational Calculus and Generalized Functions, Holt, Rinehart and Winston, New York.
55. Fadali, M. and Gnanasekaran, R. (1989) Normal matrices and their stability properties: application to 2-D system stabilization, IEEE Transactions on Circuits and Systems, CAS-36(6), pp. 873–875.
56. Feintuch, A. and Saeks, R. (1982) System Theory, Academic Press, New York and London.
57. Feng, K., Owens, D. H. and Hatonen, J. J. (2003) New connections between positivity and parameter-optimal iterative learning control, Proceedings of the 18th IEEE International Symposium on Intelligent Control, pp. 69–74.
58. Fernando, K. V. (1986) Conditions for internal stability of 2D systems, Systems and Control Letters, 7, pp. 183–187.
59. Fletcher, R. (1987) Practical Methods of Optimization, Chichester, Wiley.
60. Foda, S. and Agathoklis, P. (1989) Stability of differential multipass processes, Electronics Letters, pp. 1089.
61. Foda, S. and Agathoklis, P. (1992) Control of the metal rolling process: a multidimensional system approach, Journal of The Franklin Institute, 329(2), pp. 317–332.
62. Fornasini, E. and Marchesini, G. (1978) Doubly-indexed dynamical systems: state space models and structural properties, Mathematical Systems Theory, 12, pp. 59–72.
63. Fornasini, E. and Marchesini, G. (1980) Stability analysis of 2D systems, IEEE Transactions on Circuits and Systems, 27, pp. 1210–1217.
64. Francis, B. A. (1979) The optimal linear-quadratic time-invariant regulator with cheap control, IEEE Transactions on Automatic Control, AC-24(4), pp. 616–621.
65. Furuta, K. and Yamakita, M. (1987) The design of a learning control system for multivariable systems, Proceedings of IEEE International Symposium on Intelligent Control, pp. 371–376.
66. Gahinet, P. and Apkarian, P. (1994) A linear matrix inequality approach to H_∞ control, International Journal of Robust and Nonlinear Control, 4, pp. 421–448.
67. Galkowski, K., Rogers, E. and Owens, D. H. (1998) Matrix rank based conditions for reachability/controllability of discrete linear repetitive processes, Linear Algebra and its Applications, 275–276, pp. 201–224.
68. Galkowski, K., Rogers, E. and Owens, D. H. (1999) New 2D models and a transition matrix for discrete linear repetitive processes, International Journal of Control, 72(15), pp. 1365–1380.
69. Galkowski, K., Sulikowski, B., Paszke, W., Rogers, E. and Owens, D. H. (2001) Stability and controllability properties of extended discrete linear repetitive

processes, Proceedings of The Institute of Technical Cybernetics, Wroclaw University of Technology, 47(2), pp. 5–14.
70. Galkowski, K., Paszke, W., Rogers, E. and Owens, D. H. (2001) The state-space model of wave linear repetitive processes, Proceedings of the International Symposium on Mathematical Models in Automation and Robotics, 1, pp. 119–123.
71. Galkowski, K., Sulikowski, B., Rogers, E. and Owens, D. H. (2001) Generalised discrete linear repetitive processes, Proceedings of the International Symposium on Mathematical Models in Automation and Robotics, 1, pp. 131–135.
72. Galkowski, K., Rogers, E., Wood, J., Benton, S. E. and Owens, D. H. (2002) One-dimensional equivalent model and related approaches to the analysis of discrete non-unit memory linear repetitive processes, Circuits, Systems, and Signal Processing, 21(6), pp. 525–534.
73. Galkowski, K., Rogers, E., Xu, S., Lam, J. and Owens, D. H. (2002) LMIs – a fundamental tool in analysis and controller design for discrete linear repetitive processes, IEEE Transactions on Circuits and Systems I: Fundamental Theory and Applications, 49(6), pp. 768–778.
74. Galkowski, K., Lam, J., Rogers, E., Sulikowski, B., Paszke, W. and Owens, D. H. (2003) LMI based stability analysis and robust controller design for discrete linear repetitive processes, International Journal of Robust and Nonlinear Control, 13, pp. 1195–1211.
75. Galkowski, K., Paszke, W., Rogers, E., Xu, S., Lam, J. and Owens, D. H. (2003) Stability and control of differential linear repetitive processes using an LMI setting, IEEE Transactions on Circuits and Systems II: Analog and Digital Signal Processing, 50(9), pp. 662–666.
76. Golub, G. H. and van Loan, C. F. (1989) Matrix Computations, 2nd Edition, Baltimore, Johns Hopkins University Press.
77. Goodman, D. (1977) Some stability properties of two-dimensional linear shift-invariant digital filters, IEEE Transactions on Circuits and Systems, CAS-24(4), pp. 201–208.
78. Gramacki, A., Gramacki, J., Galkowski, K., Rogers, E. and Owens, D. H. (2002) From continuous to discrete models of linear repetitive processes, Archives of Control Sciences, 12(1-2), pp. 151–185.
79. Gunal, A. K., Sadakane, S. and Williams, E. J. (1996) Modelling of chain conveyors and their equipment interfaces, Proceedings of the Winter Simulation Conference, pp. 1107–1114, 1996.
80. Gu, G. and Lee, E. B. (1986) On stabilization independent of delay with a finite dimensional compensator, Proceedings of The American Control Conference, pp. 1229.
81. Gu, G. and Lee, E. B. (1989) Stability testing of time delay systems, Automatica, 25(5), pp. 777–780.
82. Hale, J. (1977) Theory of Functional Differential Equations, 1977, Springer-Verlag, New York.
83. Halmos, P. R. (1982) A Hilbert Space Problem Book, Springer-Verlag, New York.
84. Harley, M. (1998) Understanding repeatability and accuracy, Drives and Controls, pp. 24.
85. Hatonen, J. J., Owens, D. H. and Moore, K. L. (2004) An algebraic approach to iterative learning control, International Journal of Control, 77(1), pp. 45–54.

86. Herz, D., Jury, E. I. and Zeheb, E. (1984) Simplified analytic stability test for systems with commensurate delays, Proceedings of The Institution of Electrical Engineers, Part D, Control Theory and Applications, 131(1), pp. 52–56.
87. Hinamoto, T. (1993) 2-D Lyapunov equation and filter design based on the Fornasini-Marchesini second model, IEEE Transactions on Circuits and Systems I: Fundamental Theory and Applications, 40(2), pp. 102–110.
88. Holzhauer, R. (1996) Unit/pallet handling conveyors, Plant Engineering, 50(6), pp. 44–48.
89. Johnson, D. S. (1993) Coprimeness in Multidimensional System Theory and Computation, PhD Thesis, Loughborough University, UK.
90. Johnson, D. S., Pugh, A. C., Rogers, E., Hayton, G. E. and Owens, D. H. (1996) A polynomial matrix theory for a certain class of two-dimensional linear systems, Linear Algebra and its Applications, 241–243, pp. 669–703.
91. Jury, E. I. (1974) Inners and the Stability of Dynamic Systems, New York, Wiley.
92. Kaczorek, T. (1994) Controllability and minimum energy control of 2-D continuous-discrete linear systems, Applied Mathematics and Computer Science, 5(1), pp. 5–21.
93. Kaczorek, T. (1994) When does the local controllability of the general model of 2-D linear systems imply its local reachability?, Systems and Control Letters, 23(6), pp. 445–452.
94. Kahan, W. (1966) Numerical linear algebra, Canadian Mathematical Bulletin, 9, pp. 756–801.
95. Kailath, T. (1980) Linear Systems, Prentice-Hall, Englewood Cliffs.
96. Kalman, R. E., Falb, P. L. and Arbib, M. A. (1969) Topics in Mathematical Systems Theory, McGraw-Hill, New York.
97. Kamen, E. W. (1980) On the relationship between zero criteria for two-variable polynomials and asymptotic stability of delay differential equations, IEEE Transactions on Automatic Control, AC-25(5), pp. 983–984.
98. Khargonekar, P. P., Peterson, I. R. and Zhou, K. (1990) Robust stabilization of uncertain linear systems, IEEE Transactions on Automatic Control, 35(3), pp. 356–361.
99. Kouvaritakis, B. (1976) A geometric approach to the inversion of multivariable systems, International Journal of Control, 24, pp. 609–626.
100. Kurek, J. E. and Zaremba, M. B. (1985) Iterative learning control synthesis based on 2-D system theory, IEEE Transactions on Automatic Control, 38, pp. 121–125.
101. Langbort, C., Chandra, R. S. and D'Andrea, R. (2004) Distributed control design for systems interconnected over an arbitrary graph, IEEE Transactions on Automatic Control, 49(9), pp. 1502–1519.
102. Levy, B. C., Adams, M. B. and Wilsky, A. S. (1990) Solution and linear estimation of 2-D nearest neighbor models, Proceedings of the Institution of Electrical and Electronic Engineers, 78(4), pp. 627–641.
103. Lewis, F. L. (1992) A review of 2-D implicit systems, Automatica, 28(2), pp. 345–353.
104. Longman, R. W. (2000) Iterative learning and repetitive control for engineering practice, International Journal of Control, 73(10), pp. 930–954.
105. Lu, W.-S. and Lee, E. B. (1985) Stability analysis for two-dimensional systems, IEEE Transactions on Circuits and Systems, CAS-30(7), pp. 455–461.

106. Lu, W.-S. and Lee, E. B. (1985) Stability analysis for two-dimensional systems via a Lyapunov approach, IEEE Transactions on Circuits and Systems, CAS-32(1), pp. 61–68.
107. Lu, W.-S., Antoniou, A. and Agathoklis, P. (1986) Stability of 2-D digital filters under parameter variations, IEEE Transactions on Circuits and Systems, CAS-33, pp. 476–482.
108. Lu, W.-S. (1989) Stability robustness of two-dimensional discrete systems and its computation, IEEE Transactions on Circuits and Systems, 36(2), pp. 285–288.
109. Lu, W.-S. and Antoniou, A. (1992) Two-Dimensional Digital Filters, Volume 80 of Electrical Engineering and Electronics Series, Marcel Dekker, New York.
110. Lu, W.-S. (1994) On a Lyapunov approach to stability analysis of 2-D digital filters, IEEE Transactions on Circuits and Systems I: Fundamental Theory and Applications, 41(10), pp. 665–669.
111. Lu, W.-S. (1994) Some new results on stability robustness of two-dimensional discrete systems, Multidimensional Systems and Signal Processing, 5(4), pp. 345–361.
112. Luenberger, D. G. (1984) Linear and Nonlinear Programming, Reading, MA, Addison-Wesley.
113. MacFarlane, A. G. J. and Karcanias, N. (1975) Poles and zeros of linear multivariable systems: a survey of the algebraic, geometric and complex-variable theory, International Journal of Control, 24, pp. 33–74.
114. Martin, J. M. (1987) State-space measures for stability robustness, IEEE Transactions on Automatic Control, AC-32, pp. 509–512.
115. Melkote, H., Cloke, B. and Agarwal, V. (2003) Modelling and compensator design for self-servo writing in disk drives, Proceedings of the American Control Conference, pp. 737–742.
116. Moore, K. L. (2000) A non-standard iterative learning control approach to tracking periodic signals in discrete-time non-linear systems, International Journal of Control, 73(10), pp. 955–967.
117. Norrlof, M. and Gunnarsson, S. (2002) Time and frequency domain convergence properties in iterative learning control, International Journal of Control, 75(14), pp. 1114–1126.
118. Oberst, U. (1990) Multidimensional constant linear systems, Acta Applicande Mathematicae, 20, pp. 1–175.
119. Oberst, U. (1996) Variations on the fundamental principle for linear systems of partial differential and difference equations with constant coefficients, Applicable Algebra in Engineering, Communication and Computing, 6(4/5), pp. 211–243.
120. Owens, D. H. (1974) Feedback stability of open-loop unstable systems: contraction-mapping approach, Electronic Letters, 10, pp. 238–239.
121. Owens, D. H. (1977) Stability of multipass processes, Proceedings of The Institution of Electrical Engineers, 124(11), pp. 1079–1082.
122. Owens, D. H. (1978) Feedback and Multivariable Systems, IEE Control Engineering Series, Peter Peregrinus, London.
123. Owens, D. H. (1978) Discrete first-order models for multivariable process control, Proceedings of The Institution of Electrical Engineers, Part D, Control Theory and Applications, 126(11), pp. 525–530.

124. Owens, D. H. and Rogers, E. (1995) Frequency domain Lyapunov equations and performance bounds for differential linear repetitive processes, Systems and Control Letters, 26, pp. 65–68.
125. Owens, D. H. and Rogers, E. (1999) Stability analysis for a class of 2D continuous-discrete linear systems with dynamic boundary conditions, Systems and Control Letters, 37, pp. 55–60.
126. Owens, D. H., Amann, N., Rogers, E. and French, M. (2000) Analysis of linear iterative learning control schemes – a 2D systems/repetitive processes approach, Multidimensional Systems and Signal Processing, 11, pp. 125–177.
127. Owens, D. H. and Hatonen, J. J. (2002) A new optimality based adaptive ILC-algorithm, Proceedings of the 7th ICARV International Conference, CD Rom.
128. Owens, D. H., Rogers, E. and Moore, K. L. (2002) Analysis of linear iterative learning control schemes using repetitive process theory, Asian Journal of Control, 4(1), pp. 68–69.
129. Owens, D. H. and Feng, K. (2003) Parameter optimisation in iterative learning control, International Journal of Control, 76(11), pp. 1059–1069.
130. Padieu, F. and Su, R. (1990) An H_∞ approach to learning control systems, International Journal of Adaptive Control and Signal Processing, 4, pp. 465–474.
131. Paszke, W., Galkowski, K., Rogers, E. and Owens, D. H. (2004) Guaranteed cost control of uncertain differential linear repetitive processes. IEEE Transactions on Circuits and Systems II, 51(11) pp. 629–634.
132. Paszke, W., Galkowski, K., Rogers, E. and Owens, D. H. (2006) H_∞ control of differential linear repetitive processes, IEEE Transactions on Circuits and Systems II, 53(1), pp. 39–44.
133. Paszke, W., Galkowski, K., Rogers, E. and Owens, D. H. (2006) H_∞ and guaranteed cost control of discrete linear repetitive processes, Linear Algebra and Applications, 412, pp. 93–131.
134. Piekarski, M. S. (1977) Algebraic characterization of matrices whose multivariable characteristic polynomial is Hurwitzian, Proceedings of The International Symposium on Operator Theory, Lublock TX, USA, pp. 45–56.
135. Polderman, J. W. and Willems, J. C. (1997) Introduction to Mathematical Systems Theory, a Behavioral Approach, Springer-Verlag, Berlin.
136. Porter, W. A. and Aravena, J. L. (1984) 1-D models for m-D processes, IEEE Transactions on Circuits and Systems, CAS-31(8), pp. 742–744.
137. Ratcliffe, J. D. (2005) Iterative Learning Control Implemented on a Multi-axis System, PhD Thesis, University of Southampton, UK.
138. Ratcliffe, J. D., Hatonen, J. J., Lewin, P. L., Rogers, E., Harte, T. J. and Owens, D. H. (2005) P-type iterative learning control for systems that contain resonance, International Journal of Adaptive Control and Signal Processing, 19(10), pp. 769–796.
139. Roberts, P. D. and Becerra, V. M. (1997) An algorithm for optimal nonlinear batch process control with model-reality differences and unmatched terminal constraints, Proceedings of IFAC International Symposium Advanced Control of Chemical Processes (ADCHEM'97), pp. 517–522.
140. Roberts, P. D. (2000) Numerical investigation of a stability theorem arising from the 2-dimensional analysis of an iterative optimal control algorithm, Multidimensional Systems and Signal Processing, 11(1/2), pp. 109–204.

141. Roberts, P. D. (2000) Stability analysis of iterative optimal control algorithms modelled as linear unit memory repetitive processes, Proceedings of The Institution of Electrical Engineers, Part D, Control Theory and Applications, 147(3), pp. 229–238.
142. Roberts, P. D. and Becerra, V. M. (2002) Hierarchical control and optimisation of the Tennessee eatman process, Proceedings of 15th IFAC World Congress, CD-Rom.
143. Roberts, P. D. (2002) Two-dimensional analysis of an iterative nonlinear optimal control algorithm, IEEE Transactions on Circuits and Systems I: Fundamental Theory and Applications, 49(6), pp. 872–878.
144. Roberts, P. D. (2003) Two-dimensional analysis of a gradient method in function space optimal control algorithm, Proceedings of 42nd IEEE Conference on Decision and Control, pp. 1212–1217.
145. Roesser, R. P. (1975) A discrete state space model for linear image processing, IEEE Transactions on Automatic Control, AC-20, pp. 1–10.
146. Rogers, E. and Owens, D. H. (1989) 2D transfer functions and stability tests for differential non-unit memory linear mulitpass processes, International Journal of Control, 50(2), pp. 651–666.
147. Rogers, E. and Owens, D. H. (1989) Stability analysis for discrete linear multipass processes with non-unit memory, IMA Journal of Mathematical Control and Information, 6(4), pp. 399–409.
148. Rogers, E. and Owens, D. H. (1990) Simulation based stability tests for differential linear repetitive processes, International Journal of Control, 51(5), pp. 1051–1066.
149. Rogers, E. and Owens, D. H. (1992) Stability Analysis for Linear Repetitive Processes, Lecture Notes in Control and Information Sciences Series, Vol. 175, Springer-Verlag, Berlin.
150. Rogers, E. and Owens, D. H. (1992) New stability tests and performance bounds for differential linear repetitive processes, International Journal of Control, 56(4), pp. 831–856.
151. Rogers, E. and Owens, D. H. (1993) Stability tests and performance bounds for a class of 2D linear systems, Multidimensional Systems and Signal Processing, 4(4), pp. 355–391.
152. Rogers, E. and Owens, D. H. (1994) Output feedback control theory for discrete linear repetitive processes, IMA Journal of Mathematical Control and Information, 10(4), pp. 177–1993.
153. Rogers, E. and Owens, D. H. (1995) Error actuated output feedback control for differential linear repetitive processes, International Journal of Control, 61(5), pp. 981–998.
154. Rogers, E. and Owens, D. H. (1996) Lyapunov stability theory and performance bounds for a class of 2D linear systems, Multidimensional Systems and Signal Processing, 7(2), pp. 179–194.
155. Rogers, E. and Owens, D. H. (1997) Stability theory and performance bounds for a class of two-dimensional linear systems with interpass smoothing effects, IMA Journal of Mathematical Control and Information, 14, pp. 415–427.
156. Rogers, E., Galkowski, K., Gramacki, A., Gramacki, J. and Owens, D. H. (2002) Stability and controllability of a class of 2-D linear systems with dynamic boundary conditions, IEEE Transactions on Circuits and Systems I: Fundamental Theory and Applications, 49(2), pp. 181–195.

157. Rogers, E. and Owens, D. H. (2002) Kronecker product based stability tests and performance bounds for a class of 2D continuous-discrete linear systems, Linear Algebra and its Applications, 353(1–3), pp. 33–52.
158. Rogers, E., Zaris, P. M., Woood, J. and Pillai, H. (2006) Poles and zeros – examples of the behavioral approach applied to discrete linear repetitive processes, Multidimensional Systems and Signal Processing, 17(2–3), pp. 107–118.
159. Rosenbrock, H. H. (1970) State Space and Multivariable Theory, Wiley, New York.
160. Russell, D. L. (1978) Controllability and stabilizability theory for linear partial differential equations: recent progress and open research questions, SIAM Review 20, pp. 639–739.
161. Russell, D. L. (1981) Control canonical structure for a class of distributed parameter systems, In: Proceedings of Third IMA Conference on Control Theory, Academic Press, London, pp. 793–825.
162. Scherer, C., Gahinet, P. and Chilali, M. (1997) Multiobjective output-feedback control via LMI optimization, IEEE Transactions on Automatic Control, 42(7), pp. 896–911.
163. Siljak, D. D. (1971) New algebraic criteria for positive realness, Journal of The Franklin Institute, 291, pp. 109–120.
164. Siljak, D. D. (1975) Stability criteria for two-variable polynomials, IEEE Transactions on Circuits and Systems, CAS-22(3), pp. 185–189.
165. Stewart, G. W. (1973) Introduction to Matrix Computations, Academic Press, New York.
166. Sule, V. R. and Rogers, E. (2004) A behavioral approach to the control of discrete linear repetitive processes. Systems and Control Letters, 53(1), pp. 79–88.
167. Sulikowski, B., Galkowski, K., Rogers, E. and Owens, D. H. (2004) Output feedback control of discrete linear repetitive processes, Automatica, 40(12), pp. 2167–2173.
168. Sulikowski, B., Galkowski, K., Rogers, E. and Owens, D. H. (2005) Control and disturbance rejection for discrete linear repetitive processes, Multidimensional Systems and Signal Processing, 16, pp. 199–216.
169. Sulikowski, B., Galkowski, K., Rogers, E. and Owens, D. H. (2006) PI control of discrete linear repetitive processes, Automatica, 42(5), pp. 877–880.
170. Taylor, A. E. (1967) Introduction to Functional Analysis, Wiley, New York.
171. Trinh, H. and Fernando, T. (2000) Some new stability conditions for two-dimensional difference systems, International Journal of Systems Science, 31(2), pp. 203–211.
172. Tsoi, A. C. (1978) Recent advances in the algebraic systems theory of delay-differential equations, In Gregson, M. J. (ed). Recent Theoretical Developments in Control, Academic Press, New York and London, pp. 67–127.
173. Valcher, M. E. (2000) Characteristic cones and stability properties of two-dimensional autonomous behaviors, IEEE Transactions on Circuits and Systems I: Fundamental Theory and Applications, 47(3), pp. 290–302.
174. Weiss, L. (1970) An algebraic criterion for controllability of linear systems with time delay, IEEE Transactions on Automatic Control, 15, pp. 443–444.
175. Willson, I., Collins, W. D. and Owens, D. H. (1981) Optimal Control of Multipass Processes, Research Report, Department of Control Engineering, University of Sheffield, UK.

176. Wood, J., Oberst, U., Rogers, E. and Owens, D. H. (2000) A behavioural approach to the pole structure of 1D and nD linear systems, SIAM Journal on Control and Optimization, 38, pp. 627–661.
177. Wood, J. (2000) Modules and behaviours in nD systems theory, Multidimensional Systems and Signal Processing, 11, pp. 11–48.
178. Yamada, M., Xu, L. and Saito, O. (1999) 2D model-following servo system, Multidimensional Systems and Signal Processing, 10, pp. 71–91.
179. Youla, D. C. and Gnavi, G. (1979) Notes on nD dimensional systems, IEEE Transactions on Circuits and Systems, CAS-26, pp. 105–111.
180. Zaris, P. M. (2000) Poles and zeros of nD linear systems, PhD Thesis, Department of Electronics and Computer Science, University of Southampton, UK.
181. Zaris, P. M., Wood, J. and Rogers, E. (2001) Controllable and uncontrollable poles and zeros of nD systems, Mathematics of Control Signals and Systems, 14, pp. 281–298.
182. Zhou, K., Doyle, J. C. and Glover, K. (1996) Robust and Optimal Control, Prentice-Hall International.

Index

Lecture Notes in Control and Information Sciences

Edited by M. Thoma, M. Morari

Further volumes of this series can be found on our homepage:
springer.com

Vol. 322: Meurer, T.; Graichen, K.; Gilles, E.D. (Eds.)
Control and Observer Design for Nonlinear Finite and Infinite Dimensional Systems
422 p. 2005 [978-3-540-27938-9]

Vol. 321: Dayawansa, W.P.; Lindquist, A.; Zhou, Y. (Eds.)
New Directions and Applications in Control Theory
400 p. 2005 [978-3-540-23953-6]

Vol. 320: Steffen, T.
Control Reconfiguration of Dynamical Systems
290 p. 2005 [978-3-540-25730-1]

Vol. 319: Hofbaur, M.W.
Hybrid Estimation of Complex Systems
148 p. 2005 [978-3-540-25727-1]

Vol. 318: Gershon, E.; Shaked, U.; Yaesh, I.
H_∞ Control and Estimation of State-multiplicative Linear Systems
256 p. 2005 [978-1-85233-997-5]

Vol. 317: Ma, C.; Wonham, M.
Nonblocking Supervisory Control of State Tree Structures
208 p. 2005 [978-3-540-25069-2]

Vol. 316: Patel, R.V.; Shadpey, F.
Control of Redundant Robot Manipulators
224 p. 2005 [978-3-540-25071-5]

Vol. 315: Herbordt, W.
Sound Capture for Human/Machine Interfaces: Practical Aspects of Microphone Array Signal Processing
286 p. 2005 [978-3-540-23954-3]

Vol. 314: Gil', M.I.
Explicit Stability Conditions for Continuous Systems
193 p. 2005 [978-3-540-23984-0]

Vol. 313: Li, Z.; Soh, Y.; Wen, C.
Switched and Impulsive Systems
277 p. 2005 [978-3-540-23952-9]

Vol. 312: Henrion, D.; Garulli, A. (Eds.)
Positive Polynomials in Control
313 p. 2005 [978-3-540-23948-2]

Vol. 311: Lamnabhi-Lagarrigue, F.; Loría, A.; Panteley, E. (Eds.)
Advanced Topics in Control Systems Theory
294 p. 2005 [978-1-85233-923-4]

Vol. 310: Janczak, A.
Identification of Nonlinear Systems Using Neural Networks and Polynomial Models
197 p. 2005 [978-3-540-23185-1]

Vol. 309: Kumar, V.; Leonard, N.; Morse, A.S. (Eds.)
Cooperative Control
301 p. 2005 [978-3-540-22861-5]

Vol. 308: Tarbouriech, S.; Abdallah, C.T.; Chiasson, J. (Eds.)
Advances in Communication Control Networks
358 p. 2005 [978-3-540-22819-6]

Vol. 307: Kwon, S.J.; Chung, W.K.
Perturbation Compensator based Robust Tracking Control and State Estimation of Mechanical Systems
158 p. 2004 [978-3-540-22077-0]

Vol. 306: Bien, Z.Z.; Stefanov, D. (Eds.)
Advances in Rehabilitation
472 p. 2004 [978-3-540-21986-6]

Vol. 305: Nebylov, A.
Ensuring Control Accuracy
256 p. 2004 [978-3-540-21876-0]

Vol. 304: Margaris, N.I.
Theory of the Non-linear Analog Phase Locked Loop
303 p. 2004 [978-3-540-21339-0]

Vol. 303: Mahmoud, M.S.
Resilient Control of Uncertain Dynamical Systems
278 p. 2004 [978-3-540-21351-2]

Vol. 302: Filatov, N.M.; Unbehauen, H.
Adaptive Dual Control: Theory and Applications
237 p. 2004 [978-3-540-21373-4]

Vol. 301: de Queiroz, M.; Malisoff, M.; Wolenski, P. (Eds.)
Optimal Control, Stabilization and Nonsmooth Analysis
373 p. 2004 [978-3-540-21330-7]

Vol. 300: Nakamura, M.; Goto, S.; Kyura, N.; Zhang, T.
Mechatronic Servo System Control
Problems in Industries and their Theoretical Solutions
212 p. 2004 [978-3-540-21096-2]

Vol. 299: Tarn, T.-J.; Chen, S.-B.; Zhou, C. (Eds.)
Robotic Welding, Intelligence and Automation
214 p. 2004 [978-3-540-20804-4]

Vol. 298: Choi, Y.; Chung, W.K.
PID Trajectory Tracking Control for Mechanical Systems
127 p. 2004 [978-3-540-20567-8]

Vol. 297: Damm, T.
Rational Matrix Equations in Stochastic Control
219 p. 2004 [978-3-540-20516-6]

Vol. 296: Matsuo, T.; Hasegawa, Y.
Realization Theory of Discrete-Time Dynamical Systems
235 p. 2003 [978-3-540-40675-4]

Vol. 295: Kang, W.; Xiao, M.; Borges, C. (Eds.)
New Trends in Nonlinear Dynamics and Control, and their Applications
365 p. 2003 [978-3-540-10474-0]

Printing: Krips bv, Meppel
Binding: Stürtz, Würzburg